Engineering Crystallography: From Molecule to Crystal to Functional Form

NATO Science for Peace and Security Series

This Series presents the results of scientific meetings supported under the NATO Programme: Science for Peace and Security (SPS).

The NATO SPS Programme supports meetings in the following Key Priority areas: (1) Defence Against Terrorism; (2) Countering other Threats to Security and (3) NATO, Partner and Mediterranean Dialogue Country Priorities. The types of meetings supported are generally "Advanced Study Institutes" and "Advanced Research Workshops". The NATO SPS Series collects together the results of these meetings. The meetings are co-organized by scientists from NATO countries and scientists from NATO's "Partner" or "Mediterranean Dialogue" countries. The observations and recommendations made at the meetings, as well as the contents of the volumes in the Series, reflect those of participants and contributors only; they should not necessarily be regarded as reflecting NATO views or policy.

Advanced Study Institutes (ASI) are high-level tutorial courses to convey the latest developments in a subject to an advanced-level audience.

Advanced Research Workshops (ARW) are expert meetings where an intense but informal exchange of views at the frontiers of a subject aims at identifying directions for future action.

Following a transformation of the programme in 2006, the Series has been re-named and re-organised. Recent volumes on topics not related to security, which result from meetings supported under the programme earlier, may be found in the NATO Science Series.

The Series is published by IOS Press, Amsterdam, and Springer, Dordrecht, in conjunction with the NATO Emerging Security Challenges Division.

Sub-Series

A.	Chemistry and Biology	Springer
B.	Physics and Biophysics	Springer
C.	Environmental Security	Springer
D.	Information and Communication Security	IOS Press
E.	Human and Societal Dynamics	IOS Press

http://www.nato.int/science
http://www.springer.com
http://www.iospress.nl

Series A: Chemistry and Biology

Engineering Crystallography: From Molecule to Crystal to Functional Form

edited by

Kevin J. Roberts
School of Chemical & Process Engineering
University of Leeds
Leeds, United Kingdom

Robert Docherty
Pfizer Global R&D
Pharmaceutical Sciences, Pfizer Global R&D
Sandwich, Kent, United Kingdom

and

Rui Tamura
Graduate School of Human and Environmental Studies
Kyoto University
Sakyo-ku, Kyoto, Japan

Springer

Published in Cooperation with NATO Emerging Security Challenges Division

Proceedings of the NATO Advanced Study Institute on Molecules to Crystals to Powders: Understanding Structure Versus Function
Erice, Sicily, Italy
4–14 June 2015

Library of Congress Control Number: 2017944199

ISBN 978-94-024-1118-8 (PB)
ISBN 978-94-024-1115-7 (HB)
ISBN 978-94-024-1117-1 (eBook)
DOI 10.1007/978-94-024-1117-1

Published by Springer,
P.O. Box 17, 3300 AA Dordrecht, The Netherlands.

www.springer.com

Printed on acid-free paper

Scientific Programme Committee

Christer Aakeröy (Kansas State University, USA)
Gerard Coquerel (University of Rouen Normandy, Rouen, France)
Robert Docherty (Pfizer Worldwide R&D, UK)
Kevin J. Roberts (University of Leeds, UK)
Rui Tamura (Kyoto University, Japan)

Organising Committee

Robert Docherty (Pfizer Worldwide R&D, UK)
Annalisa Guerri (University of Florence, Italy)
Kevin J. Roberts (University of Leeds, UK)
Paola Spadon (University of Padova, Italy)
Giovanna Scapin (Merck & Co., Inc., NJ, USA)

Preface

This book is formed from selected papers and tutorials on the theme of *Engineering Crystallography: From Molecule to Crystal to Functional Form* which were presented at the 2015 International School of Crystallography. These international summer schools are renowned for bringing together scientific experts in various crystallographic fields with motivated students from around the world in an informal but engaged atmosphere. This facilitates a high level of interaction and discussion which allows the subject matter experts present to help shape the next generation of academic and industrial talent in the discipline. This was the 48th such event which was held at the Ettore Majorana Foundation and Centre for Scientific Culture in the beautiful and historic hilltop town of Erice in Sicily.

Lecturers were chosen from world experts in the fields of crystallography, solid-state chemistry, crystallisation, materials science, computational multi-scale modelling as well as particle technology and surface characterisation. Structural and modelling techniques were integrated throughout the summer school. This allowed participants to envisage how, in the next few years, computational and experimental workflows will be seamlessly integrated during the transition from molecule to crystal to function. The course consisted of plenary lectures, talks selected from poster abstracts and technologies, plus, on the final day, a facilitated question and answer session with a selected panel from the invited speakers.

Our motivation in designing the summer school was to help build bridges between the solid-state architecture, the landscape of particle and surface properties accessible from this structure and the impact of these on the function and performance of structured products. This intent is inherent in the design of this book where we have assembled the component chapters into:

- Form (inherent physical and chemical properties)
- Formation (how to prepare)
- Function (properties and performance)

Our construct of the course and the book has been shaped by a scientific storyboard created by our community over the last few decades. These include:

- The key concept of the crystal as a supramolecular assembly [1]
- Crystal engineering – the design of functional organic solids [2, 3]
- The pioneering work on stereochemical control and manipulation of nucleation and crystal growth [4]
- Towards knowledge-based approaches to crystal design [5]
- A structural perspective on the morphology and surface chemistry of molecular materials [6, 7]
- Pharmaceutical materials sciences and the materials science tetrahedron [8, 9]

The integration of this storyboard within our book is very timely, and the themes captured remain very contemporary. This is exemplified with recent conferences including the International Workshop on the Crystal Growth of Organic Materials (CGOM) [10]; Molecules, Materials and Medicines (M3) [11]; as well as the funding of the cross-sector ADDoPT (Advanced Digital Design of Pharmaceutical Therapeutics) initiative [12] in the UK. Global engagement in the area remains high with the Novartis-MIT Center for Continuous Manufacturing [13] and the Center for Structured Organic Particulate System [14] in the USA, the Synthesis and Solid State Pharmaceutical Centre (SSPC) in Ireland [15] as well as the Centre for Innovative Manufacturing in Continuous Manufacturing and Crystallisation (CMAC) in Scotland [16] reflecting the fusion of crystal design, particle engineering and innovative manufacturing paradigms.

Leeds, UK Kevin Roberts
Sandwich, Kent, UK Robert Docherty
Sakyo-ku, Kyoto, Japan Rui Tamura
March 2017

References

1. Desiraju GR (1997) The crystal as a supramolecular entity. Wiley, Chichester
2. Desiraju GR (1989) Crystal engineering: the design of organic solids. Elsevier, Amsterdam
3. Seddon KR, Zarawotko M (1999) Crystal engineering: the design and application of functional solids. NATO ASI Series, vol 539
4. Weissbuch I, Lahav M, Leiserowitz L (2003) Towards stereochemical control, monitoring, and understanding of crystal nucleation. Crystal Growth Des 3:125–150
5. Allen FH et al. (2006) Knowledge based approaches to crystal design. Cryst Eng Commun 8:11
6. Clydesdale G, Roberts KJ, Walker EM (1996) The crystal habit of molecular materials: a structural perspective, In: Gavezzotti A (ed) Molecular solid state: syntheses, structure, reactions, applications, vol 2. Theoretical aspects and computer modelling, chapter 7, pp. 203–232
7. Roberts KJ, Hammond RB, Ramachandran V, Docherty R (2016) Synthonic engineering: from molecular and crystallographic structure to the rational design of pharmaceutical solid dosage forms. In: Abramov YA (ed) Computational approaches in pharmaceutical solid state chemistry, ISBN 9781118700747, Copyright © 2016 Wiley, Inc.

8. Hancock BH, Elliot J (2006) Pharmaceutical materials science: an active new frontier in materials research. MRS Bull 31:869
9. Sun CC (2009) Materials science tetrahedron – a useful tool for pharmaceutical research and development. J Pharm Sci 98:1671–1687
10. http://www.crystalgrowth2016.co.uk/ and http://pubs.acs.org/page/cgdefu/vi/11.html
11. Almarsson O, Vadas EB (2015) Molecules, materials, medicines (M3): linking molecules to medicines through pharmaceutical material science. Crystal Growth Des 15(12):5645–5647
12. https://www.addopt.org/
13. https://novartis-mit.mit.edu/
14. http://www.csops.org/
15. http://www.sspc.ie/
16. https://www.cmac.ac.uk/

Acknowledgements

The majority of the organisational efforts that were so vital to the success of the course were championed by Annalisa Guerri and Paola Spadon. They secured the majority of the funding and co-ordinated the participant selection process in a highly diligent and professional manner. In partnership with the orange scarves team (Giovanna Scapin, Valentina Marcheselli, Vania André, Paolo Mazzeo, Matteo Lusi, Tamlyn Young, Francesco Farinella and Martin Schmidt), Annalisa and Paola created a warm welcoming atmosphere where students could engage with each other and with the invited subject matter experts. The red team Erin Davis, Fred Boyle and Fabio Nicoli played a crucial role providing audiovisual support for the lectures as well as the computing infrastructure for the tutorials. The scientific and organising committees gratefully acknowledge all of the Ettore Majorana Centre staff for their support during the event.

The editors would like to thank Ulrike Aufderhorst for directing and organising the editing of the Erice summer school proceedings and for her energetic support regarding assembling this book for publication. We would also like to thank Mauris Chen and Hayley Harding for their help.

The summer school was financed by NATO, and we gratefully acknowledge the NATO Science Committee for their continued support. Generous financial support was also received from the International Union of Crystallography, the European Crystallographic Association, the Cambridge Crystallographic Data Centre, Merck, AbbVie, Dectris, PANalytical and New York University.

The development, delivery and publication of this summer school proceeding represent a deliverable of the Advanced Digital Design of Pharmaceutical Therapeutics (ADDoPT) research project funding through the UK's AMSCI scheme, and we gratefully acknowledge them for their generous support.

Contents

Part II Formation

Part III Function

Contributors

Christer B. Aakeröy Department of Chemistry, Kansas State University, Manhattan, KS, USA

Kevin Back Pharmaceutical Sciences, Pfizer Global R&D, Sandwich, Kent, UK

Diana M. Camacho Corzo School of Chemical and Process Engineering, University of Leeds, Leeds, UK

Gerard Coquerel University of Rouen Normandy, Rouen, France

Robert Docherty Pharmaceutical Sciences, Pfizer Global R&D, Sandwich, Kent, UK

Peter J. Halfpenny School of Chemical and Process Engineering, University of Leeds, Leeds, UK

Robert B. Hammond School of Chemical and Process Engineering, University of Leeds, Leeds, UK

Jerry Y. Y. Heng Department of Chemical Engineering, Imperial College London, London, UK

Martin Iggland Institute of Process Engineering, Eidgenössische Technische Hochschule Zurich, Zurich, Switzerland

William Jones Department of Chemistry, University of Cambridge, Cambridge, UK

Alexander Krok Department of Chemical and Process Engineering, University of Surrey, Guildford, Surrey, UK

Samir A. Kulkarni Novartis-MIT Center for Continuous Manufacturing and Department of Chemical Engineering, Massachusetts Institute of Technology, Cambridge, MA, USA

Reiko Kuroda Research Institute for Science and Technology, Tokyo University of Science, Chiba, Japan

Ana Kwokal GlaxoSmithKline, R&D, Harlow, UK

Cai Y. Ma School of Chemical and Process Engineering, University of Leeds, Leeds, UK

Giovanni Maria Maggioni Institute of Process Engineering, Eidgenössische Technische Hochschule Zurich, Zurich, Switzerland

Marco Mazzotti Institute of Process Engineering, Eidgenössische Technische Hochschule Zurich, Zurich, Switzerland

Allan S. Myerson Novartis-MIT Center for Continuous Manufacturing and Department of Chemical Engineering, Massachusetts Institute of Technology, Cambridge, MA, USA

Zoltan K. Nagy School of Chemical Engineering, Purdue University, West Lafayette, IN, USA

Department of Chemical Engineering, Loughborough University, Loughborough, UK

Garry O'Connor Pharmaceutical Sciences, Kent, UK

Jose V. Parambil Department of Chemical Engineering, Imperial College London, London, UK

Chunlei Pei Department of Chemical and Process Engineering, University of Surrey, Guildford, Surrey, UK

Radoslav Y. Penchev Pharmaceutical Sciences, Kent, UK

Manomi D. Perera Department of Chemistry, Kansas State University, Manhattan, KS, USA

Jonathan Pickering School of Chemical and Process Engineering, University of Leeds, Leeds, UK

Christopher J. Price Department of Chemical and Process Engineering, University of Strathclyde, Glasgow, UK

Vasuki Ramachandran School of Chemical and Process Engineering, University of Leeds, Leeds, UK

C. Malla Reddy Department of Chemical Sciences, Indian Institute of Science Education and Research (IISER) Kolkata, Mohanpur, India

Kevin J. Roberts School of Chemical and Process Engineering, University of Leeds, Leeds, UK

Ian Rosbottom School of Chemical and Process Engineering, University of Leeds, Leeds, UK

Bhupinder Sandhu Department of Chemistry, Kansas State University, Manhattan, KS, USA

Majeed Soufian School of Chemical and Process Engineering, University of Leeds, Leeds, UK

Rui Tamura Graduate School of Human and Environmental Studies, Kyoto University, Sakyo-ku, Kyoto, Japan

Thomas D. Turner School of Chemical and Process Engineering, University of Leeds, Leeds, UK

Chuan-Yu Wu Department of Chemical and Process Engineering, University of Surrey, Guildford, Surrey, UK

About the Editors

Dr Kevin J. Roberts is the Brotherton Professor of Chemical Engineering and currently directs the EPSRC's Centre for Doctoral Training in Complex Particulate Products and Processes and the Centre for the Digital Design of Drug Products at the University of Leeds in the UK.

His research interests centre on crystallisation science and engineering. This work encompasses both fundamental and applied aspects with the latter being directed towards the needs of the pharmaceuticals, specialities, fine chemicals and nutritional product sectors. Particular focus areas include solid-state chemistry of molecular crystals; molecular and synthonic (intermolecular) modelling techniques; the use of synchrotron radiation techniques to probe condensed interfaces in situ; the development and use of process analytical techniques (PATs) for understanding, monitoring and controlling crystallisation processes; industrial crystallisation processes and their scale-up from laboratory to manufacturing; and the application of digital quality by design (dQbD) methodologies in the design and production of high-added-value structured products.

Professor Roberts gained his degree and PhD in applied physics from Portsmouth Polytechnic, UK, before taking a post-doctoral research fellowship in crystallisation and crystal characterisation at the University of Strathclyde, UK. He subsequently held a Royal Society Fellowship at the Institute of Crystallography at the Technical University of Aachen, Germany, before returning to the UK to take up a faculty position in physical chemistry at the University of Strathclyde. Later he was appointed as professor in chemical process engineering and inaugural director of the Centre for Molecular and Interface Engineering at Heriot-Watt University before moving to Leeds as head of chemical engineering in 2000. There, he has been actively involved in setting up the Institute of Particle Science and Engineering (IPSE) and the Institute of Process Research and Development (IPRD). He was awarded his DSc from the University of Strathclyde in 2004.

Over his career, he has published more than 250 peer-reviewed journal papers and 7 chapters of books.

Dr Robert Docherty He is a senior research fellow and head of materials sciences in drug product design at Pfizer in Sandwich. His group is responsible for solid form selection as well as materials and particle characterisation. A key role of this group is the integration of drug substance attributes with drug product design.

His research interests have focused on the structural aspects of the assembly of molecules to crystals, crystals to particles and particles to dosage forms. This work encompasses both fundamental and applied aspects of solid from design and particle engineering with the focus being directed towards the journey of drug molecule to crystal to particle and product performance. Specific areas covered include solid-state chemistry, molecular and materials modelling techniques and the use of state-of-the-art solid-state, particle and surface characterisation techniques. All these combine to underpin an unprecedented structural perspective of the journey from molecule to medicine.

Dr Docherty gained his degree and PhD at the University of Strathclyde and joined ICI/Zeneca in 1988 where he worked on the structure and crystallisation of dyes, pigments, biocides, electrophotographic agents, agrochemicals and pharmaceuticals. He joined Pfizer in 1999 as head of materials sciences. He holds a visiting professorship at the University of Leeds and is a fellow of the Royal Society of Chemistry.

Over his career, he has led, mentored and developed the research work of numerous PhD and post-doctoral students at the industrial/academic interface and published around 70 peer-reviewed journal papers and 6 chapters of books.

Dr Rui Tamura He is professor and councillor of the Graduate School of Human and Environmental Studies at Kyoto University in Japan and chairman of the Division of Organic Crystals in the Chemical Society of Japan.

His research fields cover synthetic and structural organic chemistry, organic crystal and liquid crystal chemistry and colloidal, magnetic and chiral chemistry. His current research interests focus on the discovery of novel complexity phenomena occurring upon the phase transition of organic crystals and in liquid crystals under non-equilibrium conditions. It should be stressed that the observed unusual phenomena cannot be reproduced in ordinary equilibrium systems. Thus far, there are two phenomena found; "preferential enrichment" is a chiral symmetry-breaking spontaneous enantiomeric resolution phenomenon observed on recrystallisation of racemic crystals, while "magneto-LC effects" refer to the generation of strong magnetic interactions induced by the application of low magnetic fields in liquid crystalline phases of organic nitroxide radical compounds.

Professor Tamura gained his PhD in the Department of Chemistry, Faculty of Science, Kyoto University, Japan, in 1980 and took two post-doctoral research fellowships in the Department of Chemistry at Colorado State University (1980–1982) and Princeton University (1982–1983), USA. He took up faculty positions in chemistry at the National Defence Academy (1983), Ehime University (1988), Hokkaido University (1995) and Kyoto University (1997) in Japan. In 2002, he was appointed as professor at Kyoto University.

Over his career, he has published more than 160 peer-reviewed journal papers, 16 chapters of books and 7 Japanese patents.

Part I
Form

Chapter 1
Crystal Science Fundamentals

Vasuki Ramachandran, Peter J. Halfpenny, and Kevin J. Roberts

Abstract The fundamentals of crystal science notably crystallography, crystal chemistry, crystal defects, crystal morphology and the surface chemistry of crystals are introduced with particular emphasis on organic crystals.

Keywords Crystallography • Crystal lattices and systems • Crystal planes and directions • Crystal defects • Crystal chemistry • Crystal morphology • Surface chemistry

1.1 Introduction

Production of many crystalline products such as pharmaceuticals are underpinned by a science-led quality by design (QbD) approach. This typically involves the need to understand, predict, manipulate and control the core processes involved in drug design, development and manufacture. The latter can be understood through a consideration of an integrated 5F materials transformation pathway developed from the target molecule's molecular structure which links and interrelates:

- Formulae (atomic and molecular structure),
- Formation (crystallisation, precipitation, spray drying, etc.),
- Form (crystal size, morphology, polymorphic form, etc.),
- Formulation (blending, granulation, compaction, etc.),
- Function (product purpose and performance).

This approach enables cross-correlation (see Fig. 1.1) for a given crystalline material's or formulated drug product's

- molecular and crystal structure,
- physical and chemical properties,
- processing behaviour,
- performance.

V. Ramachandran (✉) • P.J. Halfpenny • K.J. Roberts
School of Chemical and Process Engineering, University of Leeds, Leeds LS2 9JT, UK
e-mail: v.ramachandran@leeds.ac.uk; p.j.halfpenny@btinternet.com; k.j.roberts@leeds.ac.uk

K.J. Roberts et al. (eds.), *Engineering Crystallography: From Molecule to Crystal to Functional Form*, NATO Science for Peace and Security Series A: Chemistry and Biology, DOI 10.1007/978-94-024-1117-1_1

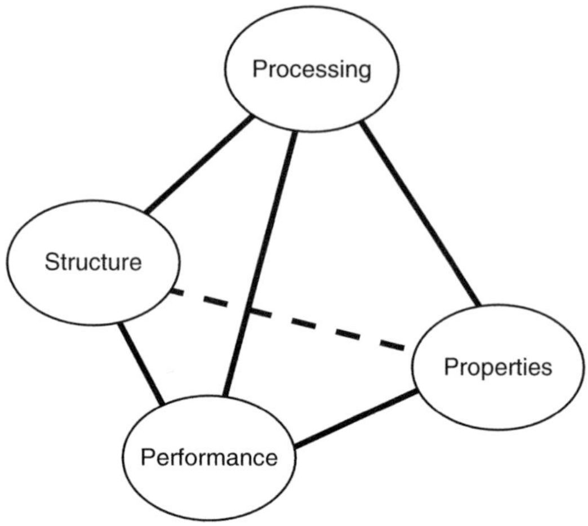

Fig. 1.1 The material science tetrahedron, after Sun [1] (Reproduced with the permission of Journal of Pharmaceutical Science)

The above holistic approach forms the overall aim of this Erice Summer School and in this brief introductory chapter the aim is to provide a short primer to the attendees for those unfamiliar with basic concepts of crystal science. The materials introduced here will underpin further chapters.

1.2 Crystals and Crystallography

1.2.1 Crystalline Solids

The principle differences between the three states of matter, solids, liquids and gases, illustrated in Fig. 1.2, lie in the separation and mobility of the atoms or molecules of which they are composed. In the gaseous state, molecules are highly mobile and separated by large distances. In liquids, molecules are in close proximity but still retain substantial mobility. While in the solid-state, molecular motion is typically limited to vibration and intermolecular distances are at a minimum

A crystal is a regular array of molecular entities, ordered in 3D where this degree of order extends up to macroscopic dimensions. A *single crystal* is the one in which this 3D order extends up to the sample's physical size. In contrast, a *polycrystalline solid* is the one where the crystal size is smaller than the actual sample size. In the latter case the sample could be aggregate of a number of smaller micro-crystalline grains such as an iron bar or a ceramic block, either of which could have a grain size of ca. 1 μm. It should be noted that the atomic level properties of both poly- and mono-crystalline solids are the same as the atomic-scale forces operate on the nm scale size, at which level the difference in nature of the molecular interactions between 1 μm^3 (micro-crystal produced in industrial crystallisation reactors) or

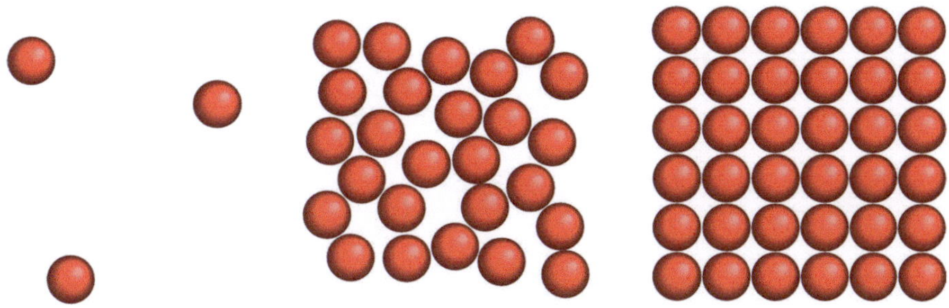

Fig. 1.2 Schematic diagram highlighting the differences between the gaseous (*left*), liquid (*centre*) and solid (*right*) states of matter

100 cm^3 (for electronic device materials) samples are negligible. Hence, the science underlying the structure and crystallisation behaviour for both of these types of materials, i.e. single crystals and polycrystalline solids is essentially the same.

Solids may be either crystalline or amorphous in nature. Crystalline solids are characterised by long-range 3D order and exhibit a periodic three-dimensional pattern in terms of the arrangement of the component atoms of the solid. The structure of amorphous solids, in contrast, closely resembles that of the liquid state. Although some short-range order may be present in amorphous solids, they are best described as disordered. An amorphous solid is thermodynamically metastable. It may transform to a crystalline structure if the kinetic barriers are not too great. However, many amorphous materials, such as glass, have retained their disordered structure for centuries. Materials with the same molecular structure can also have different crystal structures and this is referred to as polymorphism. For example, carbon forms crystals which can have one of three well known structures: graphite, diamond and buckminsterfullerene.

The structural differences between the different polymorphic forms of crystals and also those of amorphous forms of a given substance can give rise to substantial differences in both their physical and chemical properties. The relationship between structure and properties is a recurring theme which runs throughout this school.

1.2.2 Crystal Lattice

The simplest definition of a crystal is a 3D repeating pattern of atoms. The schematic given in Fig. 1.3b represents a 2D pattern of a single butterfly being repeated in the same orientation (a) which can be simply represented by an array of circles.

The 2D pattern (butterfly) is called the **motif** and the points which make up the array are called **lattice points**. One important feature of a lattice is that the

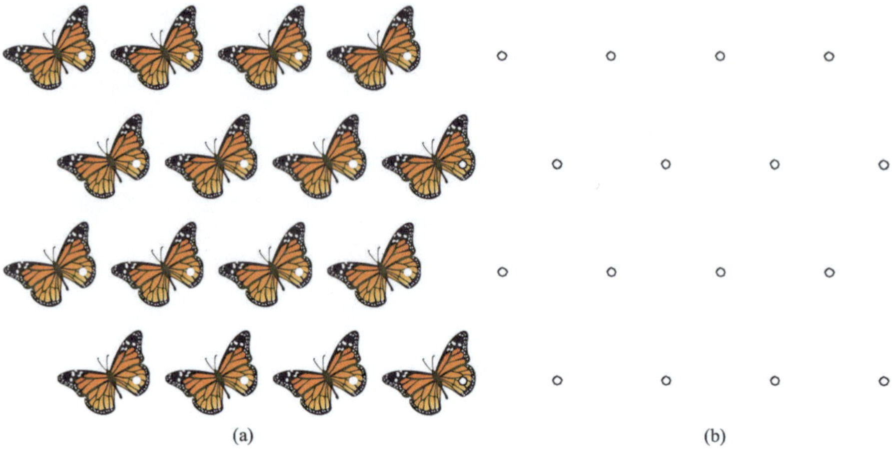

Fig. 1.3 A simple 2D pattern of a complex object (*a butterfly*) (**a**) together with its corresponding lattice (**b**) as represented by *circles*

environment of each lattice point is identical. In order to describe the scheme of repetition it is not, however, necessary to reproduce the entire lattice, but a small representative part, such as the parallelogram shown more completely in 3D in Fig. 1.4a. This still contains all the information necessary to define how the pattern is repeated and can be fully described using only the lengths of the sides (a and b) and the angle (γ) between the sides. This representative part of the lattice is referred to as a **unit cell**, as shown in Fig 1.4b.

An entity or motif sitting at the lattice point could be an atom (e.g. for metallic or elemental compounds), group of atoms (for complex atomic solid such as silicon which has 2 atoms/lattice point), collection of ions (e.g. ionic solid such as K_2SO_4 would have 2 K^+ and 1 SO_4^{2-} ions at each lattice point) or a molecule (e.g. organic compound would have complete molecule at each lattice point).

The key crystal science concepts are:

- *Crystal* = Motif (what is being repeated) + Lattice (where to repeat),
- *A more advanced definition of crystal* = Asymmetric unit (motif or what to repeat) + Space group (how to repeat about the lattice points) + Wykoff positions (where to repeat in the unit cell),
- *Crystal lattice* – infinite 3D array of points *(lattice points),*
- *Unit cell* – smallest reproducible unit within crystal lattice (Fig. 1.4),
- *Lattice direction* [uvw] – line drawn through lattice, where u, v, w are unit vectors along three non- orthogonal directions (Fig. 1.5),
- *Miller plane* (hkl) – 2D surface cut through lattice (Fig. 1.6),
- *Asymmetric unit* – atom, molecule, ions, groups of molecules.

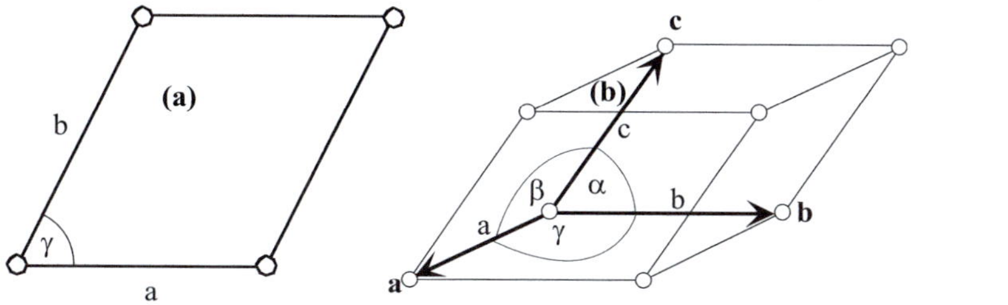

Fig. 1.4 (**a**) The unit cell and lattice parameters of the two dimensional pattern and (**b**) a three-dimensional unit cell, crystallographic axes a, b, c and the lattice parameters, a, b, c, α, β, γ

Fig. 1.5 Crystallographic directions shown by *coloured arrows*. A direction is denoted by the smallest coordinates of a point through which the *line* passes and usually the smallest integer are used

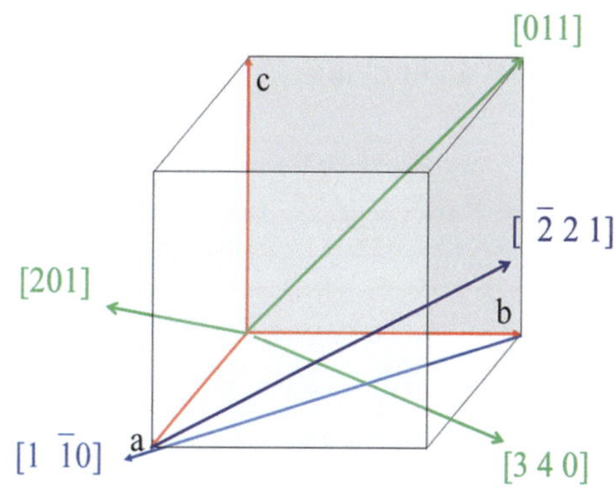

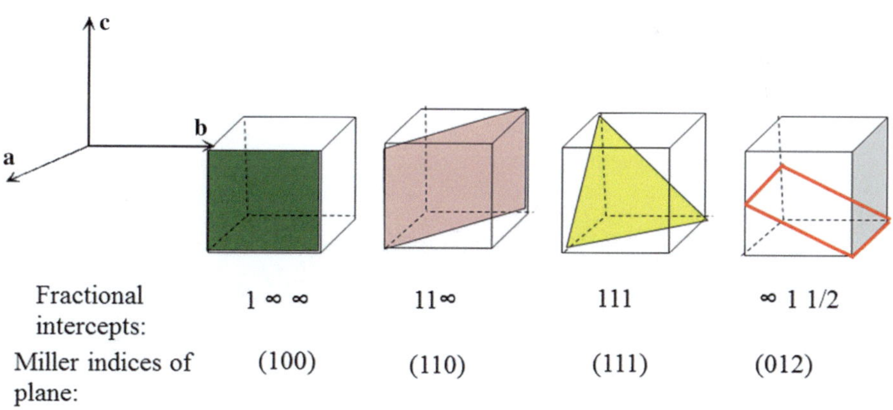

Fig. 1.6 Crystallographic planes or Miller planes are denoted by the reciprocal of the fractional coordinates of intercept of the plane with crystallographic axes

1.2.3 Crystal Systems

A vector r_{uvw} in lattice with respect to the three (integer) vectors u, v, w is defined as:

$$r_{uvw} = ua + vb + wc \tag{1.1}$$

a, b and c are lattice parameters defining the size and shape of the unit cell but are not necessarily orthogonal. Hence, this is why we additionally need three angles to define the unit cell. α – angle between b and c; β – angle between a and c, γ – angle between a and b.

The inter-relationship between a, b, c, α, β and γ gives rise to a total of 7 crystal systems as shown in Table 1.1. Examples of unit cells, Ibuprofen and D-mannitol, are illustrated in Fig. 1.7.

Table 1.1 Seven crystal systems and examples

System	Axial lengths and angles	Example
Cubic	$a = b = c, \quad \alpha = \beta = \gamma = 90°$	NaCl
Tetragonal	$a = b \neq c, \quad \alpha = \beta = \gamma = 90°$	TiO_2, Urea
Orthorhombic	$a \neq b \neq c, \quad \alpha = \beta = \gamma = 90°$	$(NH_4)2SO_4$
Rhombohedral or trigonal	$a = b = c, \quad \alpha = \beta = \gamma \neq 90°$	$CaCO_3$, α-Quartz
Hexagonal	$a = b \neq c, \quad \alpha = \beta = 90°, \gamma = 120°$	Zn
Monoclinic	$a \neq b \neq c, \quad \alpha = \gamma = 90° \neq \beta$	Paraffin, sucrose
Triclinic	$a \neq b \neq c, \quad \alpha \neq \beta \neq \gamma \neq 90°$	$CuSO_4$ $5(H_2O)$

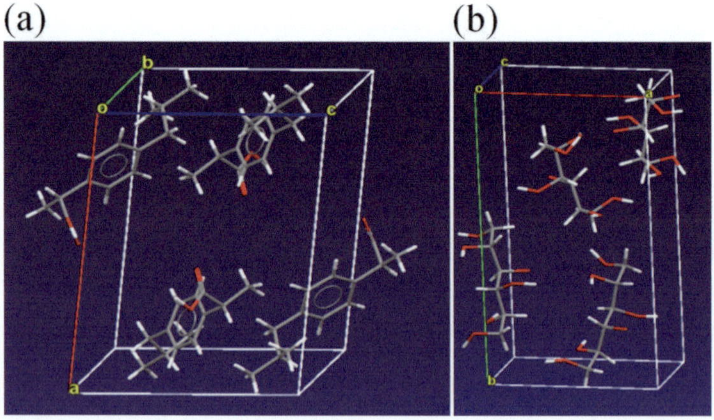

Fig. 1.7 Examples of unit cells: (**a**) Ibuprofen [2] belonging to monoclinic structure with: a = 14.67, b = 7.89, c = 10.73, β = 99.36; (**b**) D-mannitol [3] belonging to orthorhombic structure with: a = 8.67, b = 16.88, c = 5.56

1.2.4 Crystal Chemistry

We now need to add the chemical nature of the material into our concept of the crystal lattice. For a given molecule there are two dominant factors (Fig. 1.8) which influence what crystal structure a given material might adopt:

These factors affect physical and crystallographic properties. We can subdivide types of solid into five main groups of materials (see Fig. 1.9).

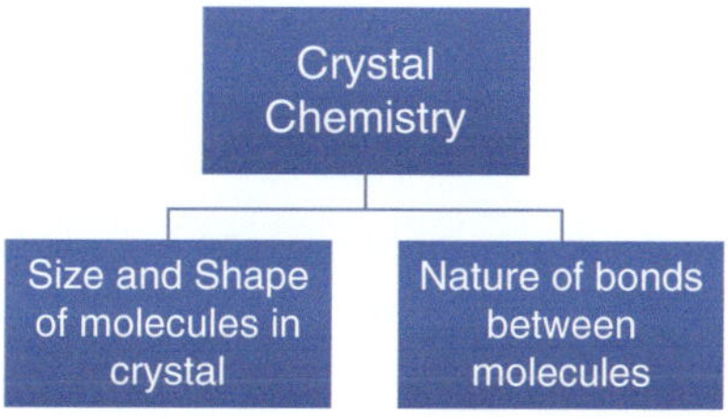

Fig. 1.8 Schematic explaining crystal chemistry with the key factors that affect the crystal structure

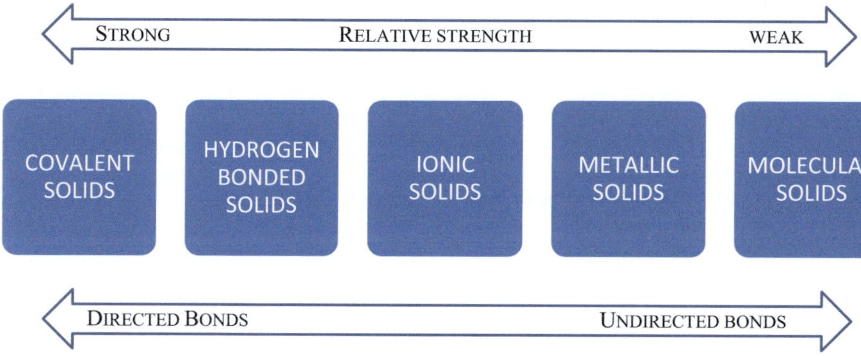

Fig. 1.9 Schematic showing the major types of inter-molecular bonding expected in the solid-state in relation to their relative strengths and their directional nature

Directed bonds tend to stop or restrict close packing which, in turn, makes for a lower density solid when compared with similar materials which have undirected bonds. The size of a molecule is reflected on the magnitude of the unit cell lattice parameters; e.g. the molecular solid benzophenone [4] ($(C_6H_5C = O)_2$) has lattice parameters of a = 10.28, b = 12.12, c = 7.99 A (orthorhombic) whilst much smaller atomic materials such as metal copper has a = 3 Å (cubic). The shape of a molecule reflects on the crystal system:

- Elemental compounds, such as metals and semiconductors, crystallising in high symmetry crystal classes such as cubic and hexagonal.
- Irregularly shaped molecules, such as organic molecules and inorganic complexes, crystallising in the lower symmetry triclinic, monoclinic classes.

The nature of bonding in any solid is of considerable importance since it influences, either directly or indirectly, a wide range of properties, notably melting point, crystal structure and mechanical properties. Most organic solids are molecular in nature. While the atoms within a molecule are held together by strong covalent bonds, the interatomic interactions are saturated and hence only weak intermolecular forces exist between molecules. Many pharmaceutical compounds are utilised in the form of salts and therefore exist as ionic species. In such cases, the role of strong ionic interactions must also be considered.

Three factors are of particular importance in assessing the influence of intermolecular bonding on the properties of organic solids:

- Strength of the interaction,
- Distance over which the interaction exerts an influence,
- Extent to which the bonding is directional.

Table 1.2 summarises the principle types of intermolecular bonding in organic solids.

Table 1.2 Types of bonding in organic solids

Type	Occurrence	Strength of interactions	Range	Directed interactions?
van der Waals	All atoms and molecules	Weak (1–10 kJ/mol)	short	No
Dipole-dipole	Between polar molecules only	Weak (3–4 kJ/mol)	long	No
Hydrogen bonding	Molecules containing O, N or F and a hydrogen bonded to one of these electronegative atoms	Strong (10–40 kJ/mol)	short	Yes
Ionic	Monatomic and molecular ions	Strong (10–50 kJ/mol)	short	No
Covalent	Intramolecular interactions between atoms	Strong	short	Yes

In organic compounds bond strengths are characterised by:

- Strong intra-molecular bonds – robust molecules,
- Weak and isotropic inter-molecular bonds – soft solids with low melting point.

The bonding directionality is un-directed except for covalent crystals and where there are H-bonds.

In summary, the main features that influence the organic solids are:

- Size of Molecules – large unit cells,
- Shape of Molecules – non-spherical shape,
- Nature of intermolecular forces – weak undirected,
- van der Waals bonds – close packed structures.

As a result, the organic crystalline solids have:

- Low melting point solids,
- Low symmetry crystallography,
- Polymorphic nature,
- Anisotropic deformation.

1.3 Defects in Crystals

So far we have assumed that the crystal lattice is perfect with no variations in crystal structure taking place on a dimensional range spanning atomic to macro-scale. Given that all molecules have thermal motion and that crystal growth is not an equilibrium process, this is an unrealistic prospect. In reality, a crystal during its formation and subsequent processing is likely to develop a range of lattice imperfections. Hence, all crystals are imperfect and can have atoms missing or in the wrong crystal lattice positions. In many respects the properties of a crystalline material are determined as much by the imperfections in the arrangement of its component atoms as by the crystal structure itself. The following discussion serves to illustrate the nature of these defects together with their importance in terms of controlling the properties and behaviour of crystalline solids.

1.3.1 Defects and Their Dimensionality with Respect to the Crystal Lattice

These collections of displaced atoms may also extend through the crystal in one, two or three dimensions. Thus the dimensionality of crystal defects beyond the atomic scale is a useful way of categorising the different types of imperfections which may exist.

Point Defects (0D) These are the most localised form of lattice defects and comprise:

- Vacancies (atoms missing from the crystal structure),
- Interstitials (atoms located in the, normally empty, voids within the structure),
- Impurities which can be foreign atoms incorporated into the crystal latter either, substitutionally, at a normal lattice site, or at an interstitial site.

Line Defects (1D) One dimensional line defects known as dislocations consist of atomic displacements and structural distortions which are confined to lie along a line through the crystal structure.

Planar Defects (2D) A variety of planar defects are known to exist within crystals. One of the most common of these is a stacking fault. These are regions of a crystal which contain either an extra layer of atoms or a missing layer of atoms. Within such regions there is a disruption to the normal stacking sequence of atoms in the structure.

Volume Defects (3D) Finally, three dimensional defects such as precipitates or inclusions are common in most crystals. Precipitates arise from the aggregation of impurity atoms whose concentration in the crystal exceeds the solubility limit due, for example, to a decrease in temperature. Inclusions, on the other hand are macro/

microscopic particles or volumes of solvent trapped inside the crystal as the growth front advances.

1.3.2 Point Defects and Their Role in Facilitating Solid State Diffusion

Point defects are equilibrium defects, i.e. due to Boltzmann statistics some atoms will always have sufficient kinetic energy to escape from their crystal lattice sites moving into *interstitial* sites or creating a *vacancy*. Without these defects we would not have any chemical reactivity in the solid state. Point defects influence almost all properties of crystalline solids, from electronic and optical properties through to mechanical behaviour. The process of solid-state diffusion, i.e. the movement of atoms through a solid is central to many physical properties and processes. It is involved in, for example, the mechanical behaviour of solids under deformation or their chemical degradation. Consider an idealised perfect crystal as shown on the left of Fig. 1.10. Because of the close packing of the atoms it would be extremely difficult for an atom of any type to move through the crystalline solid. If, however, the crystal contains vacancies (which all crystals do), atoms are able to change their positions by moving into the vacancy as shown in 4 diagrams to the right hand side of Fig. 1.10. The resulting movement of the vacancies also allows foreign atoms to move, albeit slowly, through the crystal.

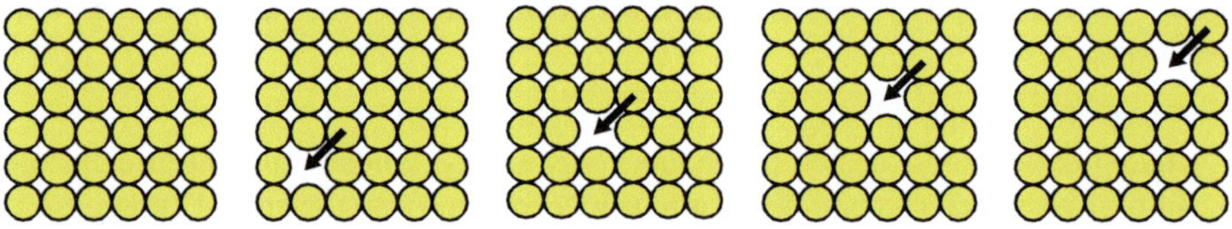

Fig. 1.10 Schematic atomistic view of a close packed crystal lattice showing how solid state diffusion processes though the movement of lattice vacancies

1.3.3 Dislocations and Their Character

The edge dislocation is perhaps the simplest type of dislocation to visualise. The geometry of an edge dislocation can be understood by considering an extra half plane of atoms inserted into an otherwise perfect crystal structure as shown in Fig. 1.11.

The edge of the extra half plane of atoms is the dislocation line. Close to the line the structure is highly distorted with significant displacement of the atoms. The direction of this distortion is called the Burgers vector and its magnitude decreases rapidly with distance from the line leaving the surrounding crystal essentially perfect. Since the chemical bonds close to the dislocation line are strained, the

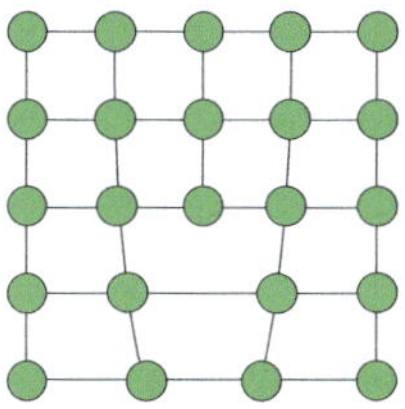

Fig. 1.11 Schematic atomistic representation showing the insertion of an extra half plane associated with the formation of an edge dislocation within the crystal lattice

activation energies for chemical reactions or dissolution can be expected to be different from the surrounding crystal. Hence, the fact that dislocations can influence on reactivity and dissolution. Note that the region below the extra half plane is in tension. This may allow the accumulation of impurities along the dislocation line which can further modify the chemical, electrical and other properties of the dislocation.

The nature and properties of a dislocation are determined by two vectors:

- Line direction (**l**)
- Burgers vector (**b**)

together with their orientation with respect to each other.

The edge dislocation, where $\underline{l}$ is perpendicular to $\underline{b}$, is not the only type of line defect. Figure 1.12 shows a second type of line defect, the screw dislocation so-called because the atomic planes around the dislocation line are helical in form. The geometry of the screw dislocation can be understood in the following hypothetical manner. A cut is made part-way through a block of crystal. The two sides are then displaced by one lattice translation with respect to each other. The dislocation line lies at the bottom of the cut. Thus, for a screw dislocation the Burgers vector and line direction are parallel. A feature of the screw dislocation, which is highly relevant to e.g. crystallisation, is the fact that it produces a partial step on the crystal surface.

The final type of dislocation is the mixed dislocation. The Burgers vector and line direction of a mixed dislocation lie at some arbitrary angle other than 0° or 90°. Visualisation of the geometry of a mixed dislocation is much more difficult than for the other two types. Their most important feature, with respect to crystal growth, is that they may also produce a surface step depending on the relative orientations of the Burgers vector with respect to the crystal face.

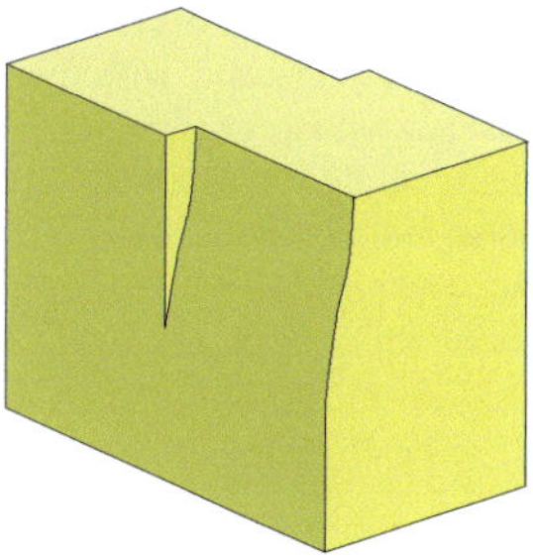

Fig. 1.12 Schematic representation of a screw dislocation highlighting that this defect generates a step on the crystal surface

1.3.4 Dislocations and Mechanical Properties

The mechanical strength of pure metals is up to five orders of magnitude lower than that predicted theoretically on the basis of the strength of bonds. This is due to the presence of dislocations which allow deformation of the metal crystals without the need for simultaneous breaking of all bonds across a deformation plane. Figure 1.13 shows this process on the atomic scale highlighting the motion of an edge dislocation through a crystal resulting in relative displacement of the two regions of the crystal above and below the dislocation line. This process of deformation via dislocation motion is known as slip. The motion of the dislocation itself in this manner is called dislocation glide. Thus, deformation of a crystalline solid can occur by the movement of dislocations through the crystal, requiring much lower stresses to initiate than would be necessary if all bonds across the plane were broken simultaneously.

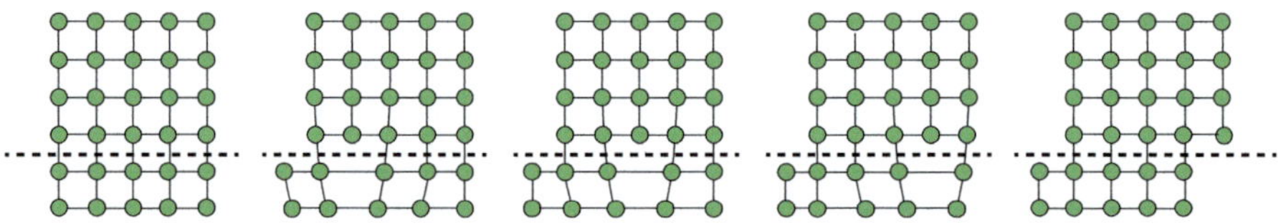

Fig. 1.13 Schematic atomistic representation of the mechanical deformation of the crystal lattice facilitated through the motion of an edge dislocation along a (*close packed*) slip plane

In simple terms the Burgers vector can be thought of as the direction in which slip has occurred. A dislocation can only move by glide in a plane which contains both its line direction and its Burgers vector. Such a plane is known as the slip plane (hkl). Thus the edge dislocation in Fig. 1.13 is confined to glide in a single slip plane indicated by the dotted line. During the mechanical deformation process layers of crystal lattice slip over each other and give rise to plastic deformation with the direction [*uvw*] of slip being along the Burgers vector.

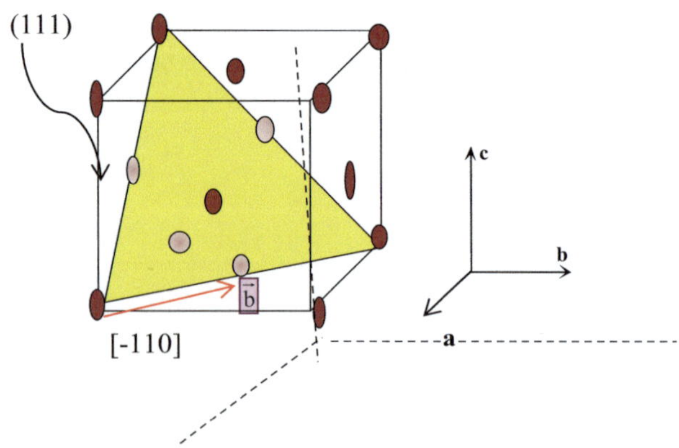

Fig. 1.14 An example of a dislocation slip system for a face centred cubic crystal structure with the likely slip plane being the most *closed packed* (111) plane with the dislocation Burgers vector being along the shortest inter-atomic contact, i.e. along the [−110] direction

The energy of a dislocation involved in this process can be approximated to:

$$E \sim G|\boldsymbol{b}|^2 \tag{1.2}$$

where G is the modulus of elasticity. From this we can see that, to minimise the dislocation energy, Burgers vectors with the shortest lengths are preferred. Figure 1.14 illustrates this for the case of a cubic face centred lattice where the shortest distance between two equivalent atom sites will be from the corner to the face centre site, i.e. a Burgers vector $\boldsymbol{b}$ = ½ a [−110] being the most likely with the likely slip plane being (111).

1.3.5 Dislocations and Chemical Reactivity

Dislocations also influence chemical reactivity. The points of emergence of dislocations on a crystal surface can be revealed by removing atoms or molecules from the surface with a suitable reagent or solvent. Enhanced reactivity and dissolution at the dislocation can result in a pit or similar feature. As we will see in later chapters, dislocations are also closely involved in the growth of crystals by providing surface steps to which atoms or molecules can more readily bind during growth.

1.4 Crystal Morphology and Surface Chemistry

1.4.1 The Nature of Crystal Surfaces

Crystals grown from solution typically exhibit regular, planar facets. Although appearing flat to the naked eye, these crystalline surfaces are rarely so at the molecular level. The various features which make up the nanoscale surface topography of the crystal face are intimately involved in the mechanisms by which crystals grow. Figure 1.15 is a schematic representation of crystal surface in which the molecules, or growth units, are represented by cubes. The principal features of interest are:

- Steps,
- Terraces between steps,
- Kink sites on the steps,
- Adsorption of molecules.

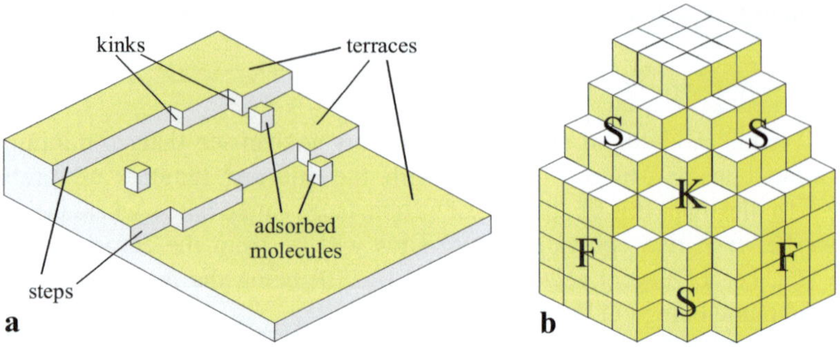

Fig. 1.15 (**a**) Steps, terraces, adsorbed molecules and kink sites on a crystal surface; (**b**) flat (F), stepped (S) and kink (K) faces

It is important to consider at what point a molecule from solution becomes truly part of the crystal. In this simplified model of a crystal surface, each cube can form a maximum of six bonds with surrounding cubes – one per face of the cube.

Examining the various sites on the schematic crystal surface, we can see that each is capable of forming a different number of bonds with an incoming molecule from the solution. A molecule adsorbed onto a terrace can form only one bond with the crystal surface. A molecule in contact with a step can form two bonds while a molecule incorporated at a kink site can form three bonds. The kink site is of particular importance since a molecule placed in such a site does not increase the surface area or, therefore, the surface energy of the crystal. A molecule can, thus, be considered to be truly part of the crystal once it has reached a kink site. Though the nature of bonding for real molecules will be substantially more complex, this simple model reproduces the key differences between the various surface sites. When the cubes of Fig. 1.15a are assembled to form a crystal, we find that three different types of crystal face occur: flat or F faces, stepped or S faces and kink or K faces as shown in Fig. 1.15b.

1.4.2 Crystal Habit

The external shape of a crystal is referred to as the crystal habit. The faceted crystals are bounded by the slowest growing faces and it follows that the crystal habit is determined by the relative growth rates of the possible faces for that crystal structure. Crystal habit is described using a variety of qualitative terms together with the Miller indices of the faces present. Figure 1.16 shows examples of (a) plate-like, (b) prismatic and (c) needle-like crystal habits.

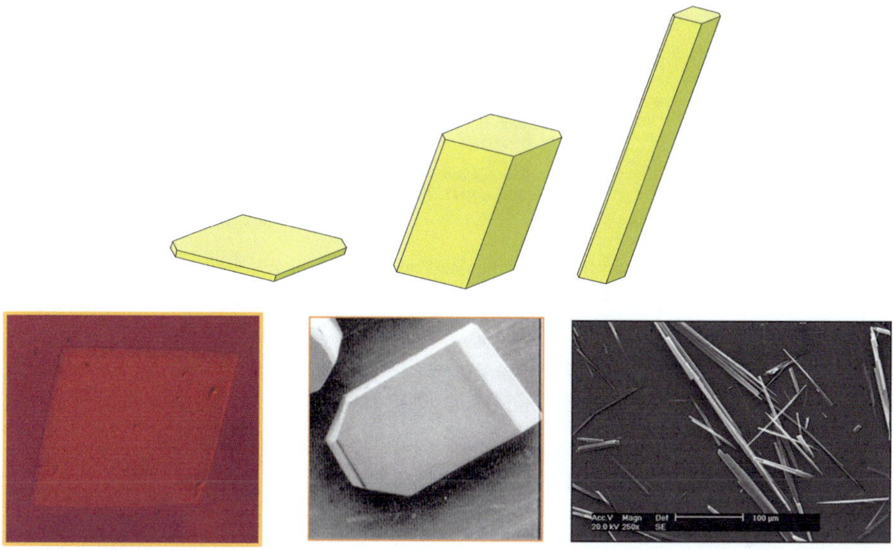

(a) (b) (c)

Fig. 1.16 Examples of crystal shapes (**a**) (*top left*) plate-like (example: methyl ester stearate [5]); (**b**) (*top centre*) prismatic (example: urea [6]); (**c**) (*top right*) needle-like (carbamazepine: nicotinamide co-crystal [7])

1.4.3 Surface Chemistry

The external shape of crystals not only has a considerable influence on the properties of the solid but also has important technological consequences, particularly in relation to handling of particulate materials.

Such differences are due to the fact that the crystal surfaces represent the termination of the bulk 3D crystal lattice and as such and depending on the crystal structure different crystal surfaces have different surface chemistry and hence interact with other crystals differently due to e.g. differences in surface wetting due to differences in their hydrophobic or hydrophilic nature. It is thus clear that the vertical and horizontal potential faces are each made up from different parts of the molecule and, therefore, contain different functional groups. As a consequence, the surface chemistry of crystals can depend strongly upon the crystal habit. This is demonstrated for the case of urea ($C = O\ NH_2$) shown in Fig. 1.17 which shows the contrasting surface chemistry exposed on the {110} surfaces which is strongly protic in nature and the {001} surface which comprises an alternative mixture of protic and carbonyl molecular sites.

In addition, the surface area for a given crystal volume depends upon the shape of the crystal. For example, the specific surface area of elongated (needle-like) crystal habit is substantially larger than that of more isometric shapes such as cubes. The combination of differences in both surface chemistry and surface area can

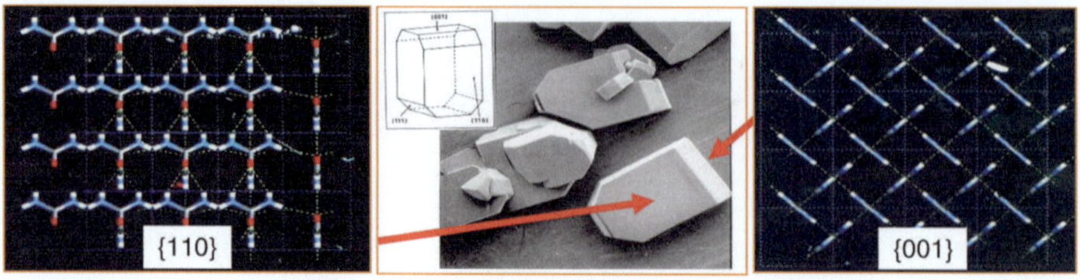

Fig. 1.17 The different surface chemistry exposed on the {110} facet (*left*) and the {001} (*right*) crystal habit faces of urea in relation to the observed crystal morphology (*centre*), after Ref. [6]

result in changes in dissolution rate of crystals as a function of crystal habit. On a technological level, crystal habit can influence a wide range of behaviour including:

- Ease of filtration of a crystal suspension,
- Crystal flow and packing characteristics,
- Crystal compression behaviour.

The latter two are of particular relevance to tabletting. Crystal shape also effects caking behaviour, i.e. the tendency of polycrystalline materials to bind together during storage.

1.5 Concluding Remarks

A brief overview of the fundamentals of crystallography and applied crystal science has been provided which will be helpful to those unfamiliar with the basic concepts underpinning the connectivity between molecular structure, the intermolecular interactions that define crystal chemistry and surface properties. The later chapters of this book will illustrate how crystal science concepts provide a useful fundamental understanding regarding how:

- crystal structure links to the physical and chemical properties of crystals,
- structure/property inter-relationship can be controlled by processing,
- structure, properties and processing can impact upon the practical performance of materials.

References

1. Sun CC (2009) Materials science tetrahedron—A useful tool for pharmaceutical research and development. J Pharm Sci 98:1671–1689
2. McConnell JF (1974) 2-(4-Isobuthylphenyl) propionic acid. $C_{13}H_{18}O_2$ ibuprofen or prufen. Cryst Struct Commun 3:73–75

3. Fronczek FR, Kamel HN, Slattery M (2003) Three polymorphs (α, β, and δ) of D-mannitol at 100 K. Acta Crystallogr Sect C Crystl Struct Commun 59:567–570
4. See e.g. Kutzke H, Klapper H, Hammond RB, Roberts KJ (2000) Meta-stable β-Phase of benzophenone independent structure determinations via X-ray powder diffraction and single crystal studies. Acta Crystallogr B 56:486–496
5. Comacho-Corzo DM, Thomas D, Muller F, Roberts KJ, Lewtas K, More I (2017) Morphology and growth kinetics of methyl stearate as a function of crystallisation environment. Crystal Growth and Design 17:563–575
6. Docherty R, Roberts KJ, Saunders V, Black S, Davey RJ (1993) Theoretical analysis of polar morphology and absolute polarity of crystalline urea. Faraday Discuss 95:11–25
7. Rahim A, Hammond RB, Sheikh AY, Roberts KJ (2013) A comparative assessment of the influence of different crystallisation screening methodologies on the solid forms of carbamazepine co-crystals. Cryst Eng Comm 15:3862–3873

Chapter 2
Molecular Structure and Chirality and Chiral Crystals

Reiko Kuroda

Abstract The term 'chiral' is defined as "geometrical figure, or group of points whose image in a plane mirror cannot be brought to coincide with itself." The obvious examples are our hands and foot, as left hand (foot) becomes right hand (foot) in a mirror, and vice-versa. In this paper we shall focus on chirality of molecules particularly in the crystalline state where molecules are densely surrounded by neighbouring molecules in an ordered manner. In the crystalline state, conformational flexibility is restricted, and chiral discrimination energy is order of magnitudes larger compared with that in the solution phase. Thus, facile optical resolution during supramolecular crystallization can be achieved. Further, co-grinding of multiple crystals of different compounds produces unique co-crystals or crystals of unique compounds. We have ascribed the origin of these interesting phenomena to the memory of the original crystal structures which is at least partially retained during the solid-state crystallization processes. These novel chiral crystalline chemistries will be introduced, mainly focused on our own work.

Keywords Chirality • Optical resolution • Chiral discrimination • Crystal co-grinding • Solid-state crystallization

2.1 Introduction

Chirality (left-right (l-r) asymmetry) is expressed throughout nature, whether microscopic or macroscopic. Some molecules are either chiral or non-chiral, and billions of them self-assemble to form macroscopic objects of either chiral or non-chiral crystals. It is intriguing how this is achieved in terms of molecular interactions. At the molecular level the biological world is totally homochiral, i.e., DNA and proteins utilize molecules of only one particular handedness. When, where, why and how was the chirality chosen on the primordial Earth?

R. Kuroda (✉)
Research Institute for Science and Technology, Tokyo University of Science, 2641 Yamazaki, Noda-shi, Chiba 278-8510, Japan
e-mail: rkuroda@rs.tus.ac.jp

K.J. Roberts et al. (eds.), *Engineering Crystallography: From Molecule to Crystal to Functional Form*, NATO Science for Peace and Security Series A: Chemistry and Biology, DOI 10.1007/978-94-024-1117-1_2

Hypotheses on the origin of life on Earth must explain the homochirality of the biological world at the molecular level. Further, as a consequence of the homochiral biological world, diastereomeric discriminations operate when chiral compounds interact with living organisms, and compounds often display different characteristics depending on the chirality, such as drug effect/side effect, taste, fragrance, etc. Thus, chirality is important in pharmaceutical, agricultural, and food industries.

2.2 Historical Aspects and Definition of Chirality

The term 'chiral' was first introduced by Lord Kelvin (William Thomson), Professor of Natural Philosophy in the University of Glasgow, in 1884 in his Baltimore lectures: I call any geometrical figure, or group of points, 'chiral', and say that it has chirality if its image in a plane mirror, ideally realized, cannot be brought to coincide with itself [1]. The word *chirality* is derived from the Greek, *χειρ* (*kheir*), "hand", a familiar chiral object.

The history of chiral stereochemistry started when Arago and later Biot discovered that there are two forms of quartz crystals, one dextrorotatory and the other laevorotatory in 1812 [2]. Before the time of Pasteur, the enantiomeric substances were termed 'physical isomers' since they appeared to differ in no respect other than the property of rotating the plane of polarized light in a clockwise or anticlockwise sense. In 1848 Louis Pasteur hand-sorted enantiomorphic crystals of sodium ammonium tartrate by distinguishing the hemihedral crystal facets, and discovered spontaneous optical resolution [3]. In his review lectures on molecular dissymmetry in 1860, Pasteur coined the term 'dissymmetric' to describe structural forms which are not superimposable by translations and rotations alone upon the corresponding mirror-image structures [4]. This is equivalent to the term 'chiral' introduced by Kelvin as described above. Pasteur noted the preferential crystallization of racemates and rare incidence of spontaneous resolution by crystallization. The second of Pasteur's discoveries is 'diastereomeric discrimination'. He noted the different chemical and physical properties of salts and other combinations of the two optical isomers ((+)-A and (−)-A) in a racemate with a single enantiomer of another chiral substance ((+)-B), i.e., (+)-A/(+)-B vs (−)-A/(+)-B. The unequal solubilities of diastereomers afford a general procedure for the optical resolution of racemates even nowadays. Pasteur also discovered the biochemical selectivity of micro-organism between the two enantiomers in a racemic mixture, exemplified by the metabolism of specifically the (+)-isomer of racemic tartaric acid by *Penicillium* moulds [3].

2.3 Molecular Chirality

Asymmetric carbon atom is a carbon atom that has four different substituents as Xabcd, and serves as a chiral center. As is well known, organic stereochemistry began with the discovery of chirality in organic compounds. Based on the number of stereoisomers of an organic compound, the stereochemistry of the carbon atom X was proposed to be tetrahedral by Le Bel [5] and van't Hoff [6]. If a molecule consists of n inequivalent asymmetric carbon atoms, there are 2^n optically-active isomers, grouped into enantiomeric pairs. If, however, the n asymmetric carbon atoms are equivalent, there are $(1/2)2^{n/2}$ inactive meso-isomers and $(1/2)2^n$ optically-active isomers, again grouped into enantiomeric pairs. The famous example is tartaric acid. Meso tartaric acid is achiral although it has two asymmetric carbon atoms (Fig. 2.1a). Similarly, there are many chiral organic compounds which are devoid of asymmetric carbon atoms. Several examples are shown in Fig. 2.1b–d. 1,1′ -Binaphthyl derivatives represent an important and interesting class of axial chirality, and helicenes are another interesting class of chiral organic compounds. If the rotation energy barrier around the central carbon-carbon bond is low, the compounds are in equilibrium of rapidly converting R an S isomers, and hence achiral. In crystals where molecular freedom is restricted, the compounds can exist as chiral entity. Metal complexes such as $[Co(en)_3]^{3+}$ (en = ethylenediamine) are chiral. When viewed along the C_3 axis, the (+)-isomer has the form of a left-handed three-bladed propeller, whereas the (−)-isomer a right-handed three-bladed propeller (Fig. 2.1e).

2.4 Results and Discussion

2.4.1 *Molecular Interactions in Racemates and Optically Active Crystals*

Spontaneous optical resolution was discovered by Pasteur on sodium ammonium tartrate, however, Scacchi obtained only one crystal type, with a holohedral form, i.e., racemate crystals in 1865 [7] (Fig. 2.2a). This is because there is a transition at 27 °C, and above the transition point, spontaneous optical resolution does not occur. Billions of molecules self-assemble to form crystals by exerting chiral discrimination. In order to understand the reason behind the formation of either chiral or racemic crystals, I have investigated the Pasteur's and Scacchi's crystal structures and compared them by making crystals above and below the transition temperature [8]. Pasteur's crystal is tetrahydrate, while Scacchi's is monohydrate. Thus, it is not possible to compare the lattice energy directly but interesting features were revealed from the X-ray crystallographic analyses. The molecular structure of the tartrate ion in the two crystals differs in the orientation of a hydroxyl group. In the active crystal, the two carboxylate groups are placed to form an intramolecular

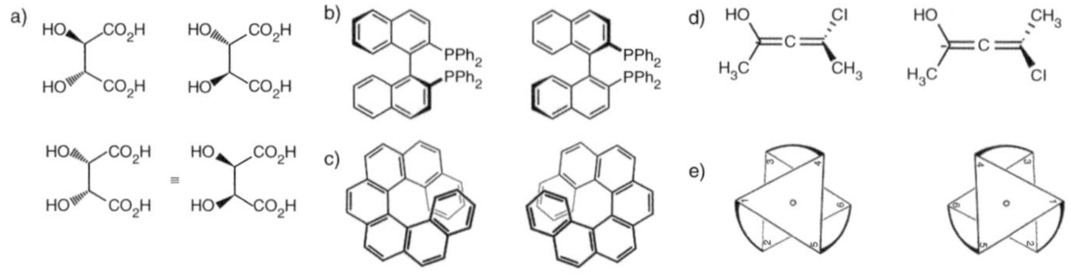

Fig. 2.1 (**a**) Optically active and *meso* tartaric acid, (**b**) BINAP, (**c**) [6]-helicene, (**d**) 4-chloropenta-2,3- dien-2-ol, and (**e**) $[Co(en)_3]^{3+}$

hydrogen bond with each hydroxyl group. In the racemic structure, only one of the hydroxyl groups of a tartrate ion forms a similar intramolecular hydrogen bond, while the other hydroxyl group is directed towards another tartrate ion and forms an intermolecular hydrogen bond. An interesting feature is that achiral water and ammonium ion play a key role in crystal chiral discrimination (Fig. 2.2b).

2.4.2 *Powerful Chiral Discrimination in Supramolecular Crystallization*

Chiral discrimination, recognition, generation and transfer in the solid phase are expected to be much stronger than in solution phase. We have achieved facile optical resolution of secondary alkyl alcohols, which is known to be quite difficult by conventional chemical methods, by simple crystallization using the target alcohols as solvent [9]. The supramolecular system is composed of (1*R*, 2*R*)-diphenyl- ethylene-diamine ((1*R*, 2*R*)-**1**) and a carboxylic acid either biphenic acid (**2**) or 2,2′-binaphthyl-3,3′- dicarboxylic acid (**3**) (Fig. 2.3). These dicarboxylic acids are not chiral in solution due to a rapid rotation around the central carbon–carbon bond, however, they can exhibit axial chirality when the rotation is restricted.

From a racemic 2-butanol solution of (1*R*, 2*R*)-**1** and **2**, colourless crystals were obtained after 1 or 2 days, which contain (1*R*, 2*R*)-**1**, **2** and 2-butanol (Fig. 2.4). In the crystal, chirality of the dicarboxylic acid is fixed to *aR*, due to chirality transfer from the chiral amine. A characteristic hydrogen bonding pattern was observed, i.e. a columnar intermolecular hydrogen-bond network around the 2_1-axis formed by the ammonium hydrogen of amine/H^+ and the carboxylate oxygen of a biphenic acid anion (Fig. 2.5).

The (*S*)-2-butanol is selectively trapped in the cavity formed by the columns, with ee (enantiomeric excess) as high as 91%. There is a hydrogen bond between the hydroxyl group of (*S*)-2-butanol and the carboxylate oxygen of a biphenic acid anion. The crystals included water which was present in the solvent, with the stoichiometry of (1*R*, 2*R*)-**1**:**2**:H_2O:2-butanol =1:1:1:1. The water molecules contribute to the maintenance of the column framework (Fig. 2.5). Interestingly, under

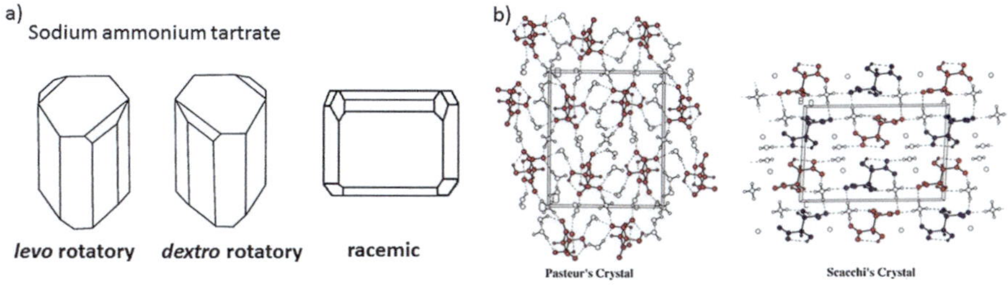

Fig. 2.2 (**a**) Crystal shape of levo, dextro and racemic sodium ammonium tartrate. (**b**) Crystal structure of Pasteur's optically active and Scacchi's racemic crystal (Ref. [8])

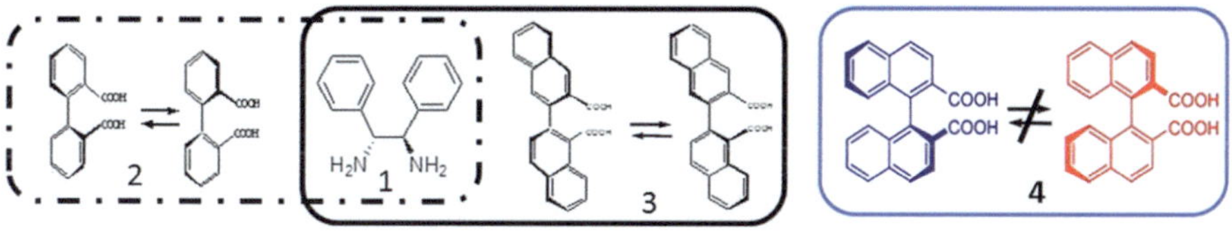

Fig. 2.3 Supramolecular crystallization was carried out between (1*R*, 2*R*)-diphenylethylene-diamine ((1*R*, 2*R*)-**1**) and achiral dicarboxylic acids **2** and **3** or chiral dicarboxylic acid **4**

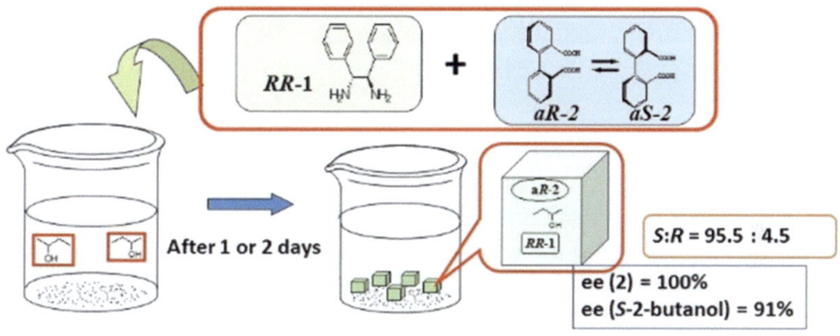

Fig. 2.4 Optical resolution of 2-butanol by supramolecular crystallization. The alcohol is used as a solvent

dry conditions using meticulously dried 2-butanol, no inclusion complex was formed and hence no optical resolution of 2-butanol was achieved.

The enantioselectivity was decreased when the length of the guest alcohol molecule increased (Table 2.1). However, we can optically resolve a variety of secondary alkylalcohols by way of changing the combination of dicarboxylic acids and the chiral diamine. The combination of (1R, 2R)-1 and 2 appears to be suitable for the resolution of small alcohols, whereas the combination of (1*R*, 2*R*)-**1** and **3** is better for larger alcohols (Table 2.1).

Successive optical resolution of rac-1,1′-binaphtyl-2,2′-dicarboxylic acid (**4**) and secondary alkylalcohols was achieved in one-pot experiment by using (1*R*, 2*R*)-**1** [10]. Unlike **2** and **3**, **4** is chiral due to the restricted rotation around the central

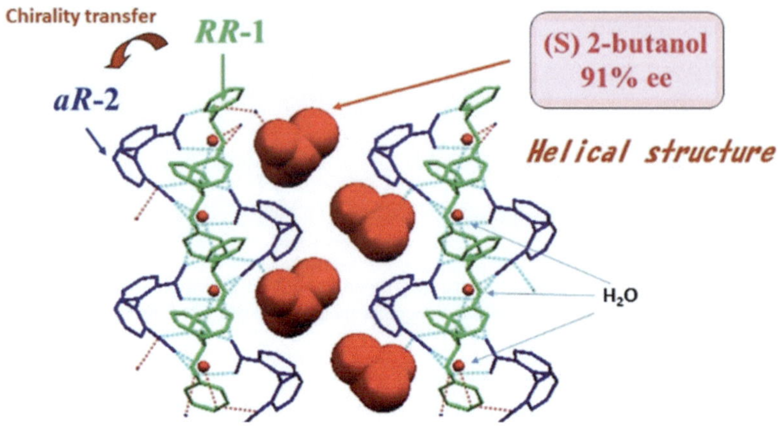

Fig. 2.5 Crystal structure of 1:1:1:1 (1*R*, 2*R*)- diphenylethylenediamine (**1**): biphenic acid (**2**): H_2O: 2-butanol complex, showing the cavity in which (*S*)-butanol is selectively trapped

Table 2.1 Optical resolution of secondary alkylalcohols by (1*R*, 2*R*)-**1**· **2** and (1*R*, 2*R*)-**1**· **3**. A.C. = absolute configuration

		A: (1*R*, 2*R*)-**1** and **2**		B: (1*R*, 2*R*)-**1** and **3**	
	alcohol	ee%	A.C.	ee%	A.C.
1	2-butanol	91	*S*	42	*S*
2	2-pentanol	61	*S*	57	*S*
3	2-hexanol	22	*R*	71	*S*
4	2-heptanol			70	*S*

carbon-carbon bond, and *aS* and *aR* forms exist. From a racemic alcohol solution of (1*R*, 2*R*)-**1** and rac-**4**, many crystals were produced after 1 day, which contained (1*R*, 2*R*)-**1** and (aS)-**4**, with as high as 98–99% ee for **4**. The solution was filtered to remove formed crystals, and the resulting filtrate was left to stand at room temperature. After 2 days, plenty of crystals were formed, which contained (1*R*, 2*R*)-**1**, (a*R*)-**4** and the solvent alcohol (Fig. 2.6). The optical purity of **4** is 94–99%, and that of alcohols is 21, 41 and 62% ee for (*S*)-2-butanol, (*S*)-2-pentanol, and (*R*)-hexanol, respectively. These ee for these alcohols are not high, but far better than those by other methods [11].

2.4.3 Chiral Discrimination in the Solid-State Crystallization – Memory of Crystal Structures

We have also found that crystal co-grinding in the total absence of liquid produces adduct crystals which are different from crystals obtained by conventional solution crystallization in terms of stereochemistry including chirality [12–17]. On grinding white racemic crystals of bis-β-naphthol (*rac*-BN) and benzoquinone (BQ), the colour of the powder changes progressively with time from very pale yellow to pink

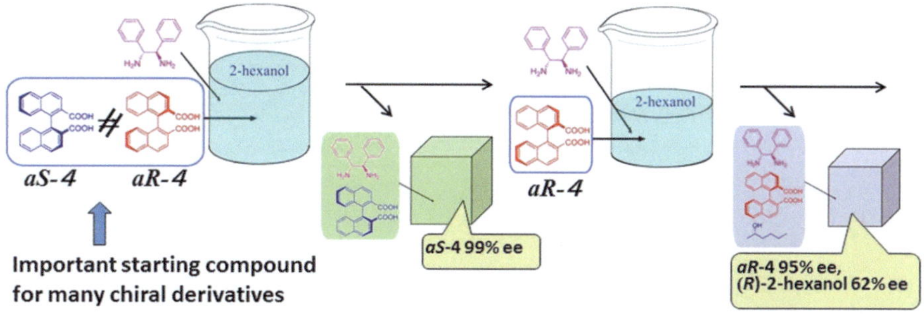

Fig. 2.6 Successive optical resolution of 1,1′-binaphtyl-2,2′-dicarboxylic acid (**4**) and 2-hexanol by crystallization

and eventually to strong red (Fig. 2.7a) [12]. The process was monitored by X-ray powder diffractometry. The powder diffraction pattern recorded after 20 min of grinding exhibited a new set of peaks in addition to the peaks corresponding to pure *rac*-BN and pure BQ crystals. After 45 min of grinding, the *rac*-BN and BQ peaks disappear completely and the new set of peaks becomes stronger (Fig. 2.7a). While leaving samples in a closed environment after 15 min co-grinding, the PXRD pattern changed continuously for 48 h (Fig. 2.7b). It should be noted that no intermediate phase is observed.

Remarkably, when adduct crystals were exposed to air, the strong red colour was lost with time and, after 23 h, the crystals eventually became white. It might be expected that the crystals would deteriorate and become amorphous, but instead, they reverted to the starting *rac*-BN crystals, as unambiguously demonstrated by the X-ray powder diffraction (Fig. 2.7a). When the adduct crystals were placed under the BQ vapour, no decomposition of the adduct was observed.

Co-grinding of crystals produced a novel crystal adduct (**I**) through crystal sheering and molecular diffusion processes in the solid state, which is structurally distinct from adducts obtained from solution (**II**) (Fig. 2.8) [12]. The composition of the two crystals is different: **I** contains 1:1.5 *rac*-BN:BQ (i.e., 1:1:3 in terms of *R*-BN:*S*-BN:BQ), whereas **II** exhibits a 1:1 ratio (1:1:2 in terms of *R*-BN:*S*-BN:BQ).

Crystal **I**: The asymmetric unit comprises one independent BN molecule and two independent BQ molecules, of which one resides at the centre of symmetry. A semi-column structure runs along the *c* axis where the aromatic planes of naphthol and BQ stack alternately. BQ is sandwiched by the naphthol rings of the *R*-BN and *S*-BN molecules, which are related by the *c*-glide. Only one of the carbonyl groups of BQ forms a hydrogen bond to the hydroxy group of a naphthol (O···O = 2.733 Å). The BQ on the centre of inversion forms a triplet structure sandwiched by the inversion-related *R*- and *S* naphthol rings. Two symmetry-related hydrogen bonds are formed between the BQ carbonyl and the hydroxy group of the nearly perpendicular naphthol ring of a BN molecule (O···O = 2.703 Å).

Crystal **II**: There are three independent molecules of BN and BQ in the asymmetric unit. The three independent BQ molecules are all sandwiched by the roughly

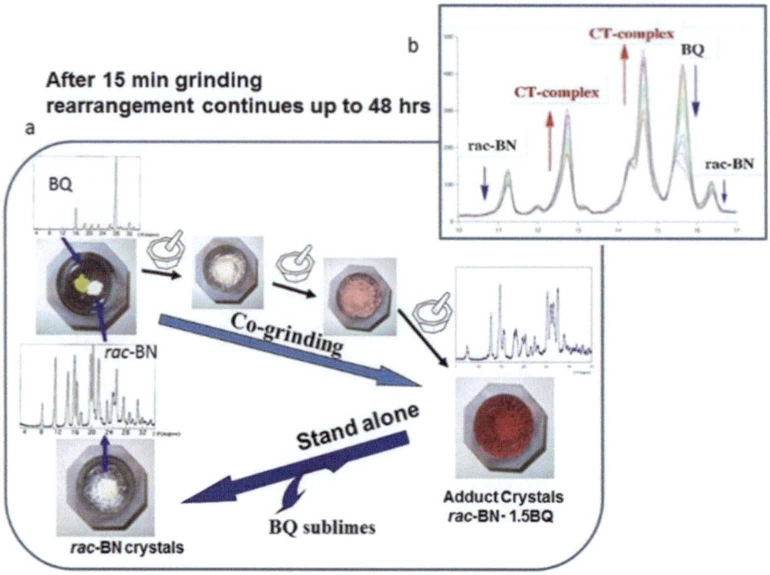

Fig. 2.7 (**a**) Co-grinding of benzoquinone(BQ) and racemic-bis-β-naphthol (*rac*-BN) crystals produced adduct crystals. *Colour* and PXRD changes. (**b**) PXRD changed continuously after 15 min co-grinding

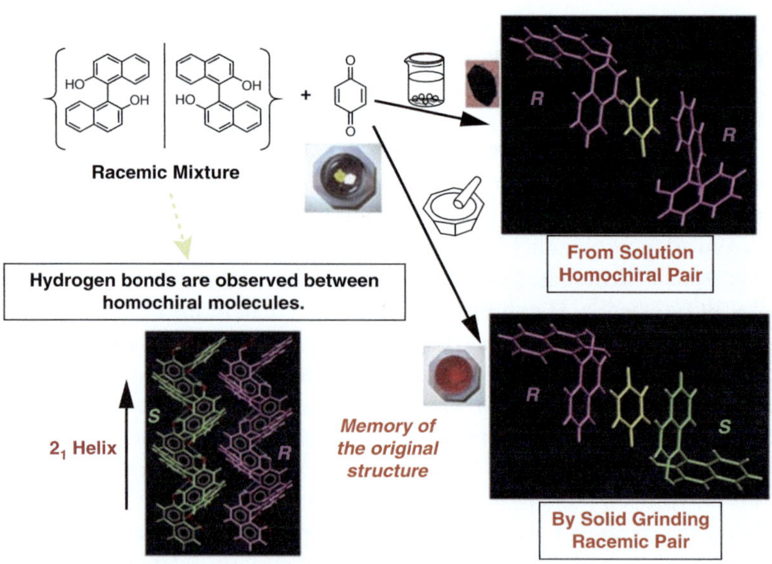

Fig. 2.8 Crystal structures of BQ:*rac*-BN adduct formed from solution and solid- state crystallization. Crystal structure of *rac*-BN is also shown [12]

parallel naphthol rings of two BN molecules. All possible hydrogen bonds are exploited between the carbonyl oxygen of BQ and the hydroxy group of the nearly perpendicular naphthol ring of BN molecules (O···O distance varies from 2.707–2.789 Å). Theoretical calculations using the ZINDO method [18] indicated that the strong red colour of **I** and **II** which appear only in the crystal phase is due to charge transfer essentially from the 1:1 BN:BQ complex [12].

The local geometry of the BN···BQ···BN triplets in **I** and **II** appears similar, however, the chirality is in sharp contrast. BQ is sandwiched by a racemic pair in **I**, but by a homochiral pair in **II**. We believe that the difference arises from the memory of original crystal structures which influences the crystallization process in the case of solid-phase crystallization. In the crystal structure of pure racemic BN (space group *Iba2*), hydrogen bonds are formed between the hydroxy groups of BN molecules related by a 2_1 screw axis, thus forming a homochiral column [19, 20]. The crystal structure of pure BQ (space group $P2_1/a$) has no hydrogen bonds [21, 22]. On co-grinding, BQ molecules may be inserted between the neighbouring opposite handed homochiral BN columns, which may encourage the formation of a hetero-chiral sandwich found in **I**. Chiral BN crystals do not form a new phase with BQ by either solid or solution crystallization. Upon complexation in the solid state, BN and BQ molecules change their crystal packing critically and form new hydrogen bonds between the two component molecules.

2.4.4 Annealing Promotes New Crystal Phase Formation in the Solid State

As seen above, diffusion and rearrangement of molecules can occur even in the solid state, and annealing promotes these. For example, co-grinding of a 1:1 mixture of Δ-$[Ru(phen)_3](PF_6)_2$ and Λ-$[Ru(phen)_3](PF_6)_2$ crystals (phen = 1,10-phenanthroline) at room temperature produced an amorphous sample (Fig. 2.9a). However, subsequent annealing treatment of the sample at 250 °C for 30 min promoted the formation of a crystal which is different from the starting crystals [23]. Δ-$[Ru(phen)_3](PF_6)_2$ and Λ-$[Ru(phen)_3](PF_6)_2$ are crystallized in the chiral space group of $P2_12_12_1$. The resultant crystals are identical to that of *rac*-$[Ru(phen)_3](PF_6)_2$ in the space group of *C2/c*.

To confirm that the formation of the racemic crystals is not due to the racemization of metal complexes in each enantiomeric component crystals, but is the result of molecular rearrangement in the solid state, we have carried out similar crystal co-grinding/annealing experiments using Δ-$[Ru(phen)_3](PF_6)_2$ and Λ-$[Os(phen)_3](PF_6)_2$.[22] Co-grinding and subsequent annealing produced quasi-racemic compound crystals. Both chiral and racemic crystals of $[Ru(phen)_3](PF_6)_2$ and $[Os(phen)_3](PF_6)_2$ are isomorphous to each other. Ru and Os complexes in chiral and racemic crystals are optically stable even at 250 °C. HPLC analyses using a chiral stationary phase have proven that the newly formed crystals indeed contain Δ-[Ru

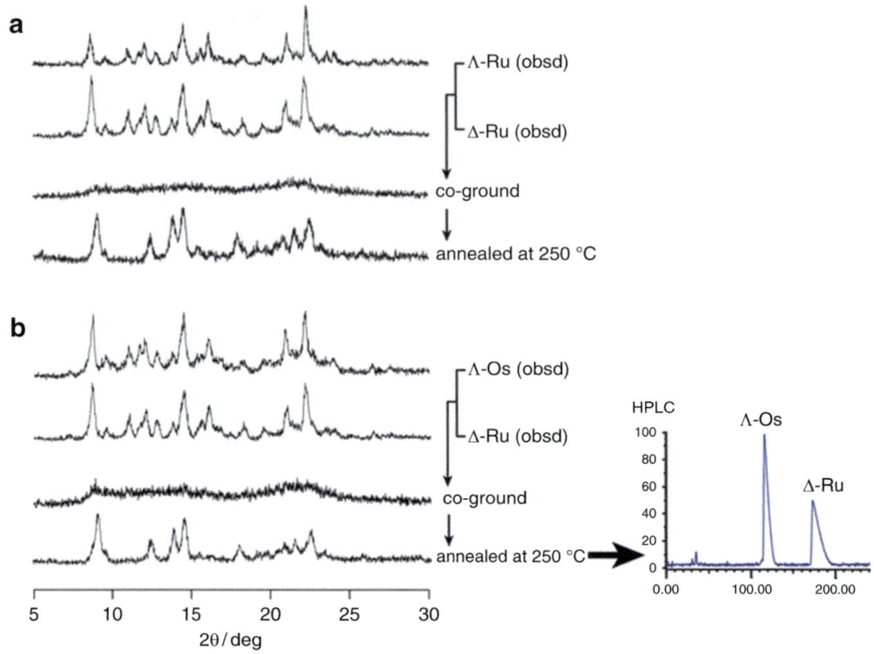

Fig. 2.9 XRPD pattern changes as a result of grinding/annealing treatment in the case of (**a**) the combination of Λ-[Ru(phen)$_3$](PF$_6$) and Δ-[Ru(phen)$_3$](PF$_6$) and (**b**) the combination of Λ-[Os (phen)$_3$](PF$_6$) and Δ-[Ru(phen)$_3$] (PF$_6$). HPLC chart after annealing at 250 °C is shown in the case of (**b**) [24]

(phen)$_3$]$^{2+}$ and Λ-[Os(phen)$_3$]$^{2+}$ in a 1:1 ratio (Fig. 2.9b). This means that annealing encourages diffusion of metal complex ions by way of recognizing their chirality to arrange the packing thereby help establishing 3-D periodicity. Annealing only and without co-grinding of the enantiomeric pair of chiral crystals did not produce racemic crystals.

We have further found that upon co-grinding/annealing of two chiral metal complexes, chirality was transferred between the complexes, if one of the metal complexes is optically labile enough to undergo epimerization in the solid state [24]. When crystals of Δ-[Ru(phen)$_3$](PF$_6$)$_2$ and *rac*-[Ni(phen)$_3$](PF$_6$)$_2$ were mixed and ground, a very broad amorphous-like powder pattern was obtained. Upon annealing the powder at 250 °C, it crystallized without melting. During this process, [Ni(phen)$_3$]$^{2+}$ de-racemized and the Λ-isomer was enriched (*ca.*50% ee). This is the result of chirality transfer between the positively charged metal complex ions, *i.e.*, from the optically stable [Ru(phen)$_3$]$^{2+}$ to optically labile [Ni(phen)$_3$]$^{2+}$. Thus, in this case, annealing promotes chirality transfer as well as diffusion and rearrangement of molecules to form new co-crystal phase.

In the case of Ru(bipy)$_3$_(ClO4)$_2$ (bipy =2,2′-bipyridine), a new anhydrous chiral phase has been prepared by mechanical grinding followed by annealing of the known chiral hydrate phase of this material complex [25].

2.4.5 Controlling the Formation of Metal Complexes in Monomeric, 1-D, 2-D or 3-D Polymeric Forms in the Solid State [26, 27]

Bis(3-cyano-pentane-2,4-dionato) (CNacac) metal complex, $[M(CNacac)_2]$, which acts as both a metal ion-like and a ligand-like building unit, forms supramolecular structures by self-assembly. Co-grinding of the metal acetates of Mn^{II}, Co^{II}, Ni^{II}, Cu^{II} and Zn^{II} with CNacacH formed a CNacac complex in all cases: mononuclear complex was formed in the cases of Mn^{II}, Cu^{II} and Zn^{II}, whereas polymeric ones were formed in the cases of Fe^{II}, Co^{II} and Ni^{II}. Subsequent annealing converted the mononuclear complexes of Mn^{II}, Cu^{II} and Zn^{II} to their corresponding polymers as a result of dehydration of the mononuclear complexes in the solid state. The resultant Mn^{II}, Fe^{II}, Co^{II}, Ni^{II} and Zn^{II} polymeric complexes had a common 3D structure with high thermal stability. In the case of Cu^{II}, a 1D polymer was obtained. The Mn^{II}, Cu^{II} and Zn^{II} polymeric complexes returned to their original mononuclear complexes on exposure to water vapour but they reverted to the polymeric complexes by re-annealing (Fig. 2.10). Co-grinding of metal chlorides with CNacacH and annealing of the mononuclear CNacac complexes prepared from solution reactions were also examined for comparison. $[Mn(CNacac)_2(H_2O)_2]$, $[M(CNacac)_2(H_2O)]$ ($M = Cu^{II}$ and Zn^{II}) and $[M(CNacac)_2]$ ($M = Mn^{II}$, Fe^{II} and Zn^{II}) formed in the solid phase are new compounds not yet reported in solution reactions. This clearly indicates the power of the combined mechanochemical/annealing method for the synthesis of varied metal coordination complexes [26].

Cd^{II} provides even more interesting stereochemistsry (Fig. 2.11) [27]. The reactions of CNacacH with Cd^{II} acetate in the solution and solid states afford different coordination polymers composed of trans-$[Cd(CNacac)_2]$ and cis-$[Cd(CNacac)_2]$ units, respectively. From a methanol solution containing CNacacH (L) and $Cd(OAc)_2{\cdot}2H_2O$ (M), a coordination polymer (Cd-1) in which trans-$[Cd(CNacac)_2]$ units are three-dimensionally linked was obtained. In contrast, two different coordination polymers, Cd-2 and Cd-3, were obtained from mechanochemical reactions of CNacacH with $Cd(OAc)_2{\cdot}2H_2O$ at M/L ratios of 1:1 and 1:2, respectively. In Cd-2, cis-$[Cd(CNacac)_2]$ units are two dimensionally linked, whereas the units are linked three-dimensionally in Cd-3. Furthermore, Cd-1 and Cd-2 converted to Cd-3 by applying an annealing treatment and grinding with a small amount of liquid, respectively, despite the polymeric structures.

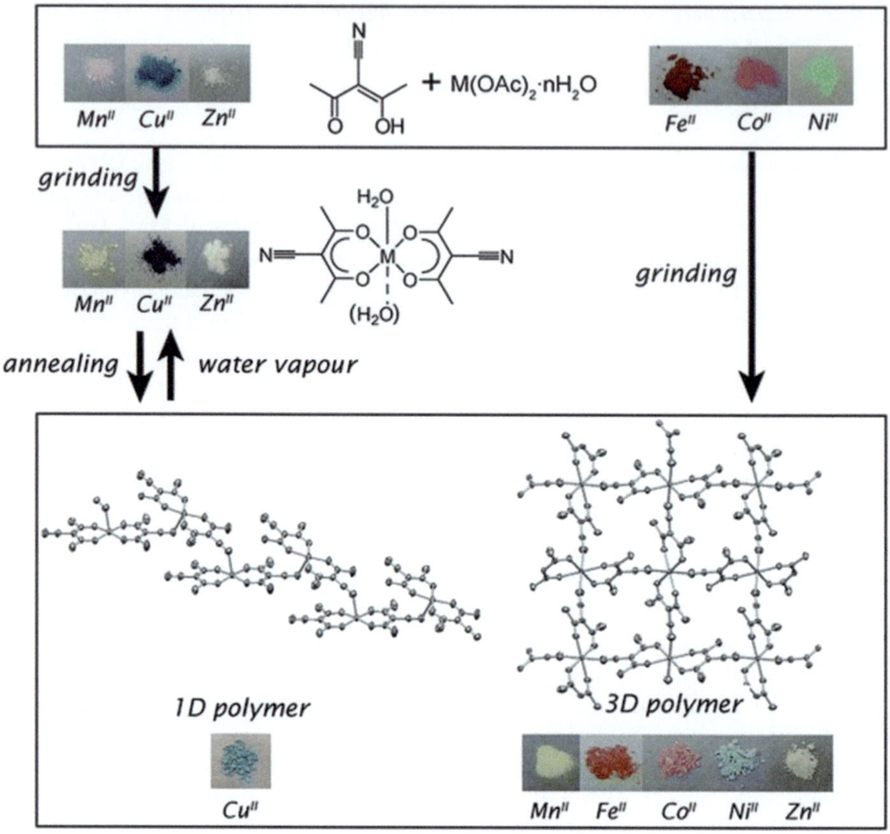

Fig. 2.10 Summary of solid-state reactions to form coordination polymers, $[M(CNacac)_2]_\infty$ (Ref. [26])

Thus,

- different structures are formed from solution and by mechanochemical reactions,
- two polymorphs are formed depending on the M/L ratio,
- structural transformation occurs in polymeric structures.

These indicate the power of the mechanochemical method in the syntheses of coordination polymers.

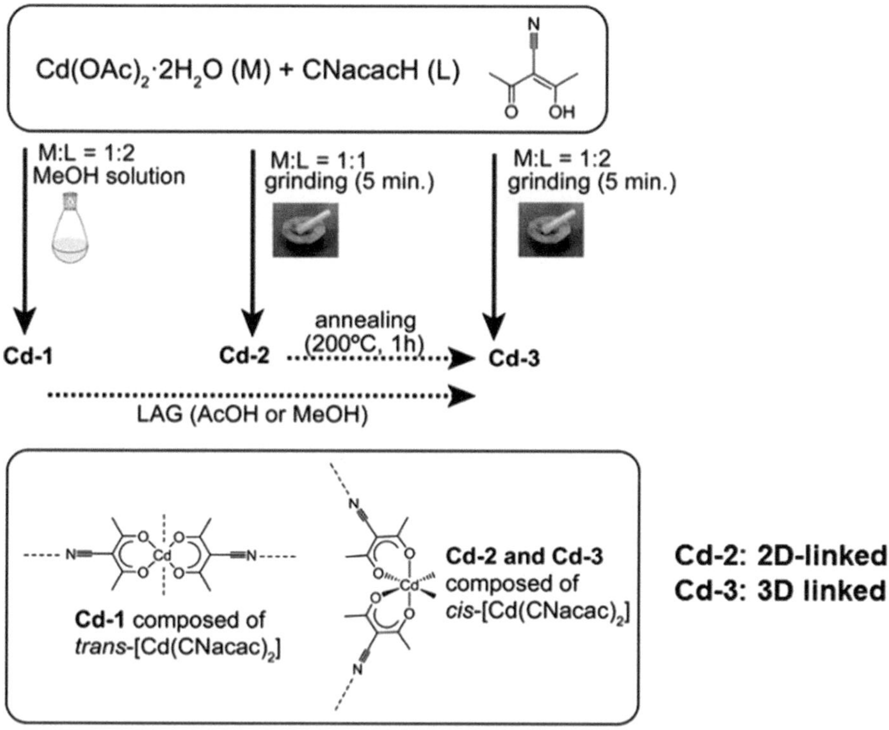

Fig. 2.11 Summary of solution and solid-state formation of $[Cd(CNacac)_2]_\infty$ (Ref. [27])

References

1. Kelvin L (1904) Baltimore lectures (1884) on molecular dynamics and the wave theory of light. Clay and Sons, London, p 449
2. Biot JB (1812) Mémoire sur un nouveau genre d'oscillations que les molécules de la lumière éprouvent, en traversant certains cristaux. Mem Inst France 1:1–32
3. Pasteur L (1848) Memoire sur la relation qui peut exister entre la forme crystalline et la composition chimique, et sur la cause de la polarisation rotatoire. C R Seances Acad Sci 26:538–538
4. Pasteur L (1860) Researches on molecular symmetry. Alembic Club Reprint, No. 14, Livingstone Edinburgh. Oeuvres 1:314–344
5. Le Bel JA (1874) Sur les relations qui existent entre les formulesatomiques des corps organiques, et le pouvoirrotatoire de leur dissolutions. Bull Soc Chim Fr 22:337
6. van't Hoff JH (1874) Arch Neerl Sci Exactes Nat 9:445, English translation given in Benfey (1963)
7. Scacchii A (1865) Del Paratartrato ammonico-sodico, Napoli, Rendiconto, IV, 250–263
8. Kuroda R, Mason SF (1981) Crystal structure of dextrorotatory and racemic sodium ammonium tartrate. J Chem Soc Dalton Trans:1268–1273
9. Imai Y, Takeshita M, Sato T, Kuroda R (2005) Efficient optical resolution of secondary alkyl alcohols by chiral supramolecular hosts. Chem Commun:3289–3291

10. Imai Y, Takeshita M, Sato T, Kuroda R (2006) Successive optical resolution of two compounds by one enantiopure compound. Chem Commun:1070–1072
11. Kassai C, Juvancz Z, Balint J, Fogassy E, Kozma D (2000) Optical resolution of racemic alcohols via diastereoisomeric supramolecular compound formation with O,O'-dibenzoyl-(2R,3R)-tartaric acid. Tetrahedron 56:8355–8359
12. Kuroda R, Imai Y, Tajima N (2002) Generation of a co-crystal phase with novel coloristic properties via solid state grinding procedures. Chem Commun:2848–2849
13. Kuroda R, Imai Y, Sato T (2001) Chirality recognition in solvent-free solid-state crystallization: chiral adduct formation by Bis-β-naphthol derivatives and benzoquinone crystals. Chirality 13:588–592
14. Imai Y, Tajima N, Sato T, Kuroda R (2002) Molecular recognition in solid-state crystallization: colored chiral adduct formations of 1,1′-Bi-2-naphthol derivatives and benzoquinone with a third component. Chirality 14:604–609
15. Kuroda R, Imai Y (2003) Spontaneous rearrangement of hydrogen bonding in a crystalline state. *Mendeleev Comm.* (dedicated to Homochirality discovered by Louis Pasteur in 1848): 148–149
16. Cheung EY, Kitchin SJ, Harris KDM, Imai Y, Tajima N, Kuroda R (2003) Direct structure determination of a multicomponent molecular crystal prepared by a solid state grinding procedure. J Am Chem Soc 125:14658–14659
17. Kuroda R, Higashiguchi K, Hasebe S, Imai Y (2004) Crystal to crystal transformation in the solid state. CrystEngComm 6:463–468
18. Ridley JE, Zerner MC (1973) An intermediate neglect of differential overlap technique for spectroscopy: Pyrrole and for azines. Theor Chim Acta 32:111–134
19. Toda F, Tanaka K, Miyamoto H, Koshima H, Miyahata I, Hirotus K (1997) Formation of Rac-compound crystal by mixing two enantiomer crystals in the solid state. Liquid carrys molecules from crystal to crystal. J Chem Soc Perkin Trans 2:1877–1885
20. Mori K, Masuda Y, Kashino S (1993) (+)-(R)- and racemic forms of 2,2′-dihydroxy-1,1-′-binaphthyl. Acta Crystallogr C49:1224–1227
21. Trotter J (1960) A three-dimensional analysis of the crystal structure of p-benzo-quinone. Acta Crystallogr 13:86–95
22. van Bolhuis F, Kiers CT (1978) Refinement of the crystal structure of p-benzoquinone at −160 °C. Acta Crystallogr B 34:1015–1016
23. Nakamura A, Sato T, Kuroda R (2004) Formation of racemic crystals of transition metal complexes by grinding 1:1 mixtures of enantiomeric crystals. Chem Commun:2858–2859
24. Kuroda R, Yoshida J, Nakamura A, Nishikiori S (2009) Annealing assisted mechanochemical syntheses of transition-metal coordination compounds and co-crystal formation. Cryst Eng Comm 11:427–432
25. Cheung EY, Fujii K, Guo F, Harris KDM, Hasebe S, Kuroda R (2011) Structural chemistry of a chiral anhydrous phase of $Ru(bipy)_3(ClO_4)_2$ established from powder X-ray diffraction analysis. Cryst Growth Des 11:3313–3317
26. Yoshida J, Nishikiori S, Kuroda R (2008) Formation of 1D and 3D coordination polymers in the solid state induced by mechanochemical and annealing treatments: bis 3-cyano-pentane-2,4-dionato metal complexes. Chem Eur J 14:10570–10578
27. Yoshida J, Nishikiori S, Kuroda R, Yuge H (2013) Three polymorphic Cd^{II} coordination polymers obtained from solution and mechanochemical reactions of 3-cyanopentane-2,4-dione with Cd^{II} acetate. Chem Eur J 19:3451–3457

Chapter 3
Supramolecular Assembly and Solid State Chemistry

Christer B. Aakeröy and Manomi D. Perera

Abstract The deceptively simple act of molecular recognition is the result of a balancing act between a variety of intermolecular interactions. Through the use of structural chemistry, interpreted against a background of calculated molecular electrostatic potential surfaces it is possible to identify binding preferences supramolecular patterns of behavior of discrete molecular species. The outcome is robust supramolecular synthetic strategies based on tunable site-specific intermolecular interactions that facilitate the preparation of co-crystals and specific solid-state motifs via selective and hierarchical self-assembly.

Keywords Hydrogen bonds • Halogen bonds • Co-crystals • Molecular recognition • Intermolecular interactions • Synthons

3.1 Introduction

Effective and successful synthetic crystal engineering demands an ability to organize and connect discrete molecular or ionic blocks into desired solid-state architectures with well-defined topologies and metrics. Such endeavors rely on relatively weak and reversible intermolecular interactions that facilitate the preparation of co-crystals and heteromeric constructions through selective and hierarchical self-assembly. This type of synthesis requires reliable structural information regarding the relative importance of the most commonly used non-covalent synthetic tools; hydrogen bonds and halogen bonds.

The most recent attempt [1] at defining hydrogen bonding arrives almost a century after Latimer and Rodebush proposed the concept of hydrogen bonding [2]. The dominant contribution in most hydrogen-bond interactions is the electrostatic component, but the hydrogen bond is partially covalent in nature, [3, 4] and induction and dispersion, in addition to induced covalency and exchange correlation from short range repulsion, all contribute to the complexity of this chemical

C.B. Aakeröy (✉) • M.D. Perera
Department of Chemistry, Kansas State University, Manhattan, KS 66506, USA
e-mail: aakeroy@ksu.edu; manomi@ksu.edu

K.J. Roberts et al. (eds.), *Engineering Crystallography: From Molecule to Crystal to Functional Form*, NATO Science for Peace and Security Series A: Chemistry and Biology, DOI 10.1007/978-94-024-1117-1_3

bond [5, 6]. Crystallographic data have been used to characterize hydrogen bonding although it is difficult to decide upon definitive hydrogen bond distances [7, 8] or energies, [9, 10] as parameters for a definition and instead the linearity of a hydrogen bond has been identified as the "discriminative attribute" [11]. Spectroscopic studies reveal that hydrogen bonds frequently result in a red-shift of X-H bands in the IR [12, 13] and a down-field shift in NMR [14] However, alternative interpretations remain as to whether these methods produce consistent changes in response to hydrogen-bond interactions [15, 16].

Halogen bonding was highlighted as a viable non-covalent interaction by Hassel, [17] and it has recently received considerable attention from the crystal engineering community and beyond [18]. The halogen bond (XB) shows fundamental similarities to the hydrogen bond, and it has been debated in ways that are reminiscent of the way in which hydrogen bonding has been described. This attention to halogen bonding is understandable given its documented importance in supramolecular synthesis, materials chemistry, biological systems and drug design [19, 20]. Halogen bond strength and effectiveness are also "tunable" through a variety of covalent modifications [21, 22]. Electron-withdrawing groups in suitable locations facilitate redistribution of electron density thereby making the halogen atom more electropositive and thus a more effective halogen-bond donor. However, electrostatic forces are not exclusively responsible for the power of halogen bonds and dispersion and induction also play a role, [23] which means that the debate about the nature of different halogen-bond interactions is very similar to that which has accompanied the hydrogen bond. While accurate energies and geometries can be determined by calculations many methods are expensive and often difficult for large halogen-bonding complexes [24, 25].

The question is how do we now develop strategies that effectively utilize the synthetic possibilities that these two interactions offer without having to resort to chance or to some supramolecular combinatorial approach [26–30]? One way of getting some answers may be through systematic structural studies where relatively simple custom-designed probe molecules, equipped with potentially competing hydrogen- and halogen-bond donor sites, are introduced to a series of molecules decorated with varying accessible acceptor sites. By examining the structural outcome of a sufficient number of experiments, it may be possible to begin to identify some of the finer details in the structural landscape that surrounds competing (or complementary) hydrogen- and halogen bonds [30–33].

3.2 Hydrogen-Bond Directed Assembly of Co-crystals

To develop supramolecular synthetic strategies based on hydrogen bonding, we must identify a series of chemical functional groups that display reliable binding preferences [34]. The molecular recognition preferences of these groups can then be explored systematically in order to establish a hierarchy of hydrogen-bond preferences [35].

Hydrogen bonds can be considered primarily electrostatic attractive forces where the hydrogen bond donor is the positively charged site and the acceptor the negatively charged group. By quantifying this charge we can develop the means for quantifying the expected relative importance of different hydrogen bond donors and acceptors [36]. Simple semi-empirical calculations (AM1) provide molecular electrostatic potential values that can be used as quantitative approximations for the charges on individual donor and acceptor groups.

According to Etter's rules, the best hydrogen bond donor forms a hydrogen bond with the best acceptor and the second best donor binds to the second best acceptor [37]. Therefore we can postulate that the donor with the highest positive molecular electrostatic potential value will preferentially bind to the acceptor with the highest negative molecular electrostatic potential value and the donor with the second highest value will bind to the acceptor with second highest value. In order to test the validity of this hypothesis we focused on five different hydrogen-bond donors, (Fig. 3.1).

To test the binding preferences of cyanooximes, acetyloximes, carboxylic acids, phenols, and amines, a series of asymmetric ditopic donor molecules decorated with a combination of these functional groups were synthesized and characterized, (Fig. 3.2). The molecular electrostatic potential values shown were calculated using semi-empirical methods.

These ligands were co-crystallized with a series of asymmetric ditopic acceptors (see Fig. 3.3 for one example) and suitable crystals obtained (from slow evaporation) were analyzed by single crystal X-ray diffraction.

The crystal structures obtained show that the best donor forms a hydrogen bond to the best acceptor and the second best donor binds to the second best acceptor (ranking based upon the values obtained from molecular electrostatic potential calculations), (Fig. 3.3).

All of the crystal structures obtained in this series display hydrogen-bond (HB) patterns and connectivities that can be rationalized in the context of preferences based on the molecular electrostatic potential calculations. Of course, even though hydrogen bonds have considerable strength and directionality, they are reversible which means that synthon polymorphism [38] and synthon crossover [39] are always possible in synthetic crystal engineering, and solvent effects can also be expected to influence the outcome (much as can be observed in conventional organic synthesis). Therefore, even though exceptions are to be expected, it is still very worthwhile to be able to identify patterns of structural behaviour, because trends clearly provide useful starting points for further studies that can validate or refine early observations.

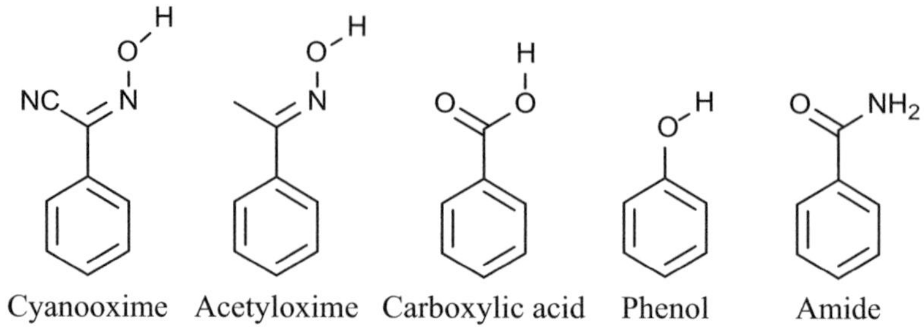

Fig. 3.1 Hydrogen bond donor groups

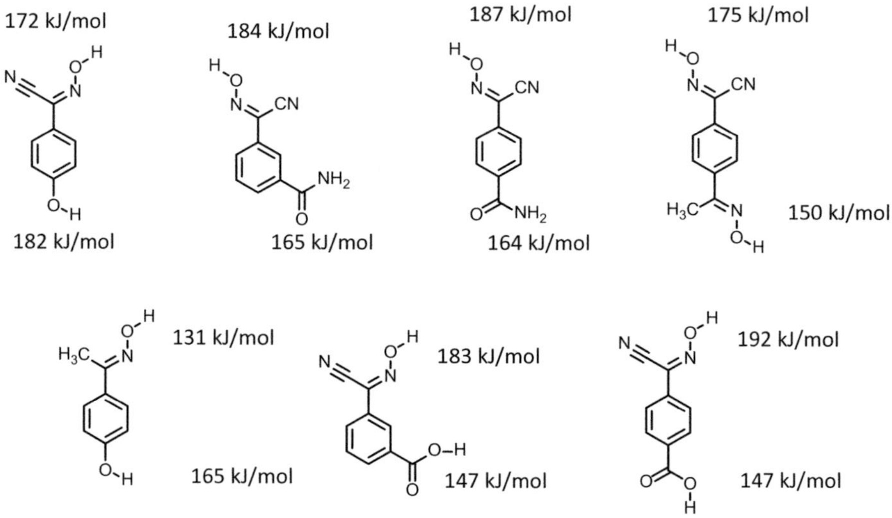

Fig. 3.2 Ditopic hydrogen-bond donors presenting two different donor sites

3.3 Expanding the Scope of Hydrogen-Bond Driven Co-Crystal Synthesis

To provide more support for a supramolecular synthetic strategy informed by calculated electrostatic molecular potential surfaces, we expanded our library of ditopic molecules with two different HB acceptor sites [40]. Our choice of building block was driven by a survey of the Cambridge Structural Database (CSD), [41] which plays a key role in offering extensive and appropriate structural information. In the search of the database, a combination of a carboxylic acid and a 2-aminopyridine based moiety yielded 27 hits whereas a combination of carboxylic acid and pyridine groups gave 202 hits [42]. The purpose of this search was to unambiguously show that carboxylic acids have the capability to bind to both 2-aminopyridine and pyridine based acceptor sites. We subsequently wanted to

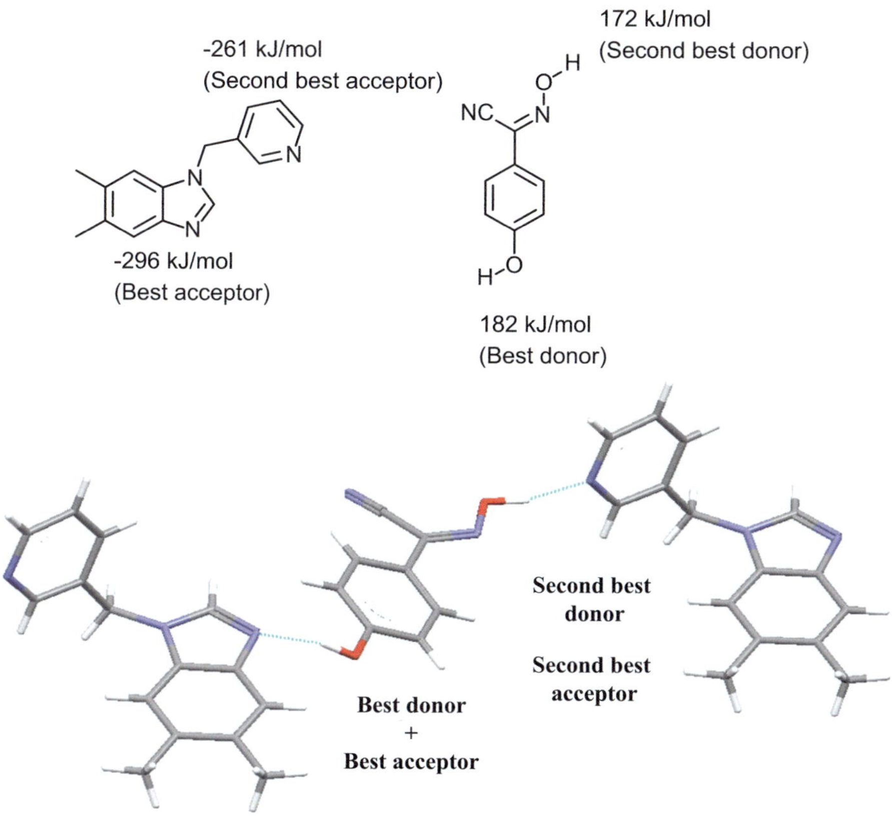

Fig. 3.3 Ranking of hydrogen-bond ability based on electrostatic potential (*top*). The best hydrogen-bond donor selects the best hydrogen-bond acceptor in the crystal structure (*bottom*)

identify any possible preference of carboxylic acids for either of the two binding sites so we chose a target molecular decorated with both functional groups. Aminopyrazine was the natural choice having 2-aminopyridine and a pyridine-type site attached to the same backbone. In order to tune the possible interaction strengths of these sites we modified the aromatic backbone with one or two bromine atoms, respectively, in order to alter the electrostatic potential at the primary hydrogen-bond acceptor site, (Fig. 3.4). We obtained ten crystal structures and 3/10 times the carboxylic acid moiety binds exclusively at the 2-amino end of pyrazine. In the remaining 7/10 there is no pronounced synthon selectivity since the carboxylic acid binds to the 2-amino and the N-4 end of pyrazine at the same time (producing discrete trimers).

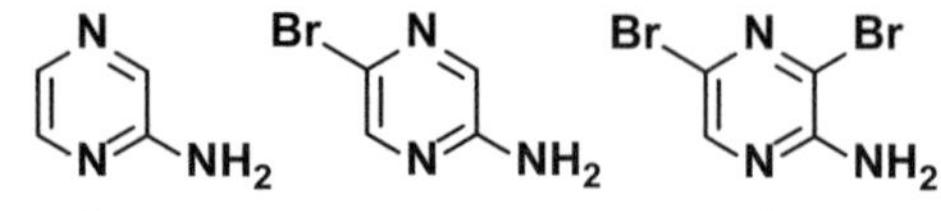

1 2 3

Fig. 3.4 Ditopic supramolecular reagents for probing intermolecular selectivity

3.4 Establishing a Hierarchy of Halogen-Bond (XB) Preferences

Etter's groundbreaking work [43] on the preparation of co-crystals using hydrogen bonds and co-crystallization reactions can be used as probes of the competition between different hydrogen-bonding interactions. Since halogen bonds are also governed to a large extent by electrostatics it is reasonable to expect that they would follow the hierarchy of interactions in a similar manner. To test our hypothesis, we designed and synthesized eight asymmetric ditopic halogen bond donor molecules, (Fig. 3.5), containing two halogen-bond donor sites with slightly different electrostatic potential value. We subsequently allowed these ditopic XB donor molecules to react with a variety of single point XB acceptors, symmetric ditopic acceptors and asymmetric ditopic XB acceptors.

According to molecular electrostatic potential surface calculations, iodine should be a better XB donor site than bromine which should make it bind preferentially to the only available acceptor pyridine site. We obtained two crystal structures, (Fig. 3.6), and in both cases the halogen bonding occurs as anticipated.

3.5 Modular Non-covalent Synthesis with Hydrogen- and Halogen Bonds

In order to refine supramolecular synthesis and to devise more robust synthetic 'reactions', it is necessary to develop supramolecular strategies that can accommodate two or more different non-covalent interactions in such a way that they are unlikely to interfere with each other. A suitable complement to widely studied hydrogen-bond based strategy could be provided by halogen bonds, which are typically formed between activated iodine- or bromine atoms (the halogen-bond donor) and an appropriate halogen-bond acceptor (electron-pair donor) such as an *N*-heterocycle. A potential problem with pairing these two interactions is that any halogen-bond acceptor can also act as a hydrogen-bond acceptor. We hypothesized that if we can choose HB and XB synthons carefully then they can operate in side-by-side in hierarchical fashion in the assembly of co-crystals. We developed a facile one-step strategy to synthesize ditopic XB/HB donor molecules; 2,3,5,6-tetrafluoro-4-halobenzoic acids (halo = iodo or bromo) [44]. We then co-crystallized these donors with the three ditopic acceptors described (Fig. 3.4). Crystal structures for all six reactions (**1–3** with the two acids) were obtained and in

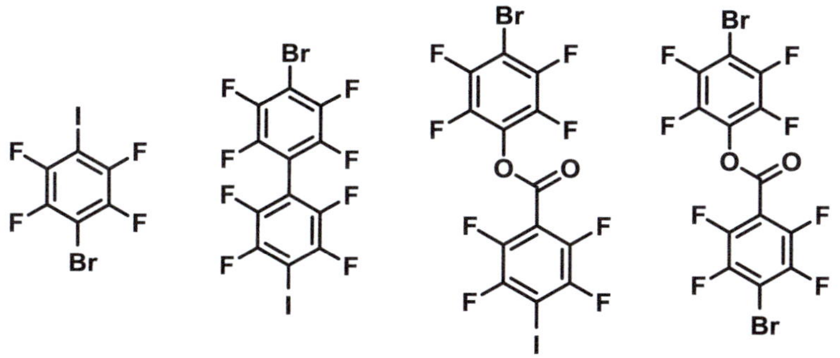

4 5 6 7

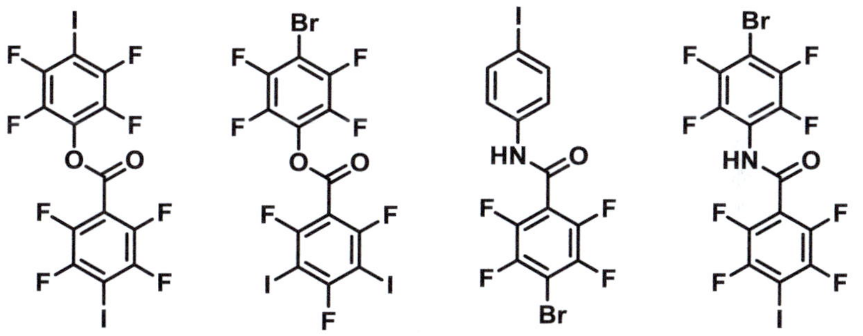

8 9 10 11

Fig. 3.5 Asymmetric ditopic halogen-bond donors

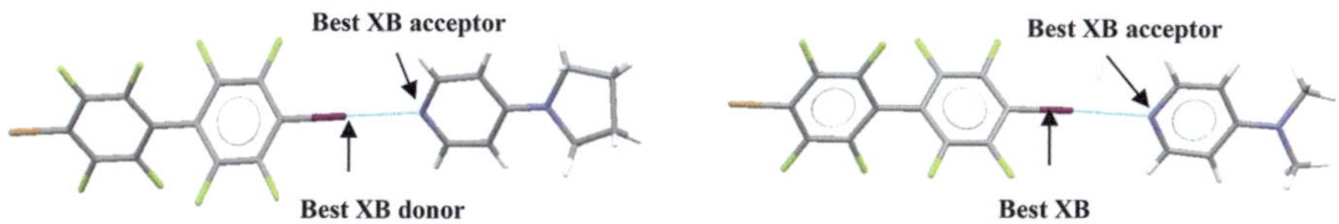

Fig. 3.6 Dimeric supramolecules constructed via hierarchical halogen bonds

each case the supermolecules were constructed via combinations of HB and XB interactions without any interference (Fig. 3.7). Moreover, the role of electrostatic potential in controlling the presence/absence of proton transfer was also emphasized as the acceptor molecule with the highest negative value was capable of abstracting a proton from both acids, leading to two salts and four co-crystals [45].

These results demonstrate how it may be possible to construct complex supramolecular assemblies with a larger number of different molecules (ternary co-crystals are still notoriously difficult to obtain) by combining interactions that can function independently of each other both at the level of molecular recognition, and at the level of overall structural control. In addition, in order to meet specific and well-defined supramolecular challenges it is necessary to employ custom-designed molecules with the appropriate functionalities which lead us to develop

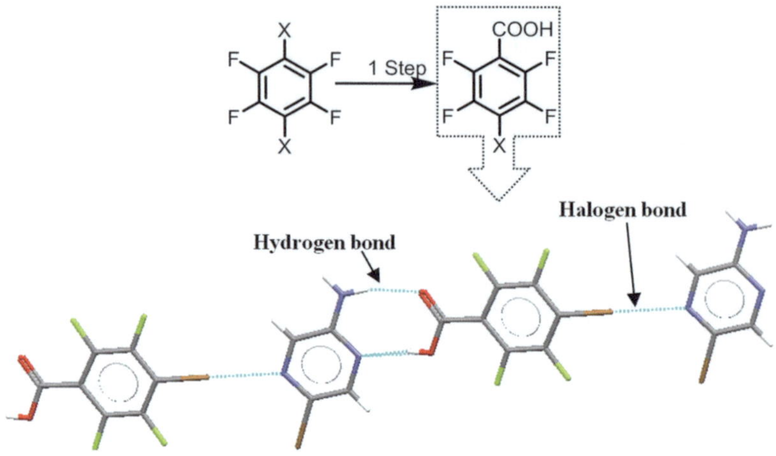

Fig. 3.7 Facile synthesis and supramolecular chemistry of HB/XB-driven multi-tasking tectons

a simple synthetic path to 2,3,5,6-tetrafluoro-4-iodobenzoic acid and 2,3,5,6-tetrafluoro-4-bromobenzoic acid, molecules which combine powerful halogen bond, and halogen bond donors, respectively.

Despite considerable recent interest in the fundamental nature and possible uses of halogen bonds, these interactions and their applications remain substantially underexplored [46]. There are relatively few reports on the logical and deliberate combination of different interactions in supramolecular synthetic and we are currently exploring the improved strategies for the targeted assembly of more complex, multimeric, molecular architectures using an expanded range of reversible intermolecular interactions [47–49].

References

1. Arunan E, Desiraju GR, Klein RA, Sadlej J, Scheiner S, Alkorta I, Clary DC, Crabtree RH, Dannenberg JJ, Hobza P, Kjaergaard HG, Legon AC, Mennucci B, Nesbitt DJ (2011) Definition of the hydrogen bond (IUPAC Technical Report). Pure Appl Chem 83:1619
2. Latimer WM, Rodebush WH (1920) Polarity and ionization from the standpoint of the Lewis theory of valence. J Am Chem Soc 42:1419–1433
3. McWeeny R (1979) Coulson's valence, 3rd edn. Oxford University Press, Oxford
4. Del Bene J, Pople JA (1970) Theory of molecular interactions. I. Molecular orbital studies of water polymers using a minimal Slater-type basis. J Chem Phys 52:4858–4866
5. Dykstra CE, Lisy JM (2000) Experimental and theoretical challenges in the chemistry of noncovalent intermolecular interaction and clustering. THEOCHEM J Mol Struct 500:375–390
6. Umeyama H, Morokuma K (1977) The origin of hydrogen bonding. An energy decomposition study. J Am Chem Soc 99:1316–1332

7. Raghavendra B, Mandal PK, Arunan E (2006) Ab initio and AIM theoretical analysis of hydrogen-bond radius of HD (D = F, Cl, Br, CN, HO, HS and CCH) donors and some acceptors. Phys Chem Chem Phys 8:5276–5286
8. Klein RA (2006) Modified van der Waals atomic radii for hydrogen bonding based on electron density topology. Chem Phys Lett 425:128–133
9. Jeffrey GA, Saenger W (1991) Hydrogen bonding in biological structures. Springer Verlag, Berlin
10. Desiraju GR, Steiner T (1999) The weak hydrogen bond. Oxford University Press, Oxford
11. Elghobashi N, Gonzalez L (2006) A theoretical anharmonic study of the infrared absorption spectra of FHF-, FDF-, OHF-, and ODF- anions. J Chem Phys 124:174308
12. Scheiner S (1997) Hydrogen bonding: a theoretical perspective. Oxford University Press, Oxford
13. Badger RM, Bauer SH (1937) Spectroscopic studies of the hydrogen bond. I. A photometric investigation of the association equilibrium in the vapor of acetic acid. J Chem Phys 5:605–608
14. Hobza P, Havlas Z (2000) Blue-shifting hydrogen bonds. Chem Rev 100:4253–4264
15. Scheiner S, Kar T (2002) Red- versus blue-shifting hydrogen bonds: are there fundamental distinctions? J Phys Chem A 106:1784–1789
16. Joseph J, Jemmis ED (2007) Steric effects on water accessability control sequence-selectivity of radical cation reactions in DNA. J Am Chem Soc 129:4620–4632
17. Hassel O (1970) Structural aspects of interatomic charge-transfer bonding. Science 170:497–502
18. Metrangolo P, Neukirch H, Pilati T, Resnati G (2005) Halogen bonding based recognition processes: a world parallel to hydrogen bonding. Acc Chem Res 38:386–395
19. Bauzá A, Quiñero D, Frontera A, Deyá PM (2011) Substituent effects in halogen bonding complexes between aromatic donors and acceptors: a comprehensive ab initio study. Phys Chem Chem Phys 13:20371–20379
20. Sarwar MG, Bojan D, Salsberg LJ, Gouliaras C, Taylor MS (2010) Thermodynamics of halogen bonding in solution: substituent, structural, and solvent effects. J Am Chem Soc 132:1646–1653
21. Riley KE, Hobza P (2008) Investigations into the nature of halogen bonding including symmetry adapted perturbation theory analyses. J Chem Theory Comput 4:232–242
22. Riley KE, Hobza P (2011) Strength and character of halogen bonds in protein-ligand complexes. Cryst Growth Des 11:4272–4278
23. Jeziorski B, Moszynski R, Szalewicz K (1994) Perturbation theory approach to intermolecular potential energy surfaces of van der Waals Complexes. Chem Rev 94:1887–1930
24. Rezac J, Riley KE, Hobza J (2012) Benchmark calculations of noncovalent interactions of halogenated molecules. J Chem Theory Comput 8:4285–4292
25. Riley KE, Hobza P (2013) The relative roles of electrostatics and dispersion in the stabilization of halogen bonds. Phys Chem Chem Phys 15:17742–17751
26. Aakeröy CB, Baldrighi M, Desper J, Metrangolo P, Resnati G (2013) Supramolecular hierarchy among halogen-bond donors. Chem Eur J 19:16240–16247
27. Aakeröy CB, Panikkattu SV, Chopade PD, Desper J (2013) Competing hydrogen-bond and halogen-bond donors in crystal engineering. CrystEngComm 15:3125–3136
28. Aakeröy CB, Panikkattu SV, DeHaven B, Desper J (2012) Establishing supramolecular control over solid-state architectures: a simple mix and match strategy. Cryst Growth Des 12:2579–2587
29. Aakeröy CB, Fasulo M, Schultheiss N, Desper J, Moore C (2007) Structural competition between hydrogen bonds and halogen bonds. J Am Chem Soc 129:13772–13773
30. Aakeröy CB, Salmon DJ (2005) Building co-crystals with molecular sense and supramolecular sensibility. CrystEngComm 7:439–448
31. Aakeröy CB, Desper J, Smith MM (2007) Constructing, deconstructing, and reconstructing ternary supermolecules. Chem Commun 38:3936–3938

32. Aakeröy CB, Desper J, Fasulo M, Hussain I, Levin B, Schultheiss N (2008) Ten years of co-crystal synthesis; the good, the bad, and the ugly. CrystEngComm 10:1816–1821
33. Aakeröy CB, Desper J, Scott BMT (2006) Balancing supramolecular reagents for reliable formation of co-crystals. Chem Commun 13:1445–1447
34. Aakeröy CB, Epa K, Forbes S, Schultheiss N, Desper J (2013) Ranking relative hydrogen-bond strengths in hydroxybenzoic acids for crystal-engineering purposes. Chem A Eur 19:14998–15003
35. Aakeröy CB, Epa KN, Forbes S, Desper J (2013) Competing hydrogen-bond donors: phenols vs. cyanooximes. CrystEngComm 15:5946–5949
36. Hunter CA (2004) Quantifying intermolecular interactions: guidelines for the molecular recognition toolbox. Angew Chem Int Ed 43:5310–5324
37. Etter MC (1990) Encoding and decoding hydrogen-bond patterns of organic compounds. Acc Chem Res 23:120–126
38. a. Mukherjee A, Desiraju GR (2011) Synthon polymorphism and pseudopolymorphism in co-crystals. The 4,4'-bipyridine-4-hydroxybenzoic acid structural landscape. Chem Commun 47:4090–4092: b. Sarma B, Sanphui P, Nangia A (2010) Polymorphism in isomeric dihydroxybenzoic acids. Cryst Growth & Des 10:2388–2399
39. Aakeröy CB, Chopade PD, Desper J (2011) Avoiding "synthon crossover" in crystal engineering with halogen bonds and hydrogen bonds. Cryst Growth Des 11:5333–5336
40. Aakeröy CB, Chopade PD, Rajbanshi A, Ganser C, Desper J (2012) Exploring the structural landscape of 2-aminopyrazines via co-crystallizations. CrystEngComm 14:5845–5853
41. Allen FA (2002) The Cambridge Structural Database: a quarter of a million crystal structures and rising. Acta Crystallogr Sect B Struct Sci 58:380–388
42. CSD search carried out on ConQuest Version 1.12 (2010)
43. Etter MC, Frankenbach GM (1989) Hydrogen-bond directed cocrystallization as a tool for designing acentric organic solids. Chem Mater 1:10–12
44. Aakeröy CB, Chopade PD, Ganser C, Desper J (2011) Facile synthesis and supramolecular chemistry of hydrogen bond/halogen bond-driven multi-tasking tectons. Chem Commun 47:4688–4690
45. Aakeröy CB, Fasulo ME, Desper J (2007) Cocrystal or salt: does It really matter? Mol Pharm 4:317–322
46. Aakeröy CB (2015) Is there any point in making co-crystals? Acta Crystallogr Sect B 71:387–391
47. Aakeröy CB, Spartz CL, Desper J (2015) Systematic study of halogen bonding versus hydrogen bonding within supramolecular synthetic systems. IUCrJ 2:498–510
48. Aakeröy CB, Wijethunga TK, Desper J (2015) Molecular electrostatic potential dependent selectivity of hydrogen bonding. New J Chem 39:822–828
49. Aakeröy CB, Wijethunga TK, Desper J, Đaković M (2015) Crystal engineering with iodoethynylnitrobenzenes: a group of highly effective halogen-bond donors. Cryst Growth Des 15:3853–3861

Chapter 4
Solid Form Landscape and Design of Physical Properties

Christer B. Aakeröy and Bhupinder Sandhu

Abstract Non-covalent interactions are relatively weak and reversible which means that it is particularly difficult to identify trends and patterns and to establish a hierarchy of molecular-recognition efficiency in a competitive situation, but it is important to note that the seemingly simple act of molecular recognition is achieved by balancing a range of non-covalent forces, and the synthesis of more complex architectures requires a systematic and targeted application of hierarchical self-assembly. In this contribution we illustrate how robust and reliable supramolecular synthetic strategies can be translated into the deliberate design of molecular solids, notably co-crystals, with tunable or desired function. We will focus our attention on energetic materials, pharmaceutically active compounds, and cavitands for host-guest applications.

Keywords Co-crystals • Halogen bonding • Hydrogen bonding • Supramolecular synthesis • Solid-state properties • Molecular recognition

4.1 Introduction

Organic crystalline solids are crucial to the synthesis and processing of specialty chemicals and materials that impact all aspects of modern society. The chemical composition and structure of a molecular solid determine its physical properties, yet most molecular solids are homomeric and we have little control over the way in which individual building blocks organize themselves in a crystalline material [1]. The deceptively simple act of molecular recognition is achieved by balancing a range of non-covalent forces, and the synthesis of more complex architectures requires a systematic and targeted application of hierarchical self-assembly. This means that we need to develop one-pot synthetic methodologies that rely on binding and organization of multiple molecules that take place at distinct levels of gradually decreasing strength. If we are to accomplish this, we need to access and establish

C.B. Aakeröy (✉) • B. Sandhu
Department of Chemistry, Kansas State University, Manhattan, KS 66506, USA
e-mail: aakeroy@ksu.edu; bhupindersandhu90@ksu.edu

K.J. Roberts et al. (eds.), *Engineering Crystallography: From Molecule to Crystal to Functional Form*, NATO Science for Peace and Security Series A: Chemistry and Biology, DOI 10.1007/978-94-024-1117-1_4

versatile supramolecular design strategies based on tunable site-specific intermolecular interactions, and by addressing specific questions about how relatively simple molecules prefer to bind to each other, we may be able to devise reproducible and reliable links between molecular structure and practical methodologies for directed non-covalent synthesis [2].

The importance of the solid-state, and the connections between structure and activity are felt particularly strong in areas such as pharmaceuticals, electronic materials, pigments and dyes, energetic materials, toxins, foodstuff, and detergents. Co-crystals represent new classes of compounds where bulk physical properties such as thermal stability, solubility, mechanical strength or optical response may be amenable to fine-tuning by making controllable alterations to the crystalline lattice that "houses" the active molecular species. The links between crystal structure, morphology and solid-state properties may also offer opportunities for improving processing, performance and shelf-life of a wide range of specialty chemicals [3]. Consequently, an ability to control and change the crystalline environment of a material without altering molecular properties would be of enormous significance to manufacturers and consumers alike.

4.2 Hydrogen-Bond Directed Assembly of Co-crystals

In order to develop versatile supramolecular synthetic strategies that can be applied in hydrogen-bond (HB) driven assembly [4] of co-crystals it is necessary to identify building blocks that display reliable binding preferences in the presence of a range of chemical functionalities. We may realize this goal by assembling a library of custom-designed ditopic molecules (with two different HB donors/acceptors) and then allowing them to react with probe molecules in order to identify a ranking of synthons within an intermolecularly competitive framework. The design of the ditopic molecules, as well as the observed binding preferences, can then be rationalized in the context of a mainly electrostatic view of intermolecular interactions (Fig. 4.1).

4.3 Supramolecular Strategy and Covalent Building Blocks

The success of 'conventional' covalent synthesis is often measured in terms of selectivity, versatility, and yield, and the quality of a supramolecular reagent can be assessed along similar lines. First, "selectivity" translates to the ability to assemble supermolecules with predictable connectivity based upon principles of molecular recognition. Second, "versatility" means that the supramolecular reagent should be able to operate effectively under different reaction conditions (*e.g.* change in solvents and temperature). Third, "yield", in a supramolecular sense, translates to frequency of occurrence of a particular synthon in the presence of potentially

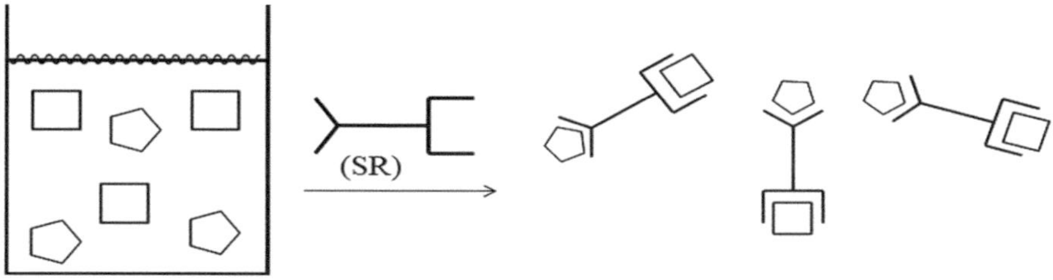

Fig. 4.1 A one-pot supramolecular reaction between two different molecules and a ditopic supramolecular reagent (SR) resulting in 1:1:1 ternary supermolecules

disruptive intermolecular interactions. With this in mind, we can evaluate any crystal engineering strategy geared towards the assembly of heteromeric crystalline solid using quantitative and transferrable metrics.

4.4 How Do We Design, Engineer and Build a Molecular Crystal?

At the core of crystal engineering resides a search for practical avenues that allow for the construction of crystalline materials from discrete molecular building blocks using non-covalent interactions [5, 6]. To complete this search we need to develop reliable and versatile means for the synthesis of molecular materials with specific and tunable properties [7], such as materials that can perform chemical separations, or that have non-linear optical [8], magnetic [9], or catalytic properties [10]. - Non-linear optical properties require individual molecules to be organized in a non-centrosymmetric arrangement with appropriately aligned dipole moments, in magnetic materials molecules and ions must be positioned so that communication between spins is facilitated, and an effective host-material must be able to selectively entrap molecules or ions. Therefore, if we wish to build functional materials, we need to control *how* molecular building blocks recognize each other, exchange information and subsequently assemble architectures with precisely defined metrics.

4.5 Energetic Materials

Supramolecular synthesis of solid-state architectures, crystal engineering, is achieved by directional and site-specific intermolecular interactions [11]. Since our first paper on co-crystals was published in 1996 [12], we have developed robust and transferable practical strategies for the synthesis of co-crystals of molecules containing a wide-range of chemical functionalities [13], and we have also demonstrated how physical properties of active ingredients can be 'dialed-in' through a careful selection of co-crystallizing agents based on extensive synthetic, structural, and spectroscopic efforts [14]. In this section we will shift our focus onto functionalities that impart function and performance of energetic materials. When it comes to the deliberate design and construction of co-crystals of energetic materials, the synthesis is primarily dictated by the intermolecular chemistry of nitro and, to a lesser extent, amino groups. Unfortunately, the -NO_2 moiety is one of the least active chemical functionalities from the point of view of non-covalent interactions. In fact, nitro groups are frequently thought of as nothing more than a space-filling entity without an ability to influence solid-state assembly in a meaningful or predictable manner [15]. Although, examples of co-crystals of energetic materials have been reported [16], we do not yet have access to versatile and reliable synthetic strategies for the targeted assembly of co-crystals, new solid forms, of energetic materials of strategic and commercial importance.

Ethylenedinitramine (**EDNA**) is a known energetic material which requires attention partly due its chemical instability originating with its two highly acidic protons. In order to stabilize EDNA, a co-crystallization approach targeting the acidic protons using a series of co-crystallizing agents with suitable hydrogen-bond acceptors was employed [3]. A systematic co-crystallization study of **EDNA** demonstrated that the acidic protons in the energetic material can be successfully targeted with suitable hydrogen-bond acceptors. Six of the eight co-crystals synthesized were characterized using single-crystal diffraction and the outcome was predictable supramolecular motifs based upon N-H···N and N-H···O structure-directing hydrogen bonds. The co-formers also act as "supramolecular protecting groups" resulting in a reduced chemical instability/corrosiveness which is otherwise detrimental to the storage and processability of **EDNA**. The co-crystal of **EDNA** and 1,2-bis(4-pyridyl)¬ethylene was recognized as a more thermally stable alternative to **EDNA** while the co-crystal of **EDNA** and pyrazine N,N′-dioxide showed comparable detonation strengths (and improved chemical stability) compared with EDNA. The co-crystal of EDNA and 4,4′-bipyridine was found to be about 50% less impact sensitive than pure EDNA. Thermal properties impact sensitivity, and detonation velocities and pressure could also be modified and altered with a degree of predictability since the structural consistency throughout this series unearthed some correlations between molecular structure/property of the pure co-former and the energetic properties of the resulting co-crystal. This clearly suggests that systematic co-crystallizations may allow us to fine-tune properties that are important for storage, handling, and processing, with minimal negative impact on the performance of the targeted substance.

4.6 Constructing Molecular Capsules via Hydrogen- and Halogen Bonds

Directional self-assembly processes provide invaluable corner stones of many synthetic strategies designed for bringing together two or more discrete molecular building blocks into capsular structures [17]. Nano-sized molecular capsules with well-defined inner cavities can find applications related to sensing of small molecules and ions [18], new reaction chambers [19], regioselective control of reactivity [20], stabilization of reactive intermediates [21], chiral receptors [22], gas encapsulation [23], and social isomerization [24]. The use of non-covalent reversible self-assembly processes can offer particular opportunities for 'catch-and-release' of a guest with potential applications in controlled drug delivery and for enabling completely new reactions [19].

In order to provide more options for directed assembly of supramolecular capsules, there is a need for reliable and effective synthetic methods for covalently decorating the upper rim of cavitand-based host-structures with appropriate functional groups. We have designed and synthesized different cavitands of significantly different depth and interior volume functionalized with four aldoxime groups capable of forming homomeric or heteromeric capsules through hydrogen bonding. The rim of the cavitands can be decorated with aldoxime groups through 'Solvent-Assisted Grinding', a greener process which radically limits the amount of organic solvent that needs to be employed and subsequently disposed of. The reactions are achieved quantitatively using few drops of solvents, within few minutes, without any purification [25].

4.7 From Cups to Capsules

To test the reliability and versatility of the findings from our systematic studies of hydrogen bonds involving small molecules, we sought to assemble larger molecules into discrete capsular structures. This was achieved through cavitands decorated with four self-complementary acetamidopyridine moieties [16]. We avoided polymers by utilizing 'supramolecular chelation', producing the desired dimeric capsule (Fig. 4.2).

Supramolecular chelation was again instrumental in the formation of a discrete species as opposed to an unwanted polymer. A halogen-bonded capsule was also obtained [26] *via* directed assembly of a rigid tetra(3-pyridyl) cavitand and a flexible tetra(4-iodotetrafluorophenyl)calix[4]arene. The pyridyl nitrogen atoms from one cavitand molecule interact with the iodine atoms of a single calixarene molecule through short and directional I···N halogen bonds (Fig. 4.3). The flexibility of the ethylenedioxy moieties on the calixarene platform results in positional flexibility of the iodotetrafluorobenzene sites which, coupled with a supramolecular

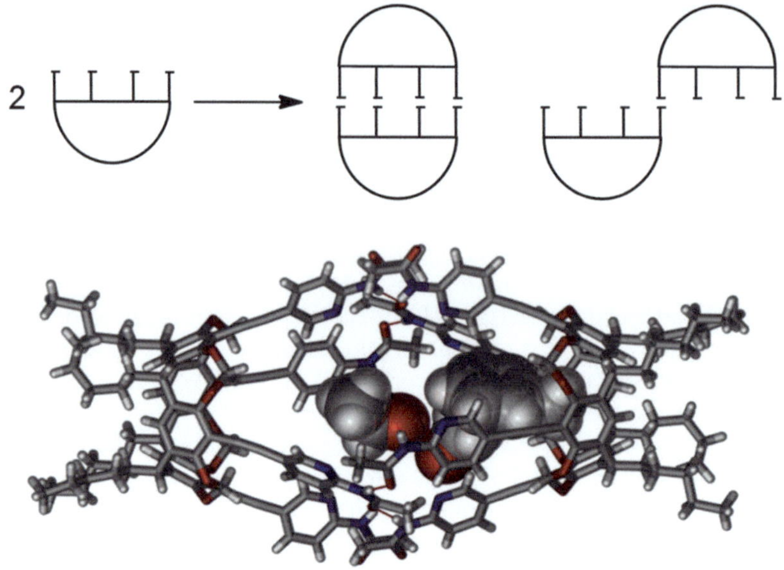

Fig. 4.2 Two possible outcomes of the assembly of self-complementary cavitands; **(a)** dimers (*top left*) and **(b)** polymers (*top right*). Supramolecular chelation leads to dimer formation (*bottom*) [16]

chelating effect, allow for an effective partner-induced geometric fitting between four nitrogen atoms on the cavitand and four iodine atoms on the calixarene.

We also employed isothermal titration calorimetry to show that a targeted pentameric hydrogen-bonded architecture (one tetra-pyridyl cavitand and four 4-nitrobenzoic acids) which had been isolated in the solid-state, was present in solution in acetonitrile with the same stoichiometry and with an enthalpy of binding of approximately −45 kJ/mol [27].

4.8 Synthesis and Solubility of Co-crystals of Pharmaceuticals

The vast majority of active pharmaceutical ingredients (API's) exists as solids at room temperature and the particular solid form defines many of the API's physical properties such as solubility, thermal and mechanical stability, and particle morphology (which influences downstream processability and formulation). Furthermore, the API's physicochemical properties also govern pharmacokinetic (PK) properties such as bioavailability, absorption, and distribution. A major challenge in this field is the lack of versatile technologies that can alter/improve key physicochemical properties of API's, and while there are many drug-delivery technologies that can lengthen the time a drug remains in the human body or how it

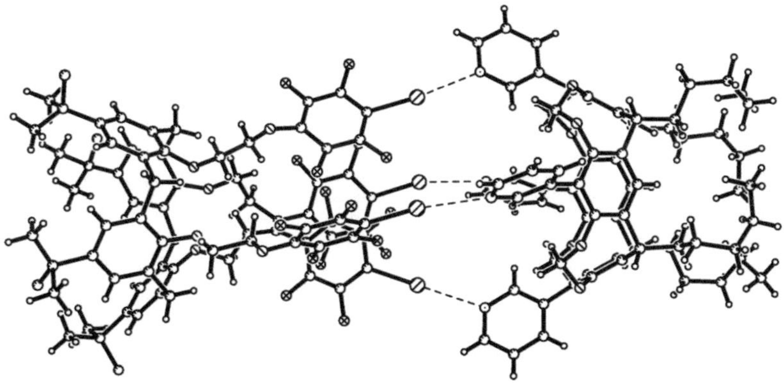

Fig. 4.3 A dimeric capsule constructed by four N···I halogen bonds between the calixarene on the *left* and the cavitand on the *right*

is delivered, there are few options to control solubility. Particle size reduction may improve the kinetic solubility of the API in some cases, however it may also introduce additional technical challenges, such as agglomeration and instability, in formulation and manufacturing as well as higher cost in production. Instead, it would be more beneficial if properties such as solubility or chemical stability of the solid-form of the API could be engineered at the molecular level [14]; access to such expertise and technologies would have dramatic and long-term impact on the pharmaceutical industry.

We have previously [14] presented a systematic structure-property study of a series of co-crystals of hexamethylenebisacetamide, **A**, a compound with potential capabilities for inhibiting proliferation of lung cancer cells, and which may also have therapeutic uses in the treatment of myelodysplastic syndrome (MDS) and resistant acute myelogenous leukemia (AML) [28]. Our strategy was to synthesize infinite API···diacid···API···diacid chains using the well-known COOH···py hydrogen-bond based synthons [29], and these chains were subsequently going to be arranged into 2-D layers as a result of API-based self-complementary amide···amide hydrogen bonds [30]. All diacids used in this study (Fig. 4.4), are generally regarded as safe by FDA. The main objectives of this study were (i) to determine if a series of co-crystals with the desired structural consistency could be synthesized; (ii) to establish if melting point of the resulting co-crystals could be correlated with the co-crystallizing agent and; (iii) to establish if solubility could be correlated with some feature of the co-crystallizing agent.

Co-crystallization reactions were carried out between **A** and **1–5**, and IR spectroscopy was used to screen all resulting solids for co-crystal formation. We were also able to grow crystals suitable for single crystal X-ray diffraction of all products. In **A1** (the 1:1 co-crystal of **A** and **1**) a primary interaction between the pyridyl moiety and carboxylic acid was observed resulting in 1-D chains which, in turn were organized into layers via interchain, N-H···O hydrogen bonds. Similar

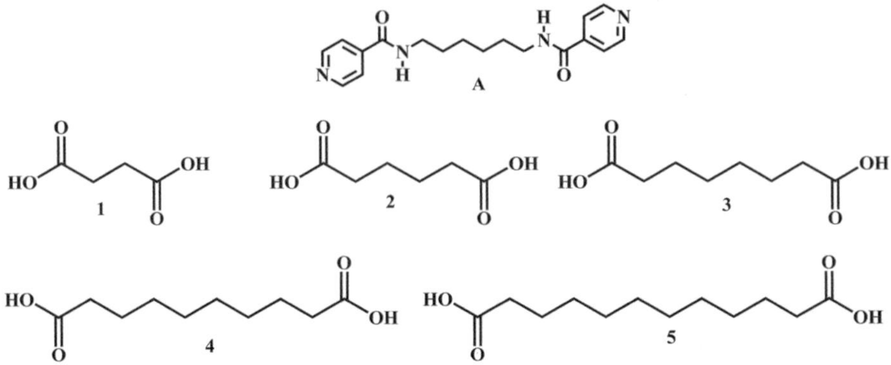

Fig. 4.4 Target API and dicarboxylic acids

structures were obtained with longer chain acids where again O-H···N acid···py and amide···amide hydrogen bonds generate 2-D layers (Fig. 4.5).

Having achieved the required structural consistency, we subsequently examined whether the thermal stability of these co-crystals could be correlated with any molecular feature of the five co-crystallizing agents (Fig. 4.6). The data clearly show that the melting points of these five crystalline solids are directly related to the melting points in the dicarboxylic acids. The highest melting co-crystal contains the dicarboxylic acid with the highest melting point (the lowest melting acid produces the lowest melting co-crystal) which demonstrates that the thermal stability of the five solid forms of this API can be modulated in a predictable manner over a considerable range, 148–188 °C (the melting point of the API itself is 181–182 °C).

Although thermal stability is an important issue, solubility is a key factor, and we therefore also determined the aqueous solubilities for **A1–A5** (Fig. 4.7).

Although the solubilities of the five co-crystals of the API do not produce a linear correlation as was the case with the thermal stabilities, the trend in physico-chemical properties of the co-crystals can certainly be rationalized in terms of the aqueous solubilities of the dicarboxylic acids. The co-crystals of the longer chain diacids, which are less polar and more hydrophobic in nature, show decreased aqueous solubility compared to that of the API itself. The fact that we retain structural consistency throughout this series of co-crystals (they all display similar 2-D layers), makes it possible to relate bulk physical performance of the co-crystals to a specific features of each molecular co-former.

4.9 Endnote

By addressing specific questions about how relatively simple molecules prefer to bind to each other, we will attempt to establish reproducible and reliable links between molecular function and directed non-covalent synthesis. The ability, (a) to

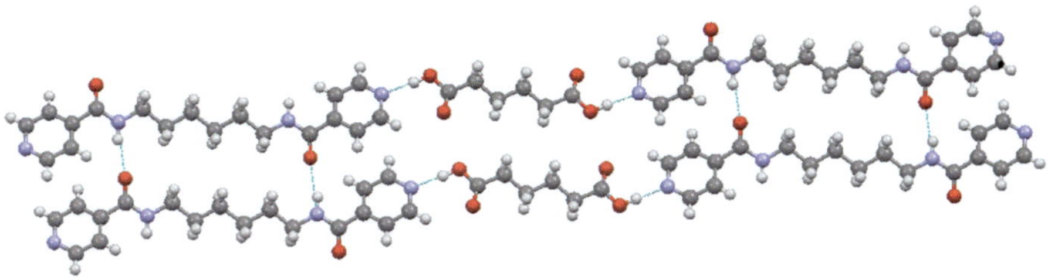

Fig. 4.5 2-D sheet in **A2** generated through O-H···N and N-H···O hydrogen bonds

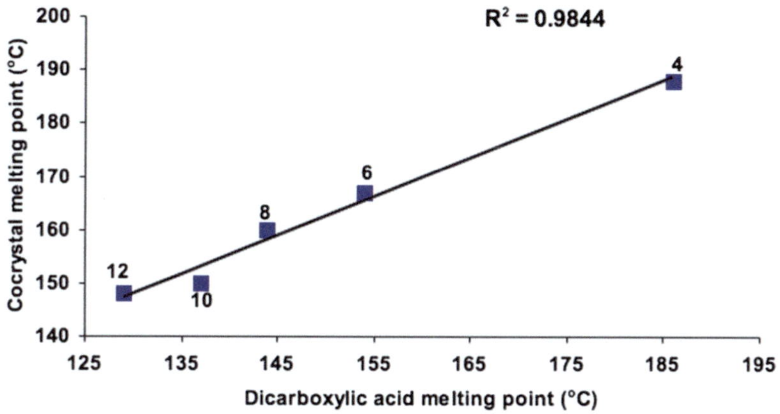

Fig. 4.6 Thermal stabilities for **A1–A5**

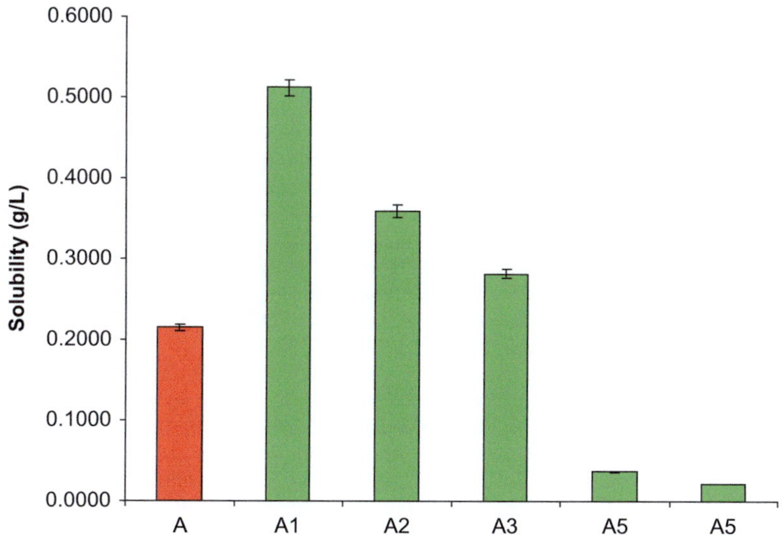

Fig. 4.7 Aqueous solubilities of **A** and **A1–A5**

position molecules exactly where we want them to be; (b) to construct heteromolecular architectures with desirable metrics; and (c) to translate intermolecular communication into blueprints for materials design and for constructing viable biological mimics, represent significant long-term goals of interest to a wide range of scientists.

References

1. Aakeröy CB, Spartz CL, Desper J (2015) Systematic study of halogen bonding versus hydrogen bonding within supramolecular synthetic systems. IUCrJ 2:498–510
2. Aakeröy CB (2015) Is there any point in making co-crystals? Acta Crystallogr Sect B 71:387–391
3. Aakeröy CB, Wijethunga TK, Desper J (2015) Crystal engineering of energetic materials: Co-crystals of ethylenedinitramine (EDNA) with modified performance and improved chemical stability. Chem Eur J 21:11029–11037
4. (a) Braga D, Desiraju GR, Miller JS, Orpen AG, Price SL (2002) Innovation in crystal engineering. Cryst Eng Comm 4:500–509; (b) Angeloni A, Crawford PC, Orpen AG, Podesta TJ, Shore BJ (2004) Does hydrogen bonding matter in crystal engineering? Crystal structures of salts of isomeric ions. Chem -A Eur J 10:3783–3791
5. (a) Lehn JM (1990) Perspectives in supramolecular chemistry—from molecular recognition towards molecular information processing and self-organization. Angew Chem Int Ed Engl 29:1304; (b) Desiraju GR (1989) Crystal engineering: the design of organic solids. Elsevier, Amsterdam; (c) Steed JW, Atwood JL (2000) Supramolecular chemistry: an introduction. Wiley, Chichester
6. Aakeröy CB (1997) Crystal engineering: strategies and architectures. Acta Crystallogr B53:569
7. (a) Desiraju GR (2001) Chemistry beyond the molecule. Nature 412:397; (b) Coronado E, Galan-Mascaros JR, Gomez-Garcia CJ, Laukhin V (2000) Coexistence of ferromagnetism and metallic conductivity in a molecule-based layered compound. Nature 408:447
8. (a) Muthuraman M, Masse R, Nicoud JF, Desiraju GR (2001) Molecular complexation as a design tool in the crystal engineering of noncentrosymmetric structures. Ideal orientation of chromophores linked by O–H···O and C–H···O hydrogen bonds for nonlinear optics. Chem Mater 13:1473; (b) König O, Burgi HB, Armbruster T, Hulliger J, Weber T, (1997) A study in crystal engineering: structure, crystal growth, and physical properties of a polar perhydrotriphenylene inclusion compound. J Am Chem Soc 119:10632
9. (a) Kahn O (2000) Chemistry and physics of supramolecular magnetic materials. Acc Chem Res 33:647–657; (b) Miller JS (2000) Organometallic- and organic-based magnets: new chemistry and new materials for the new millennium. Inorg Chem 39:4392
10. (a) Aoyama Y (1998) Functional organic zeolite analogues. Topics Curr Chem198:131; (b) Bassani DM, Darcos V, Mahony S, Desvergne JP (2000) Supramolecular Catalysis of Olefin [2 + 2] Photodimerization. J Am Chem Soc 122:8795
11. (a) Aakeröy CB, Chopade PD, Ganser C, Desper J (2011) Facile synthesis and supramolecular chemistry of hydrogen bond/halogenbond-driven multi-tasking tectons. Chem Commun 47:4688–4690; (b) Aakeröy CB, Desper J, Fasulo M, Hussain I, Levin B, Schultheiss N (2008) Ten years of co-crystal synthesis; the good, the bad, and the ugly. CrystEngComm 10:1816–1821; (c) Aakeröy CB, Desper J, Helfrich BA, Metrangolo P, Pilati T, Resnati G, Stevenazzi A (2007) Combining halogen bonds and hydrogen bonds in the modular assembly of heteromeric infinite 1-D chains. Chem Commun 4236–4238; (d) Desiraju GR (2007) Crystal engineering: a holistic view. Angew Chem Int Ed 46:8342–8356; (e) Moulton B, Zaworotko MJ (2001) From molecules to crystal engineering: supramolecular isomerism and

polymorphism in network solids. Chemical Reviews 101:1629–1658; (f) Thomas JM (2011) Crystal engineering: origins, early adventures and some current trends. CrystEngComm 13:4304–4306

12. Aakeröy CB, Cooke TI, Nieuwenhuyzen M (1996) The crystal structure of the molecular cocrystal L-malic acid L-tartaric acid (1/1). Supramol Chem 7:153–156
13. (a) Aakeröy CB, Beatty AM, Nieuwenhuyzen M, Zou M (2000) Organic Assemblies of 2-pyridones with dicarboxylic acids. Tetrahedron 56:6693–6699; (b) Aakeröy CB, Beatty AM, Helfrich BA (2001) "Total synthesis" supramolecular style: design and hydrogen-bond-directed assembly of ternary supermolecules. Angew Chem Int Ed Engl 40:3240–3242; (c) Aakeröy CB, Beatty AM, Helfrich BA (2002) A high-yielding supramolecular reaction. J Am Chem Soc 124:14425–14432; (d) Aakeröy CB, Beatty AM, Helfrich BA, Nieuwenhuyzen M (2003) Combining halogen bonds and hydrogen bonds in the modular assembly of heteromeric infinite 1-D chains. Cryst Growth Des 3:159–165; (e) Aakeröy CB, Salmon DJ (2005) Building co-crystals with molecular sense and supramolecular sensibility. Cryst Eng Comm 7:439–448
14. (a) Aakeröy CB, Forbes S, Desper J (2009) Using cocrystals to systematically modulate aqueous solubility and melting behavior of an anticancer drug. J Am Chem Soc 131: 17048–17049; (b) Aakeroy CB, Fasulo ME, Desper J (2007) Cocrystal or salt: does it really matter? Mol Pharm 4:317–322; (c) Aakeröy CB, Grommet AB, Desper J (2011) Co-crystal screening of diclofenac. Pharmaceutics 3:601–614; (d) Aakeröy CB, Forbes S, Desper J (2012) The effect of water molecules in stabilizing co-crystals of active pharmaceutical ingredients. Cryst Eng Comm 14:2435–2443
15. Robinson JMA, Philp D, Harris KDM, Kariuki BM (2000) Weak interactions in crystal engineering—understanding the recognition properties of the nitro group. New J Chem 24:799–806
16. Aakeröy CB, Rajbanshi A, Desper J (2011) Hydrogen-bond driven assembly of a molecular capsule facilitated by supramolecular chelation. ChemCommun 47:11411–11413
17. Rebek J Jr (2005) Simultaneous encapsulation: molecules held at close range. Angew Chem Int Ed 44:2068
18. Castellano RK, Craig SL, Nuckolls C, Rebek J Jr (2000) Detection and mechanistic studies of multicomponent assembly by fluorescence resonance energy transfer. J Am Chem Soc 122:7876
19. Chen J, Rebek J Jr (2002) Selectivity in an encapsulated cycloaddition reaction. Org Lett 4:327
20. (a) Fujita M, Umemoto K, Yoshizawa M, Fujita N, Kusukawa T, Biradha K (2001) Molecular paneling via coordination. Chem Commun 509; (b) Yoshizawa M, Kusukawa T, Fujita M, Yamaguchi K. (2000) Ship-in-a-bottle synthesis of otherwise labile cyclic trimers of siloxanes in a self-assembled coordination cage. J Am Chem Soc 122:6311; (c) Kang JM, Santamarìa J, Hilmersson G, Rebek J Jr (1998) Self-assembled molecular capsule catalyzes a diels−alder reaction. J Am Chem Soc 120:7389; (d) Chen J, Körner S, Craig SL, Rudkevich DM, Rebek J Jr (2002) Chemical amplification with encapsulated reagents. Nature 99:2593–2596
21. (a) Kusukawa T, Fujita M (1999) "Ship-in-a-bottle" formation of stable hydrophobic dimers of cis-azobenzene and -stilbene derivatives in a self-assembled coordination Nanocage. J Am Chem Soc 121:1397; (b) Körner SK, Tucci FC, Rudkevich DM, Heinz T, Rebek J Jr (2000) A self-assembled cylindrical capsule: new supramolecular phenomena through encapsulation. Chem Eur J 6:187
22. Scarso A, Shivanyuk A, Hayashida O, Rebek J Jr (2003) Asymmetric environments in encapsulation complexes. J Am Chem Soc 125:6239
23. Ajami D, Rebek J Jr (2008) Gas behavior in self-assembled capsules. Angew Chem Int Ed 47:6059
24. Ajami D, Rebek J Jr (2007) Adaptations of guest and host in expanded self-assembled capsules. Proc Natl Acad Sci U S A 104:16000
25. Aakeröy CB, Chopade PD (2011) Oxime decorated cavitands functionalized through solvent-assisted grinding. Org Lett 13:1

26. Aakeröy CB, Rajbanshi A, Metrangolo P, Resnati G, Parisi MF, Desper J, Pilati T (2012) The quest for a molecular capsule assembled via halogen bonds. CrystEngComm 14:6366–6368
27. Aakeröy CB, Chopade PD, Quinn CF, Desper J (2014) Structure and thermodynamics of a multimeric cavitand assembly. CrystEngComm 16:3796–3801
28. (a) Andreeff M, Stone R, Michaeli J, Young CW, Tong WP, Sogoloff H, Ervin T, Kufe D, Rifkind RA, Marks PA (1992) Hexamethylene bisacetamide in myelodysplastic syndrome and acute myelogenous leukemia: a phase II clinical trial with a differentiation- inducing agent. Blood 80:2604–2609; (b) Callery SP, Egorin MJ, Geelhaar LA, Nayar BS (1986) Identification of metabolites of the cell-differentiating agent hexamethylene bisacetamide in humans. Cancer Res 46:4900–4903; (c) Siegel DS, Zhang X, Feinman R, Teitz T, Zelenetz A, Richon VM, Rifkind RA, Marks PA, Michaeli J (1998) Hexamethylene bisacetamide induces programmed cell death (apoptosis) and down-regulates BCL-2 expression in human myeloma cells. Proc Natl Acad Sci USA 95:162–166
29. (a) Shan N, Zawarotko MJ (2008) The role of cocrystals in pharmaceutical science. Drug Discovery Today 11:440–446; (b) Aakeröy CB, Beatty AM, Helfrich BA, Nieuwenhuyzen M (2003) Do polymorphic compounds make good cocrystallizing agents? A structural case study that demonstrates the importance of synthon flexibility. Cryst Growth Des 3:159–165; (c) Aakeröy CB, Desper J, Helfrich BA (2004) Heteromeric intermolecular interactions as synthetic tools for the formation of binary co-crystals. CrystEngComm 6:19–24; (d) Aakeröy CB, Desper J, Urbina JF (2005) Syntheses and crystal structures of versatile supramolecular reagents based upon [(Benzimidazol-1-yl)methyl]-benzamides. Cryst Growth & Design 5:1283–1293
30. Lauher JW, Fowler FW, Goroff NS (2008) Single-crystal-to-single-crystal topochemical polymerizations by design. Acc Chem Res 41:1215–1229

Chapter 5
Design of Physical Properties and Solid Form Design

Robert Docherty and Kevin Back

Abstract The selection of the solid form for development is a key milestone in the conversion of a new chemical entity into a drug product. An understanding of the materials science of a new active pharmaceutical is crucial at the interface of medicinal chemistry and pharmaceutical development. The physical and chemical properties of a new chemical entity that impact product performance and product robustness are strongly influenced by the solid state structure of the active pharmaceutical ingredient. Product performance can only be assured when the new chemical entity is delivered to the patient in a chemically and physically stable solid form. In this chapter we will attempt to integrate progress with cutting edge computational tools in academia to the best current industrial practices so that the medicinal chemist and pharmaceutical scientist can transform the journey from molecule to medicine.

Keywords Solid-form informatics • Cambridge structural database • Solid-form selection • Crystal packing and solvation • Low solubility • Crystallography

5.1 Introduction

Selection of the commercial solid-form and associated crystallization process is one of the key milestones in the development of any new drug molecule. It is critical not only from a drug-substance manufacturing standpoint but also from a drug-product functionality perspective (bioperformance, processing, and stability). The regulatory landscape associated with the solid form and particle attributes of the active pharmaceutical ingredient (API) and dosage form development has been described previously [1, 2]. The issues associated with the emergence of an unexpected solid form [3, 4] and the importance of intellectual property around crystallization process-design and polymorphs has also been well documented [5, 6]. The progress of automation and structural informatics technologies, which allow development

R. Docherty (✉) • K. Back
Pharmaceutical Sciences, Pfizer Global R&D, Ramsgate Road, Sandwich, Kent CT13 9NJ, UK
e-mail: Robert.Docherty@pfizer.com; Kevin.Back@pfizer.com

K.J. Roberts et al. (eds.), *Engineering Crystallography: From Molecule to Crystal to Functional Form*, NATO Science for Peace and Security Series A: Chemistry and Biology, DOI 10.1007/978-94-024-1117-1_5

scientists to search and identify the solid form with optimal properties, has also been reported [7].

In 1987 the Nobel Prize for chemistry was awarded to Cram, Lehn and Pedersen for their work on supramolecular chemistry. Since then publications [8–10] have charted the evolution of Pharmaceutical Materials Science. More recently the importance of the Materials Science tetrahedron in depicting the relationships between internal structure, particle properties and the processing and performance of a drug product has been described [10]. The industrial perspective on engineering pharmaceutical materials has also been highlighted [11, 12]. Pharmaceutical Materials Science has emerged as a foundation of Quality by Design (QbD) with solid form, crystallization and particle engineering being core elements linking the drug-product functional form to the final steps of the synthetic pathway of the API.

Whilst increasing interest in the crystallisation of pharmaceutical entities within academia has resulted in substantial progress over the last decade, the challenge for the pharmaceutical scientist in tackling the crystallisation of highly complex, new chemical-entities remains a significant one. These complex organic structures are exacting because:

- Increasing molecular complexity results in a complicated solid form space (salts, cocrystals, polymorphs, hydrates and solvates)
- Different crystal faces exhibit different surface chemistry and consequently interactions with solvents, inherent process impurities and excipients.
- Different solid forms may have different chemical and physical stabilities, biopharmaceutical properties and drug product processing behaviour.

Whilst traditionally, the solid form selection process has focused on achieving an appropriate degree of bioavailability, increasing emphasis is now being placed on selection of solid forms which have optimal functionality such as chemical and physical stability, mechanical behaviour, surface properties and particle shape. Given such a perspective, this chapter outlines some recent progress on the application of emerging structural computational technologies as foundation elements of the modern QbD strategy for the development and manufacture of advanced functional particulate products (Fig. 5.1). This chapter highlights the opportunity for structural tools to build the bridge across the chemical, analytical and formulation disciplines. These relationships, combined with institutionalised corporate knowledge of formulation design practices provide a clear route map to a fully integrated, holistic product design process consistent with the emerging QbD philosophy realised through the use of increasingly sophisticated particle and surface chemistry design tools. This enables the journey from molecule to crystal to functional form and product performance.

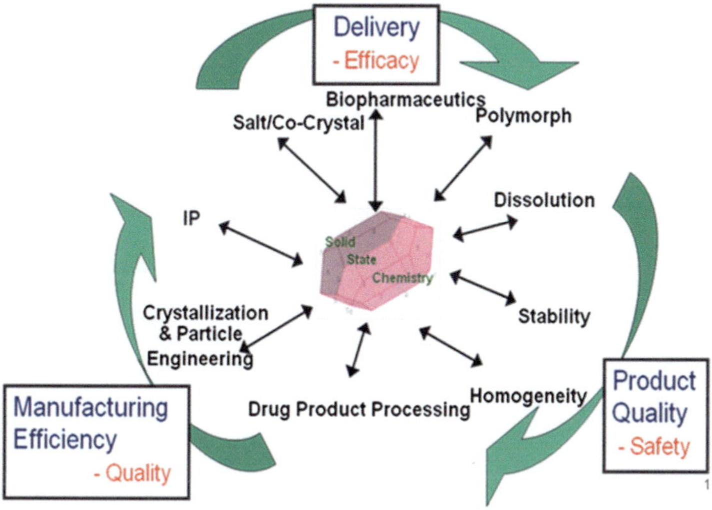

Fig. 5.1 The importance of the solid state in impacting product safety, efficacy and quality

5.2 The Crystal Chemistry and Crystal Packing of Drug Molecules

Molecules can essentially be regarded as impenetrable systems whose shape and volume characteristics are governed by the molecular conformation and the radii of the constituent atoms. The atomic radii are essentially exclusion zones in which no other atom may enter except under special circumstances, such as bonding [13]. Figure 5.2 shows a comparison between a ball and stick and van der Waals (space-fill) representation of paracetamol [14].

The structures and crystal chemistry of molecular materials are often classified into different categories according to the type of intermolecular forces present [15, 16]. A number of factors are of particular importance in assessing the influence of intermolecular bonding on the physico-chemical properties of organic solids [17, 18] including:

- the size and shape of the molecular entities that make up the structure
- the strength of the intermolecular interaction
- the distance over which the interaction exerts an influence
- the extent to which the interaction is directional or not

Organic molecules in general and drug molecules in particular are found in only a limited number of low symmetry crystal systems. The general uneven shapes of molecular structures tend to result in unequal unit cell parameters [19]. A further consequence of their unusual shape is that organic molecules prefer to adopt space groups which have translational symmetry elements, as this allows the most efficient spatial packing of the protrusions of one molecule into the gaps left by

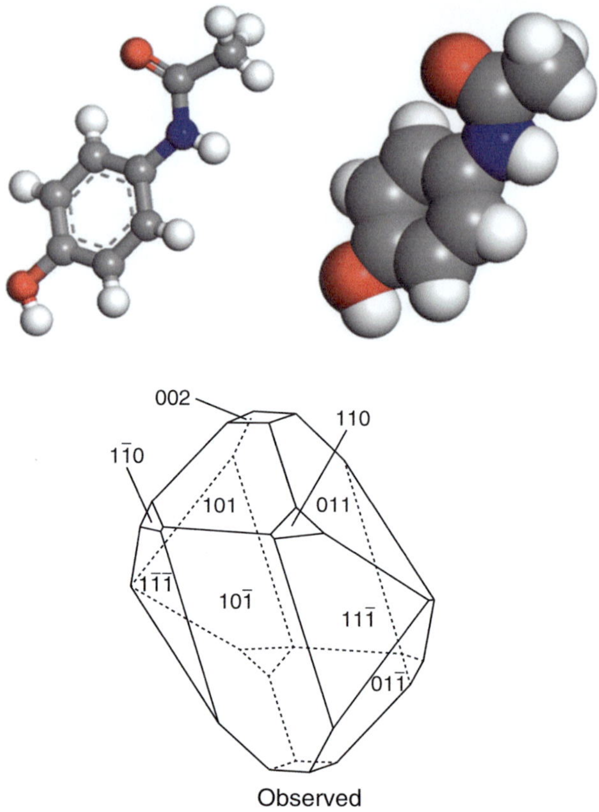

Fig. 5.2 The molecule and the crystal morphology of paracetamol. The molecular structure of paracetamol (*top*) is shown in ball and stick and space-fill. The crystal morphology (*bottom*) shows the observed crystal faces labelled with the corresponding Miller indices

the packing arrangements of its neighbours [20]. These tendencies are reflected in an analysis of the Cambridge Crystallographic Database where the vast majority of the organic structures reported prefer the lower symmetry triclinic, monoclinic and orthorhombic crystal systems [21].

It is clear that the crystal chemistry of organic molecules differs from that of highly symmetrical inorganic systems. The uneven molecular shapes and directional bonding influence the result of the crystal packing. These directing features can be more prominent in drug molecules, as many have multiple hydrogen bonding donors and acceptors. The trend to increasing molecular weight for new drug candidates can also result in more anisotropic chemical structures. Understanding the role the molecular structure plays in defining the crystal structures and hence physical properties is key to risk assessing polymorphism and to predicting performance, as will be described later in this chapter.

5.3 Solid Form Informatics

The Cambridge Crystallographic Data Centre (CCDC) has developed tools specifically designed to assess the likelihood of the formation of specific intermolecular interactions for a given molecule. Such utilities are provided in the Solid Form Module of *Mercury* [22]. These include the *Hydrogen Bond Propensity* [23, 24] tool, which applies a statistical analysis to structures in the Cambridge Structural Database (CSD) in order to determine quantitatively the likelihood of hydrogen-bond formation between individual functional groups of the target molecule. A chart, plotting all possible combinations of hydrogen-bond donors with acceptors for the target molecule, ranked in terms of propensity and coordination score highlights where the observed solid form resides in a landscape of possible hydrogen-bond networks. Such a tool can be used to assess the likelihood of polymorphism and stability through differing coordination modes of hydrogen bonding [25].

The investigation of the potential polymorphism of Maraviroc was a very early example of how structural informatics principles influenced the solid form selection experiments carried out during the pre-clinical development phase of this new drug candidate. The first form of Maraviroc crystallised (Form A) exhibited a three-centred intermolecular hydrogen-bond architecture involving the amide and triazole groups as shown in Fig. 5.3. The two N···N donor-acceptor distances were observed to be 3.239(5) and 3.221(5) Å in this arrangement. A ConQuest contact search established that these distances were longer and atypical of similar interactions observed in the CSD at the time (CSD version 5.26).

This prompted a further more detailed examination of the geometry of such interactions by creating a bespoke Isostar plot using a combination of ConQuest and Isogen. A ConQuest search query was constructed with the aim of finding molecules with functional groups that reflected the hydrogen-bond donor and acceptor capacity of secondary amide and triazole functions, respectively. In addition, the search was directed to find those molecules that exhibited a packing arrangement that brought the two groups in close contact but with an arbitrary geometry. Analysis of such structures was then be used to define the favoured geometry for the interaction of the two groups as a model for the more specific secondary amide-triazole interaction [25].

The first step of the analysis was to view all of the contacts together on a scatter plot, as in Fig. 5.3 where the position of the probe group for each contact is displayed relative to a normalised central-group position. From this plot it was immediately apparent that the NH probe groups were clustered in a small area, with the hydrogen atoms pointing in the direction of one central group N atom. This was direct evidence that a preferential orientation of the two groups is observed in the CSD, which typically results in an N–H···N hydrogen bond interaction. It was also clear from this plot that a number of other NH probe groups are spread at random relative to the central group. The diffuse nature of their arrangement indicated that

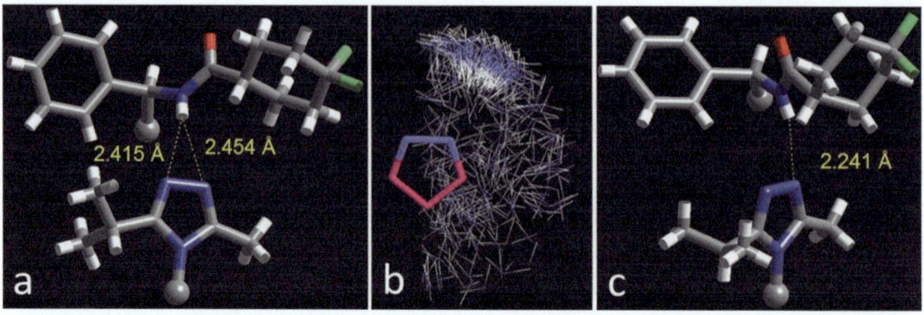

Fig. 5.3 (**a**) The secondary amide-triazole hydrogen bond motif observed in Form A of maraviroc. The motif is bifurcated with N···N donor-acceptor distances of 3.239(5) & 3.221 (5) Å. (**b**) A scatter plot of all the NH probe groups relative to the five-membered ring central group is described. The NH groups are concentrated into a cluster indicating the presence of a hydrogen-bond interaction. (**c**) The second form crystallised with stronger, linear H-bonding, and has been determined to be the more stable form

these were contacts that occurred through close packing of molecules rather than any structure-directing intermolecular interaction.

An inspection of Fig. 5.3 revealed that the hydrogen-bond geometry exhibited for Form A was clearly unusual when compared to other structures in the CSD. This was taken as an indication that a more thermodynamically stable form might yet exist, utilising the more commonly observed amide-triazole hydrogen bond geometry and hence further crystallisation experiments were initiated. The more stable polymorph Form B was quickly identified. The crystal structure of Form B was indeed found to exhibit an amide-triazole hydrogen bond, although this time as a more usual two-centred interaction using just one of the triazole nitrogen acceptor atoms (Fig. 5.3). The N···N interaction distance of 3.075 Å was within the typical range.

5.4 Intermolecular Interactions, Crystal Packing (Lattice) Energies

In order to understand the principles which govern the wide variety of solid state properties and structures of drug molecules it is important to describe both the energy and direction of interactions between molecules. As a result of the pioneering work in the development of atom-atom intermolecular potentials [21, 26], it is now possible to interpret intermolecular packing effects in organic crystals in terms of interaction energies. The basic assumption of the atom-atom method is that the interaction between two molecules can be considered to simply consist of the sum of the interactions between the constituent atom pairs.

$$E_{latt} = 1/2 \sum_{k=1}^{N} \sum_{i=1}^{n} \sum_{j=1}^{n'} V_{kij} \tag{5.1}$$

The lattice energy E_{latt} (often referred to as the crystal binding or cohesive energy) for molecular materials can be calculated by summing up all the interactions between a designated central molecule and all the surrounding molecules. Hence, if there are n atoms in the central molecule and n' atoms in each of the N surrounding molecules then the lattice energy can be calculated using Eq. 5.1 [21, 26].

Each atom-atom interaction pair (V_{kij}) consists of an attractive and repulsive dispersive interaction which can be described by a van der Waals representation, together with an electrostatic interaction and, in some cases (particularly for pharmaceuticals), a hydrogen bonding potential. The former two are broadly speaking undirected interactions whilst the latter is not. On a per atom basis, the hydrogen bond is much stronger than say a dispersive interaction but the latter involve many more atomic interactions and so, for molecules such as pharmaceuticals where the molecular weight is relatively high, contributions from the undirected van der Waals interaction tend to dominate the lattice energy.

The use of these potentials has been validated by comparing the theoretical values against the known crystal structures and experimentally measured lattice energies (sublimation enthalpies) [27, 28]. A particular advantage of the calculated energy is that it can be broken down into specific interactions along particular crystallographic directions and further partitioned onto the constituent atom-atom and/or group contributions. This is the key link between the intrinsic molecular structure and the crystal packing, allowing a profile of the important interactions to be built up within families of compounds. This approach permits the discussion between the medicinal chemist and the pharmaceutical scientist in optimising the design of molecular features, and with the pre-formulation scientist working on the optimisation of the physical properties for the intended dosage form. A number of papers have highlighted the impact of this increased understanding in recent years, including the design of features to disrupt crystal packing and therefore enhance solubility [29, 30]. Table 5.1 shows the breakdown of intermolecular interactions for the two polymorphs of *para*-aminobenzoic acid determined by a deconstruction of the lattice energy onto various structural features [31].

Recently a simple, robust Quantitative Structure Property Relationship (QSPR) model for enthalpies of sublimation (and hence crystal lattice energies) of a large, chemically diverse set of organic components was published. This model differs from previous models of enthalpy of sublimation in its applicability to more chemically diverse molecules and in its higher accuracy. The most important descriptors [32, 33] used for predicting sublimation enthalpy were the hydrophilicity, the fractional charged partial surface area, the water-excluded molecular volume and the polar surface area. These descriptors mainly represent dispersion, van der Waals, and electrostatic interactions that contribute to crystal packing stability thus increasing the sublimation enthalpy of organic compounds. A variation on this

Table 5.1 Deconstruction of the key structural features of the two polymorphs of *para*-aminobenzoic acid. Data distilled from information in [31]

α	Lattice energy contributions	β
−24.45	Lattice energy (kcal/mol)	−22.73
12.3	NH_2 (%)	23.8
39.4	C_6H_6 (%)	42.5
48.3	COOH (%)	33.7
7	Number of key interactions	8
71	Percentage of lattice energy from key interactions	75
30	Cluster size for lattice energy convergence	35

model using a machine learning algorithm utilizing recursive partitioning and linear regression has been developed [33]. This methodology combines supervised and 'unsupervised' learning methods (instance based predictions) and has proven very effective in handling large datasets. An additional benefit is that the model can be rapidly adapted as we build refinements to the model including specific data sets. Our prediction based on the test data set is shown below in Fig. 5.4 and the R^2 for the training set is 0.97 (Root Mean Squared Error (RSME) = 5.7 kJ/mol) and this is close to the original published model with R^2 of 0.95. It is used further in later sections around solvation and crystal packing.

5.5 Crystal Structure Prediction

The *ab initio* generation of reliable solid state structural details and properties through computational methods based only on molecular descriptors remains a major scientific goal. The methods being developed for structure prediction usually involve the stages of generating, clustering and refining trial structures. Final refinement of the potential structures is carried out minimising the lattice energy with respect to the unit cell dimensions **(a, b, c, α, ß and γ)**.

Despite the inherent difficulties, predictions from first principles have been the subject of much elegant investigation through the last decade with increasing application to pharmaceutical compounds [34, 35]. Proponents of these methods have now become so confident in their approaches that they are prepared to engage in blind tests to assess the predictability of their methods [36]. Figure 5.5 shows a typical energy density diagram from a polymorph prediction run. Through comparing the computationally predicted solid form landscape with the experimentally known solid form data on a new candidate, it is possible to guide the solid form screening work needed. If the current solid form is consistent with the thermodynamically stable form and energetically distinct (red coloured square in bottom right hand side in Fig. 5.5) then confidence in the current form should be high and experimental screening work minimal. If the current solid form is more consistent/ closely matched with those less stable structures (blue diamonds at the top left of

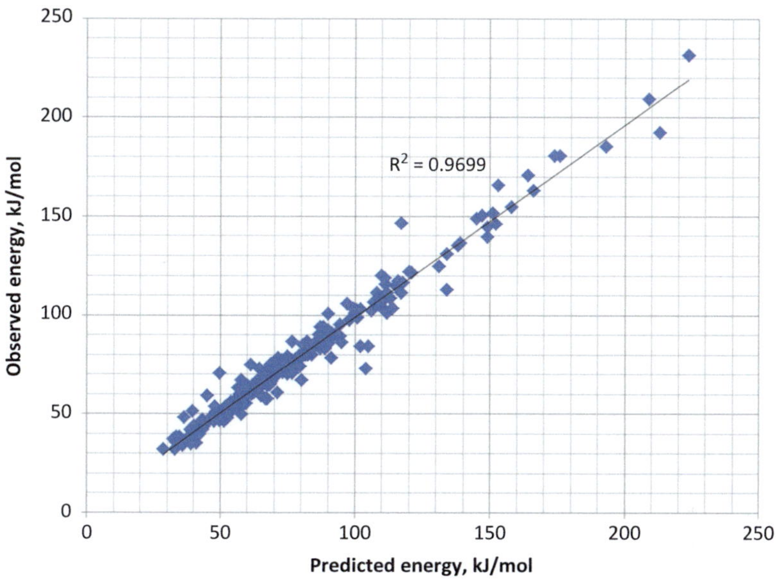

Fig. 5.4 Observed vs. predicted lattice energy for QSPR model from Ref. [33] based on the elegant models developed in Ref. [32]. Values are quoted in kJ/mol to be consistent with Ref. [32]

Fig. 5.5) then this is a situation where more expansive experimental screening should be undertaken to ensure solid form space has been effectively explored. This is an early example of successful prediction for the steroid progesterone [37]. These tools and technologies have been applied more recently to allow the rapid acceleration of an oncology candidate to the patient post promising clinical results [38].

5.6 Addressing Low Solubility – Deconvolution of Solvation and Crystal Packing

It has been reported that over 75% of drug development candidates have low solubility. An increasing trend towards low solubility is a major issue for drug development as the formulation of low solubility compounds can be problematic. From a pharmaceutical perspective, the crystalline solid is usually the final form of choice when developing a new chemical entity (NCE) into a product and consequently it would be of great value to be able to accurately predict the intrinsic solubility of crystalline drug molecules. Despite tremendous efforts [39–41], a definitive accurate and comprehensive approach to predicting solubility has proven elusive.

The precise interplay between solubility and other molecular properties has been at the hub of physical chemistry in drug discovery and development and the subject of much debate over the last couple of decades. Recent literature [30, 40] has seen

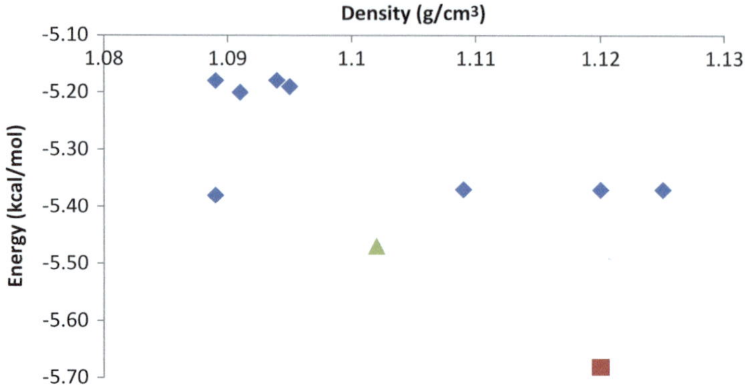

Fig. 5.5 The packing energy/density plot for a polymorph prediction run, plotted from data in Table 6 of Ref. [37]. Each *dot* represents a potential crystal structure. The *blue diamonds* represent higher energy forms. The most stable forms are those with the lowest energy and highest density – the *green triangle* matches the β form, the red square the low energy α form. This demonstrated that the α form is the thermodynamically stable form, distinct from the other possible structures

numerous solubility models proposed, generally based on hydrophobicity in combination with other descriptors. The General Solubility Equation (GSE) [41] succinctly links solubility with hydrophobicity and melting point.

Lately there have been several semi-empirical attempts to probe the changes in solubility as a function of structural changes in NCE structures in specific classes as well as systematic approaches to looking at matched molecular pairs to determine improved solubility as a function of structural changes and inferred crystal packing disruption [42]. Two recent papers highlight the potential of these approaches and the attempts to build scientific bridges across the two communities. The first paper overviews the optimisation of the solubility of an antibiotic drug using structural modifications to disrupt the crystal lattice which was limiting the solubility [43] and the second describes the use of co-crystals to optimise the dissolution rate of a psychotropic drug with known dissolution challenges [44]. The tools described in this section provide capabilities that can allow an unprecedented deconstruction of the importance of molecular solvation and crystal packing on solubility.

The grid methodology proposed elsewhere [33] and highlighted in this section provides a standard framework that can allow a systematic deconstruction of the importance of solvation and crystal packing on solubility. Harnessing this capability will hopefully lead to the creation of a 'design by first intent' strategy, at the Discovery/Development interface, for solid forms with tailored physicochemical attributes (see Fig. 5.6). Additionally the grid opens up the opportunity to explore changes in solubility at fixed solvation values using pairs of polymorphs [45]. Matched molecular pairs, where a single group on a molecule is changed, give the opportunity to step across both solvation and lattice energy changes.

- Quadrant I – solubility not limited by solvation or packing
- Quadrant II – solubility limited by solvation

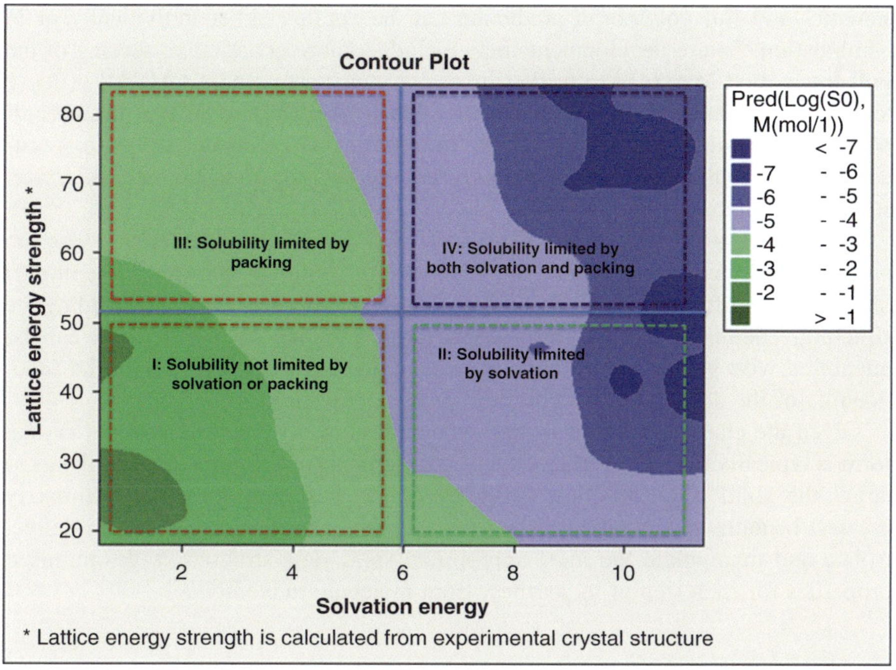

Fig. 5.6 Solvation packing grids based on mixing energies and full lattice energy calculations (*top*) and estimated lattice energy based on the QSPR approach [32, 33]. The solvation energies are in kcal/mol and the lattice energies in kJ/mol to be consistent with Ref. [32]

- Quadrant III – solubility limited by packing
- Quadrant IV – solubility limited by both solvation and packing

5.7 Summary and Future Outlook - Solid Form Design

The ICH Q9 [46] Quality Risk Management guidelines define risk as a combination of the probability of occurrence and the severity of the impact. The ICH Q6a guidelines [47] consolidate this risk framework into a decision tree on polymorphs. The first two decision points on this framework remain the key questions that need to be addressed by the development scientist along with one additional main question - how many experiments do I need to execute to ensure robustness.

- Decision Point 1: PROBABILITY- "can different polymorphs be formed?"
- Decision Point 2: IMPACT – "do the forms have different properties? (e.g. solubility)"

In attempting to provide a greater definition of the probability of a new form appearing tools ranging from quantum chemistry analysis, hydrogen bonding statistics and full polymorph prediction can be applied either individually or in combination. Future developments may include a more quantitative scoring of the Full Interaction Map [48] to further its use as a risk assessment tool. All of these tools point towards the potential for the data in the CSD to inform the user of possible structures for a new molecule: informatics based crystal structure speculation, a tool that would complement current methodologies in the field of crystal structure prediction.

In this chapter we have described how new approaches can help assess the probability of new forms. When combined with institutional knowledge the impact of potentially different structures can begin to be determined. By understanding the structural chemistry and the biopharmaceutics the risk of a new form can be quantified with greater rigour and experimental plans shaped to design the functionality of the dosage form accordingly [49].

Given the challenges industry has encountered [3, 4] the current target crystal form is typically the most stable. One of the elements of the summer school was to target the solid state structure which gives the best properties and ultimately greatest benefits to the patient. The vision for the future is to be able to define, isolate and manipulate the most appropriate solid state structure with optimised properties for each step of its journey, from molecule to medicine.

References

1. DeCamp WH (2001) Amer Pharm Rev 4(3):70
2. Byrn SR, Pfeiffer R, Stowell JG (2002) Amer Pharm Rev 5(3):92
3. Chemburkar SR, Bauer J, Deming K, Spiwek H, Patel K, Morris J, Henry R, Spanton S, Dziki W, Porter W, Quick J, Bauer P, Donaubauer J, Narayanan BA, Soldani M, Riley D, McFarland K (2000) Dealing with the impact of ritonavir polymorphs on the late stages of bulk drug process development. Org Process Res Dev 4:413
4. Erdemir D, Lee AY, Myerson AS (2007) Polymorph selection: the role of nucleation, crystal growth and molecular modeling. Curr Opin Drug Discov Devel 10:746
5. Glaxo Inc. and Glaxo Group Limited versus Novopharm Limited, No 5:94-CV-527-BO(1) 931 F. Supp. 1280 and 96-1466 DCT 94-CV-527
6. Cabri W, Ghetti P, Pozzi G, Alpegiani M (2007) Polymorphisms and patent, market, and legal battles: cefdinir case study. Org Process Res Dev 11:64
7. Storey R, Docherty R, Higginson PD (2003) Integration of high throughput screening methodologies and manual processes for solid form selection. Amer Pharm Rev 6(1):100
8. Thayer A (2007) Chem Eng News 85:17
9. Desiraju GR (1997) The crystal as a supramolecular entity. Wiley, Chichester
10. Sun CC (2009) Materials science tetrahedron—a useful tool for pharmaceutical research and development. J Pharm Sci 98(5):1671
11. Chow K, Tong HYH, Lum S, Chow AHL (2008) Engineering of pharmaceutical materials: an industrial perspective. J Pharm Sci 97:2855
12. Hancock BH, Elliot J (2006) Pharmaceutical materials science: an active new frontier in materials research. MRS Bull 31:869
13. Phillips FC (1963) An introduction to crystallography, 3rd edn. Longmans, London
14. Nichols G, Frampton CJ (1998) Physicochemical characterization of the orthorhombic polymorph of paracetamol crystallized from solution. J Pharm Sci 87:684

15. Wright JG (1987) Molecular crystals. Cambridge University Press, Cambridge
16. Etter MC (1990) Encoding and decoding hydrogen-bond patterns of organic compounds. Acc Chem Res 23(4):120
17. Taylor R, Kennard O (1984) Hydrogen-bond geometry in organic crystals. Acc Chem Res 17:320
18. Taylor R, Kennard O (1982) Crystallographic evidence for the existence of C-H···O, C-H···N and C-H···Cl hydrogen bonds. J Chem Soc 104:5063
19. Docherty R, Jones W (1997) In: Jones W (ed) Organic molecular solids: properties and applications, Chapter 3. CRC Press, London
20. Brock CP, Dunitz JD (1994) Towards a grammar of crystal packing. Chem Mater 6:1118
21. Kitaigorodsky AI (1973) Molecular crystals and molecules. Academic Press, New York
22. Macrae CF, Edgington PR, McCabe P, Pidcock E, Shields GP, Taylor R, Towler M, van de Streek J (2006) Mercury: visualization and analysis of crystal structures. J Appl Crystallogr 39:453–457
23. Galek PTA, Allen FH, Fábián L, Feeder N (2009) Knowledge-based H-bond prediction to aid experimental polymorph screening. CrystEngComm 11:2634–2639
24. Galek PTA, Chisholm JA, Pidcock E, Wood PA (2014) Hydrogen-bond coordination in organic crystal structures: statistics, predictions and applications. Acta Crystallogr B 70:91–105
25. Feeder N, Pidcock E, Reilly AM, Sadiq G, Doherty CL, Back KR, Meenan P, Docherty R (2015) The integration of solid-form informatics into solid-form selection. J Pharm Pharmacol 67(6):857–868
26. Williams DE (1966) Nonbonded potential parameters derived from crystalline aromatic hydrocarbons. J Chem Phys 45:3770
27. Lifson S, Hagler AT, Dauber P (1979) Consistent force field studies of intermolecular forces in hydrogen-bonded crystals. 1. Carboxylic acids, amides, and the C=O···H- hydrogen bonds. J Amer Chem Soc 101:5111
28. Momany FA, Carruthers LM, McGuire RF, Scherega HA (1974) Intermolecular potentials from crystal data. III. Determination of empirical potentials and application to the packing configurations and lattice energies in crystals of hydrocarbons, carboxylic acids, amines, and amides. J Phys Chem 78:1595
29. Scott JS, Birch AM, Brocklehurst KJ, Broo A, Brown HS, Butlin RJ, Clarke DS, Davidsson Ö, Ertan A, Goldberg K, Groombridge SD, Hudson JA, Laber D, Leach AG, MacFaul PA, McKerrecher D, Pickup A, Schofield P, Svensson PH, Sörme P, Teague J (2012) Use of small-molecule crystal structures to address solubility in a novel series of G protein coupled receptor 119 agonists: optimization of a lead and in vivo evaluation. J Med Chem 55:5361
30. Ishikawa M, Hashimoto Y (2011) Improvement in aqueous solubility in small molecule drug discovery programs by disruption of molecular planarity and symmetry. J Med Chem 54:1539
31. Roberts KJ, Rosbottom I, Docherty R (2015) The solid state, surface and morphological properties of p-aminobenzoic acid in terms of the strength and directionality of its intermolecular synthons. CrystEngComm 17(30):5768–5788
32. Salahinejad M, Le TC, Winkler DA (2013) Capturing the crystal: prediction of enthalpy of sublimation, crystal lattice energy, and melting points of organic compounds. J Chem Inf Model 53:223–229
33. Docherty R, Pencheva KP, Abramov Y (2015) Low solubility in drug development. De-convoluting the relative importance of solvation and crystal packing. J Pharm Pharmacol 67(6):847–856
34. Payne RS, Rowe RC, Roberts RJ, Charlton MH, Docherty R (1999) Potential polymorphs of aspirin. J Comput Chem 20:262–273
35. Day GM, Motherwell WDS, Jones W (2007) A strategy for predicting the crystal structures of flexible molecules: the polymorphism of phenobarbital. Phys Chem Chem Phys 9:1693–1704
36. Bardwell DA et al (2011) Towards crystal structure prediction of complex organic compounds - a report on the fifth blind test. Acta Crystallogr B 67:535–551

37. Payne RS, Roberts RJ, Rowe RC, Docherty R (1999) Examples of successful crystal structure prediction: polymorphs of primidone and progesterone. Int J Pharm 177:231–245
38. Abramov YA (2013) Current computational approaches to support pharmaceutical solid form selection. Org Process Res Dev 17:472–485
39. Llinàs A, Glen RC, Goodman JM (2008) Solubility challenge: can you predict solubilities of 32 molecules using a database of 100 reliable measurements? J Chem Inf Model 48 (7):1289–1303
40. Hill AP, Young RJ (2010) Getting physical in drug discovery: a contemporary perspective on solubility and hydrophobicity. Drug Discov Today 15:648–655
41. Ran Y, Yalkowsky SH (2001) Prediction of drug solubility by the general solubility equation (GSE). J Chem Inf Comput Sci 41:354–357
42. Leach AG, Jones HD, Cosgrove DA, Kenny PW, Ruston L, MacFaul P, Wood JM, Colclough N, Law B (2006) Matched molecular pairs as a guide in the optimization of pharmaceutical properties; a study of aqueous solubility, plasma protein binding and oral exposure. J Med Chem 49:6672–6682
43. Tehler U, Fagerberg JH, Svensson R, Larhed M, Artursson P, Bergström CA (2013) Optimizing solubility and permeability of a biopharmaceutics classification system (BCS) class 4 antibiotic drug using lipophilic fragments disturbing the crystal lattice. J Med Chem 56 (6):2690–2694
44. Zhang T, Yang Y, Wang H, Sun F, Zhao X, Jia J, Liu J, Guo W, Cui X, Gu J, Zhu G (2013) Using dissolution and pharmacokinetics studies of crystal form to optimize the original iloperidone. Cryst Growth Des 13:5261–5266
45. Pudipeddi M, Serajuddin A (2005) Trends in solubility of polymorphs. J Pharm Sci 94 (5):929–939
46. ICH Q9 Quality Risk Management (2005) http://www.ich.org/fileadmin/Public_Web_Site/ICH_Products/Guidelines/Quality/Q9/Step4/Q9_Guideline.pdf
47. ICH Q6A Specifications: Test procedures and acceptance criteria for new drug substances and new drug products: chemical substances. (1999) http://www.ich.org/fileadmin/Public_Web_Site/ICH_Products/Guidelines/Quality/Q6A/Step4/Q6Astep4.pdf
48. Wood PA, Olsson TS, Cole JC, Cottrell SJ, Feeder N, Galek PT, Groom CR, Pidcock E (2013) Evaluation of molecular crystal structures using full interaction maps. Cryst Eng Comm 15:65–72
49. Roberts KJ, Hammond RB, Ramachandran V, Docherty R, Chapter 7 in computational approaches in pharmaceutical solid state chemistry, Edited by Abramov YA, Wiley-Blackwell

Chapter 6
Modelling Route Map: From Molecule Through the Solution State to Crystals

Robert B. Hammond

Abstract An introductory review is presented which places molecular scale modelling within the context of the multi-scale modelling paradigm. The link between constitutive equations for material properties and atomistic calculations is highlighted. The molecular modelling toolkit is described together with approaches for calculating the shapes of crystals and for exploring interactions between crystal surfaces and molecules in the solution environment. Similarly, approaches for the molecular modelling of solutions are described.

Keywords Molecular modelling • Molecular mechanics • Crystal habit • Morphology • Molecular dynamics • Multi-scale

6.1 Introduction

In this paper I will try to address aspects of these three questions. Why model? What can we model? How can we model it? Possible reasons for modelling include enhancing our understanding and interpretation of experimental observations, determining material properties that are difficult to measure by experiment and reducing the level of empiricism in the design of experiments. The main theme of this course naturally takes us across multiple length scales and times scales. In a review article [1] James Elliott gives a materials scientist's view of multiscale modelling identifying two basic approaches (i) hierarchical, where separate models are run at the different scales and coupled via the exchange of parameter values and (ii) hybrid, where models are computed concurrently over spatially-separated regions whilst exchanging information periodically. An example of the later is coupling quantum mechanics with classical molecular dynamics and a finite element model to investigate how the effects from an oxidation reaction on the (111) surface of silicon propagate through the crystal lattice [2]. However at the present time the hierarchical approach is the more common. Later we will consider some of

R.B. Hammond (✉)
School of Chemical and Process Engineering, University of Leeds, Leeds LS2 9JT, UK
e-mail: r.b.hammond@leeds.ac.uk

K.J. Roberts et al. (eds.), *Engineering Crystallography: From Molecule to Crystal to Functional Form*, NATO Science for Peace and Security Series A: Chemistry and Biology, DOI 10.1007/978-94-024-1117-1_6

the parameters that molecular scale modelling can be used to calculate and how these are coupled to modelling approaches at the meso and process scales.

6.1.1 Possible Approaches to Modelling

On the one hand atoms and on the other hand crystals: our journey may start from either place.

So the research question might be, when manufacturing a formulated product which contains solid particles, how will those particles respond to being conveyed to the die or subjected to compression in a tablet press? In trying to tackle such a problem, a materials scientist who is interested in the mechanical properties of a crystalline material may adopt a perspective in which the crystalline state is regarded as a continuum. In this approach one considers the inherent symmetry of the crystal (without invoking atoms) and the way the crystal responds to the application of forces. Hence the mathematical description of the mechanical behaviour of the crystalline material is through a set of constitutive equations [3]. A volume element of a material may be subjected to a uniaxial stress (force per unit area), denoted σ_{xx}, which may produce a uniaxial strain (change in length per unit length and hence dimensionless), denoted ε_{xx}. Eq. 6.1 denotes how the strains are related linearly to the stresses through the compliance constants of the material S_{ijkl} where i, j, k, l = 1, 2, 3 and indicate the orthogonal directions selected.

$$\varepsilon_{ij} = S_{ijkl}\sigma_{kl} \tag{6.1}$$

Alternatively the stresses are related linearly to the strains in terms of the stiffness constants of the material C_{ijkl} as given in Eq. 6.2.

$$\sigma_{ij} = C_{ijkl}\varepsilon_{kl} \tag{6.2}$$

In terms of notation, a shear stress applied to a material, which has components in two orthogonal directions, is often denoted τ_{ij} where $i \neq j$ (equivalent to σ_{kl} for $k \neq l$ and σ_{ij} for $i \neq j$ respectively in Eqs. 6.1 and 6.2) while shear strain, which again has components in two orthogonal directions, is often denoted γ_{ij} where $i \neq j$ (equivalent to ε_{ij} for $i \neq j$ and ε_{kl} for $k \neq l$ respectively in Eqs. 6.1 and 6.2). Fortunately, through an appropriate choice of orthogonal reference-axes the eighty-one compliance constants and stiffness constants are reduced to just twenty-one unique constants in the most general case (triclinic crystal system) and the linear, constitutive equations can be written in terms of column vectors with six elements and symmetric 6 × 6 matrices as shown in Fig. 6.1.

If the material is isotropic then there are only two unique material constants C_{11} and C_{12} and these can be related to Young's modulus and Poisson's ratio for the material. Later we will return to how molecular modelling can be used to compute the material constants required to evaluate such constitutive equations. This

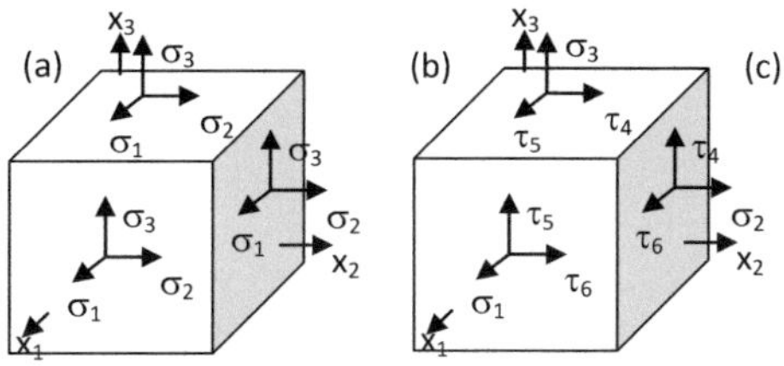

(c)
$$\begin{pmatrix}\sigma_1\\ \sigma_2\\ \sigma_3\\ \tau_4\\ \tau_5\\ \tau_6\end{pmatrix} = \begin{bmatrix} C_{11} & C_{12} & C_{13} & C_{14} & C_{15} & C_{16}\\ & C_{22} & C_{23} & C_{24} & C_{25} & C_{26}\\ & & C_{33} & C_{34} & C_{35} & C_{36}\\ & & & C_{44} & C_{45} & C_{46}\\ & sym & & & C_{55} & C_{56}\\ & & & & & C_{66}\end{bmatrix} \begin{pmatrix}\varepsilon_1\\ \varepsilon_2\\ \varepsilon_3\\ \gamma_4\\ \gamma_5\\ \gamma_6\end{pmatrix}$$

Fig. 6.1 (**a**) The forces on the faces of a unit cube in a homogeneously stressed body, (**b**) using a simplified notation, (**c**) the constitutive equations expressed in vector and matrix form for the most general case

continuum based approach is useful because Finite Element (FE) analysis allows it to be applied to predict the behaviour of real materials. Alternatively when we have assemblies of particles then the Discrete Element Method (DEM) can be employed to understand how forces are transmitted through the powder. In DEM, contact mechanics are used with a spring-dashpot model but again the model equations require us to know material constants such as the surface energy for specific particle-surfaces.

6.1.2 Atomistic or Molecular Perspective in Modelling

What is molecular modelling?

- A broad set of approaches for investigating the chemical and physical properties of matter where matter is regarded as an assembly of discrete particles i.e. atoms or molecules.
- Hence molecular modelling can treat any material in any physical state: solid, liquid or gas (fluid i.e. in supercritical region).
- Models can be used, for example, to compute the system energy, interactions between matter and incident electromagnetic radiation, properties of individual phases such as density, viscosity, structure, chemical reactivity and rates of reactive processes.

Modelling approaches are often classified using the descriptors: first principles (ab initio), semi-empirical and empirical. This way of classifying molecular modelling methods is based on considering:

- the amount of information/data used that comes directly from experimental measurements,
- the use of analogy when modelling the physics i.e. whether a simplified physical model is used (which matches what is observed).

The hierarchy of methods is also characterised by:

- the types of energy considered (potential/kinetic),

- whether temperature is considered,
- whether the phenomenon being modelled is treated as time dependent or independent.

For processes involving making and breaking of covalent bonds or significant changes in electron density distribution, then generally employ first principles (ab initio) quantum mechanics through the application of Hartree Fock equations (Sects. 6.3.4, 6.3.5, and 6.3.6) or electronic Density Functional Theory (DFT) (Sect. 6.3.8). Similarly, but where lower precision is required, semi-empirical quantum mechanical approaches can be used. These methods are partially parameterised against experimental data. For processes that are mostly determined by non-bonded interactions can use empirical, atomistic force-field potentials to describe the behaviour of molecules.

Methods that don't consider the influence of temperature but only consider potential energy i.e. not kinetic energy and not free energy and entropy are good for comparing the relative stability of solid state structures (polymorph prediction), calculating surface energies of facetted particles and calculating the shape of crystalline particles.

Methods that take account of temperature in some form but not directly kinetic energy (momentum) include Monte Carlo (MC) [4] simulations which can be used to explore potential energy surfaces. In MC simulations particles can escape from local energy-minima. A series of trials are performed in which the adjustments to particle positions are determined by random numbers. There is a probability to accept trial moves which increase the overall potential energy of the system. The probability to accept a move is weighted by the Boltzmann factor exp.($-\Delta E/\beta T$) where ΔE is the energy change associated with a trial move, β is a suitable scaling constant and T a pseudo temperature. One possible evolution for a system with time is implied by an MC simulation but the pathway is not defined rigorously for example by solving Newton's equations of motion.

Methods that take account of temperature and kinetic energy (momentum) and follow a deterministic trajectory in phase space (three positions and three momenta per particle) include classical Molecular Dynamics (MD) simulations. Here solve Newton's equations of motion for all the particles (molecules) in the system at the scale of individual atoms. The approach may be coarse-grained to treat groups of atoms as the centres of mass (rather than every individual atom) in the case of, for example, biopolymers. From the trajectory (history) file produced in a simulation one may extract ensemble averages, radial distribution functions describing local structure in solutions and relate simulation results to bulk properties via approaches from statistical thermodynamics.

6.2 Molecular Coordinates

The text [5] provides a good overview of the topics covered in this section. The chemical structure of a component is only partially encoded by its chemical or molecular formula which might be thought of as a one-dimensional descriptor for a material. Hence in the case of molecules which have structural isomers (i.e. a common molecular formula), we generally differentiate between isomers by using a two-dimensional representation of the molecular structure which is itself a molecular model. A number of approaches have been developed to encode structural information within an alphanumeric character-string. One is called SMILES (Simplified Molecular Interpretive Line Entry System), for example the SMILES string C1CCCCC1 encodes the molecule cyclohexane whereas c1ccccc1 encodes benzene. Molecular modelling programs such as INTERCHEM [5] can generate three-dimensional coordinates for the atoms in molecules by interpreting a SMILES string. The procedure uses look-up tables of bond lengths and angles to create the molecular framework and then applies a molecular mechanics approach (see Sect. 6.4) to optimise the molecular geometry. Alternatively a two-dimensional sketch of a molecular structure, made on a regular grid, can be converted to a three-dimensional structure in a similar way. Aspects of stereochemistry can be encoded in the sketch by using wedges to indicate whether an atom lies above or below the plane.

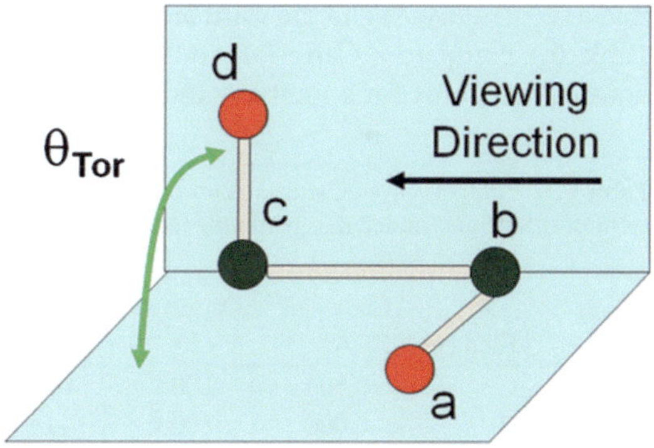

Fig. 6.2 Definition of a Dihedral angle

6.2.1 *Dihedral (Torsion) Angles*

For a molecule, see Fig. 6.2, when viewing along the bond direction from atom b to atom c the torsion angle is the magnitude of the rotation needed to bring the plane containing atoms b, c and d into coincidence with the plane containing atoms a, b and c.

Coincidence means that atom a (proximal) eclipses atom d (distal) along the viewing direction. If the smaller rotation required, when moving the distal atom d, is clockwise then the sense of the torsion angle is positive and vice versa.

Here the torsion angle a-b-c-d θ_{Tor} is $-90°$ (negative) as the smaller rotation to achieve coincidence of planes abc and bcd is anticlockwise when viewed from atom b towards atom c, $-180° < \theta_{Tor} \leq 180°$.

6.2.2 *Cartesian and Internal Coordinates*

There are two common approaches used to describe the positions of the N atoms in a molecule. Option one is to use a vector of Cartesian-coordinate triples (x_i, y_i, z_i), or often for describing the positions of atoms in crystals, fractional-coordinate triples (a_i, b_i, c_i). Equivalent sets of Cartesian coordinates can always be computed from fractional coordinates by adopting a convention to interrelate the directions of the basis vectors in the two systems (see Appendix A1 from reference (5)). However, Cartesian and fractional coordinates cannot be directly related to aspects of molecular geometry such as covalent-bond lengths, bond angles and dihedral angles which are referred to as internal coordinates. The so called Z-matrix is a list of internal coordinates which allows the positions of all the atoms to be determined. In the general case for a molecule of N atoms, 3 N – 6 internal coordinates will be required to specify the atomic positions. The three rigid-body translations and three rigid-body rotations are removed by placing the first numbered atom at the origin of the underlying Cartesian coordinate system, the second atom on a specified Cartesian axis (usually Z) and the third atom in a specified Cartesian plane (usually ZX). Table 6.1 compare a Cartesian coordinate and Z-matrix specification of the positions of the atoms for a methane molecule.

Table 6.1 Comparison of Cartesian coordinates and internal coordinates (Z-matrix) to specify the atom positions and molecular geometry for a methane molecule

Atom number	Element	Cartesian X (Å)	Cartesian Y (Å)	Cartesian Z (Å)	Bond length (Å)	Bond angle (°)	Dihedral angle (°)
1	C	0.000	0.000	0.000	None	None	None
2	H	0.000	0.000	1.089	1.089	None	None
3	H	1.027	0.000	−0.363	1.089	109.471	None
4	H	−0.513	−0.889	−0.363	1.089	109.471	120.000
5	H	−0.513	−0.889	−0.363	1.089	109.471	−120.000

6.3 Electronic Structure Calculations for Molecules

The texts [6–11] provide a good overview of the topics covered in this section. First principles molecular modelling approaches are methods based on the application quantum mechanics. Here we begin by considering a single atom in a vacuum which consists of a central nucleus, containing protons and neutrons, surrounded by a 'diffuse cloud' of electrons. Mathematically, one can only treat exactly the case where there is just one electron for example H, He^{+}, Li^{2+} etc. To deal with more than one electron can apply the Self Consistent Field, SCF, approach described in Sect. 6.3.6.

The electrons in isolated atoms are found to occupy atomic orbitals. When atoms come together to form an isolated molecule the atomic orbitals combine to form molecular orbitals. Hence atomic orbitals form a basis on which to compute molecular orbitals. This approach to calculating molecular orbitals is known as the Linear Combination of Atomic Orbitals or LCAO approximation.

When describing atoms and molecules it is generally considered that protons, neutrons and electrons are the smallest subdivision of the 'atom' required to understand chemical processes. Crudely all chemistry is to do with where electrons reside in space as a function of time. Need a frame of reference to answer the question '*Where are the electrons?*' The atomic nucleus is three orders of magnitude more massive than its associated electrons so consider the nucleus to be at rest when seeking to describe where the electrons are situated. This is the Born Oppenheimer approximation.

6.3.1 Quantum Mechanics and the Schrödinger Equation

Modelling Electrons Fundamental Principles

- Energy and momentum are quantised

$E = h\upsilon$	E, photon energy, υ, photon frequency
$c = \upsilon\lambda$	c speed of light, λ, photon wavelength
$p = h/\lambda = h\upsilon/c$	p photon momentum (de Broglie)

- The size of the quantum, Planck's constant, h, is very small. Try working out the EM wavelength of a bus with a mass of 7500 kg travelling at 10 m/s.
- In quantum mechanics, QM, address the 'where question' in terms of a mathematical function called the wave function, Ψ.
- Ψ has to be interpreted in terms of probability theory.
- The Hamiltonian Operator, H, operates on the wavefunction to yield the total energy of the system E.

The Schrödinger Equation can be written as:

$$H\Psi = E\Psi \quad (6.3)$$

H was first worked out for 'classical' Newtonian systems and later adapted in quantum theory. For one dimension and considering the time independent Schrödinger Equation

$$\mathrm{H} = -\hbar^2/2\mathrm{m}\left(\mathrm{d}^2/\mathrm{dx}^2\right) + \mathrm{V}(\mathrm{x}) \tag{6.4}$$

where $\hbar = h/2\pi$ and m is the mass of the particle. The first term is associated with the kinetic energy the second term, V(x), with the potential energy.

6.3.1.1 Particle in One-Dimensional Potential Energy Well

The simplest problem to treat with QM is a particle in a one-dimensional energy well where $V(x) = 0$ in the well and $V(x) = \infty$ everywhere else. The Schrödinger Equation for this system can be written as Eq. 6.5. Here the subscript index n is introduced to distinguish between different members of the set of possible solutions. General solutions for the wavefunction are given by Eq. 6.6 where $p_n = (2mE_n)^{1/2}$ this is true since the particle only has kinetic energy. Noting outside the energy well $V(x) = \infty$ when $0 > x$, $x > a$ whereas $V(x) = 0$ when $0 \leq x \leq a$, boundary conditions can be imposed on the possible solutions to the Schrödinger Equation. First, the wavefunction must vanish at the boundaries of the well i.e. there must be nodes at the boundaries $\Rightarrow C = 0$. The second normalisation condition is that the product of the wavefunction, Ψ, with its complex conjugate, Ψ*, integrated over all space must equal to one (c.f. case of a continuous probability distribution) this is expressed in Eq. 6.7. Hence the energy of the particle in the 1-D energy well is quantised and depends on the quantum number n where n = 1, 2, 3.., the corresponding wavefunctions for these states have n-1 nodes lying between the boundaries of the cell.

$$\left(\frac{d^2}{dx^2}\right)\Psi_n = -\left(\frac{2mE_n}{\hbar^2}\right)\Psi_n \tag{6.5}$$

$$\Psi_n(x) = C\cos\left(\frac{p_n x}{\hbar}\right) + D\sin\left(\frac{p_n x}{\hbar}\right) \tag{6.6}$$

$$\begin{aligned} &\textstyle\int_0^a \Psi_n^*(x)\Psi_n(x)dx = 1 \\ &\text{hence } \textstyle\int_0^a D^2 \sin^2\left(\frac{n\pi x}{a}\right) = \frac{1}{2}aD^2 \end{aligned} \tag{6.7}$$

$$\begin{aligned} &\Rightarrow D = \left(\tfrac{2}{a}\right)^{\frac{1}{2}} \\ &\Rightarrow \Psi_n = \left(\tfrac{2}{a}\right)^{\frac{1}{2}} \sin\left(\frac{n\pi x}{a}\right), \quad E_n = \frac{n^2h^2}{8ma^2} \end{aligned} \tag{6.8}$$

6.3.1.2 The Hydrogen Atom and Atomic Orbitals

For the quantum mechanical description of the hydrogen atom, the Hamiltonian operator is given by Eq. 6.9. Spherical-polar coordinates are used to solve the Schrödinger Equation for the hydrogen atom where μ is the reduced mass, m_e mass of an electron, m_p mass of a proton, ε_0 is the permittivity of free space, ∇^2 is the Laplace operator (spherical polar coordinates), e is the charge on an electron. Can factorise the wave function, $\Psi_{nlm}(r,\theta,\phi)$, in terms of a radial, $R_{nl}(r)$, and spherical, $Y_{lm}(\theta,\phi)$, function, hence $\Psi_{nlm}(r,\theta,\phi) = R_{nl}(r)Y_{lm}(\theta,\phi)$. The solutions to the equation for the radial and spherical parts of the wave function define the atomic orbitals for the hydrogen atom. There are now three quantum numbers required, n, l and m.

$$H = -\left(\frac{\hbar^2}{2\mu}\right)\nabla^2 - \frac{e^2}{4\pi\varepsilon_0 r} \quad \text{where } \mu = \frac{m_e m_p}{m_e + m_p} \tag{6.9}$$

$$E_{nlm} = -\left(\frac{\mu e^4}{32\pi^2\varepsilon_0^2\hbar^2 n^2}\right) \text{ where } n = 1, 2, 3\ldots \tag{6.10}$$

The exact solution for the system energy is given by Eq. 6.10. Notice the energy quantisation depends on a single quantum number n. Since the 2s and 2p orbitals have the same Principle Quantum Number, n, this implies they are degenerate however this is only true for the hydrogen atom. For atoms with more than one electron the degeneracy is split because of the interactions between different electrons in the same atom. Although there is no exact analytical solution to the Schrödinger Equation for atoms with more than one electron, approximate methods can yield results that are in excellent agreement with experimental, spectroscopic observations. One approach to this is based on variation theory described briefly below. Finally, it turns out that there is a fourth quantum number required, the spin quantum number s, where s = ±½, to completely specify the quantum state of electrons in atoms and molecules.

6.3.2 Electrons in Atoms and Molecules: Pauli Exclusion Principle

The quantum state of an electron is fully specified by FOUR quantum numbers n, l, m and s. The Pauli Exclusion Principle states that no two electrons in a system can exist in the same quantum state. This implies that no two electrons can have all four corresponding quantum numbers with the same values. In atoms or molecules which have more than one electron, the total wavefunction for the system can be expressed as the product of one-electron wavefunctions.

However, we need to take account of electron spin so introduce the concept of a spin orbital in which the spatial part of the one-electron wavefunction is multiplied by a spin function denoted α or β where α represents the spin-up state and β represents the spin-down state. Just as the one-electron wavefunctions that are solutions to the SE for the hydrogen atom form an orthonormal set the opposite spin-state functions are orthogonal to each other as indicated by Eq. 6.11. Here the ds is an element of 'volume' in the space spanned by the spin coordinates. So for spin orbitals rather than integrating over a volume element dv in three spatial elements dx, dy and dz. we integrate over all space $d\tau = dvds$. For a helium atom in its ground state, with two electrons in the 1 s orbital, it might seem that the wavefunction can be written as $\psi = \phi_{1s}\alpha(1)\phi_{1s}\beta(2)$ where the labels in brackets identify the two different electrons in the atom. However, the total wavefunction for a system of two or more electrons must be antisymmetric with respect to the exchange (permutation) of any pair of electrons since electrons are fermions having half-integral spin. For example in Eq. 6.12 the labels 1 and 2 denote the two different electrons in an atom of helium. Note that if the electron labels are exchanged the requirement that $\Psi \rightarrow -\Psi$ is satisfied. Note also that the correct form of the wavefunction can be expressed as a determinant, the so called Slater determinant. In a similar way the wavefunction of a hydrogen molecule, H_2, containing two electrons, can be derived from the one-electron wavefunctions for an H_2^+ molecular ion which contains a single electron (see below).

$$\int \alpha{*}\alpha\, ds = \int \beta{*}\beta\, ds = 1 \text{ and} \int \alpha{*}\beta\, ds = \int \beta{*}\alpha\, ds = 0 \qquad (6.11)$$

$$\begin{aligned} \psi &= \frac{1}{\sqrt{2}}\phi_{1s}\alpha(1)\phi_{1s}\beta(2) - \frac{1}{\sqrt{2}}\phi_{1s}\alpha(2)\phi_{1s}\beta(1) \\ \text{or } \psi &= \frac{1}{\sqrt{2}}\begin{vmatrix} \phi_{1s}\alpha(1) & \phi_{1s}\beta(1) \\ \phi_{1s}\alpha(2) & \phi_{1s}\beta(2) \end{vmatrix} \end{aligned} \qquad (6.12)$$

6.3.3 Introduction to Variation Theory

Variation theory defines an approach to improve the accuracy of a trial solution to the SE when the trial solution is approximate rather than exact. Hence, if E_0 is the lowest eigenvalue (ground state energy) found when the Hamiltonian operator operates on the exact wavefunction then for a trial wavefunction Ψ_{trial} the Inequality (13) is true. The trial solution will provide an over-estimate of the exact ground-state energy. Therefore, to obtain an improvement in the estimate to the exact wavefunction the strategy is to minimise the quantity ε given by Eq. 6.14. So we now have a strategy for improving our approximation for the exact wavefunction and ground state energy which can be exploited using computation.

$$\int d\tau \psi^*_{trial} \widehat{H} \psi_{trial} \geq E_0 \int d\tau \psi^*_{trial} \psi_{trial} \tag{6.13}$$

$$\varepsilon = \frac{\int d\tau \psi^*_{trial} \widehat{H} \psi_{trial}}{\int d\tau \psi^*_{trial} \psi_{trial}} \tag{6.14}$$

6.3.4 Linear Combination of Atomic Orbitals to Give Molecular Orbitals

If the trial wavefunction is expressed as a linear combination of other wavefunctions, Eq. 6.15, then seek to minimise the sum of integrals given by Eq. 6.16 subject to the constraint given by Eq. 6.17.

$$\Psi_{trial} = \sum\nolimits_m c_m \Psi_m \tag{6.15}$$

$$\sum\nolimits_{n,m} \int d\tau \Psi^*_m \widehat{H} \Psi_n c^*_m c_n \tag{6.16}$$

$$\sum\nolimits_{n,m} \int d\tau \Psi^*_m \Psi_n c^*_m c_n = 1 \tag{6.17}$$

Then the minimum energy is the minimum route of the determinant given by Eq. 6.18 known as the secular determinant.

$$\det|H_{mn} - E_m S_{mn}| = 0 \text{ where } H_{mn} = \int d\tau \psi^*_m \widehat{H} \psi_n \quad S_{mn} = \int d\tau \psi^*_m \psi_n \tag{6.18}$$

The coefficients that define the weightings in the linear combination may be obtained from the simultaneous equations defined by Eq. 6.19

$$\sum_n c_n (H_{mn} - E_m S_{mn}) = 0 \tag{6.19}$$

6.3.4.1 Example of Molecular Orbital Method: Hydrogen Molecular Ion H_2^+

The hydrogen molecular ion is the analogue of the hydrogen atom because it represents the simplest molecular system that contains only a single electron. So in this case we do not need to consider interactions between electrons. Use the idea of a *linear combination of atomic orbitals* (LCAO) to approximate the true wavefunction for the molecule.

- Can write $\Psi(1) = c_a a(1) + c_b b(1)$
- Here c_a and c_b are constants that need to be determined
- $a(1) = \Psi_{1s(a)}(r_a)$ and $b(1) = \Psi_{1s(b)}(r_b)$ wavefunctions for the 1 s atomic orbitals on hydrogen atom a and b respectively.

Can write down a determinant that yields the energy, E, of the system and find the coefficients c_a and c_b (Fig. 6.3).

Note for the elements on the main diagonal of the determinant S = 1. Whereas for H_2^+ all the integrals that must be computed are over a single electron, for a hydrogen molecule also need to compute two-electron integrals and account for electron spin. Although this increases the complexity of computations the basic approach remains the same

$$S = \int d\tau_1 a(1)b(1) \text{ (the overlap integral)}$$
$$x = \int d\tau_1 a(1)\hat{H}a(1) \text{ (a Coulomb integral)}$$
$$y = \int d\tau_1 a(1)\hat{H}b(1) \text{ (a resonance integral)}$$

$$\begin{vmatrix} x-E & y-ES \\ y-ES & x-E \end{vmatrix} = 0$$

$$E = \frac{x \pm y}{1 \pm S}$$

$$c_a = \pm c_b = \pm \frac{1}{\sqrt{2}(1 \pm S)^{\frac{1}{2}}}$$

$$\Psi_{bonding} = \frac{a(1)+b(1)}{\sqrt{2}(1+S)^{\frac{1}{2}}} \quad \Psi_{antibonding} = \frac{a(1)-b(1)}{\sqrt{2}(1-S)^{\frac{1}{2}}}$$

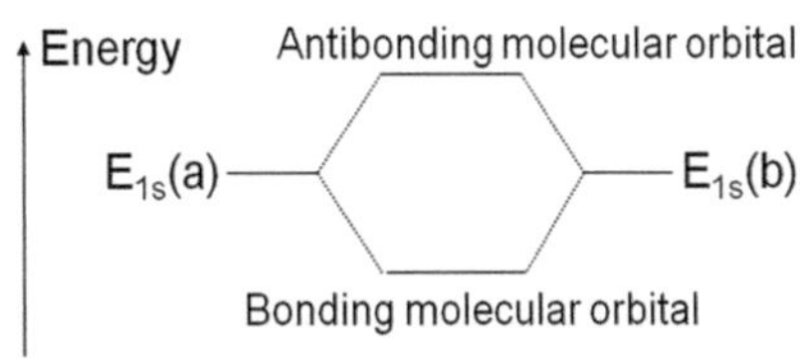

Fig. 6.3 Simple molecular orbital description of the hydrogen molecular ion H_2^+

6.3.5 *Basis Sets for Molecular Orbitals*

To compute the ground state energy for a molecule and expand the molecular wavefunction in terms of one-electron spin orbitals, a set of mathematical functions is required to represent the orbitals. If we take the example of a water molecule then, as a minimum, the atomic orbitals of all the shells with occupied atomic orbitals must be considered. For the oxygen atom this means we consider the 1 s, 2 s, $2p_x$, $2p_y$, $2p_z$ atomic orbitals. The atomic orbitals for an oxygen atom are solutions to the SE for the H atom (Eq. 6.20 for the 1 s orbital where r_1 is the position vector of the electron and R_A the position vector of the nucleus) but where the exponent in the exponential has been optimised using a variational approach. Together with the 1 s orbitals on the hydrogen atoms we then have seven atomic orbitals which are used in linear combinations to compute the molecular orbitals. The atomic orbitals form a basis set from which to compute the molecular orbitals. In this case it is the minimal basis set since using any fewer atomic orbitals would not make sense chemically. Unfortunately this particular exponential form which defines what are referred to as Slater-type orbitals or STOs is not ideal from a mathematical standpoint for carrying out the necessary integrations. For practical purposes Gaussian-type orbitals are used instead of STOs. To enable Gaussian type orbitals to be used, linear combinations of Gaussians are employed to represent a single STO. For more complete information about basis sets for molecular orbital calculations see, for example Section 2.7 of reference 6.

$$\Psi_{1s}^{H}(r_{1A}) = \frac{1}{\sqrt{\pi}} e^{-|r_1 - R_A|} \tag{6.20}$$

6.3.6 *Self Consistent Field Approach*

The self-consistent field approach is a method for implementing variation theory to perform energy minimisation when using a particular basis set to construct the molecular orbitals. The approach is iterative in that it requires an initial guess at the orbitals. The method of Lagrange multipliers is used to add a set of constraints which are only satisfied (and become equal to zero) when the system energy is minimised. Throughout the procedure the requirement that the orbitals form an orthonormal set is maintained. When the positions of the atomic nuclei are changed so as to optimise the molecular geometry by minimising the ground-state energy, a self-consistent field calculation is performed for every step-change in the molecular geometry. The procedure is stopped when the change in energy between successive steps becomes smaller than a set threshold-value or when the energy gradients with respect to the parameters being refined (e.g. the Cartesian coordinates of the atomic nuclei) are sufficiently close to zero.

6.3.7 *Charge Distribution in Molecules*

It is important to know the distribution of electronic charge in molecules particularly when modelling intermolecular interactions in condensed phases such as liquid solutions and in crystalline solids. To a first approximation this can be achieved by representing the charge distribution as a series of poles with their centres located at the centre of mass of the molecule (although the choice of location is arbitrary). The first term in the series is a monopole which is a single charge of a specified sign and magnitude. If the molecule is charge neutral then this monopole will have zero magnitude. The second term in the series is a dipole, which is a vector quantity. The direction is from the centre of a positive charge to the centre of a negative charge having the same magnitude. The dipole moment, μ, is the product of the magnitude of the vector and the magnitude of the charge. The third term in the series is the quadrupole which may be envisaged as an alternating arrangement of two positive and two negative charges, all of the same magnitude, placed at the four corners of a square. As the quadrupole moment is a tensor quantity its definition is more complex than the dipole moment's.

Although these so called multipoles form an infinite series, trying to represent the charge distribution with a single series placed at the centre of mass of the molecule is often not sufficiently accurate. A much better computational approach is to calculate many multipole expansions, usually one per atom of the molecule, and to derive these from the ground state wavefunction which is a direct measure of the electron density distribution. Eq. 6.21 shows how the electron probability density, ρ(r), can be computed from the wavefunction for a molecule containing N electrons. The electrostatic potential is the interaction energy between a unit-positive test charge, placed at a point with position vector, R_p, and the molecule and is denoted $V(R_p)$. Eq. 6.22 gives an expression for the electrostatic potential where R_n in the summation term is the position vector of an atomic nucleus with charge Z_n. Atomistic potentials which use a molecular mechanics description of molecules (see Sect. 6.4) often take just the first term for each series. Hence a monopole is located, for example, at the position of the nucleus for every atom in the molecule. The sum of these monopole charges is zero if the molecule is charge neutral. The set of charges is computed so as to minimise the differences, in a least-squares sense, between the electrostatic potential calculated from the molecular wavefunction, on a specified surface of the molecule, and that calculated from the monopole charges placed at the positions of the atomic nuclei. Eq. 6.23 summarises the procedure in which the electrostatic potential is evaluated for a grid of points which define the surface of the molecule. The terms q_n are the set of fitted charges placed at the locations of the atomic nuclei.

$$\rho(\mathbf{r_1}) = N \int \ldots \int d\mathbf{r_2} \ldots d\mathbf{r_N} |\Psi(\mathbf{r_1}, \mathbf{r_2}, \ldots \mathbf{r_N})|^2 \tag{6.21}$$

$$V(\mathbf{R_p}) = -\int d\mathbf{r}\rho(\mathbf{r})|\mathbf{R_p} - \mathbf{r}|^{-1} + \sum_{n=1}^{N} Z_n |\mathbf{R_p} - \mathbf{R_n}|^{-1} \tag{6.22}$$

$$\min \sum_{p=1}^{grid} \left[V(\mathbf{R_p}) - \sum_{n=1}^{n=N} q_n |\mathbf{R_p} - \mathbf{R_n}|^{-1} \right]^2 \tag{6.23}$$

Sets of charges derived in this way are called ElectroStatic Potential (ESP) charges. Sometimes restraints are applied so that the magnitudes of individual charges cannot become too large. The charges are then known as Restrained ElectroStatic Potential (RESP) charges. The surface used is often that defined by the van der Waals radii of the atoms that constitute the molecule or these radii scaled by a factor. When greater accuracy is required in the treatment of electrostatic interactions between molecules the fitting procedure is extended to derive sets of dipoles, quadrupoles and higher multipoles. Sometimes the placement of the multipoles is decoupled from the positions of the nuclei of the atoms in the molecule however, this increases still further the computation complexity when trying to use the multipoles in practical calculations.

6.3.8 Electronic Density Functional Theory

A different approach for determining the ground-state energy and electron density distribution is called Density Functional Theory (DFT). A 'functional' is simply a function of a function so, for example, the system energy is a functional of the Hamiltonian operator which is itself a function. By extension it turns out that the energy is a functional of the electron density ρ(r) and this is written as the energy density functional E[ρ(r)]. The energy density functional has three key properties:

1. It has a unique value for a given electron density so $E[\rho_1] \neq E[\rho_2]$ if $\rho_1 \neq \rho_2$ which means if the exact electron density is known the exact energy can be determined.
2. It obeys the variational principle, $E[\rho] \geq E[\rho_{exact}]$.
3. It can be obtained via a one-electron functional as shown in Eq. 6.24.

$$E[\rho] = \int d\mathbf{r}\rho(\mathbf{r})\nu(\mathbf{r}) \text{ where } \nu(\mathbf{r}) = \frac{\partial E[\rho(\mathbf{r})]}{\partial \rho(\mathbf{r})} \quad (6.24)$$

The one-electron functional, υ(r), can in turn be expressed in terms of orbitals that are analogous to those used in the molecular orbital approach. In fact the basis sets used for DFT calculations are the same. The main challenge in the approach is that the mathematical form of the one-electron functional is not known. To mitigate this difficulty, the energy density functional is split into the sum of three terms, the kinetic energy T[ρ], the potential energy due to the electron-nuclear attraction, $V_{ne}[\rho]$, and the potential energy due to electron-electron repulsion, $V_{ee}[\rho]$. Some contributions to these energy terms are known, and those which are not known are grouped together and called the exchange-correlation potential. A description of the history of the development of useful exchange-correlation energy functionals is beyond the scope of these brief notes. However, some major advances in the theory between 1988 and 1993 led to the development of an accurate functional B3LYP which is still widely employed today.

It is possible to apply DFT to periodic systems and to rewrite the appropriate Hamiltonian in terms of cell-periodic parameters in reciprocal space. The equivalent of the repeating unit, the unit cell, is direct space is called the Brillouin Zone in reciprocal space. An appropriate choice of basis set to model electrons in crystals is a set of plane waves to model the distribution of electron density. One of the issues is whether to include the core electrons explicitly in the calculations as in the full electron code WIEN2k or to treat just valence electrons and use a pseudo potential to represent the effect of the core-shell electrons as in the Cambridge Serial Total Energy Package (CASTEP).

6.3.9 Intermolecular Interactions

The key intermolecular interactions fall into two categories, specific and non-specific interactions. Non-specific interactions include the following types.

Ion-dipole interactions:- a molecule with a permanent dipole moment will be orientated by any electric field it encounters. An ionic species generates an electric field locally and a permanent dipole on a molecule will be orientated so as to oppose the field. The interaction energy $E_{ion/dipole}$ is proportional to the product of the charge on the ion Q with the magnitude of the dipole moment μ divided by the distance from the centre of the ion to the mid-point of the dipole d, hence $E_{ion/dipole}$ α Qμ/d.

Dipole-dipole interactions:- again a permanent dipole on one molecule produces an electric field that interacts with the permanent dipole on a second molecule.

The molecules tend to orientate themselves in a position of minimum energy so that the dipoles are either orientated end-to-end or where the molecules are large side-by-side, see Fig. 6.4.

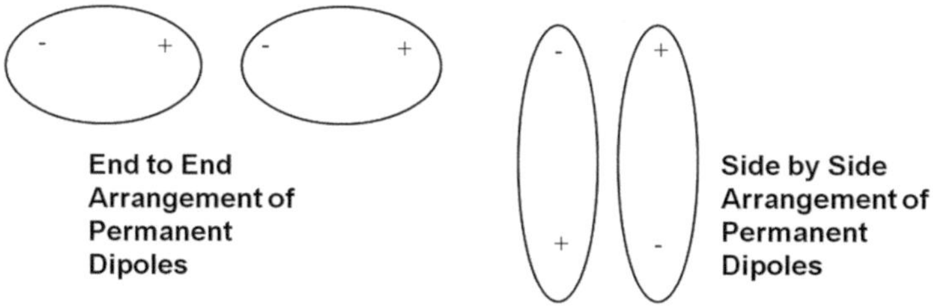

Fig. 6.4 Types of dipole interactions

The interaction energy is given by the expression:

$$E_{dipole/dipole} = -\frac{C}{R^6} \text{ where } C = \left(\frac{2}{3kT}\right)\left(\frac{\mu_1\mu_2}{4\pi\varepsilon_0}\right)^2 \qquad (6.25)$$

Here R is the distance between the centres of the dipoles, μ_1 and μ_2 are the respective dipole moments, ε_0 the relative permittivity of free space, k the Boltzmann constant and T the absolute temperature. Note that the strength of this type of interaction in inversely proportional to the absolute temperature as increased Brownian motion tends to disrupt the alignment of the permanent dipoles and hence to reduce the strength of the interactions.

Dipole-induced dipole:- the electric field of a molecule with a permanent dipole moment will induce a temporary dipole moment on any neighbouring molecule that doesn't otherwise have a permanent dipole moment. The induced dipole will tend to align with the permanent dipole of the molecule that induced it. Since the effect is transitory, with interactions constantly forming and reforming, increasing the temperature has little effect on the strength of these induced interactions. The key property of the molecule without a permanent dipole moment in determining the strength of the interaction is the molecular polarizability. The greater the polarizability, the larger the temporary dipole moment that can be induced and hence, the greater the strength of the interaction.

The interaction energy is given by the expression below where α_2' is the polarizability volume of the second molecule and where E is the electric field strength.

$$E_{dipole/induceddipole} = \frac{-C}{R^6} \text{ where } C = \frac{\mu_1^2 \alpha_2'}{4\pi\varepsilon_0} \tag{6.26}$$

Induced-dipole induced-dipole interactions:- the final type of interaction is known as the dispersion force. It originates from random fluctuations in the distribution of electrons around molecules leading to the formation of a transitory dipole moment. This transitory dipole in turn induces a dipole on a neighbouring molecule leading to a net attractive force. The interaction energy is given by the expression where I_1 and I_2 are the ionization energies of the two molecules.

$$E_{induced/induced} = \frac{-C}{R^6} \text{ where } C = \left(\frac{3}{2}\alpha_1' \alpha_2'\right)\left(\frac{I_1 I_2}{I_1 + I_2}\right) \tag{6.27}$$

In contrast to non-specific interactions, specific intermolecular interactions include:

Hydrogen-bonding interactions:- of particular importance here are molecules in which a hydrogen atom is bonded to either an oxygen or a nitrogen atom. Such molecules can act as hydrogen-bond donor molecules and if they are acting as a solvent are referred to as protic solvents. Hydrogen bonding capability has a very strong influence on solubility for solute molecules that have hydrogen-bond acceptor and donor atoms.

Electron pair donor-acceptor interactions:- Here one molecule acts as the electron pair donor, having a pair of electrons in a high energy molecular orbital, and the other molecule acts as a electron pair acceptor having a low energy unoccupied molecular orbital. Some solvent molecules such as water can act both as electron pair acceptors and donors. This type of specific interaction is often associated with a large interaction energy.

6.4 Modelling Molecules but Neglecting Electrons: Molecular Mechanics

The texts [12, 13] provide an overview of the material presented in this section. There are many circumstances in which it is sufficient to compute the relative potential energy of two or more molecular or crystal structures without requiring the calculations to be very precise. Then we can draw an analogy between a molecular structure and a purely mechanical system. In the mechanical representation we disregard the positions of the electrons completely and compute the total potential energy based on the positions of the atomic nuclei which are the centres of mass.

6.4.1 *Terms in Molecular Mechanics Potential*

The following sections introduce the different energy terms that are used, typically, in a molecular mechanics representation of a molecule.

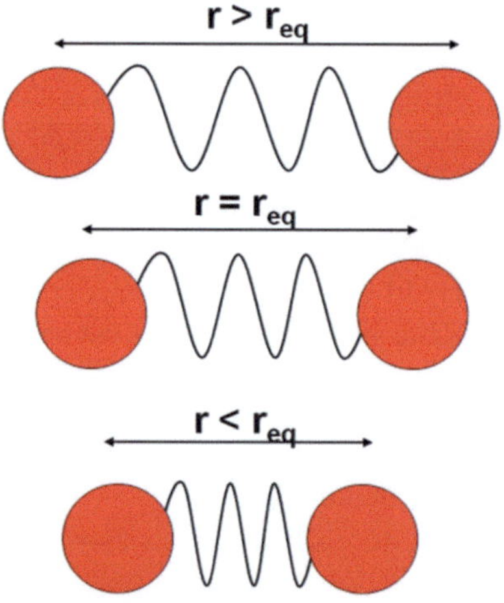

Fig. 6.5 Harmonic bond potential

6.4.1.1 Harmonic Bond Potential and Morse Potential

Bond Energy: when two atoms are directly chemically bonded via a covalent bond there is an optimal separation distance between the nuclei of the bonded atoms. When the nuclei are separated by this optimum distance the potential energy is zero. Increasing or reducing the separation is like stretching or compressing a spring. The potential energy increases according to a Harmonic Potential. The potential energy term is given by Eq. 6.28 where k_{ab} is force constant associated with a particular bond involving atoms a and b with units of energy per square Angstrom, r_{ab} is the actual distance between atoms a and b, r_{0ab} is the equilibrium distance between atoms a and b.

$$E_{Bond} = \frac{1}{2}k_{ab}(r_{ab} - r_{0ab})^2 \tag{6.28}$$

Note this is a two-body potential as the value of the potential energy depends on the position of two independent atoms or 'bodies' (Fig. 6.5).

For larger displacements the behaviour of a real covalent bond deviates from the harmonic potential. The potential energy increases more rapidly on compression of the bond (a harder wall potential) and on extension of the bond the dissociation limit is reached. This 'real' behaviour is more accurately modelled using a Morse Potential which is given by Eq. 6.29 where E_{0ab} is the dissociation energy of bond between atoms a and b, k_{ab} an asymmetry parameter for the bond a-b, and r_{0ab} equilibrium length of bond a-b (Fig. 6.6).

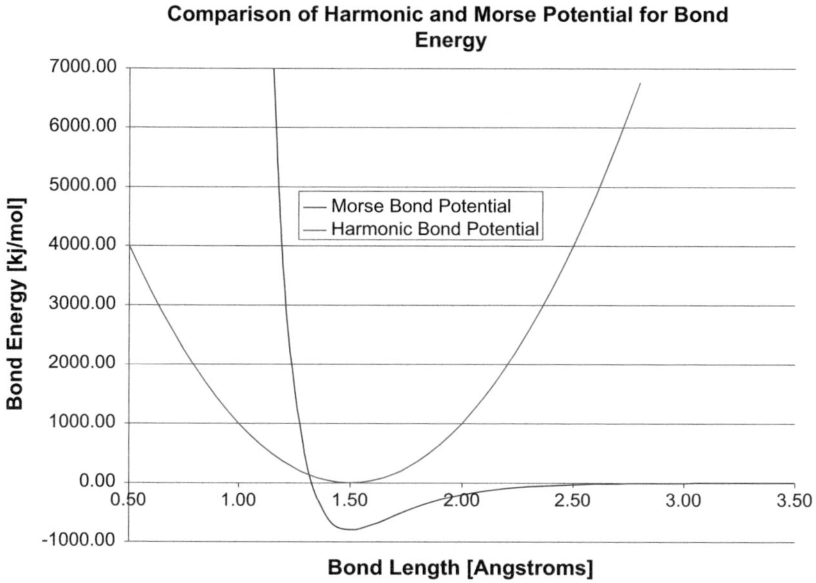

Fig. 6.6 shows a comparison of the harmonic and Morse bond potentials. Note the minimum energy is zero for the harmonic potential but equal to minus the bond dissociation energy for the Morse potential

$$E_{Morse} = E_{0ab}\left[\{1 - \exp(-k_{ab}(r_{ab} - r_{0ab}))\}^2 - 1\right] \tag{6.29}$$

6.4.1.2 Harmonic Bond Angle Potential

In a similar way the potential energy associated with bond angle flexing is given by Eq. 6.30.

$$E_{Angle} = \frac{1}{2} k_{abc} (\theta_{abc} - \theta_{0abc})^2 \tag{6.30}$$

Here k_{abc} is a force constant associated with a particular bond angle involving atoms a, b and c. The units of the force constant are energy per square radian, $\theta_{\mathbf{abc}}$ is the actual bond angle between atoms a and c about b, $\theta_{\mathbf{0abc}}$ is the equilibrium bond angle between atoms a and c about b. This is an example of a three body potential i.e. the potential energy depends on the positions of three independent atoms. Fig. 6.7 indicates the analogy between increasing and decreasing the bond angle and stretching and compressing a spring.

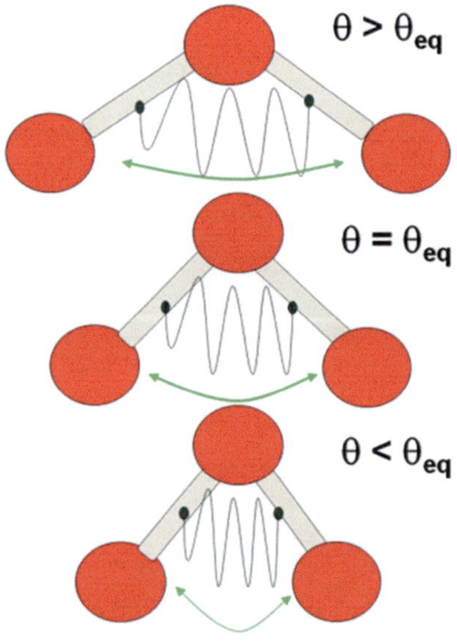

Fig. 6.7 Harmonic angle bend potential

6.4.1.3 Dihedral Angle Potential

The dihedral angle was defined in Sect. 6.2.1. Since the electrons are neglected in a molecular mechanics representation, loss of orbital overlap due to changing a dihedral angle is not explicit. Taking the molecule ethene as an example, it is necessary to have an energy penalty-function to penalise any rotation about the carbon-carbon double bond which would take two of the hydrogen atoms out of the plane of the molecule. The form of the potential energy function used is given by Eq. 6.31 where θ_{tor} is the dihedral angle relating the distal and proximal atoms about a common covalent bond and the terms V_j are energies. Often just a single term is used in the sum to calculate the energy. This is a four-body potential as it requires the coordinates of four atoms to evaluate the energy.

$$E_{Torsion} = \frac{1}{2}\sum_{j} V_j(1 - \cos(j\theta_{tor})) \tag{6.31}$$

6.4.1.4 Non-bonded Intra-molecular Potential

Pairs of atoms in a molecule that are not directly connected by a covalent bond (1–2 interactions) or both connected by covalent bonds to a common atom (1–3 interactions) are considered to be non-bonded. The potential energy associated with non-bonded atom-atom interactions is usually modelled using a Buckingham potential, Eq. 6.32, or Lennard-Jones type 12–6 potential Eq. 6.33.

$$E_{Buckingham} = A_{ij}\exp\left(-\frac{r_{ij}}{\rho_{ij}}\right) - \frac{C_{ij}}{r_{ij}^6} \tag{6.32}$$

$$E_{12-6} = \frac{A_{ij}}{r_{ij}^{12}} - \frac{B_{ij}}{r_{ij}^6} \tag{6.33}$$

In both cases the second term represents the attractive dispersion force and the first term the short range repulsions due to overlap of electron clouds. A_{ij}, B_{ij}, C_{ij} and ρ_{ij} are energy parameters that depend on the specific pair of atoms and r_{ij} is their separation distance.

When atoms with different electronegativities are covalently bonded this results in the polarisation of the covalent bond and hence a partial separation of charge leading to electrostatic interactions between atoms. The simplest approach for modelling the electrostatic interactions is to define a set of point charges (monopoles) that coincide in position with the nuclei of the constituent atoms of the molecule. Then the electrostatic or Coulombic interaction between a pair of atoms is given by Eq. 6.34 where ε_0 is the permittivity of free space, q_i and q_j are the point charges on atoms i and j and r_{ij} is the separation distance between atoms i and j.

$$E_{Coulombic} = \frac{1}{4\pi\varepsilon_0}\frac{q_i q_j}{r_{ij}} \tag{6.34}$$

6.4.1.5 Evaluation of the Total System Eenergy

The energy terms in the molecular mechanics description are pairwise additive so the total energy is given by Eq. 6.35.

$$E_{Total} = \sum_{n_{bonds}} E_{bond} + \sum_{n_{angles}} E_{angle} + \sum_{n_{torsions}} E_{torsion} + \sum_{n_{nonbond}} E_{nonbond} \tag{6.35}$$

- Take the example of a molecule of butane (Fig. 6.8).
- 10 C-H bond energy terms
- 3 C-C bond energy terms
- 8 H-C-H bond angle terms
- 14 H-C-C bond angle terms
- 2 C-C-C bond angle terms
- How many torsion angle terms?
- How many non-bonded atom-atom interaction terms?

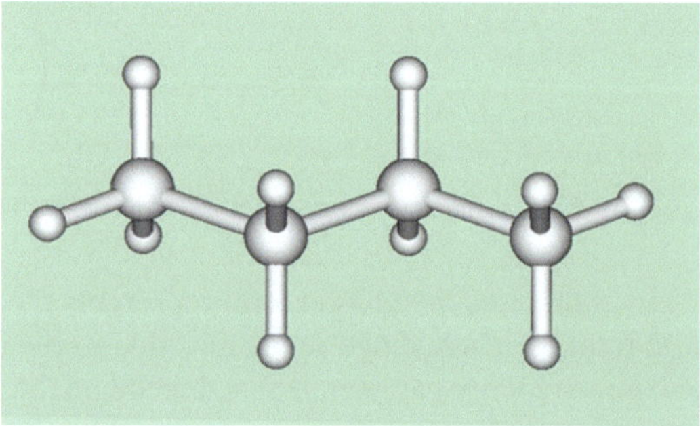

Fig. 6.8 Molecular model of butane

6.4.2 *Optimising Molecular Structure by Energy Minimisation*

Energy minimization is an important concept in molecular modelling. It provides the means for locating an optimum molecular geometry. Every atom in a molecule introduces three independent Cartesian variables with respect to which all non-zero potential energy terms in the potential-energy summation must be minimised. Take the example of a diatomic molecule. In a molecular mechanics representation the potential energy can be expressed as a function of a single variable, the distance between the atoms, r_{12}. The conditions for minimum energy are given by Eq. (Inequality) 6.36.

$$\left(\frac{dE_{bond}}{dr_{12}}\right) = 0 \text{ and } \left(\frac{d^2E_{bond}}{dr_{12}^2}\right) > 0 \tag{6.36}$$

Now $E_{bond} = f^n(r_{12})$ and $r_{12} = f^n(x_1,y_1,z_1,x_2,y_2,z_2)$ where x_1, y_1 and z_1 are the Cartesian coordinates of the first atom and x_2, y_2 and z_2 are the Cartesian coordinates of the second atom. Hence employ the chain rule to differentiate the expression for the bond energy with respect to the Cartesian directions (see Eqs. 6.37).

$$r_{12}^2 = (x_1 - x_2)^2 + (y_1 - y_2)^2 + (z_1 - z_2)^2$$
$$\left(\frac{\partial E_{bond}}{\partial x_1}\right) = \left(\frac{dE_{bond}}{dr_{12}}\right)\left(\frac{\partial r_{12}}{\partial x_1}\right) \quad \text{here} \quad \left(\frac{\partial r_{12}}{\partial x_1}\right) = \frac{(x_1 - x_2)}{r_{12}} \quad \text{hence}$$
$$\left(\frac{\partial E_{bond}}{\partial x_1}\right) = \left(\frac{\partial E_{bond}}{\partial x_2}\right) = \ldots = \left(\frac{\partial E_{bond}}{\partial z_2}\right) = 0 \text{ at energy minimum} \tag{6.37}$$

To characterise a stationary point as a minimum must generate the so called Hessian matrix of second order derivatives of the energy with respect to pairs of the independent Cartesian coordinates. In this case the form of the Hessian matrix is given by Eq. 6.38.

$$\begin{pmatrix} \left(\frac{\partial^2 E}{\partial x_1^2}\right) & \left(\frac{\partial^2 E}{\partial x_1 \partial y_1}\right) & \cdots & \left(\frac{\partial^2 E}{\partial x_1 \partial z_2}\right) \\ \left(\frac{\partial^2 E}{\partial y_1 \partial x_1}\right) & \left(\frac{\partial^2 E}{\partial y_1^2}\right) & \cdots & \vdots \\ \vdots & \vdots & \ddots & \vdots \\ \left(\frac{\partial^2 E}{\partial z_2 \partial x_1}\right) & \cdots & \cdots & \left(\frac{\partial^2 E}{\partial z_2^2}\right) \end{pmatrix} \tag{6.38}$$

The Hessian matrix must be diagonalized to yield a set of eigenvalues and eigenvectors (in this case six of each). The resulting eigenvectors are referred to as the normal coordinates of the molecule. For the case of a diatomic molecule at its equilibrium geometry diagonalization of the Hessian matrix will yield five zero eigenvalues and one positive eigenvalue thus characterising the stationary point as a minimum in the potential energy.

There are a number of different numerical approaches for performing potential energy minimisation. Two such methods are outlined below.

Steepest Descents: here there are a number of cycles in which the Cartesian coordinates of the atoms are updated according to the following formula Eq. 6.39 after n cycles. Here c is a constant that determines the size of the step that is taken between cycles in the minimisation procedure. The Steepest Descent method does not take account of the curvature of the potential energy surface that is described by the 3 N independent Cartesian coordinate variables where N is the number of atoms in the molecule.

$$x_i^{n+1} = x_i^n - c\left(\frac{\partial E_{Total}}{\partial x_i}\right)\Bigg|_{x_i = x_i^n} \tag{6.39}$$

In the Newton-Raphson approach the Hessian matrix of second derivatives is employed to scale the energy gradients and hence modify the step taken between cycles in the method. The greater the curvature of the energy surface in a particular Cartesian coordinate direction the smaller the scaling factor applied to that component of the energy gradient. The Cartesian coordinates of the atoms are updated according to the formula in Eq. 6.40 after n cycles.

$$x_i^{n+1} = x_i^n - c\left(\frac{\partial^2 E_{Total}}{\partial x_i^2}\right)^{-1}\Bigg|_{x_i=x_i^n}\left(\frac{\partial E_{Total}}{\partial x_i}\right)\Bigg|_{x_i=x_i^n} \text{ or in vector notation} \qquad (6.40)$$

$$\mathbf{q}_{n+1} = \mathbf{q}_n - c\mathbf{H}_n^{-1}\mathbf{g}_n$$

6.4.3 Potential Energy Surfaces (Example of Single Molecule of Pentane)

We have seen that molecular mechanics can be applied to calculate the potential energy of an isolated molecule if the positions in 3D space of all the atoms are known. Often want to map the change in potential energy when one or more parameters describing the geometry of the molecule are changed. A typical parameter is a torsion angle describing the *conformation* of a molecule. Take the example of two torsion angles that describe the conformation of the molecule n-pentane.

Two torsion angles τ_1 and τ_2 describing the conformation of the n-pentane molecule are shown in Fig. 6.9.

We can construct a potential energy surface by calculating the total potential energy of a n-pentane molecule on a two-dimensional grid of points defined by τ_1 and τ_2.

There are two approaches to calculate the potential energy at a grid point: (i) keep the geometry of the molecule fixed apart from changing the torsion angles τ_1 and τ_2, (ii) minimise the potential energy subject to the constraint that the torsion angles τ_1 and τ_2 must maintain their specified values. Can either use a contour plot (2D) or a surface plot (3D) to display the potential energy as a function of the torsion angles τ_1 and τ_2.

Figure 6.10 (left) shows an example of a contour plot displaying different potential energies as coloured regions. Note there are several minima with similar energies. Also note the line of mirror symmetry bottom left hand corner to top right hand corner. The right hand plot in Fig. 6.10 is similar, but now the potential energy is plotted as a surface. The potential energy is plotted on the z axis.

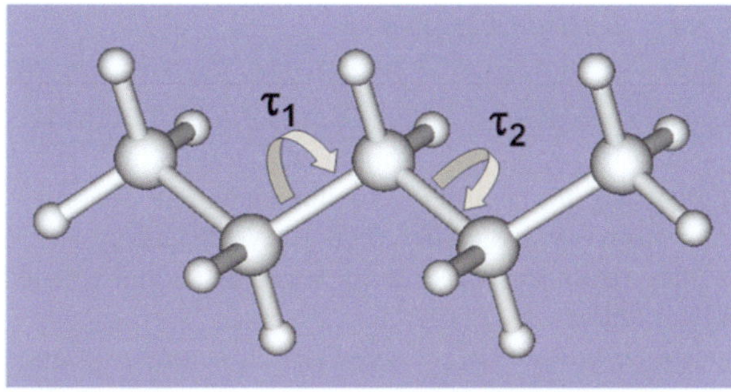

Fig. 6.9 Definition of two torsion angles in normal pentane

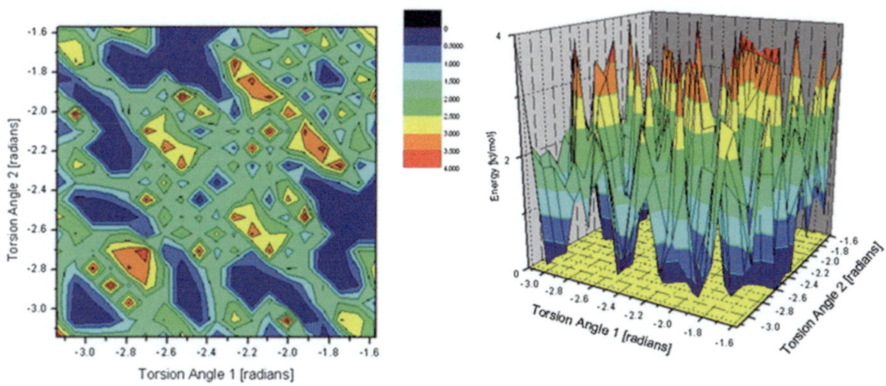

Fig. 6.10 Representations of a potential energy surface for variation of two torsion angles in n-Pentane (*left*) a contour plot, (*right*) a 3-D representation

6.5 Molecules Assembled in Crystals

Texts [5, 14, 15] provide background reading for the topics presented in this Section. The smallest group of atoms that in combination with the appropriate space group symmetry-operations is needed to generate the fractional coordinates of all the atoms in the unit cell of a crystal is called the asymmetric unit. Often for molecular crystals the asymmetric unit consists of one molecule of the material from which the crystal is formed. However sometimes the asymmetric unit contains more than a single molecule for example:

- two or more molecules of the same component where the internal arrangements of the separate molecules (conformations) are different and not related by the space group symmetry,
- two or more different molecules as in a hydrate, solvate or co-crystal,
- a molecular ion and a counter ion as in a salt.

6.5.1 Lattice Energy

As a starting point it is important to establish a working definition of lattice energy that can be employed in molecular modelling. The lattice energy can, to a first approximation, be regarded as the change in potential energy when the constituent ions, atoms or molecules of the crystal lattice, initially at rest and separated by infinite distances, are brought together to their respective locations at rest in the crystal lattice.

Notice the lattice energy, E_{Latt}, is actually an energy change or difference, E, accompanying a transformation between two hypothetical states conducted reversibly at the absolute zero of temperature. Although both the initial state, a perfect gas at absolute zero, and the final state, a perfect crystal at absolute zero, cannot be realised exactly in any practical experiment, conceptually they are quite easy to understand as points of reference. The real physical process to which this energy change most closely corresponds is sublimation, the formation of a vapour directly from a solid. Accordingly, lattice energies calculated using molecular modelling approaches, E_{Latt}, are often compared with experimentally determined heats of sublimation, H_{Sub}. To make a comparison between a calculated lattice energy and experimentally determined heat of sublimation at a finite temperature, T, for a material, a correction must be made for the relative change in enthalpy due to the difference between the heat capacities of the solid and gaseous states. Hence, the correction is given by Eq. 6.41. The calculated lattice energy and sublimation enthalpy measured at temperature T are related by Eq. 6.42.

$$\begin{aligned} &\Delta H_{Sub}^{T=T} = \Delta H_{Sub}^{T=0} + \int_0^T \left(C_P^{gas} - C_P^{solid}\right) dT \\ &\text{where } C_P^{gas} - C_P^{solid} = \left(6.\frac{1}{2}R + R\right) - 6.R = -2R \end{aligned} \tag{6.41}$$

$$\Delta H_{Sub}^{T=T} = -E_{Latt} - 2RT \text{ since } \Delta H_{Sub}^{T=0} = -E_{Latt} \tag{6.42}$$

For some materials the heat capacities, at constant pressure, of the solid and gaseous states are known from experiment as a function of temperature. These experimental data allow the heat of sublimation at any temperature to be calculated directly.

The lattice energy is therefore a measure of the cohesive energy. The procedure for calculating a lattice energy is as follows. Using the fractional coordinates of the atoms in the asymmetric unit of the crystal structure, apply the space group symmetry operations to generate the fractional coordinates of all the atoms in the unit cell. These coordinates enable the distances between all atom pairs to be calculated. Assign potential energy types A_i, B_i and point charges q_i to all the atoms. Using a computer programme perform the energy summation indicated in Eq. 6.43.

$$U_{non-bonded} = \frac{1}{2}\sum_{i=1}^{n1}\sum_{j=1}^{n2}\frac{A_{ij}}{r_{ij}^{12}} - \frac{B_{ij}}{r_{ij}^{6}} + \frac{q_i q_j}{r_{ij}} \tag{6.43}$$

The first sum is over all atoms i in the central molecule the second sum is over all atoms j in all the other molecules. If there is more than one molecule in the asymmetric unit then there is a further term in the summation over all the different molecular entities. This number corresponds to the value of Z prime for the crystal structure.

The total potential or cohesive energy of a molecular crystal can be regarded as the sum of a very large number of similar interactions between pairs of atoms. Hence quite different periodic arrangements of the same type of molecule can have similar energies. This is why different solid-state structures or 'polymorphs' are particularly observed for molecular materials.

Crystal polymorphs are different periodic arrangements of the same molecular species in distinct crystal lattices. If the main molecular species is combined with one or more additional molecular species as in a hydrate or solvate then the different crystal phases are referred to as pseudo-polymorphs. Crucially different polymorphs have distinct physical properties such as solubility and bioavailability. If for a pair of polymorphs with melting points T_1 and T_2, ($T_1 > T_2$) respectively one polymorph is more thermodynamically stable at all temperatures below the melting point T_1 then the polymorphs are *monotropes* (related monotropically). Alternatively, if the stability order swaps over at a temperature $T_3 < T_1$ then the polymorphs are *enantiotropes* (related enantiotropically). At the transition temperature T_3 the free energy of the two polymorphs is the same. Since for monotropes the relative thermodynamic stability is independent of temperature, very accurate calculations of lattice energy at zero Kelvin can be used to rank the stability order. In addition, a significant amount of research effort has been devoted to the prediction of crystal polymorphs from first principles. It is highly beneficial to know whether a particular polymorph is the most thermodynamically stable form since, if a pathway is available, a metastable form will tend to transform with concomitant changes to key physical properties.

6.5.2 *The Shape of Crystals as Grown*

The shape of a crystal depends on the relative rate of growth perpendicular to each surface of the crystal during the growth process. Every surface on a crystal can be defined by a Miller plane and its associated Miller index. The Miller plane that describes a crystal surface is the plane (or set of planes) exactly parallel to that surface. Hence through the definition of the Miller plane a direct link is made between the crystal surface observed at the bulk, macroscopic scale and the atomic structure of the crystal surface at the atomic, nano-scale. To predict crystal shape need a theory of how the relative rate of growth perpendicular to each crystal surface relates to properties of the Miller plane describing the surface.

6.5.2.1 Bravais Friedel Donnay Harker Approach

Firstly Bravais and later Friedel, Donnay and Harker proposed that the rate of crystal growth is inversely proportional to the spacing between consecutive Miller Planes in the family of Miller Planes described by the Miller Index. This is the Bravais Friedel Donnay Harker or BFDH approach to predict crystal shape.

Table 6.2 BFDH calculation for primitive cubic crystal

Miller index	d_{hkl}	d_{hkl}/a	a/d_{hkl}
{100}	a	1.00000	1.00000
{110}	a√(1/2)	0.70711	1.41421
{111}	a√(1/3)	0.57735	1.73205
{200}	a/2	0.50000	2.00000
{210}	a√(1/5)	0.44721	2.23607

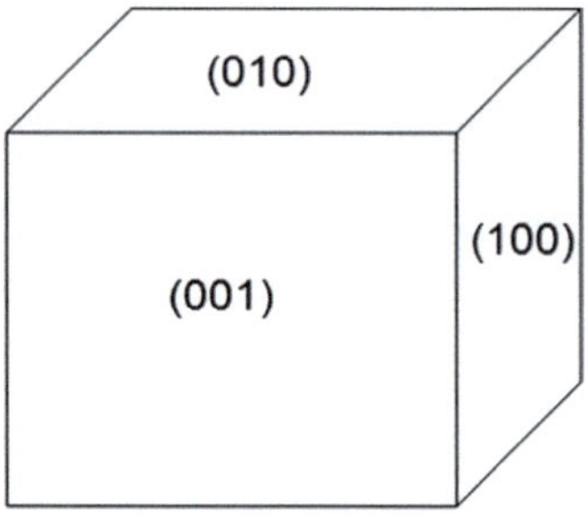

Fig. 6.11 BFDH crystal shape prediction for primitive cubic crystal

Table 6.3 BFDH calculation for body centred cubic crystal

Miller index	d_{hkl}	$d_{hkl}/a\sqrt{(½)}$	$\sqrt{(½)}a/d_{hkl}$
{110}	a√(1/2)	1.00000	1.00000
{200}	a/2	0.70711	1.41421
{211}	a√(1/6)	0.57735	1.73205
{222}	a√(1/12)	0.40825	2.44949

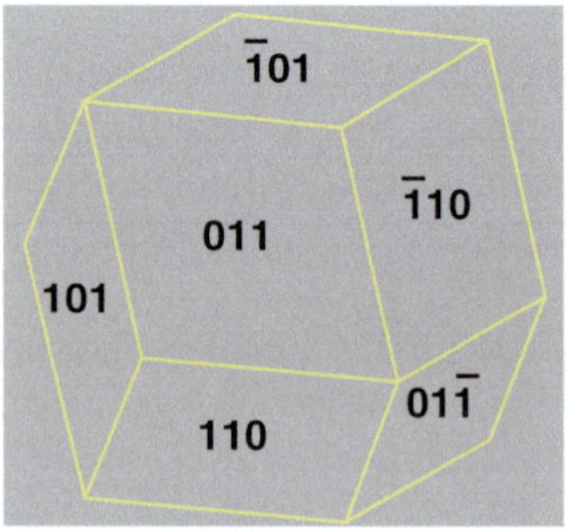

Fig. 6.12 BFDH crystal shape prediction for body centred cubic crystal

For a primitive, cubic lattice with lattice parameter a, the inter-planar spacing d_{hkl} for a plane with Miller index (hkl) is given by Eq. 6.44.

$$d_{hkl} = a\sqrt{\frac{1}{h^2 + k^2 + l^2}} \tag{6.44}$$

The relative rate of crystal growth perpendicular to a face (hkl) is given by (a/d_{hkl}) so the relative distances from the centre of a crystal to each face along the direction perpendicular to the face are also given by (a/d_{hkl}). Once the relative distances from the centre of a crystal to each face have been defined a model of the crystal shape can be constructed. For the case of a primitive cubic crystal the calculation is summarised in Table 6.2. Hence for a primitive cubic crystal the predicted shape is shown in Fig. 6.11. In the BFDH predicted crystal shape only a single form {100}, consisting of six individual faces, is predicted to appear. For a lattice containing additional lattice points, e.g. a body centred lattice, there are rules to determine which X-ray reflections are extinguished by total destructive interference. Donnay and Harker proposed that these extinctions should be included when considering the external surfaces of crystals. This is extended to include the effect of space group symmetry-elements such as a screw axis and glide plane. For a body centred lattice, extinctions are governed by the formula $h + k + l = 2n$ where n is an integer. So for a body centred cubic lattice, rather than a primitive cubic lattice, we do not include the set of crystal faces {100}, {111} or {210} when considering the relative rates of growth perpendicular to the crystal surfaces. The corresponding calculation for a body centred cubic crystal is summarised in Table 6.3 and the predicted crystal shape shown in Fig. 6.12. The BFDH approach takes no account of inter-atomic forces in predicting relative growth rates and may not always be accurate in predicting crystal shape.

6.5.2.2 Attachment Energy Model

The attachment energy model for predicting crystal shape (crystal morphology) states that the rate of growth perpendicular to a crystal surface (hkl) is proportional to the attachment energy for that surface. Hence the relative rates of growth perpendicular to the growing surfaces that enclose (or bound) a crystal can be calculated from the attachment energies for the surfaces. Knowing the relative rates of growth allows the relative centre to surface distances of all the crystal surfaces to be defined and this gives a model of the shape of a crystal (Fig. 6.13).

A slice is a slab of crystal bounded by two adjacent planes (hkl) in the crystal lattice. Hence the slice thickness is d_{hkl} The crystal is envisaged as growing by the addition of crystal slices onto an existing crystal surface (hkl). The slice energy, E_{sl}, is the summation of all the atom-atom interaction energies between the atoms in a *central molecule* in the slice and all the atoms in all the other molecules contained

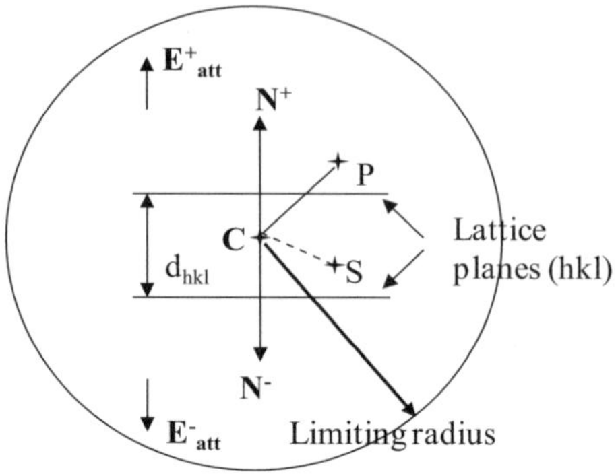

Fig. 6.13 Method for calculating slice and attachment energies

within the same slice. The attachment energy, E_{att}, is the summation of all the atom-atom interaction energies between the atoms in a *central molecule* in the slice and all the atoms in all the other molecules outside the same slice.

The attachment energy can also be interpreted at the energy released when a slab of crystal of thickness d_{hkl} is incorporated into the crystal lattice. Hence the sum of the slice energy and attachment energy is equal to the crystal lattice energy. The action of centring and/or elements of space group symmetry, screw axes and glide planes, may reduce the thickness of the growth slab on a particular surface of a crystal. For example in the monoclinic space group $P2_1$ the growth slab on the (010) surface is separated by adjacent (020) planes and not adjacent (010) planes. Hence the correct slice thickness is d_{020} and not d_{010} (i.e. the slice thickness is halved).

The attachment energy model for predicting crystal shape (crystal morphology) states that the rate of growth perpendicular to a crystal surface (hkl) is proportional to the attachment energy for that surface. Hence the relative rates of growth perpendicular to the growing surfaces that enclose (or bound) a crystal can be calculated from the attachment energies for the surfaces. Knowing the relative rates of growth allows the relative centre to surface distances of all the crystal surfaces to be defined and this gives a model of the shape of a crystal. The attachment energy model, unlike the BFDH approach, does take account of the intermolecular forces operating in the crystal.

Generally get good agreement between observed and predicted crystal shapes provided (i) the growth environment does not play a big role, (ii) there is no relaxation of surface molecular species with respect to the bulk crystal structure. Figure 6.14 compares the crystal shape predicted for ascorbic acid using the attachment energy model with crystals grown from solution.

The surface energy (in a vacuum), γ_{hkl}, of a specific surface of a crystal (hkl) can be estimated from the attachment energy according to Eq. 6.45. Here Z is the number of asymmetric units per unit cell, d_{hkl} is the inter-planar spacing, V_{cell} is the unit cell volume, N_a is Avogadro's number and E_{att} is the attachment energy of the slice bounded by adjacent planes (hkl).

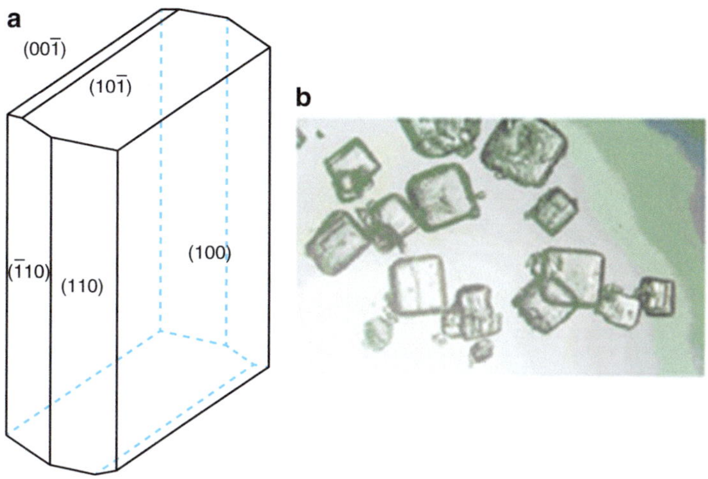

Fig. 6.14 (**a**) Crystal shape for ascorbic acid predicted by Attachment Energy Model (**b**) Crystals grown from solution

$$\gamma_{hkl} = \frac{ZE_{att}^{hkl} d_{hkl}}{2V_{cell}N_a} \tag{6.45}$$

6.5.3 Examining Individual Crystal Surfaces

Section 6.5.2 explained the concept of a crystal slice which is needed to calculate the slice and attachment energy for predicting the shape of crystalline particles. The slab of crystal that makes up a slice can also be used to create a model of a particular surface of a crystal which is likely to be engaged in the process of crystal growth from solution. During growth a surface expresses certain functional groups into the solution phase depending on the crystallographic orientation defined by the Miller Index of the surface. So surfaces may have different propensities to interact with other molecules in the solution phase, principally solvent molecules but also impurity species which may be structurally similar to the host molecule. Hence the rate of crystal growth can be modified on a face specific basis through these interactions with solvent and impurity species.

A useful approach for assessing such interactions is grid-based systematic searching. In the systematic-searching procedures employed in this context a single molecule of the host-crystal material, the solvent or an impurity molecule is referred to as a probe molecule. Molecular models of all the crystal surfaces of interest are produced in turn. A grid of points is generated in space lying adjacent to the crystal surface of interest. The locations of the grid-points are calculated based on the crystallographic parameters for a specific surface. A model of a slab of the

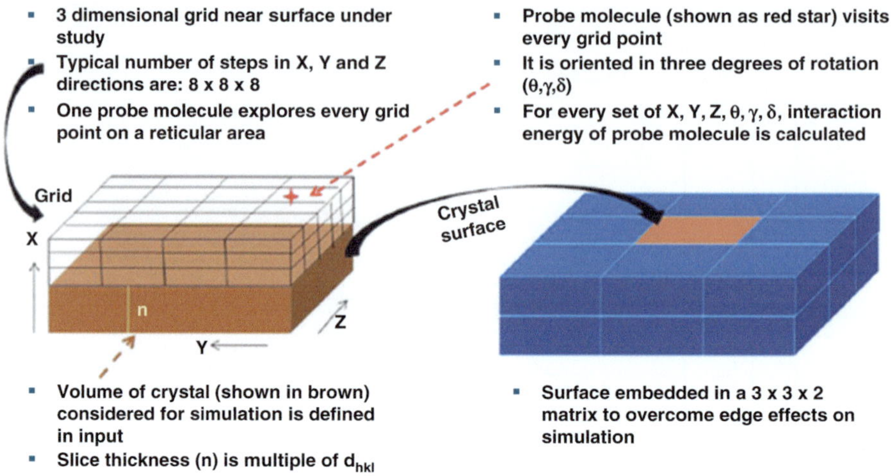

Fig. 6.15 Description of grid used in systematic searching of a crystal surface by a probe molecule

crystal terminated by the surface of interest is produced. In this procedure the direction normal to the surface is aligned with the Cartesian X-axis. To achieve this the unit cell is automatically redefined by the software. In plane, fractional coordinates are retained to describe atomic positions in the Y, Z plane and Cartesian coordinates are used to describe atomic positions in terms of the perpendicular distance to the surface (X coordinate). Hence the search utilises the periodic conditions in-plane to confine the area searched within a single reticule of the surface. The size of the slab employed is designed to be large enough to avoid edge effects in calculating the interaction energy between the probe and the molecules in the crystal-slab. Hence the probe search is usually confined to the central reticule of an array which is at least 3 × 3. Figure 6.15, taken from a recent publication [16], shows details of the systematic-search grid. The principle axes are calculated for the probe molecule (which is treated as a rigid body) and the orientation of the probe molecule is described by three angles. The entire search-path required to explore the possible interactions between the probe molecule and the crystal surface has a total of six degrees of freedom.

6.6 Modelling Molecules in Solution

Following on from the discussion of potential energy surfaces in Sect. 6.4.3, when describing a system that contains, for example, N n-pentane molecules, using classical mechanics, at a temperature above absolute zero we do not expect all the molecules to have the same potential energy. The molecules will also have kinetic energy by virtue of their motion, and the total energy of the system of N molecules depends on the way the molecules are distributed between all the possible energy states available to the system.

So need to introduce the concept of an ensemble.

- Suppose we have M systems that contain the same number of molecules, N, have the same volume, V, and are at the same temperature T.
- Every system in the ensemble has the same set of allowed total-energy states E_1, $E_2, \ldots, E_j, \ldots E_{max}$.
- The M systems in the ensemble can have different distributions over these allowed energy states.

If n_j is the number of systems in the ensemble with energy E_j the total energy of the ensemble E_{total} is given by Eq. 6.46, and the total number of systems M is given by $M = \sum_j n_j$

$$E_{Total} = \sum_j n_j E_j \tag{6.46}$$

The ensemble in which all the systems have the same number of molecules N, volume, V and temperature T is called the canonical ensemble. Can now introduce the partition function Z(V, T, N) for the canonical ensemble as Eq. 6.47. Here E_j is the energy of state j of a system within the ensemble, k_B is the Boltzmann constant, T is absolute temperature. The probability of finding a system of molecules, chosen at random from the canonical ensemble, in a particular energy state E_j is p_j and is given by the Eq. 6.48.

$$Z = \sum_{\text{all states j}} e^{-E_j/k_B T} \tag{6.47}$$

$$p_j = \frac{e^{-E_j/k_B T}}{\sum_{\text{all states j}} e^{-E_j/k_B T}} = \frac{e^{-E_j/k_B T}}{Z} \tag{6.48}$$

If a system consists of non-interacting molecules in the gas phase, can express the canonical ensemble partition function, Z, in terms of the product of single-molecule partition functions z_i. Hence, it is possible to use quantum mechanics to find mathematical equations to evaluate the molecular canonical ensemble partition function and hence the canonical ensemble partition function for ideal gases. This allows all the thermodynamic State Functions to be calculated through Eq. 6.49 since A = U – TS where A is the Helmholtz Free Energy and U is the internal energy, T the absolute temperature and S the entropy. However, we will not consider this approach further here because most systems of interest contain interacting molecules and it is not possible to calculate Z directly for such systems. What is required instead is a method to simulate a system of interacting molecules, in thermodynamic equilibrium at a given temperature and volume, so that the way the molecules sample the available energy states reflects the free energy differences between the states. One approach is to use Molecular Dynamics simulations as described in the next section.

$$A = -k_B T \ln Z \tag{6.49}$$

6.6.1 Molecular Dynamics

In Molecular Dynamics (MD) simulations the trajectories of all the molecules are determined from the inter-atomic forces by solving Newton's equations of motion. This is in contrast to Monte Carlo MC simulations in which a move is determined by generating random numbers and the trajectory is stochastic (random) not deterministic as for MD. In classical MD a molecular mechanics, atomistic potential is used as the basis for calculating the forces on the atoms. For example in MD the velocities of the atoms, $\underline{v}$, can be updated according to Eq. 6.50 where Δt is the time step, m the atomic mass and $\underline{f}(t)$ the force at time t. Following the calculation of the new set of velocities, these are used to update the positions of the atoms $\underline{r}$ as indicated in Eq. 6.51. This particular algorithm for calculating the Molecular Dynamics trajectory is known as the Verlet Leapfrog algorithm. There are several other approaches that can be used. At the beginning of an MD simulation the first set of displacements of the atoms is generated using random numbers with the step sizes scaled to the appropriate temperature. These displacements generate forces from which the initial set of velocities is calculated. Unless the seed used for random number generation is changed, identical trajectories are calculated if the coordinates of the atoms and all details of the force field remain the same.

$$\underline{v}\left(t + \frac{1}{2}\Delta t\right) = \underline{v}\left(t - \frac{1}{2}\Delta t\right) + \Delta t \frac{\underline{f}(t)}{m} \tag{6.50}$$

$$\underline{r}(t + \Delta t) = \underline{r}(t) + \Delta t \, \underline{v}\left(t + \frac{1}{2}\Delta t\right) \tag{6.51}$$

Typically MD simulations are run in one of three different thermodynamic ensembles

1. Constant N, V and E
2. Constant N, V and T
3. Constant N, p and T

A typical time step employed in an MD simulation is one femto-second, 1×10^{-15} s, so that to simulate a nano-second of real time requires one million time steps to be computed. Initially the simulation is in an equilibration period after which the trajectories of all the atoms are recorded (production period). The trajectory file typically stores the atom positions, velocities and computed system-energy terms at every time step.

6.6.2 Calculating Free Energy Differences

The shape or conformation of a molecule which is optimal when it packs with copies of itself (as mediated by the space group symmetry and simple translations) to create a crystal may not be the shape favoured in a solution. In the latter case the solute molecules will seek an optimal balance of their interactions with solvent and other solute molecules. If we want to determine the free energy differences between different solute conformations in a solution using explicit solvent molecules then the result will depend on the effective composition of the solution. This is reflected in the relative numbers of solute and solvent molecules used in a MD simulation.

Frequently the simulation is set up to mimic 'infinite dilution' so that a single solute-molecule and large number of solvent molecules are used. Periodic boundary conditions are usually employed so if the simulation box is, for example, a cube a molecule which leaves the box through a surface is mapped across to the corresponding position on the opposite surface. The edge length of the simulation box must be greater than the cut-off distance used in the atom-atom summation so that there is no direct interaction between the solute molecule and its copies by translation in the adjoining cells. Although there is only one solute molecule, we make use of the ergodic hypothesis which states that the probability distribution of molecular conformations computed for a large number of molecules at an instant in time is the same as the probability distribution for a single molecule over a sufficiently long period of time.

It is possible to estimate whether a sufficiently long period of time has been computed in an MD simulation to obtain the correct probability distribution. The total MD trajectory is divided into two, equal parts in terms of elapsed time and the relative-probabilities of pairs of different conformations are calculated separately. The relative probabilities should be the same for the first and second portions of the trajectory.

6.6.3 Free Energy Perturbation and Potential of Mean Force

If two conformations of interest are not sufficiently well sampled in an MD trajectory over a reasonable period of simulation time then there are further approaches which can be employed to obtain the free energy difference between the conformations. For example if we want to compare two conformers A and B of a solute molecule in a given solvent but, the solute stays in conformation A at each time step, the conformation is changed to B and the potential energy calculated. During this procedure the positions of the atoms in the solvent molecules are not changed. Hence the simulation is allowed to resume with the solute in conformation A. For each time step the difference in potential energy between the conformations is determined and the mean value of the term $\exp(-(E_B-E_A)/RT)$ is calculated. The free energy difference between the solute conformations in solution is given by Eq. 6.52.

$$\Delta G_{BA} = -RT \ln \left\langle \exp\left(-\frac{E_B - E_A}{RT} \right) \right\rangle \tag{6.52}$$

However, if the difference in the overall shape of the molecule between conformer A and B is large, it is usually necessary to sample some intermediate states of the molecule lying between the two conformers. In effect this procedure defines a trajectory is phase space or a 'reaction coordinate'. A series of separate pairwise comparisons is needed so, for example, if two conformers intermediate between A and B are selected, labelled AB1 and AB2, three separate MD simulations are needed. The first compares conformer A and conformer AB1, the second compares conformer AB1 and AB2 and the third compares conformer AB2 and B. From the three MD simulations a free energy profile is constructed which is known as a Potential of Mean Force (PMF).

6.6.4 *Umbrella Sampling and Biasing Potential*

An alternative way to generate a PMF is by a technique known as umbrella sampling. In this approach a biasing potential is used in conjunction with the normal atomistic potential which causes the MD trajectory to sample intermediate structures along a reaction coordinate. So if a reaction coordinate joining two conformers is described by changes to a dihedral angle, τ, then to sample conformers in a region with dihedral angles close to a value τ_1, a biasing potential of the form given in Eq. 6.53 is used where the constant k is chosen to have a minimum value of $E_{\tau 1}$ when the dihedral angle has the value τ_1. A series of MD simulations are run for dihedral angle values τ_1, τ_2, τ_3 etc. making sure that there is overlap in the ranges of dihedral angle sampled between adjacent regions of the PMF as it is mapped out. The free energy change, as a function of τ, is calculated directly from the number of instances of a conformer with a dihedral angle τ found in the trajectory file. This gives directly the probability $p(\tau)$ and $\Delta G(\tau) = -RT\ln(p(\tau))$. Since the k values vary, there will be discontinuities between the segments of the PMF produced by each MD simulation. However the curves can be matched up in the regions of overlap to produce a continuous PMF.

$$E_{\tau 1} = E + k(\tau - \tau_1) \tag{6.53}$$

6.6.5 Implicit Solvent Models

Rather than using molecular dynamics simulations, a less expensive computational approach to compute, for example, a solvation free energy is to use an implicit solvent model. Here the solvent molecules that surround a solute molecule are treated as a dielectric continuum (or polarisable continuum) which is specified by a single parameter, the dielectric constant. The surface charge density of the solute molecule polarizes the continuum and thereby generates an apparent surface charge density, σ(**r**), at the boundary of the cavity which has to be created within the dielectric continuum to accommodate the solute molecule. The apparent surface charge density is calculated from the Poisson equation, Eq. 6.54, where V(**r**) is the electrostatic potential of the solute molecule and n(r) a unit vector normal to the solute solvent interface. Analytical solutions for Eq. 6.54 exist for only simple cases such as a point charge Q in a spherical cavity of radius r for which the apparent surface charge density is given by Eq. 6.55. Substitution of this expression into an equation for the energy of the solute molecule in solution gives the Born solvation energy. Another approach for solving Eq. 6.54 is to divide the surface of the cavity in the dielectric continuum into elements or tesserae, perform surface integrations numerically over these elements and then add up the contributions. This is usually the method applied when *ab initio* QM calculations, employing a high level of theory, are used to describe the solute molecule. There are two versions of the approach commonly used, the polarisable continuum model (PCM) and the COSMO model. A part of the calculation is empirical because the contribution from the dispersion force is not included generally. Hence, appropriate atomic radii to use to define the surface of the solute molecule are parameterised from experimental solvation energies. Importantly, separate parameterisations are used which match the level of theory employed.

$$\sigma(\mathbf{r}) = \frac{1}{4\pi}\left(1 - \frac{1}{\varepsilon}\right)\nabla V(\mathbf{r}) \cdot n(\mathbf{r}) \tag{6.54}$$

$$\sigma(\mathbf{r}) = -\frac{1}{4\pi}\left(1 - \frac{1}{\varepsilon}\right)\frac{Q}{r^2} \tag{6.55}$$

References

1. Elliott JA (2011) Novel approaches to multiscale modelling in materials science. Int Mater Rev 56(4):207–225
2. Ogata S, Lidorikis E, Shimojo F, Nakano A, Vashishta P, Kalia RK (2001) Hybrid finite-element/molecular-dynamics/electronic-density-functional approach to materials simulations on parallel computers. Comput Phys Commun 138:143–154
3. Nye JF (1985) Physical properties of crystals: their representations by tensors and matrices. Oxford University Press, Oxford

4. Landau DP, Binder K (2000) A guide to Monte Carlo simulations in statistical physics. Cambridge University Press, Cambridge
5. Bladon P, Gorton J, Hammond RB (2011) Molecular modelling computational chemistry demystified. RSC Publishing, Cambridge
6. Jensen JH (2010) Molecular modeling basics. CRC Press, Boca Raton
7. Höltje HD, Sippl W, Rognan D (2003) Molecular modeling basic principles and applications. WILEY-VCH, Weinheim
8. Heine T, Joswig JO, Gelessus A (2003) Computational chemistry workbook: learning through examples. WILEY-VCH, Weinheim
9. Moore E (ed) (2002) Molecular modelling and bonding. RSC Publishing, Cambridge
10. Atkins PW (1983) Molecular quantum mechanics. Oxford University Press, Oxford/New York
11. Richards WG, Cooper DL (1985) Ab initio molecular orbital calculations for chemists. Oxford Science Publications, Oxford
12. Hinchliffe A (2003) Molecular modelling for beginners. Wiley, Hoboken
13. Leach AR (2001) Molecular Modelling: principles and applications. Prentice Hall, Harlow/Munich
14. Gavezzotti A (ed) (1997) Theoretical aspects and computer modeling of the molecular solid state. Wiley, Chichester
15. Hammond C (2009) The basics of crystallography and diffraction. International Union of Crystallography/Oxford University Press, Oxford
16. Ramachandran V, Murnane D, Hammond RB, Pickering J, Roberts KJ, Soufian M, Forbes B, Jaffari S, Martin GP, Collins E, Pencheva K (2015) Formulation pre-screening of inhalation powders using computational atom–atom systematic search method. Mol Pharm 12:18–33

Chapter 7
Crystal Growth and Morphology of Molecular Crystals

Ian Rosbottom and Kevin J. Roberts

Abstract Organic molecular crystals are often the main active ingredient in pharmaceutical drug products. The crystal morphology of these materials plays a significant role in their ease of separation from the mother solution phase, physical behaviour during downstream unit processes and their dissolution profiles and delivery of the active ingredient to the patient. Molecular modelling can be used to predict crystal morphologies, in terms of the strengths of their internal intermolecular interactions and their external crystallisation environment, hence providing a guide to the experimental conditions required to produce a pre-defined crystal morphology.

Here, the use of calculations of intermolecular interaction strength, nature and directionality in the prediction and analysis of morphologies is reviewed. These calculations are in terms of the intermolecular interactions within the crystal structure, along with the solute/solvent interactions at the crystal/solution interface, to predict an equilibrium morphology and habit modification that can be governed by the crystallisation solvent.

Keywords Morphology • Crystal growth • Molecular modelling • Atomistic simulations • Attachment energies • Habit modification

7.1 Introduction

The morphology of organic crystals can play a substantial role in how a material behaves during the downstream unit processes, such as filtration, milling, compression, etc. associated with the manufacturing of pharmaceutical and fine chemical products containing organic crystalline material. For example, needle-like crystal morphologies can often cause issues due to their shape, causing them to pass through filters and often break during compression [1]. Though the nucleation transition pathway defines many properties that are displayed by a molecular

---

I. Rosbottom • K.J. Roberts (✉)
School of Chemical and Process Engineering, University of Leeds, Leeds LS2 9JT, UK
e-mail: i.rosbottom@leeds.ac.uk; k.j.roberts@leeds.ac.uk

K.J. Roberts et al. (eds.), *Engineering Crystallography: From Molecule to Crystal to Functional Form*, NATO Science for Peace and Security Series A: Chemistry and Biology, DOI 10.1007/978-94-024-1117-1_7

crystal, it is thought that the crystal growth process has the strongest influence on the final crystal morphology [2]. The crystal growth of the material from solution can be strongly affected by the intermolecular interactions within the structure, along with those formed between solute or solvent molecules and the crystalline surface defined by the Miller Planes (*hkl*) [3]. Hence, both the internal crystal structure and the external growth conditions (temperature, supersaturation, solvent, etc.) can strongly influence the crystal growth and morphology of organic crystalline materials [4–10], and both should be considered when trying to predict and control the final crystal morphology.

Molecular modelling has been used to predict the morphology of crystals, with these models including properties such as interplanar spacing, symmetrical considerations [11–13] and the relationship between the surface terminated intermolecular interactions and face-specific relative crystal growth rates [14–16], through calculations of the strength of the intermolecular interactions within the crystal structure. The intermolecular interactions are often calculated using interatomic potentials, derived from experimental data. This method has been expanded to calculate surface roughening transitions through the connected nets approach and α-factor calculations [17–22]. High performance computing techniques, such as molecular dynamics, have also been utilised to calculate solvent effects on crystal morphologies [6, 23–26]. This chapter focusses on the prediction of the crystal morphology of organic materials from a calculation of the strength and directionality of the intermolecular interactions within the crystal structure, and identifying those that are exposed at the growth surfaces and govern the relative face-specific growth rates, through the attachment energy theory [15, 16]. The analysis of the surface chemistry is also used to qualitatively explain some examples of solvent dependent habit modification, along with examples of how calculations of solvent interactions with the exposed solute molecules at the crystal surfaces can also be used to predict such solvent dependent crystal morphologies.

7.2 Equilibrium Morphology and the Wulff Plot

In 1875, Gibbs postulated that the three-dimensional shape of a crystal will be the one in which the total free energy is at a minimum, according to Eq. 7.1.

$$\Sigma_i \gamma_i A_i \tag{7.1}$$

Whereby γ_i and A_i are the surface energy and surface area respectively of the ith face. In 1901, Wulff developed this theory such that the equilibrium crystal habit would consist of crystal faces whose distance from the origin, in a specific direction defined by a Miller plane (*hkl*), is proportional to the specific surface energies of the faces and the crystal growth rates away from the nucleation centre. An example of a Wulff plot is shown in Fig. 7.1.

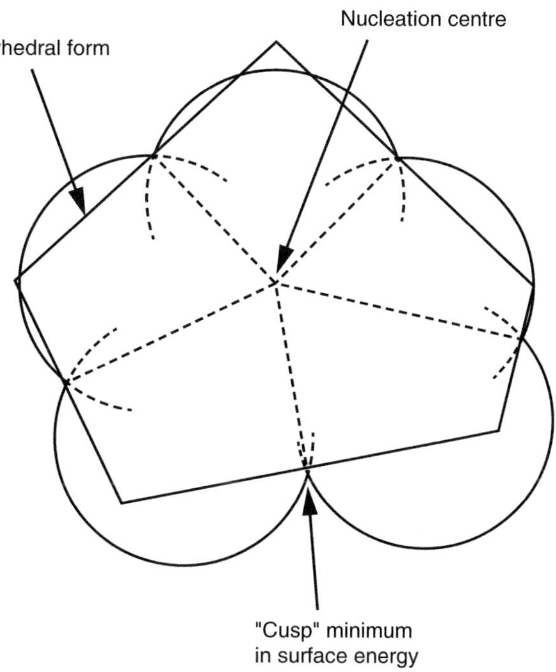

Fig. 7.1 Diagram of the Wulff plot of the equilibrium morphology of a crystal [27]

Molecular drawing programs such as Shape [28] and Mercury [29] can take crystallographic information and relative crystal growth rates, expressed as the nucleation centre to crystal face distance, and construct a Wulff plot of the resulting crystal morphology.

The anisotropic nature of molecular crystals often results in significantly varied surface chemistry at the crystal faces, as defined by specific Miller planes (*hkl*). This variation in surface chemistry can result in considerably different face-specific growth rates that are due to the variation in the strength and nature of the intermolecular interactions which govern the growth of the different crystal surfaces. Such surface chemistry anisotropy can often produce anisotropic crystal morphologies, such as needles. Indeed, it is often observed that the slow growing surfaces of a crystal dominate the final morphology and fast growing faces are small, or can 'facet out' and not be present in the final morphology, demonstrated schematically in Fig. 7.2.

Since this variation in the strength and nature of the intermolecular interactions at each crystal surface can impact on the crystal morphology, a calculation of the strength and directionality of the intermolecular interactions within a crystal structure, and at the crystal surfaces, can be used to predict and understand the crystal morphologies of organic materials.

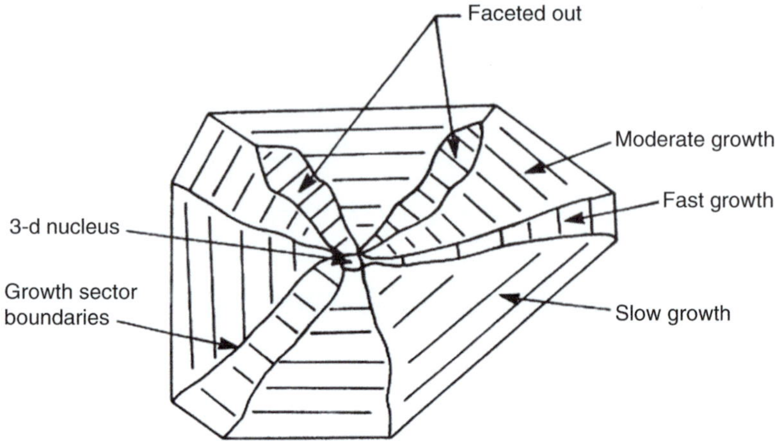

Fig. 7.2 Representation of a cut through a particulate grain boundary, showing that the final three-dimensional morphology is created from a series of two dimensional growth processes. Here, the 3D morphology dominated by slow growing faces [30]

7.3 Intermolecular Interactions (Synthonic Engineering)

Whereas the synthesis of molecules involves the making and breaking of covalent bonds, the synthesis and design of molecular crystals with pre-defined physical properties involves the making and breaking of solvent/solvent, solute/solvent and solute/solute intermolecular interactions (synthons) [27, 30–35]. This approach is the basis for the emerging field of synthonic engineering.

The internal structures of molecular crystals are often dominated by 'weak' intermolecular interactions such as hydrogen bonds (H-bonds), π-π stacking interactions and less specific van der Waals interactions (vdW). The molecules pack in arrangements that maximise these interaction energies, whilst still being in relatively energetically favourable conformations. An example of the different types of intermolecular synthons that can be found in the crystal structure of a relatively small molecule organic crystal, α-para aminobenzoic acid (*p*ABA), is shown in Fig. 7.3.

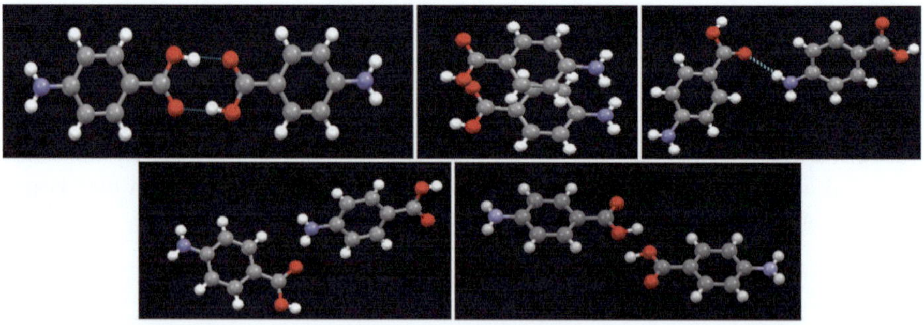

Fig. 7.3 Different types of synthons found in the α-*p*ABA, demonstrating the different types of hydrogen bonding and van der Waals interactions that can be formed [36]

The fully satisfied intermolecular interactions that are present within the bulk of the material are known as 'intrinsic synthons', whilst the unsaturated intermolecular interactions exposed at the surface of a material are known as 'extrinsic synthons'. The nature, strength and directionality of these synthons can direct the physico-chemical properties of a crystalline material, such as crystal growth, polymorphism and purity. Therefore, through calculation of the strength of both the intrinsic and the extrinsic synthons present at the surface of molecular crystals, these properties can be predicted.

Often the strength of intermolecular interactions are calculated using empirical interatomic-potentials, which can be derived from experimental data, and the intermolecular interaction between two molecules can be calculated from summing all of the constituent interatomic interactions between pairs of molecules (atom-atom approach [37]). As computing power has increased, the sophistication of *ab initio* calculations of molecular energies has developed. Density functional theory (DFT) calculations simulate the electronic structure of atoms and molecules based on electron density, though these calculations are less adept at modelling the long range dispersion interactions present in molecular crystals. However, the development of dispersion corrections has improved the capability of these simulations to account for the dispersion effects of neighbouring molecules [38, 39]. Despite this, these calculations are often computationally expensive and time consuming and, as of yet, are unproven to offer significantly more accurate results than empirical interatomic potentials, with respect to crystal morphology prediction. A more detailed description of interatomic potentials is provided in Chaps. 6 and 11 [40, 41].

7.3.1 Lattice Energies

The lattice energy can be calculated as the sum of all of the intermolecular interactions between an origin molecule in a central unit cell, and all of the other molecules in the origin and surrounding unit cells, up to a distance where the interactions become negligible. Further details of how these calculations are carried out using the HABIT [34] and VisualHabit [42] packages are described in Chap. 11 [41]. The calculated lattice energy can be compared to experimentally obtained sublimation enthalpies (ΔH_{sub}) to test the suitability of an interatomic potential. The 'experimental lattice energy' (V_{exp}) is related to the sublimation enthalpy through Eq. 7.2:

$$V_{exp} = -\Delta H_{sub} - 2RT \tag{7.2}$$

where the 2RT (R is the ideal gas constant and T is the temperature) term is the correction for the difference between the gas phase enthalpy and the vibrational contribution to the crystal enthalpy. Good agreement between the sublimation enthalpy and calculated lattice energy suggests that the potential is suitable to predict the strength of the interatomic interactions, and hence predict the strength of the intermolecular interactions.

There are generalised potential parameters available in the literature that have previously provided good predictions of the strength of intermolecular interactions between organic molecules [43–47]. These potentials are parameterised to handle a range of molecules, and as such there can be some doubt over the absolute values of the intermolecular interactions that these potentials calculate. An alternative is to optimise potentials against known crystal structures, experimental data and ab initio energy calculations to create a potential that is suited to a specific molecule or system [48–50]. Though in many cases this improves the accuracy of the calculation of intermolecular interaction strength, the process can be time consuming. Considering that the morphology is affected by the relative strength of the intermolecular interactions, rather than the absolute values, changing the potential often does not affect the crystal morphology prediction [51]. Optimisation of the potential is more common within the crystal structure prediction community, where the absolute values of intermolecular interactions are far more important [50, 52–54]. Hence, if the interatomic potential gives a good approximation of the lattice energy, then the approximation of the intermolecular interactions can be deemed to be acceptable for crystal morphology prediction and surface chemistry analysis.

The HABIT [34] program calculates the lattice energy within a user defined sphere, and the lattice energy can be calculated as a function of the size of the sphere, to examine how the lattice energy increases to convergence as more molecular units are added around a central molecule. When the lattice is fully constructed and the lattice energy stops increasing when the limiting radius of the sphere is increased, the lattice energy has converged [41]. How the lattice energy changes as a function of size can give an indication of how the stability of nano-crystals changes with size, and the influence the different coordination spheres of molecules have on the lattice stability. The converged lattice energy can also give an indication of the polymorphic stability of macro sized crystals.

An example of the lattice energy convergence calculation of ibuprofen, using the Dreiding potential, [46] is shown in Fig. 7.4, along with the increase in energy as the radius of the sphere of calculation is increased. From this it can be seen that the first coordination sphere of interactions are often the most important interactions to stabilise the lattice energy.

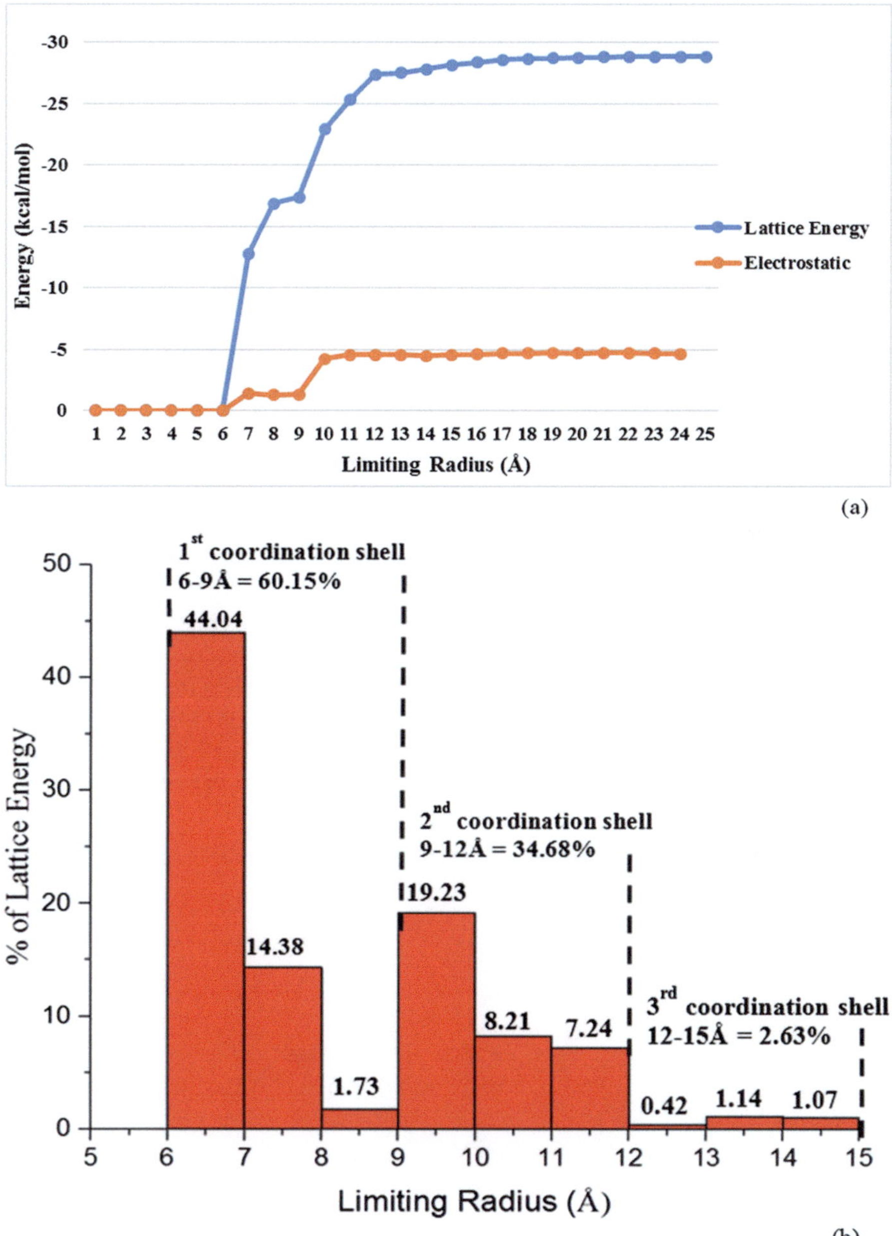

Fig. 7.4 (**a**) Lattice energy of ibuprofen as a function of increasing limiting radius of the sphere of calculation; (**b**) the % of the lattice energy added with the increase in radius of the sphere, highlighting the amount of energy in 3 coordination shells [55]

The fully converged calculated lattice energy of −28.86 kcal/mol (Fig. 7.4a) was in good agreement with the experimental lattice energy, calculated from the heat of sublimation for ibuprofen of 30.10 kcal/mol [56]. This suggests that the Dreiding potential provides an acceptable reproduction of the strength and nature of the synthons within the crystal structure. The coulombic interactions contributed a relatively low amount to the lattice energy, perhaps reflecting that the majority of the molecule is apolar in nature, due to only the H-bonding COOH group containing significantly electronegative atoms to contribute to the polar nature of the molecule.

Figure 7.4b shows that the largest contribution to the lattice energy was found to be from the interaction between the central molecule and the molecules closest to it, occurring between 6 Å and 7 Å from the origin molecule, with over 44% of the lattice energy held within those nearest neighbours. This highlights the importance of the nearest neighbours for stabilising the crystal lattice energy of molecular crystals, such as ibuprofen.

7.3.2 Bulk Synthon Analysis

Careful characterisation of the bulk synthons can provide information on how they can direct physical properties and stabilise the crystal lattice. The strongest synthons in the α-*p*ABA crystal structure are characterised below from a calculation of the strength of the interactions using the Momany potential [43] (Fig. 7.5).

The synthons described in Table 7.1(a) and (b) are calculated as all the interactions between the molecules within one asymmetric unit in the central unit cell,

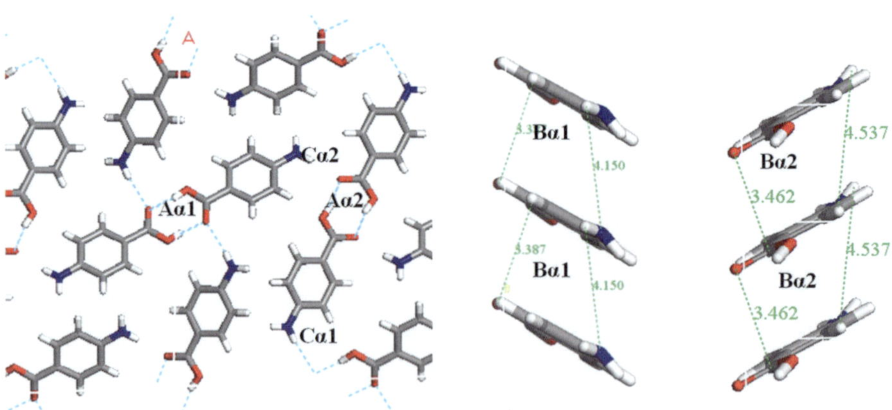

Fig. 7.5 The strongest synthons in the α-*p*ABA structure labelled on the α packing diagram. Strongest synthons are a combination of H-bonding interactions (**Aα** and **Cα**) and π-π stacking (**Bα**) indicating that both types of interactions are important in the formation of the α-*p*ABA structure. Interactions are tabulated in Table 7.1 [36]

Table 7.1 Seven strongest intermolecular interactions in α-*p*ABA based upon Lattice energy summations centred on the two crystallographically independent molecules 1(a) and 2 (b). Distance (column 3) reflects centre of mass to centre of mass of the molecules involved in the interaction (herein and after). Total energy of interaction is energy of the bond multiplied by the multiplicity. The strongest interatomic interaction type and the functional group contribution to the individual pairwise interaction energies are also displayed [57]

Bond	Multiplicity	Distance (Å)	Intermole-cular energy (kcal/mol)	Percentage contribution to lattice energy	Dominating interatomic interaction type	COOH % Contribution to interaction	C6H4 % Contribution to interaction	NH2 % Contribution to interaction
(a)								
Aα2	1	8.34	−5.63	22.9	H-bond	96.7	3.6	−0.4
Bα2	2	3.86	−2.66	21.7	π-π stacking	14.5	72.6	13.0
Cα2	1	7.90	−2.27	9.3	H-bond	41.7	20.7	37.6
Dα2	1	7.76	−1.21	4.9	H-bond	38.8	26.1	35.1
Fα2	2	6.94	−1.9	7.7	vdW	80.8	20.0	−0.9
Total			−16.33	66.5				
(b)								
Aα1	1	8.23	−5.67	23.1	H-bond	96.4	4.0	−0.4
Bα1	2	3.86	−2.68	21.8	π-π stacking	14.5	72.6	13.00
Cα1	1	7.90	−2.27	9.3	H-bond	41.7	20.7	37.6
Dα1	1	7.75	−2.00	8.2	H-bond	38. 8	26.1	35.1
Eα1	2	7.96	−2.26	9.2	vdW	79.9	21.0	−0.9
Total			18.65	71.6				

with all of the other molecules within the central and surrounding unit cells of the crystal structure. Since α-*p*ABA has two crystallographically independent molecules in the asymmetric unit, the interactions from both molecules are described in Table 7.1(a) and (b). The strongest synthonic interactions in α-*p*ABA were calculated to be the OH…O H-bonding interactions between the carboxylic acid groups (Aα1 and 2), which contributed approximately 23% of the lattice energy, and approximately 96% of the synthon interaction strength was found to come from the COOH group. This synthon was calculated to be approximately 3 kcal/mol stronger than any other synthon within the structure. The second strongest synthonic interaction was found to be the π-π stacking interactions between the phenyl ring structures (Bα1 and 2), which were calculated to have over 70% of the synthon interaction strength coming from the phenyl ring group. Since there are unbroken π-π stacking chains running through the structure, each molecule can form two of these interactions (one above and one below the plane of the ring), therefore these interactions contribute approximately 22% of the lattice energy.

7.3.3 Functional Group Lattice Energy Contributions

The contribution of each atom to the calculated lattice energy can be identified, which can be useful in terms of understanding which functional groups on the molecular structure are important in stabilising the lattice. The breakdown of the lattice energy of α-*p*ABA onto the functional groups of its molecular structure is shown in Fig. 7.6.

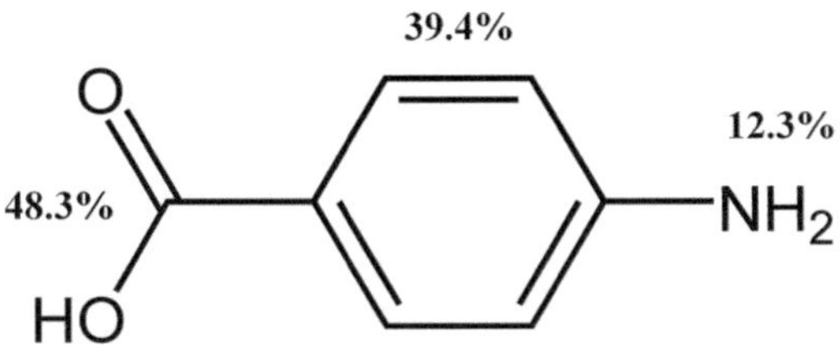

Fig. 7.6 Breakdown of the lattice energy of α-*p*ABA, highlighting the contribution of the functional groups of the *p*ABA molecular structure to the lattice energy

Figure 7.6 suggests that the COOH and phenyl ring groups are the major contributors to the stabilisation of the lattice energy of α-*p*ABA. Although hydrogen bond interactions can play the most vital role in stabilising the crystal lattice of small molecule organic crystals, here it is shown that the interactions arising between the phenyl ring groups play a greater role than the interactions arising from the NH2 group, even though this group forms an NH…O H-bond in the structure.

7.4 Crystal Morphology Prediction

7.4.1 Identification of Morphologically Important Surfaces

Early relationships of interplanar spacing to morphological importance, linked with lattice geometry, led to the Bravais, Friedel, Donnay and Harker model (BFDH) [11–13, 27, 58]. This model is still used to identify the morphologically dominant faces (*hkl*). The rule suggests that, after allowances have been made for space group symmetry, the crystallographic forms with greatest interplanar spacing d_{hkl} will be the most morphologically important at the surface of the crystalline particle [18]. Figure 7.7 shows an example of how the repetition of a molecular structure along a certain direction can impact on the morphological importance of a surface.

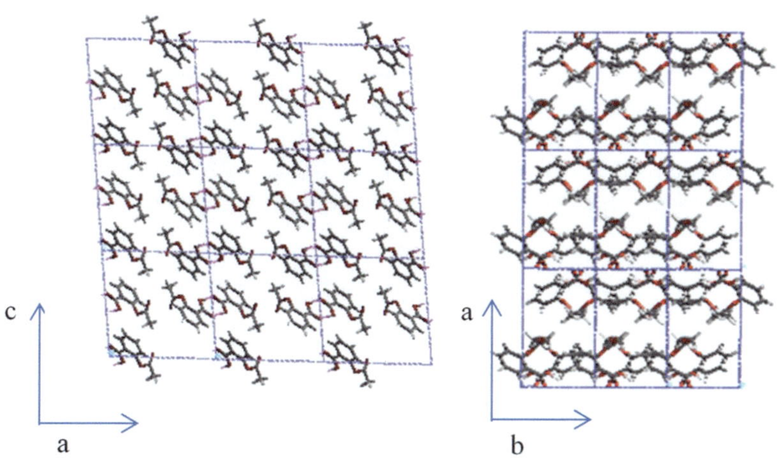

Fig. 7.7 Aspirin {0 0 1} form (*left*) and {1 0 0} form (*right*) showing how the structure is repeated in the [0 0 1] direction but not the [1 0 0], resulting in reduction of the morphological importance of the {0 0 1} but not the {1 0 0} form [59]

Figure 7.7 shows how, in the case of aspirin with a P21/n space group, the crystal structure repeats within a unit cells length along the [0 0 1] direction, but in the [1 0 0] direction this does not happen. Hence, the morphological importance of the {0 0 1} form is reduced to {0 0 2}, as the effective interplanar spacing is $d_{hkl}/2$. In contrast, the {1 0 0} form remains morphologically important. Computer programs, such as Morang [58], can calculate the interplanar spacing in specific crystallographic directions and therefore efficiently predict the likely morphologically important surfaces.

7.4.2 Attachment Energy Calculations

The BFDH rule does not take into account the effect of the intermolecular interactions on the face specific crystal growth rates, and hence their effects on the crystal morphology. This was recognised by Hartman and Perdok in 1955, with the derivation of the 'periodic bond chain' theory and first use of the term 'attachment energy', to describe the strength of the interactions formed when adding another growth layer of a thickness d_{hkl} [15].

Once the lattice energy (E_{cr}) has been calculated, it can then be partitioned into slice (E_{sl}) and attachment (E_{att}) energies for faces predicted to be morphologically important by the BFDH rule. The interrelationship between the lattice, slice and attachment energies is given by Eq. 7.3.

$$\boldsymbol{E_{cr} = E_{sl} + E_{att}} \tag{7.3}$$

The slice energy is the summation of all the interactions between the central molecule and all other molecules within the slice boundary, whereas the attachment energy is the summation of all the interactions between the central molecule and molecules outside the slice boundary. This process can be calculated using the HABIT program [34], and a molecular visualisation of how the interactions are divided into those which contribute to the slice, and those which contribute to the attachment energies, is shown in Fig. 7.8.

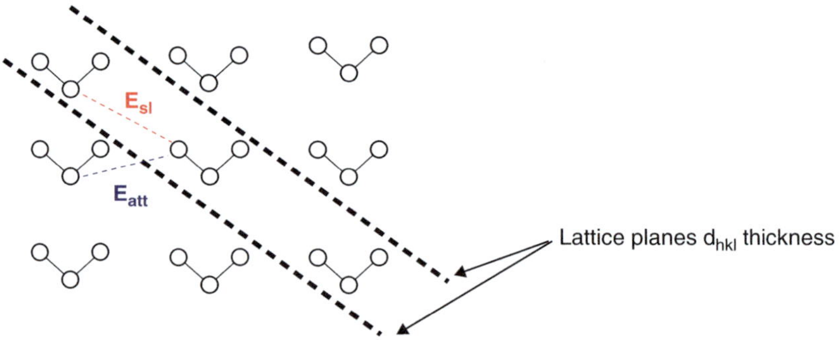

Fig. 7.8 Example crystal structure of water, demonstrating how the attachment energy is calculated at the molecular level. Intermolecular interactions that are within the slice contribute to the slice energy, whilst the interactions that form across the slice boundary contribute to the attachment energy

In this model it is assumed that the attachment energy can be taken to be proportional to the growth rate, according to Eq. 7.4 [15, 16].

$$\boldsymbol{R \propto E_{att}} \tag{7.4}$$

The attachment energies can be converted into centre to face distances and then used to construct a Wulff plot prediction of the crystal morphology. Generally, attachment energy morphological predictions give the best match to the morphology of crystals grown in the vapour phase or at low driving force in solution, where the surfaces often grow by a Burton, Cabrera, Frank (BCF) or Birth and Spread (B&S) growth mechanism, as these conditions are closest to the equilibrium conditions in which the attachment energy morphology is predicted [16]. The attachment energy morphology predictions of anthracene, bi-phenyl and succinic acid are shown in Fig. 7.9, together with the comparison of these predictions to the observed morphologies of these crystals.

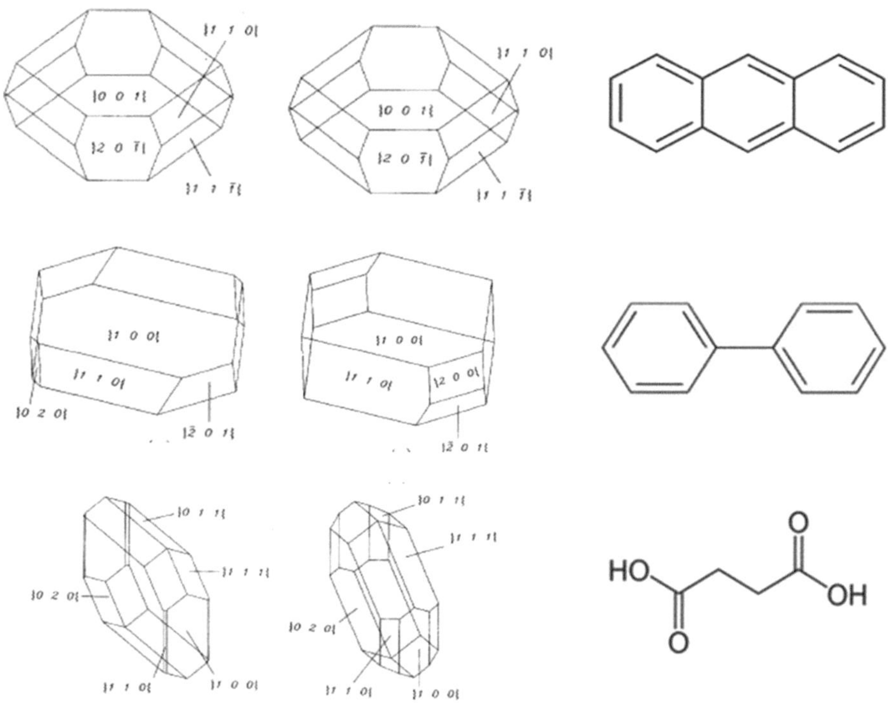

Fig. 7.9 Observed (*left*) and attachment energy calculated (*right*) morphologies for anthracene (*top*), biphenyl (*middle*) and succinic acid (*bottom*). Molecular structures shown on the right hand side [33]

Figure 7.9 shows how the attachment energy model can give a good prediction of the shape of molecular crystals with varied molecular chemistry. The morphological prediction of anthracene almost completely matched the experimental crystal morphology. Biphenyl and succinic acid gave a good match to the general shape of the experimental morphologies. However, the biphenyl simulation shows an under estimation of the (−2 0 1) surface, whilst the succinic acid simulation shows a slightly thinner morphology and over-estimates the size of the (1 1 1) surface, compared to the experimental morphologies. The morphologies predicted for these materials are all somewhat equant, suggesting that the surfaces of these materials grow by stable crystal growth mechanisms, such as BCF or B&S,

whereby the attachment energy theory can effectively predict the relative growth rates of faces growing by such mechanisms [16]. Expanding on this idea, it is perhaps not surprising that the best match of the predicted, to the experimental, morphology is anthracene. This probably due to be due to this molecule only forming weak vdW interactions and being relatively inflexible, probably resulting in stable, slow crystal growth.

7.5 Influence of Solution Environment on Crystal Morphology

Since the majority of industrial crystallisation takes place in the solution phase, it is important to consider the how the intermolecular interactions between the solvent and the morphologically important crystal surfaces can impact upon the relative growth rates of the morphologically important crystal faces.

7.5.1 Surface Chemistry Analysis

Characterisation of the extrinsic synthons, which can be identified from an attachment energy calculation, can provide information as to how the crystallisation solvent may interact with these exposed, unsaturated functional groups at a particular crystal surface, and hence impact on the individual growth rates of the surfaces, and the resultant crystal morphology. The predicted morphology of α-*p*ABA is shown in Fig. 7.10, together with a comparison to the observed morphology for crystals experimentally grown from ethanol solutions.

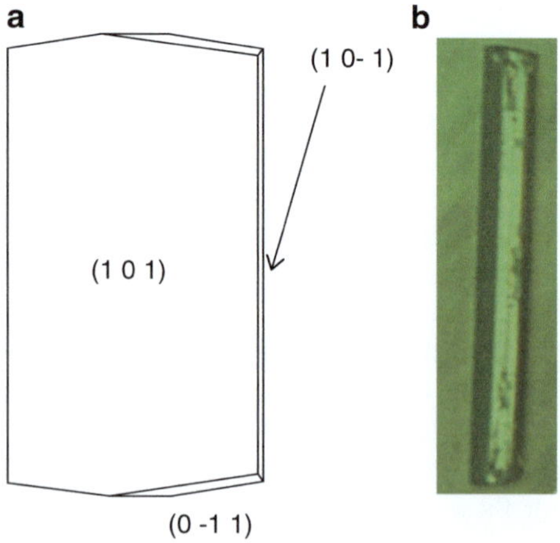

Fig. 7.10 (**a**) Attachment energy crystal morphology prediction of α-*p*ABA; (**b**) α-*p*ABA crystal grown from EtOH solution [36]

Figure 7.10 shows that the attachment energy prediction of α-*p*ABA was in general agreement with the crystals produced from ethanol, however, the experimental crystals were found to be much more needle-like. Hence, the synthons which contribute to the attachment energies of the morphologically important surfaces of α-*p*ABA were determined to characterise how the molecular surface chemistry impacts on the crystal growth and morphology of the material (Fig. 7.11).

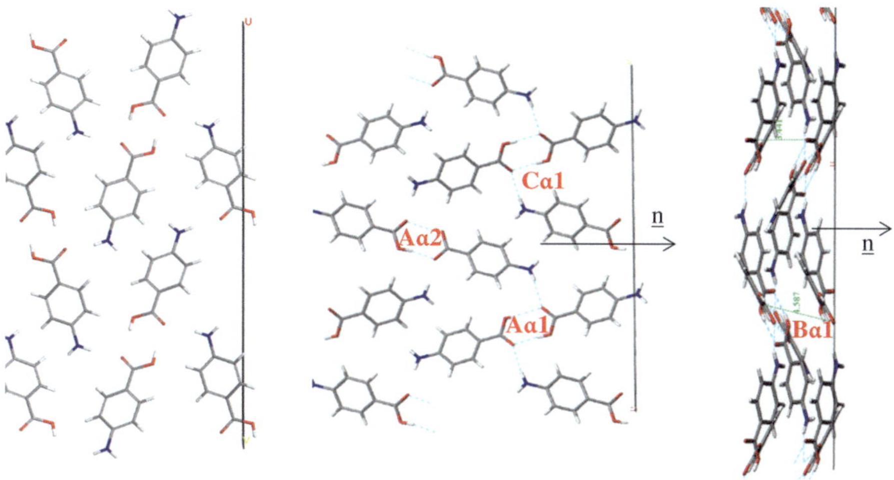

(1 0 1) (1 0 -1) (0 1 -1)

Fig. 7.11 Ideally cleaved surfaces of the major *(1 0 1)*, *(1 0–1)* and *(0 1–1)* faces of α-*p*ABA, with the strongest synthons from the bulk structure (given in Table 7.1), which were calculated to contribute to the growth of the *(1 0–1)* and *(0 1–1)* surfaces shown with the *red* labels [36]

The extrinsic synthon analysis revealed that the (1 0 1) surface (Fig. 7.11 (left)) was found to have none of the strong bulk synthons (Table 7.1) contributing to the attachment energy, and hence the growth of this surface was predicted to be very slow. This was consistent with this being the slow growing and hence dominant crystal surface within the α-*p*ABA crystal morphology. In contrast, the faster growing side (1 0–1) surfaces were found to be dominated by the OH…O H-bonding interactions and the fastest growing face (Fig. 7.11 (centre)) was calculated to be dominated by the π-π stacking interactions between the close packed phenyl rings.

Figure 7.10b shows that the experimental crystal morphology as crystallised from EtOH solutions, which was found to be significantly more extended along the b-axis, in comparison to the predicted morphology. This needle-like morphology is consistently observed from polar and protic solvents [60–62]. Hence, it is likely the growth of the side (1 0–1) surface (Fig. 7.11 (right)) is significantly more hindered than the growth of the capping (0 1–1) surface in polar solvents such as ethanol. This is probably due to the ethanol molecules more strongly interacting with the exposed COOH groups at the (1 0–1) surface, due to the ethanol being able to form

H-bonds with the carbonyl and acid groups. In comparison, since it is mainly the phenyl ring structures exposed on the capping (0 1–1) surface, it is likely that ethanol does not strongly interact with these groups and that ethanol does not hinder growth on this surface, hence explaining the needle-like morphology which crystallises from ethanol solutions.

7.5.2 Solvent Induced Modification of Crystal Morphology

Figure 7.12 shows how the morphology of benzophenone is strongly modified when crystallised from toluene solutions, in comparison to the morphology when crystallised from most other common solvents. The size of the (0 2 1) surface is greatly enhanced, whereas other surfaces, such as the (0 2 0) surface, are observed to grow out.

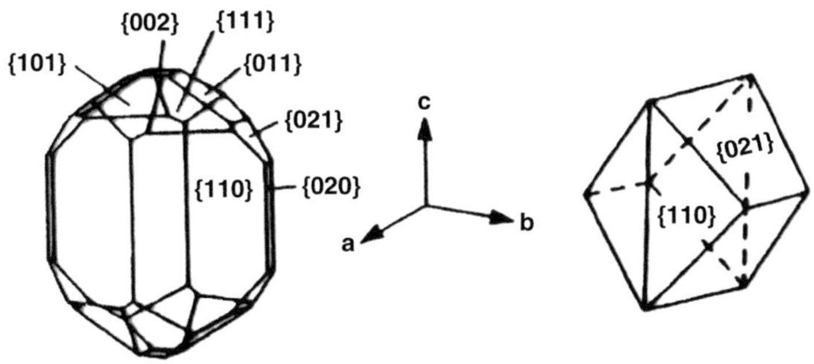

Fig. 7.12 Benzophenone morphology from most solvents (*left*) and from toluene (*right*) [2]

Comparing how toluene interacts with the (0 2 1) and (0 0 2) (Fig. 7.13) surfaces can rationalise why the (0 2 1) surface is greatly enhanced in size, compared to the other faces, when benzophenone is crystallised from toluene.

The toluene molecule is observed to incorporate much more easily into the growth layer of the (0 2 1) surface compared to the (0 2 0). Therefore, the de-solvation of the (0 2 1) surface is likely to be much slower and will have the effect of slowing down the growth rate of this surface, hence increasing the surface area. Compared to the (0 2 0), where toluene incorporation into the growth layer is much more difficult, growth is less likely to be hindered and hence the face grows out and is not observed at the surface of benzophenone crystals grown from toluene.

Crystal morphology modification has been well studied, particularly in the cases of problematic needle-like morphologies [6, 63–66] that can be difficult to process and result in pipe blockages. Therefore, there is a need to be able to effectively predict the experimental environments that can produce crystal morphologies that do not produce problems during downstream manufacturing processes.

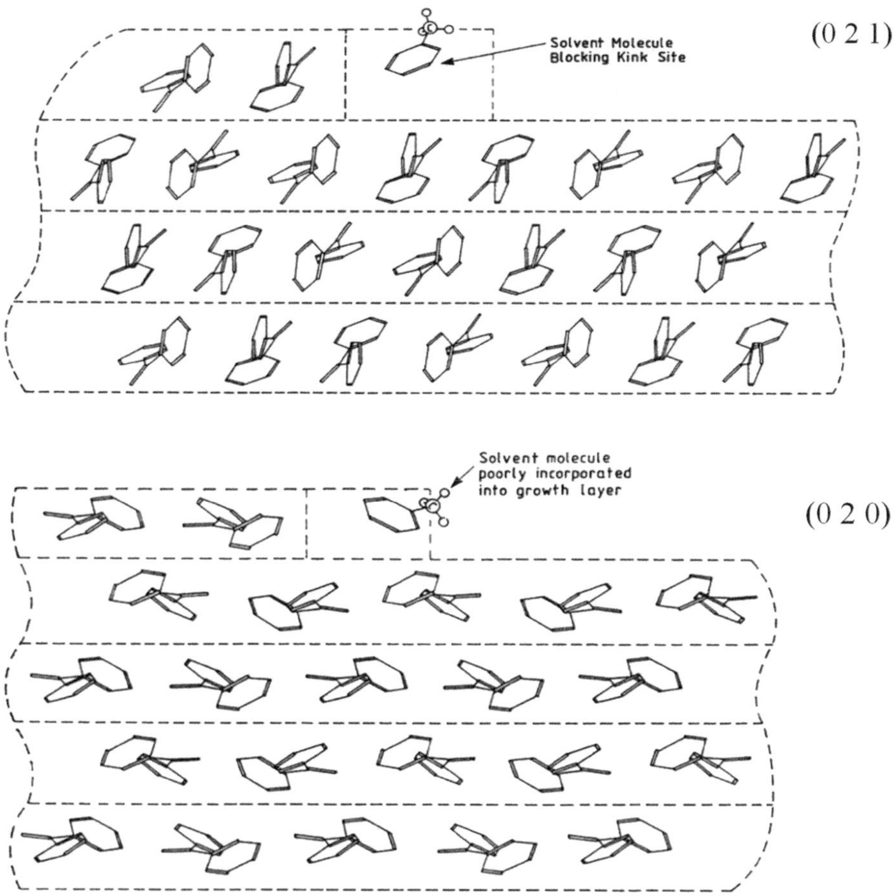

Fig. 7.13 Predicted solvent molecule incorporation at the *(0 2 1)* surface and *(0 2 0)* surfaces of benzophenone [2]

7.5.3 Grid-Based Methods for Predicting Solvent Induced Morphologies

Computationally efficient grid-based methods can be used to calculate solute and solvent binding energies on crystal habit faces, hence allowing the prediction of solvent induced morphologies [40].

The SystSearch grid search program [67] creates a three-dimensional grid near the surface of a crystal where a probe molecule explores every point to find the most favourable energetic interaction point at where the probe molecule interacts with the crystal surface.

In addition to translating across the grid, the probe molecule is also allowed to explore rotational degrees of freedom at each point to find the most stable geometry. The method uses the intermolecular energy calculations to identify the most stable molecular orientation from the search, along with its interaction energy. Additionally, the distribution of all the energies found from the grid search can be plotted, which can give a broader picture of the probe to surface interactions. This can allow prediction as to how solute or solvent molecules will incorporate at a given surface.

The binding energy of aspirin, ethanol and water molecules to the morphologically important surfaces of an aspirin crystal [5] were calculated using grid-based methods. This data was then used to calculate a solution-modified attachment energy for crystals of aspirin grown in ethanol:water mixtures. Figure 7.14 shows a comparison of the vacuum-state morphology prediction, solution mediated morphology prediction and experimentally crystallised aspirin.

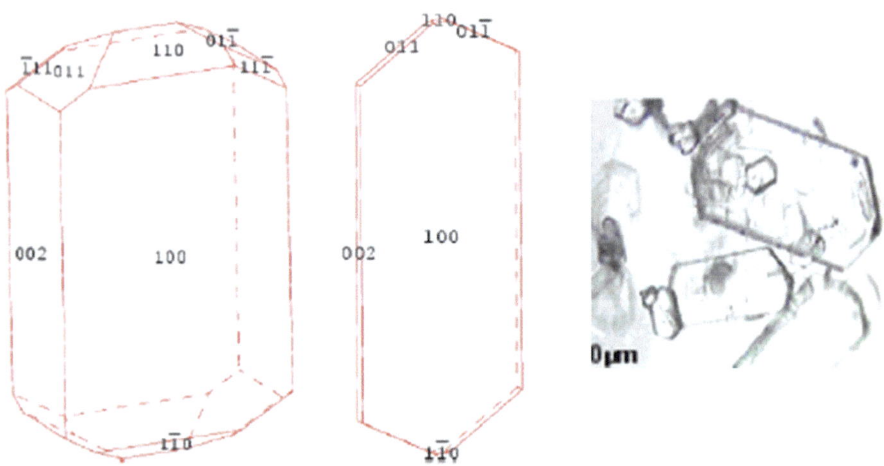

Fig. 7.14 (*Left*) Aspirin vacuum attachment energy morphology prediction. (*Middle*) Aspirin solution mediated morphology prediction in 62:38 ethanol: water mixed solvent. (*Right*) Experimentally grown aspirin crystals from 62:38 ethanol:water mixes [5]

The solution mediated morphology showed excellent agreement to the experimental crystals. Compared to the vacuum state morphology prediction, it is clear that the growth of the (1 0 0) surface has retarded significantly compared to the other major surfaces. Examination of this surface revealed that it has exposed carbonyl and hydroxyl groups at the surface, which hence are expected to strongly interact with the surrounding solvent and slow down the de-solvation at this surface (Fig. 7.15).

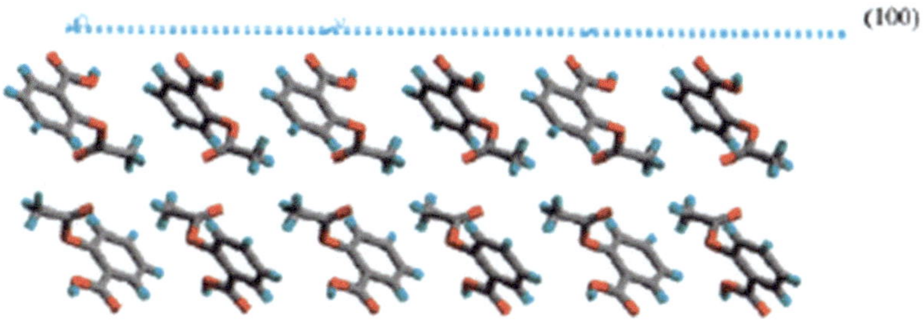

Fig. 7.15 Cleared crystal surface of aspirin (*1 0 0*) surface

7.6 Conclusions

The prediction of synthonic strength and directionality can be used to aid in the prediction of crystal stability, growth and morphology. The calculation of lattice energies and the bulk intrinsic synthons can give information on the key synthons that stabilise the crystal structure. In turn, the unsaturated extrinsic synthons at the morphologically important surfaces can provide a molecular understanding as to how a crystal grows and how this affects the final morphology, as well as how the crystal may interact with the surrounding crystallisation environment.

Morphological prediction using the attachment energy theory gives a good prediction for crystals grown at low driving forces. The identification of the extrinsic synthons of α-pABA allowed for a qualitative explanation for the needling observed in polar and protic solvents, and this method can be expanded to other materials that produce undesirable crystal morphologies.

In addition, grid-based methods have provided evidence of their potential application for predicting a crystal morphology as a function of external conditions, with the modification of the aspirin crystal morphology prediction to accurately resemble the crystal morphology produced in ethanol/water mixtures. However, these methods have only been applied to a relatively small number of case studies, where, although the results show promise, they must be applied to a higher number of studies, especially with respect to applying these methods to crystals that show kinetic anisotropic morphologies.

References

1. MacLeod CS, Muller FL (2012) On the fracture of pharmaceutical needle-shaped crystals during pressure filtration: case studies and mechanistic understanding. Org Process Res Dev 16:425–434
2. Roberts KJ, Sherwood JN, Yoon CS, Docherty R (1994) Understanding the solvent-induced habit modification of Benzophenone in terms of molecular recognition at the crystal/solution Interface. Chem Mater 6:1099–1102

3. Ramachandran V, Halfpenny PJ, Roberts KJ (2017) Crystal science fundamentals, Chapter 1. In: Roberts KJ, Docherty R, Tamura R (eds) Engineering crystallography: from molecule to crystal to functional form Springer Advanced Study Institute (ASI) Series, 2017, in press
4. ter Horst JH, Geertman RM, van Rosmalen GM (2001) The effect of solvent on crystal morphology. J Cryst Growth 230:277–284
5. Hammond RB, Pencheva K, Ramachandran V, Roberts KJ (2007) Application of grid-based molecular methods for modeling solvent-dependent crystal growth morphology: aspirin crystallized from aqueous ethanolic solution. Cryst Grow Des 7:1571–1574
6. Chen J, Trout BL (2010) Computer-aided solvent selection for improving the morphology of needle-like crystals: a case study of 2,6-Dihydroxybenzoic acid. Crystal Grow Des 10:4379–4388
7. Hod I, Mastai Y, Medina DD (2011) Effect of solvents on the growth morphology of dl-alanine crystals. CrystEngComm 13:502–509
8. Li Destri G, Marrazzo A, Rescifina A, Punzo F (2011) How molecular interactions affect crystal morphology: the case of haloperidol. J Pharm Sci 100:4896–4906
9. Punzo F (2013) Unveiling the role of molecular interactions in crystal morphology prediction. J Mol Struct 1032:147–154
10. Singh MK, Banerjee A (2013) Role of solvent and external growth environments to determine growth morphology of molecular crystals. Cryst Growth Des 13:2413–2425
11. Bravais A (1886) Etudes Crystallographiques Gauthiers Villars, Paris
12. Friedel G (1907). Bulletin De La Societe Francaise De Mineralogie Et De Crystallographie 30:326
13. Donnay JDH, Harker D (1937) A new law of crystal morphology extending the law of bravais. Am Mineral 22:446–467
14. Volmer M, Weber A (1926) Germ-formation in oversaturated figures. Zeitschrift Fur Physikalische Chemie--Stochiometrie Und Verwandtschaftslehre 119:277–301
15. Hartman P, Perdok WG (1955) On the relations between structure and morphology of crystals. I. Acta Crystallogr 8:49–52
16. Hartman P, Bennema P (1980) The attachment energy as a habit controlling factor: I theoretical considerations. J Cryst Growth 49:145–156
17. Bennema P, van der Eerden JP (1977) Crystal growth from solution: development in computer simulation. J Cryst Growth 42:201–213
18. Human HJ, Van Der Eerden JP, Jetten LAMJ, Odekerken JGM (1981) On the roughening transition of biphenyl: transition of faceted to non-faceted growth of biphenyl for growth from different organic solvents and the melt. J Cryst Growth 51:589–600
19. Davey R (1982) The role of additives in precipitation processes. In: Jancic SJ, de Jong EJ (eds) Industrial crystallization 81. North-Holland Publishing Co, Amsterdam
20. Jetten LAMJ, Human HJ, Bennema P, Van Der Eerden JP (1984) On the observation of the roughening transition of organic crystals, growing from solution. J Cryst Growth 68:503–516
21. Davey R (1986) The role of the solvent in crystal growth from solution. J Cryst Growth 76:637–644
22. Jackson KA (1958) Mechanisms of growth. In: ASF M (ed) Liquid metals and solidification. American Society for Metals, Cleveland
23. Walker EM, Roberts KJ, Maginn SJ (1998) A molecular dynamics study of solvent and impurity interaction on the crystal habit surfaces of epsilon-caprolactam. Langmuir 14:5620–5630
24. Winn D, Doherty MF (1998) A new technique for predicting the shape of solution-grown organic crystals. AICHE J 44:2501–2514
25. Li J, Tilbury CJ, Joswiak MN, Peters B, Doherty MF (2016) Rate expressions for kink attachment and detachment during crystal growth. Cryst Growth Des 16:3313–3322
26. Tilbury CJ, Green DA, Marshall WJ, Doherty MF (2016) Predicting the effect of solvent on the crystal habit of small organic molecules. Cryst Growth Des 16:2590–2604

27. Docherty R, Clydesdale G, Roberts KJ, Bennema P (1991) Application of Bravais-Friedel-Donnay-Harker, attachment energy and Ising-models to predicting and understanding the morphology of molecular-crystals. J Phys D-Appl Phys 24:89–99
28. Dowty E (1980) Computing and drawing crystal shapes. Am Mineral 65:465–472
29. Bruno IJ, Cole JC, Edgington PR, Kessler M, Macrae CF, McCabe P, Pearson J, Taylor R (2002) New software for searching the Cambridge structural database and visualizing crystal structures. Acta Crystallogr B 58:389–397
30. Roberts KJ, Docherty R, Bennema P, Jetten L (1993) The importance of considering growth-induced conformational change in predicting the morphology of Benzophenone. J Phys D-Appl Phys 26:B7–B21
31. Desiraju G, Vittal JJ, Ramanan A (1989) Crystal engineering: the design of organic Solids. Elsevier, Amsterdam
32. Desiraju GR (1995) Supramolecular Synthons in crystal engineering—a new organic synthesis. Angew Chem Int Ed Engl 34:2311–2327
33. Docherty R, Roberts KJ (1988) Modeling the morphology of molecular-crystals - application to Anthracene, biphenyl and Beta-succinic acid. J Cryst Growth 88:159–168
34. Clydesdale G, Docherty R, Roberts KJ (1991) HABIT - a program for predicting the morphology of molecular crystals. Comput Phys Commun 64:311–328
35. Etter MC (1991) Hydrogen bonds as design elements in organic chemistry. J Phys Chem 95:4601–4610
36. Rosbottom I, Roberts KJ, Docherty R (2015) The solid state, surface and morphological properties of p-aminobenzoic acid in terms of the strength and directionality of its intermolecular synthons. CrystEngComm 17:5768–5788
37. Kitaygorodsky A (1961) Interaction curve of non-bonded carbon and hydrogen atoms and its application. Tetrahedron 14:230–236
38. Zhao Y, Truhlar DG (2006) A new local density functional for main-group thermochemistry, transition metal bonding, thermochemical kinetics, and noncovalent interactions. J Chem Phys 125
39. Grimme S (2011) Density functional theory with London dispersion corrections. Wiley Interdiscip Rev: Comput Mol Sci 1:211–228
40. Hammond RB (2017) Molecular Modelling Route Map, Chapter 6. In: Roberts KJ, Docherty R, Tamura R (eds) Engineering crystallography: from molecule to crystal to functional form Springer Advanced Study Institute (ASI) Series, 2017, in press
41. Pickering J, Hammond RB, Ramachandran V, Soufian M, Roberts KJ (2017) Synthonic Engineering Modelling Tools for Product and Process Design, Chapter 11. In: Roberts KJ, Docherty R, Tamura R (eds) Engineering crystallography: from molecule to crystal to functional form Springer Advanced Study Institute (ASI) Series, 2017, in press
42. Roberts KJ, Hammond RB, Ramachandran V, Docherty R (2016) Synthonic engineering: from molecular and crystallographic structure to the rational design of pharmaceutical solid dosage forms. In: Abramov Y (ed) Computational approaches in pharmaceutical solid state chemistry. Wiley, New Jersey
43. Momany FA, Carruthers LM, McGuire RF, Scheraga HA (1974) Intermolecular potentials from crystal data. III. Determination of empirical potentials and application to the packing configurations and lattice energies in crystals of hydrocarbons, carboxylic acids, amines, and amides. J Phys Chem 78:1595–1620
44. Lifson S, Hagler AT, Dauber P (1979) Consistent force-field studies of Inter-molecular forces in hydrogen-bonded crystals .1. Carboxylic-acids, amides, and the C=O...H- hydrogen-bonds. J Am Chem Soc 101:5111–5121
45. Nemethy G, Pottle MS, Scheraga HA (1983) Energy parameters in polypeptides. 9. Updating of geometrical parameters, nonbonded interactions, and hydrogen bond interactions for the naturally occurring amino acids. J Phys Chem 87:1883–1887
46. Mayo SL, Olafson BD, Goddard WA (1990) Dreiding - a generic force-field for molecular simulations. J Phys Chem 94:8897–8909

47. Wang JM, Wolf RM, Caldwell JW, Kollman PA, Case DA (2004) Development and testing of a general amber force field. J Comput Chem 25:1157–1174
48. Price SL, Wibley KS (1997) Predictions of crystal packings for uracil, 6-azauracil, and allopurinol: the interplay between hydrogen bonding and close packing. J Phys Chem A 101:2198–2206
49. Gale JD, Henson NJ (1994) Derivation of interatomic potentials for Microporous Aluminophosphates from the structure and properties of Berlinite. J Am Chem Soc Faraday T 90:3175–3179
50. Price SL, Leslie M, Welch GWA, Habgood M, Price LS, Karamertzanis PG, Day GM (2010) Modelling organic crystal structures using distributed multipole and polarizability-based model intermolecular potentials. Phys Chem Chem Phys 12:8478–8490
51. Brunsteiner M, Price SL (2001) Morphologies of organic crystals: sensitivity of attachment energy predictions to the model intermolecular potential. Cryst Grow Des 1:447–453
52. Ouvrard C, Price SL (2004) Toward crystal structure prediction for conformationally flexible molecules: the headaches illustrated by aspirin. Cryst Growth Des 4:1119–1127
53. Day GM, Cooper TG, Cruz-Cabeza AJ, Hejczyk KE, Ammon HL, Boerrigter SXM, Tan JS, Della Valle RG, Venuti E, Jose J, Gadre SR, Desiraju GR, Thakur TS, van Eijck BP, Facelli JC, Bazterra VE, Ferraro MB, Hofmann DWM, Neumann MA, Leusen FJJ, Kendrick J, Price SL, Misquitta AJ, Karamertzanis PG, Welch GWA, Scheraga HA, Arnautova YA, Schmidt MU, van de Streek J, Wolf AK, Schweizer B (2009) Significant progress in predicting the crystal structures of small organic molecules - a report on the fourth blind test. Acta Crystallogr Sect B: Struct Sci Cryst Eng Mater 65:107–125
54. Price SL (2009) Computed crystal energy landscapes for understanding and predicting organic crystal structures and polymorphism. Acc Chem Res 42:117–126
55. Nguyen TTH, Rosbottom I, Hammond RB, Marziano I, Roberts KJ (2017) Crystal morphology and interfacial stability of RS-Ibuprofen in relation to its molecular and synthonic structure. Crystal Grow Des (Accepted)
56. Bunyan J, Shankland N, Sheen D (1991) Solvent effects on the morphology of ibuprofenAIChE Symposium Series, pp 44–54
57. Rosbottom I (2015) The influence of the intermolecular Synthons on the molecular aggregation, polymorphism, crystal growth and morphology of p-Aminobenzoic acid crystals from solution. School of Chemical and Process Engineering University of Leeds, Leeds
58. Docherty R, Roberts KJ, Dowty E (1988) Morang — a computer program designed to aid in the determinations of crystal morphology. Comput Phys Commun 51:423–430
59. Olusanmi D, Roberts KJ, Ghadiri M, Ding Y (2011) The breakage behaviour of aspirin under quasi-static indentation and single particle impact loading: effect of crystallographic anisotropy. Int J Pharm 411:49–63
60. Gracin S, Rasmuson ÅC (2004) Polymorphism and Crystallization of p-Aminobenzoic acid. Cryst Growth Des 4:1013–1023
61. Sullivan RA, Davey RJ (2015) Concerning the crystal morphologies of the alpha and beta polymorphs of p-aminobenzoic acid. Cryst Eng Comm 17:1015–1023
62. Toroz D, Rosbottom I, Turner TD, Corzo DMC, Hammond RB, Lai X, Roberts KJ (2015) Towards an understanding of the nucleation of alpha-para amino benzoic acid from ethanolic solutions: a multi-scale approach. Faraday Discuss 179:79–114
63. Panina N, van de Ven R, Janssen FFBJ, Meekes H, Vlieg E, Deroover G (2009) Study of the needle-like morphologies of two beta-Phthalocyanines. Cryst Grow Des 9:840–847
64. McArdle P, Hu Y, Lyons A, Dark R (2010) Predicting and understanding crystal morphology: the morphology of benzoic acid and the polymorphs of sulfathiazole. Cryst Eng Comm 12:3119–3125
65. Lovette MA, Doherty MF (2013) Needle-shaped crystals: causality and solvent selection guidance based on periodic bond chains. Cryst Grow Des 13:3341–3352

66. Walshe N, Crushell M, Karpinska J, Erxleben A, McArdle P (2015) Anisotropic crystal growth in flat and Nonflat systems: the important influence of van der Waals contact molecular stacking on crystal growth and dissolution. Crystal Grow Des 15:3235–3248
67. Hammond RB, Pencheva K, Roberts KJ (2006) A Structural−kinetic approach to model face-specific solution/crystal surface energy associated with the Crystallization of acetyl salicylic acid from supersaturated aqueous/ethanol solution. Cryst Grow Des 6:1324–1334

Chapter 8
Determining Surface Energetics of Solid Surfaces

Jerry Y.Y. Heng

Abstract The most fundamental description of an interface is the thermodynamic free energy per unit area, γ_{ij} governs a range of solid-solid, solid-liquid and solid-vapour interactions. The determination of the solid-vapour free energy is not straight-forward. In this chapter, conventional experimental techniques such as; sessile drop contact angle measurements, capillary rise and Wilhelmy plate is presented. Models for the determination of surface energy from contact angles are discussed. Furthermore, recent developments in inverse gas chromatography and dynamic vapour sorption approaches for the evaluation of the solid surface free energy are also described.

Keywords Contact angle • Surface energy • Wettability • Inverse gas chromatography

8.1 The Contact Angle

The most fundamental description of an interface is the thermodynamic free energy per unit area, γ_{ij}. γ_{ij} can be defined in terms of the Gibbs free energy as Eq. 8.1:

$$\gamma_{ij} = \left[\frac{\partial G}{\partial A}\right]_{T,P,N_i} \tag{8.1}$$

γ_{ij} can be calculated by measuring the contact angles of various reference liquids and analysing the data with an appropriate theory.

Liquid droplets on ideally smooth, flat, isotropic, rigid and non-deformable surfaces in thermodynamic equilibrium with its own vapour will form a sessile drop and possesses a unique profile (Fig. 8.1). This droplet profile was first observed and a detailed description was put forward by Thomas Young in 1805 [1]. At the

J.Y.Y. Heng (✉)
Department of Chemical Engineering, Imperial College London, South Kensington Campus, London SW7 2AZ, UK
e-mail: jerry.heng@imperial.ac.uk

K.J. Roberts et al. (eds.), *Engineering Crystallography: From Molecule to Crystal to Functional Form*, NATO Science for Peace and Security Series A: Chemistry and Biology, DOI 10.1007/978-94-024-1117-1_8

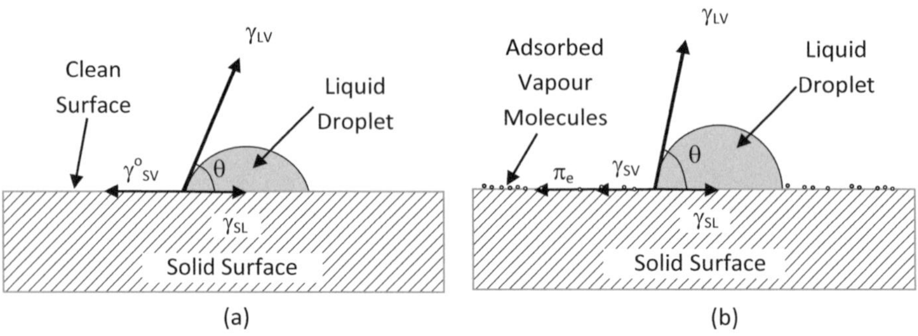

Fig. 8.1 Equilibrium contact angles on (**a**) clean surface and (**b**) surface with adsorbed vapour molecules and the balance of forces at the three phase contact points

solid-liquid-vapour interface, the angle formed by the tangent of the liquid-vapour interface and the solid substrate is defined as the contact angle, θ_e. By balancing the forces at the three phase point, along the horizontal axis, the surface energy of the solid can be obtained. The contact angle is the manifestation of the solid surface energy, and is often the preferred method of choice for its determination. The Young equation is written as:

$$\boldsymbol{\gamma}^{\mathbf{o}}{}_{SV} = \boldsymbol{\gamma}_{SL} + \boldsymbol{\gamma}_{LV} \cdot \cos\theta_e \tag{8.2}$$

where $\boldsymbol{\gamma}^{o}{}_{SV}$, $\boldsymbol{\gamma}_{SL}$ and $\boldsymbol{\gamma}_{LV}$ are the solid-vapour, solid-liquid and liquid-vapour interfacial energies, respectively. The superscript nought represents a bare clean surface, free of any adsorbed molecules. In the case of solid surfaces with adsorbed vapour molecules, the spreading pressure, π_e may be appreciable (see DVS section below).

8.2 Determining Surface Energies by Contact Angle Measurement

Numerous techniques, depending on the nature of the material, have been established for determining the surface energies of solid materials. Conventional methods include sessile drop, capillary rise in a powder bed, air-pressure techniques, Wilhelmy plate, sedimentation volume, film flotation and vapour probe techniques. Most of these traditional methods of characterising solid materials rely on the use of a packed powder bed. More recently, the use of atomic force microscopy has been employed.

8.2.1 Sessile Drop

The sessile drop method is a simple, straight forward procedure which is sensitive to the outermost molecular layers of the solid surface, the first 10 Å (1 nm) of a surface [2]. A liquid drop is placed on a surface and a tangent aligned at the three-phase point to obtain the contact angle as shown in Fig. 8.1. Measurements are made optically and the use of computer programmes to fit drop profile dimensions using various models is now routine, via a number of commercial instruments.

A flat solid sample area of only a few square millimetres is required and small quantities of probe liquid are sufficient to obtain contact angle data. The probe liquid should be non-reactive with the solid surface and have a low volatility. In order to obtain the advancing contact angle, θ_a, the liquid front has to advance to wet fresh surface. This condition is achieved by increasing the drop volume slowly, then subsequently decreasing the volume to obtain the receding contact angle, θ_r.

Whilst this method is relatively straighforward, a surface which is smooth, flat and homogenous is ideally required. Without these requirements, the applicability of this method is further complicated by the existence of the contact angle hysteresis, which is defined as the difference between the θ_a and the θ_r. This phenomenon occurs mainly due to roughness and heterogeneity of the surface.

8.2.2 Capillary Rise

Washburn, Bartell and Osterhof [3] were pioneers on characterising the wettability of powders. Their work involves measuring the liquid penetration rate into a powder bed, or the pressure required to halt the penetration of such a liquid front.

A column packed with a sample powder is brought into contact with the surface of a liquid, and the rate the liquid penetrates or wicks into the powder bed is measured. The liquid should be non-reactive with the sample and a range of liquids with differing surface tension should be used. Characterisation of powders by capillary rise method is primarily based on the Washburn equation as shown below:

$$h^2 = \frac{r_{eff}\gamma_{LV}\cos\theta}{2\eta} \cdot t \tag{8.3}$$

where h is the distance travelled by the liquid in time t and r_{eff} the effective interstitial pore radius. γ_{LV} and η are the surface tension and viscosity of the liquid respectively. The model is derived in conjunction with the Hagen-Poisseuille equation and based on the assumption of a laminar flow of liquid, negligible gravitational effects and constant cylindrical pore geometry. By measuring the mass gained, the Washburn equation can be expressed as:

$$\frac{m^2}{t} \cdot \frac{\eta}{\rho^2} = K_C \gamma_{LV} \cos\theta \tag{8.4}$$

with K_C the capillary constant, representing the geometry of the porous network. The capillary constant, K_C is calculated from the complete wetting case where contact angle, $\theta = 0^\circ$ while the contact angles measured for the partial wetting case are calculated based on the K_C obtained.

However, the reproducibility of column packing and poorly defined effective pore radius have often raised speculation on the accuracy of the experiment. Swelling of the powder bed also affects the effective pore radius yielding higher contact angles compared to the results of Wilhelmy and sessile drop method. Contact angles derived via the Washburn equation incorporate numerous assumptions, all of which are based on idealised conditions and may not be applicable to real particulate materials, especially for porous materials. Furthermore, penetration of liquid into the powder bed might cause swelling of the particles in the packed bed resulting in a sample temperature change and/or altering the effective capillary radius of the bed.

8.2.3 *Wilhelmy Plate*

Wilhelmy (1863) developed a method to measure the downward force exerted on a vertical plate when it is brought into contact with liquid, and related this force to the contact angle between the plate and the liquid [4]. By measuring the wetting force of liquids with known surface tensions, γ_{LV}, and a known perimeter of the plate, p_p, the contact angle of the solid surface can be calculated by:

$$F_W = p_p \gamma_{LV} \cdot \cos\theta \tag{8.5}$$

However, when the plate is immersed in the liquid, a reduction in force is measured due to the buoyancy force, notably in the case of a thick vertical plate. The contact angle can then be calculated by:

$$F_W = p_p \gamma_{LV} \cdot \cos\theta - Vg\Delta\rho \tag{8.6}$$

where V is the volume of displaced fluid and $\Delta\rho$ the density difference between the solid plate and the liquid and g is the gravitational force constant.

Dynamic contact angles can be obtained by advancing the plate or cylinder into the liquid to obtain θ_a and then retrieving the plate to obtain θ_r. The use of the Wilhelmy plate method might seem favourable and simple. However, the method works best with a smooth plate with constant or known perimeter. The plate should also have the same compositions on all surfaces. This technique has been usefully applied to fibres and thin films. Dynamic contact angles of powders can be obtained by either compacting powder into a flat plate or by adhering the powder particles to a glass slide coated with an adhesive.

8.2.4 Inverse Gas Chromatography

Inverse gas chromatography (IGC) is a simple, yet versatile and robust technique for studying the physicochemical properties of particulate and fibrous materials. This technique was first developed in the 1950s, and early work focused on characterisation of catalysts, adsorbents and polymers. IGC is simply the inverse use of a conventional gas chromatography (GC), in which a column is packed with an *unknown* solid sample and *known* vapour probes are injected into the column via an inert carrier gas. The retention time of the probe molecules is recorded by a GC detector, allowing the retention volume and then the partitioning co-efficient for the solid-vapour interaction to be determined. From this primary data, a wide range of physicochemical properties of solid materials such as surface energies, acid-base functionality of surfaces, diffusion kinetics, solubility parameters, surface heterogeneity and phase transition temperatures may be determined [5].

At infinite dilution, the retention behaviour of organic vapour probes has been successfully measured using IGC to study the surface energies of pharmaceutical solids. With infinite dilution injections, only a small amount of probe (adsorbate) is injected and adsorption normally occurs in the linear Henry region of the adsorption isotherm. The retention behaviour of these injection pulses yields a net retention volume, $V_R{}^0$ which is a fundamental surface thermodynamic property of the solid-vapour interaction process. From this parameter, a wide range of surface properties, including surface energies can be derived [6]. For infinite dilution IGC in the Henry's law region, the chromatogram peaks obtained are symmetrical (Gaussian). The net retention volume, $\boldsymbol{V_R}{}^o$ is given by:

$$V_R^o = \frac{j}{m} \cdot F \cdot (t_R - t_o) \cdot \frac{T}{273.15} \tag{8.7}$$

where T is the column temperature in Kelvin (K), F is the carrier gas exit flow rate at standard temperature and pressure (STP), t_R is the retention time for adsorbing probe and t_o is the mobile phase hold-up time (dead-time) and j is the James-Martin correction factor, which corrects the retention time for the pressure drop along the column bed.

$\boldsymbol{V_R}{}^o$ can be related to the free energy of adsorption by Eq. 8:

$$\Delta G^o = RT \ln V_R^o + K = N_A \cdot a_m \cdot W_A \tag{8.8}$$

where R is the universal gas constant, K a constant, N_A the Avogadro's number, a_m the cross sectional surface area of the adsorbed molecule and $\mathbf{W}_A$ is the work of adhesion. It is from the retention volume that a range of thermodynamic properties can be determined, including surface energy. The dispersive surface energy is typically determined using the Schultz approach or the Dorris-Gray approach, and acid-base using the Gutmann concept or van Oss-Good-Chaudhary approach (based on monopolar probes) [7].

IGC has been used to study a range of materials; catalysts, food and cosmetics including pharmaceuticals. The robustness of the IGC technique is exemplified by the increasing wide scope and range of applications including for pharmaceutical solids. This information could be used to evaluate performance or stability of products as well as suitability of pharmaceutical delivery systems.

8.2.5 Dynamic Vapour Sorption

The reduction in solid surface energy due to the adsorption of a vapour phase is known as the equilibrium spreading pressure, π_e. The work of adhesion as described by the Young-Dupre assumes a negligible spreading pressure ($\pi_e = 0$). Such an assumption is widely accepted for the case of partial wetting when contact angles, θ are greater than 10°. However, for liquids which completely wet a solid ($\theta = 0^o$), the spreading pressure may be significant and work of adhesion can be described as follows:

$$W_A = 2\gamma_{LV} + \pi_e = 2\sqrt{\gamma_S^d \gamma_L^d} + 2\sqrt{\gamma_S^p \gamma_L^p} \tag{8.9}$$

From the Gibbs surface tension equation, the spreading pressure can be calculated by:

$$\pi_e = \frac{RT}{\sigma} \int \Theta \cdot d\ln p \tag{8.10}$$

where σ is the specific surface area and Θ is the total amount adsorbed.

Adamson and co-workers [8], in their pioneering work employed the sensitivity of the ellipsometry technique for measurements of vapour film thickness to calculate the π_e.

The introduction of a highly sensitive and accurate flow control for determination of sorption isotherms rapidly, has enabled such measurements of higher precision to be obtained. In turn, properties such as the surface energy of powders can also be determined with confidence. Therefore, by measuring the adsorption isotherm of the vapour for a completely wetting liquid, the π_e can be determined and hence W_A, enabling calculation of the surface energy of the solid.

8.2.6 Atomic Force Microscopy

The basic principle of the Atomic Force Microscopy is to allow a small probe to scan across a sample of interest, and permits the measurement of surface features by observing the probe-sample interactions, which is capable of imaging features down to the atomic scale.

The adhesion of a particle on a solid surface occurs via a range of complex intermolecular forces. The two most commonly used models for the approximation of such interactions are the Johnson-Kendall-Roberts (JKR) [9] and Derjaguin-Muller-Toporov (DMT) [10] approaches. The measurement of the adhesion force, F_{Adh} by AFM has been utilised by several investigators to study surface energetics of pharmaceutical solids.

8.3 Determining Surface Energy from Contact Angles

8.3.1 Zisman's Approach

Partial wetting ($\theta > 0^o$) and non-wetting ($\theta > 90^o$) occurs when a liquid with a higher surface tension interacts with a solid of lower surface energy, and wetting ($\theta = 0^o$) occurs for the reverse case. It can be expected that contact angles, θ be greater for liquids of higher surface tension, and vice versa. The hypothetical scenario when a liquid just about wets the solid ($\theta \to 0^o$) was introduced by Zisman as the concept of the critical surface energy, γ_C [11]. At this critical point, the solid surface energy is thought to be equivalent to the surface tension of liquid. Contact angle of near 0° may be difficult to measure, consequently the γ_C is typically extrapolated from a plot of $\cos\theta$ versus γ_{LV} to $\cos\theta = 1$ ($\theta = 0^o$). A smooth trend in the cosine of the contact angle versus liquid surface tension was observed. This relationship can be described by Eq. 8.11:

$$\cos\theta = 1 - b_Z(\gamma_{LV} - \gamma_{SV}) \tag{8.11}$$

where b_Z is the slope of the regression line.

8.3.2 Component's Approaches

Fowkes proposed that surface energy of a solid can be expressed as a summation of a number of independent surface energy components, each arising for specific type of intermolecular forces [12]. The surface energies are normally grouped into the dispersive and non-dispersive components as follows:

$$\boldsymbol{\gamma}_{SV} = \boldsymbol{\gamma}_{SV}^{d} + \boldsymbol{\gamma}_{SV}^{h} + \boldsymbol{\gamma}_{SV}^{dp} + \boldsymbol{\gamma}_{SV}^{i} + \boldsymbol{\gamma}_{SV}^{m} = \boldsymbol{\gamma}_{SV}^{d} + \boldsymbol{\gamma}_{SV}^{nd} \tag{8.12}$$

where $\boldsymbol{\gamma}^{d}{}_{SV}$ and $\boldsymbol{\gamma}^{nd}{}_{SV}$ are the dispersive and non-dispersive components, respectively. $\boldsymbol{\gamma}^{h}{}_{SV}$, $\boldsymbol{\gamma}^{dp}{}_{SV}$, $\boldsymbol{\gamma}^{i}{}_{SV}$ and $\boldsymbol{\gamma}^{m}{}_{SV}$ are the surface energy components due to hydrogen bonding, dipole-dipole interactions, induction force and metallic interactions respectively. Dispersive forces (due to London forces) exist for all molecules whilst non-dispersive forces occur only between molecules with permanent dipole moments. Fowkes also assumed that for interfacial interactions, thermodynamic terms such as the work of adhesion, $\mathbf{W}_A$ could likewise be approximated as a simple sum of the independent terms, each relating to a specific type of intermolecular interactions.

A range of semi-empirical approaches were subsequently developed by Fowkes and other workers which allows the various components of $\mathbf{W}_A$, dispersive or acid-base interactions for example, to be calculated.

The non-bonding interactions between molecules are made up by the London, Debye and Keesom forces. Fowkes estimated the interaction between purely dispersive molecules to be a geometric mean based on Berthelot's principle. By using a purely dispersive liquid ($\gamma^{nd}{}_{LV} = 0$), the interfacial interactions can be described as:

$$\gamma_{SL} = \gamma_{SV} + \gamma_{LV} - 2\sqrt{\gamma_{SV}^{d}\gamma_{LV}^{d}} \tag{8.13}$$

Combining Eq. 8.13 with the Young's equation yields:

$$\cos\theta = 2\sqrt{\gamma_{SV}^{d}} \cdot \frac{1}{\sqrt{\gamma_{LV}^{d}}} - 1 \tag{8.14}$$

By plotting cos θ versus $1/(\gamma^{d}{}_{LV})^{1/2}$ with an intercept of -1, the dispersive energy for a solid is obtained from the slope of the line.

Owens and Wendt [13] extended Fowkes' equation by grouping the non-bonding London, Debye and Keesom interactions into a similar term which they called dispersive (also recognised as the Lifshitz-van der Waals interactions) while the remaining terms were grouped as the polar contributions.

Wu proposes a harmonic mean relationship for the interfacial tension could describe low surface energy surfaces more accurately, based primarily on the mathematical fits. Arithmetic, quadratic and anharmonic means were also proposed by others but did not gain popularity.

8.3.3 Acid–Base Approaches

van Oss et al. proposed an acid-base approach by the combination of London, Debye and Keesom interactions into the apolar Lifshitz-van der Waals component,

γ^{LW} and an acid-base component, γ^{AB} [14, 15]. However, the short-range interactions were divided into acid, γ^{+} and base, γ^{-} parameters. They assumed that the acid-base interactions between phases were dissimilar, where the acid part of one phase interacts with the basic part of the other. The total surface energy is given by:

$$\gamma_i = \gamma_i^{LW} + 2\sqrt{\gamma_i^{+}\gamma_i^{-}} \tag{8.15}$$

where i denotes solid or liquid phase, and γ_i^{+} the electron acceptor (Lewis acid) and γ_i^{-} the electron donor (Lewis base) components. Combining Eq. 8.15 with the Young's equation yields:

$$\gamma_{LV}(1 + \cos\theta) = 2\sqrt{\gamma_{SV}^{LW}\gamma_{LV}^{LW}} + 2\sqrt{\gamma_{SV}^{+}\gamma_{LV}^{-}} + 2\sqrt{\gamma_{SV}^{-}\gamma_{LV}^{+}} \tag{8.16}$$

Three liquids with known surface tension components, of which at least two are polar (with γ^{+} and γ^{-}), are needed to calculate the surface energy components of solid. Whilst this approach yields the most information of a surface, liquid properties of different scales proposed by van Oss et al. (γ_{H2O}^{+}:$\gamma_{H2O}^{-} = 1$) and Della Volpe and Siboni (γ_{H2O}^{+}:$\gamma_{H2O}^{-} = 6.5$) [16] yield results differing in magnitude, whilst the order or acidity/basicity is often unchanged.

The van Oss et al. scale utilised water as reference with γ^{+}: γ^{-} ratio equal to unity. Evaluation of surface energies with the van Oss scale often predicts an excess basic property. Della Volpe proposed a different ratio suggesting that water was not amphoteric but acidic. The acid-base approach of van Oss et al. assumes that W_A^{AB} is always greater than zero while the Chang-Chen model allows for both attractive and repulsive interactions [17].

8.3.4 Equation of State Models

In contrast to the components approach, Neumann et al. [18] proposed an equation of state approach. The equation of state approaches stipulates that interaction between two phases is governed solely by liquid and solid surface energies, and that the measurable contact angle is only a function of these two variables. As such, surface energies are not made up of different components and contact angle measurements do not reflect local surface chemical information. This approach was modelled purely from a thermodynamic perspective. Employing Berthelot rule, a geometric mean combining rule of the potential energy parameters of like-pair interactions, the unlike-pair interactions can be estimated by

$$\varepsilon_{SL} = \sqrt{\varepsilon_{SS}\varepsilon_{LL}}$$

Li et al. proposed a modification to the geometric mean rule, to take into account the over-predictions of unlike-pair interactions, by introducing an empirical parameter [19]. This parameter should be a decreasing function to equal unity when potential energy parameters of like-wise pairs are equal. The interfacial tension is therefore described as:

$$\gamma_{SL} = \gamma_{SV} + \gamma_{LV} - 2\sqrt{\frac{\gamma_{SV}}{\gamma_{LV}}} \cdot e^{-\beta(\gamma_{LV}-\gamma_{SV})^2} \quad (8.17)$$

where β is an empirical constant with an average value of 1.057 x 10^{-4} (m^2 mJ^{-1})2. Combining Eq. 8.17 with the Young's equation yields:

$$\cos\theta = -1 + 2\sqrt{\frac{\gamma_{SV}}{\gamma_{LV}}} \cdot e^{-\beta(\gamma_{LV}-\gamma_{SV})^2} \quad (8.18)$$

The equation of state approach enables calculation of the surface energy of a solid by measuring the contact angle of a single liquid, with a known surface tension, on a solid surface.

8.3.5 *Non-ideal Surfaces*

The classification of an ideal surface is one which is molecularly smooth and homogenous while the liquid is non-reactive to the solid surface. However, in the case of real solid surfaces, which are often far from the idealised case, experimental observations have revealed that various stable regimes exist resulting in a maximum and a minimum contact angle corresponding to the advancing, θ_a and receding, θ_r contact angles respectively. The difference between θ_a and θ_r is known as hysteresis.

The analysis of the effects of surface roughness on contact angles by Wenzel simply takes into account the additional area to be wetted by the probe liquid on a rough surface compared to a smooth surface [20]. Wenzel proposed a roughness factor r_W which is defined as the ratio of the total surface area to be wetted due to roughness, to the projected surface area. Wenzel proposed a model to include this roughness effect on the equilibrium contact angle as shown:

$$r_W \cdot \cos\theta_Y = \cos\theta_W \quad (8.19)$$

where θ_Y is the Young equilibrium contact angle and θ_W is the measurable Wenzel contact angle due to the surface roughness.

For heterogenous surfaces, the Cassie-Baxter equation describes the apparent contact angle as a function of the surface area fractions of each component. For a two component surface where the liquid fully wets the entire surface, the Cassie-Baxter equation is written as [21]:

$$\cos\theta_i = \alpha_1 \cos\theta_{i1} + \alpha_2 \cos\theta_{i2} \tag{8.20}$$

where α_1 and α_2 are the surface area fractions of components 1 and 2, respectively while *i* denotes that the contact angle could be for the advancing or receding case.

Contact angle hysteresis is evident in cases where heterogeneity, physical or chemical. It has been suggested that the hysteresis may be reproducible, and hence may actually represent the surface chemistry of the solid. Chibowski used this viewpoint to propose a method of evaluating the solid surface energy from θ_a and θ_r data. In the model, it is assumed that the receding angle represents the surface energy of a film left behind from the advancing drop. Detailed development of this model is found elsewhere [22]. The equation proposed is as follows:

$$\gamma_{SV} = \gamma_{LV} \cdot (\cos\theta_r - \cos\theta_a)\frac{(1+\cos\theta_a)^2}{(1+\cos\theta_r)^2 - (1+\cos\theta_a)^2} \tag{8.21}$$

8.4 Concluding Remarks

Unlike liquid surfaces, the direct surface energy measurement for solids is not viable as solid surfaces are rigid. Almost all indirect approaches involve the use of known vapours, liquids or solids as probes for the solid materials under investigation. The knowledge of the interaction between the solid and the external probe may be analysed via an appropriate model in terms of properties of the solid of interest. Care should be taken in selection of method and model for the determination of surface energetics.

References

1. Young T (1805) An essay on the cohesion of fluids. Philos Trans R Soc Lond 95:65–87
2. Kwok DY, Neumann AW (1999) Contact angle techniques and measurements. In: Milling AJ (ed) Surface characterization methods: principles, techniques and applications. Marcel Dekker Inc., New York
3. Bartell FE, Osterhof HJ (1927) Determination of the wettability of a solid by a liquid. Ind Eng Chem 19:1277–1280
4. Heertjes PM (1971) The determination of interfacial tensions with the Wilhelmy plate method. Chem Eng Sci 26:1479–1480
5. Papirer E, Balard H (1999) In: Pefferkorn E (ed) Interfacial phenomena in chromatography. Marcel Dekker Inc, New York
6. Conder JR, Young CL (1979) Physicochemical measurement by gas chromatography. Wiley, Chichester
7. Ho R, Heng JYY (2013) A review of inverse gas chormatography and its development as a tool to characterize anisotropic surface properties of pharmaceutical solids. KONA Powder Part J 30:164–180

8. Adamson AW (1968) An adsorption model for contact angle and spreading. J Colloid Interface Sci 27:180–187
9. Johnson KL, Kendall K, Roberts AD (1971) Surface energy and the contact of elastic solids. Proc R Soc Lond A 324:301–313
10. Derjaguin BV, Muller VM, Toporov YP (1975) Effect of contact deformations on the adhesion of particles. J Colloid Interface Sci 53:314–326
11. Ellison AH, Zisman WA (1954) Wettability studies on nylon, polyethylene terephthalate and polystyrene. J Phys Chem 58:503–506
12. Fowkes FM (1962) Determination of interfacial tensions, contact angles, and dispersion forces in surface by assuming additivity of intermolecular interactions in surfaces. J Phys Chem 66:382
13. Owens DK, Wendt RC (1969) Estimation of the surface free energy of polymers. J Appl Polym Sci 13:1741–1747
14. van Oss CJ, Good RJ, Chaudhury MK (1988) Additive and nonadditive surface tension components and the interpretation of contact angles. Langmuir 4:884–891
15. van Oss CJ (1994) Interfacial forces in aqueous media. Marcel Dekker Inc., New York
16. Volpe CD, Siboni S (1997) Some reflections on acid–base solid surface free energy theories. J Colloid Interface Sci 195:121–136
17. Chen F, Chang WV (1991) Applicability study of a new acid base interaction model in polypeptides and polyamides. Langmuir 7:2401–2404
18. Neumann AW, Good RJ, Hope CJ, Sejpal M (1974) An equation-of-state approach to determine surface tensions of low-energy solids from contact angles. J Colloid Interface Sci 49:291–304
19. Li D, Neumann AW (1992) Equation of state for interfacial tensions of solid-liquid systems. Adv Colloid Interf Sci 39:299–345
20. Wenzel RN (1936) Resistance of solid surfaces to wetting by water. Ind Eng Chem 28:988–994
21. Cassie ABD (1948) Contact angles. Discuss Faraday Soc 3:11–16
22. Chibowski E (2003) Surface free energy of a solid from contact angle hysteresis. Adv Colloid Interf Sci 103:149–172

Chapter 9
Crystal Effects Influencing the Course of Organic Solid State Reactions: Perfect, Imperfect and Surface Effects

William Jones

Abstract Chemical reactivity in the solid state is an important topic with regards to the exploitation of organic solids. The relatively immobile nature of the molecules in a crystal can lead to highly selective conversions and be exploited in the area of chemical synthesis. It can also, using crystal engineering methods, be used to provide chemical stability, isolating potentially reactive molecules in an environment where conformational and configurational changes required for conversion become impossible. While such effects will be present in both amorphous and crystalline materials this chapter reviews the implications for crystalline solids. It considers the role of the perfect lattice (as revealed by conventional single crystal diffraction methods) as well as the nature and implications of defects (planar, linear and point as well as the terminating faces of a crystal) as possible nucleation sites for transformations to occur.

Keywords Defects • Reactivity • Topochemistry • AFM • Electron microscopy

9.1 Introduction

The way in which reactions occur in organic solids has long been of interest. Significant progress in terms of how crystal packing influences reactivity, however, required the development of structure determination using single crystal X-ray diffraction in a suitably routine way so that systematic studies of related solids could be performed. The most cited example is that of the cinnamic acids which was developed in the 1960s. The main conclusion was that reaction could occur in the (perfect) crystal if molecules were suitably arranged for diffusionless reaction. The phrase "topochemical control" was coined to describe such processes. Shortly afterwards, however, reactions which apparently contravened this interpretation

W. Jones (✉)
Department of Chemistry, University of Cambridge, Lensfield Road,
Cambridge CB2 1EW, UK
e-mail: wj10@cam.ac.uk

K.J. Roberts et al. (eds.), *Engineering Crystallography: From Molecule to Crystal to Functional Form*, NATO Science for Peace and Security Series A: Chemistry and Biology, DOI 10.1007/978-94-024-1117-1_9

were identified and the hypothesis developed that perhaps structural imperfections within the crystal might be the sites at which "non-topochemical" reaction took place. Less well developed has been an assessment of how reactions at the surface of a crystal might differ from that of the bulk (perfect or imperfect). In a similar manner to the need for routine surface studies required the development of a technique with better resolution than that provided by optical microscopy. The development of surface probe microscopy and in particular atomic force microscopy has allowed progress in this area.

9.2 Topochemical Control – Perfect Packing Dominates

In a series of papers, Schmidt, Cohen and colleagues [1, 2] established that the photochemical reactivity of crystals of cinnamic acid and various cinnamic acid derivatives was a function of the molecular registry as revealed by single-crystal diffraction studies. The crystal structures of numerous substituted cinnamic acids can be categorised into three general types – α, β and γ. Some derivatives can adopt multiple different crystal structures – for example, o-ethoxy-trans-cinnamic acid is polymorphic and crystallises in all three structural types. In each of the structure types, the molecules pack in one dimensional stacks and form pairwise hydrogen bonding interactions across centres of symmetry. Within the stacks the molecules lie parallel with a distance of the order of 3.5 Å between molecular planes. These three types differ, however, in the angle between this normal and the stack axis, in the repeat distance along this axis and in the extent and type of overlap between adjacent molecules in the stack. The β-type resulted in the mirror dimer; the α-type the centrosymmetric dimer; whilst the γ-type was light stable – see Fig. 9.1.

As an extension of this lattice controlled dimerization of cinnamic acids the technique of solid-state polymerisation was developed. Hirschfeld and Schmidt [3] suggested that the necessary condition for matrix controlled polymerisation is that no significant disruption of the molecular positions occurred on polymerisation. They referred specifically to crystals of molecules with two potentially reactive centres which were so oriented that they could rotate in place to link up with their neighbours without any linear displacement of the molecular centres – see Fig. 9.2. Two groups of materials that explored this concept were the diacetylenes and divinyl monomers. In the case of divinyl monomers, an important example was that of 2,5-distyrylpyrazine, which readily polymerised in the solid state [4] – with the final crystalline photo-product having a molecular weight several times greater than the solution grown polymer. This could be rationalised on the basis of packing within the perfect lattice. The work of Wegner and co-workers [5] demonstrated that solid state polymerisation of diacetylenes (thermal and photolytic) could also be rationalised on the basis of crystal packing in the perfect lattice, the polymerisation proceeding to high molecular weight products with little change in lattice parameters.

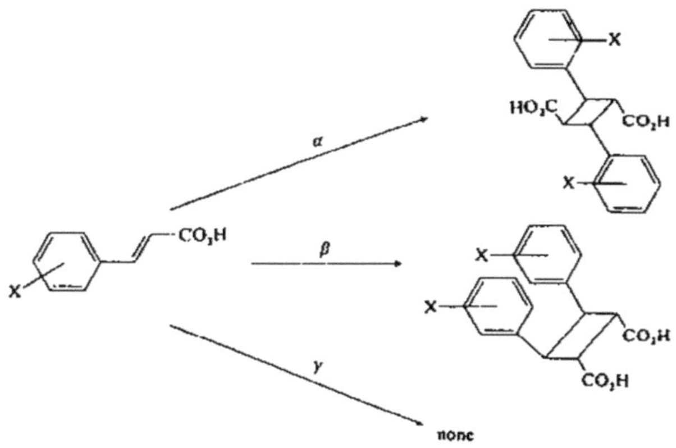

Fig. 9.1 Dimerisation of cinnamic acids

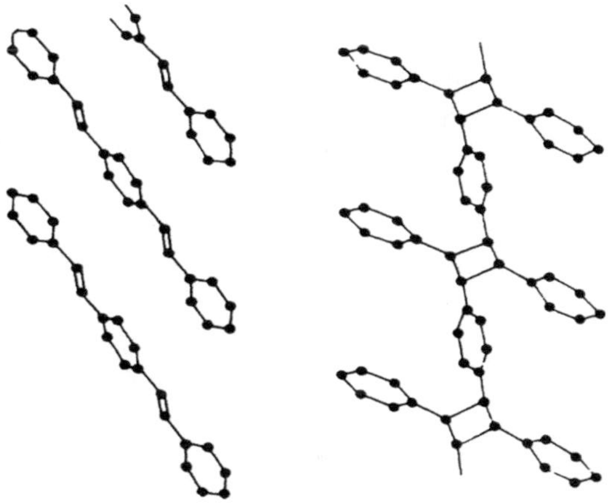

Fig. 9.2 Polymerisation reaction

9.3 Non-topochemical Reactions – Types of Defects in Organic Crystals and Their Possible Role

A recognised challenge, however, was found in the abnormal solid-state photochemical reactions of anthracene and 9-cyanoanthracene (amongst others) [6]. In these systems, analysis of the perfect lattice would suggest stability as molecules are inappropriately arranged to react (either by orientation or separation). Despite this, reactions did in fact take place. One explanation was that those reactions which gave the "non-topochemical" product occurred preferentially at defect sites in the crystal – see Figs. 9.3 and 9.4 [7–10].

This was supported by the work of Thomas and his group [11, 12] who separately etched and photodimerized matched halves of cleaved melt-grown anthracene crystals (Fig. 9.4). They found that the degree of correspondence between etch pits and dimerization centres was usually greater than 85%. For 9-cyananoanthracene the observed product was further rationalised in terms of

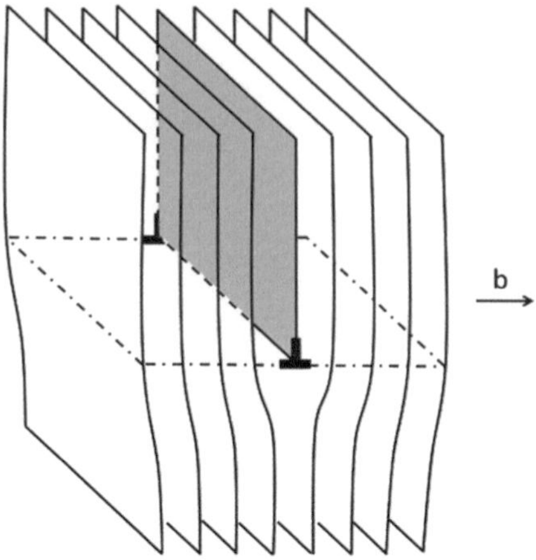

Fig. 9.3 Representation of an edge dislocation

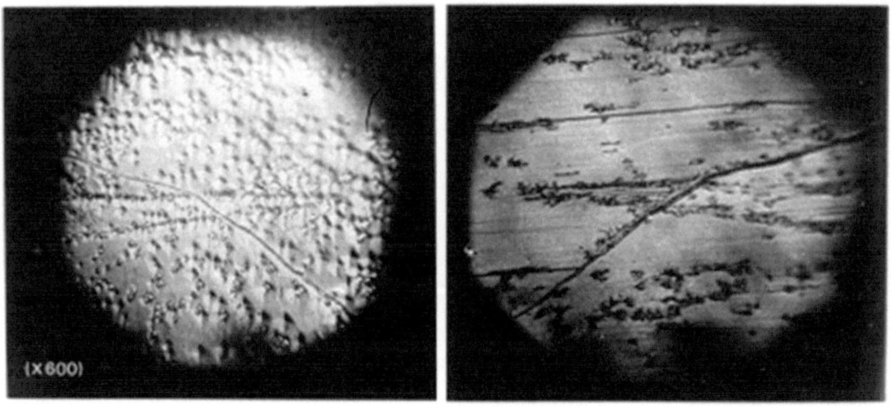

Fig. 9.4 Defect and product formation

stacking faults in the crystals which were generated by the dissociation of [100] (221) dislocations (Fig. 9.5) [11].

Recent micro-Raman studies of the photodimerisation of 9-cynanoanthracene has demonstrated that there is a delay between the onset of chemical reaction and the appearance of the product crystals – the initial reaction can occur within the crystal but there is no diffraction evidence until a reconstructive phase transition occurs [13].

The presence of crystalline imperfections in a solid may lead both to local changes in topology at the imperfection and to changes in energy owing to any strain set up or relieved by the imperfection [9]. The chemical potential of a molecule associated with an imperfection will be different to that of molecules in the perfect lattice. In order to anticipate the role that imperfections will play in the properties of an organic solid, it will be necessary to establish completely their character. Some salient properties of imperfections in general, and in particular

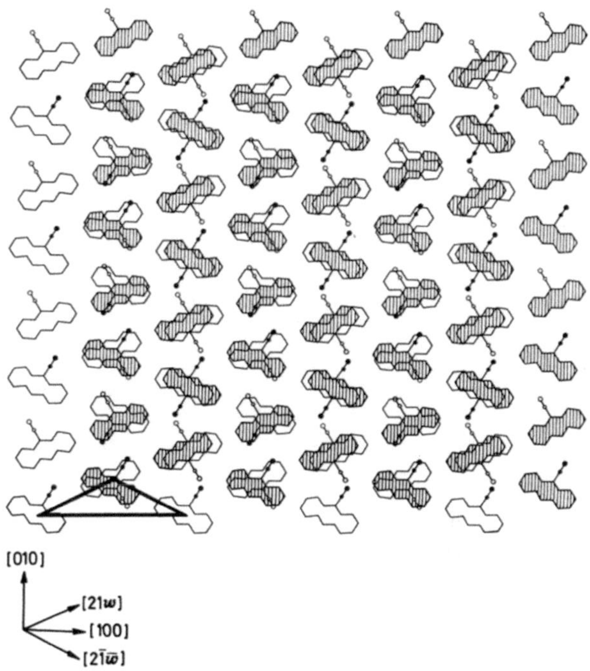

Fig. 9.5 Stacking fault in 9-cyanoanthracene leading to appropriate orientation for reaction [11]

point defects, linear (dislocations) and planar defects (e.g. stacking faults) are now given. We also note that the concept of a volume defect might apply where solvent occlusion is also present.

Structural imperfections can be classified into three general types based on geometry: point, line and planar.

(i) Point defects: In a pure material there are several types of point defects. Vacancies exist where atoms or molecules are missing from their normal positions in the crystal lattice. Molecules may also occupy non-lattice (interstitial) sites (Fig. 9.6). At temperatures above 0 K there is always present an equilibrium number of vacancies in a crystal with the number increasing as the temperature increases. During crystal growth, a quantity of these defects in excess of the equilibrium number may be introduced either through rapidly quenching from high temperatures, through plastic deformation or through radiation damage.

Line Imperfections: Line imperfections, or dislocations, are linear regions of elastic distortion caused by atoms that are slightly displaced from their equilibrium lattice positions. The two basic types of dislocations are the edge dislocation and the screw dislocation [9]. An example of an edge dislocation was given in Fig. 9.3. If the dislocation line AB were imagined to bend in the slip plane, then it will be possible for the dislocation line to run parallel to the direction of the Burgers vector

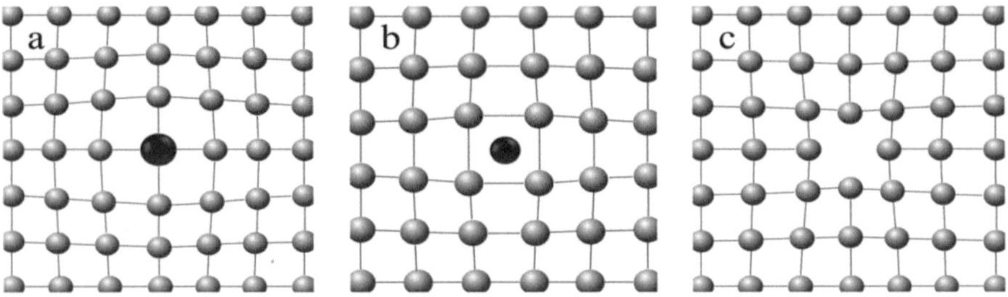

Fig. 9.6 Various types of point defects in an idealised atomic-array

of the dislocation. This condition describes the screw dislocation (Fig. 9.7). The Burgers vector of a screw dislocation is parallel to the line of the dislocation. The structural error that arises in the screw dislocation is a spiral ramp of displaced atoms that advances one Burgers vector per circuit of the dislocation.

(ii) Planar imperfections: Planar imperfections are any surfaces across which the atom positions in a perfect crystal are not preserved. Two common planar faults found in crystals are twin planes and stacking faults. Another common defect will be grain boundaries – the region between randomly oriented domains within the crystal. Twin planes separate the two parts of a twinned crystal, the parent and its twin – see Fig. 9.8.

Transmission electron microscopy (TEM) is particularly powerful for imaging dislocations in organic crystals [14]. As an example see Fig. 9.9.

9.4 Atomic Force Microscopy and Comparison of Surface Reactivity Compared to Bulk

AFM is a member of the family of scanning tunnelling microscopy (STM) in that it relies on a physical probe to explore the surface of interest. [16, 17] The AFM "feels" atomic forces, such as Van der Waals or electrostatic interactive forces between the tip and the surface, by the bending of the cantilever on an AFM probe. See Figs. 9.10 and 9.11. This surface sensitivity allows molecular height effects to be observed [17].

Besides imaging individual surfaces, it is also crucial to understand the surface changes of molecular crystals during chemical reactions or in response to environmental alterations. One example of *ex situ* imaging is offered by the study of (caffeine)·(oxalic) acid and (caffeine)·(malonic) acid cocrystals, which were demonstrated by Trask et al. to be stable in 98% RH and 75% RH respectively by PXRD [18]. Cassidy et al. monitored the surface response of these cocrystals by *ex situ* AFM after exposure to these humidities (Fig. 9.12) [19]. Caffeine and oxalic acid

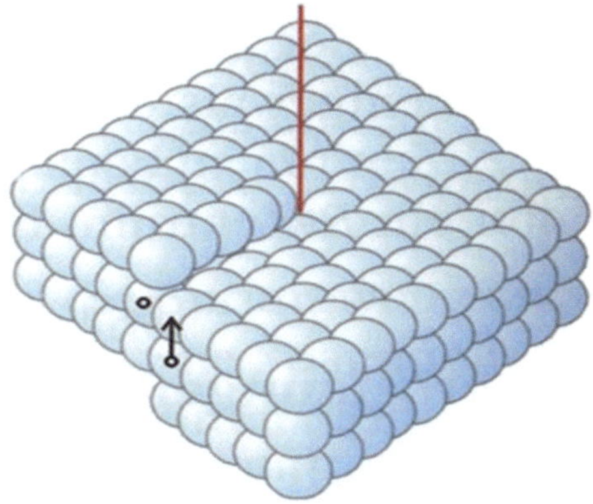

Fig. 9.7 Schematic of a screw dislocation

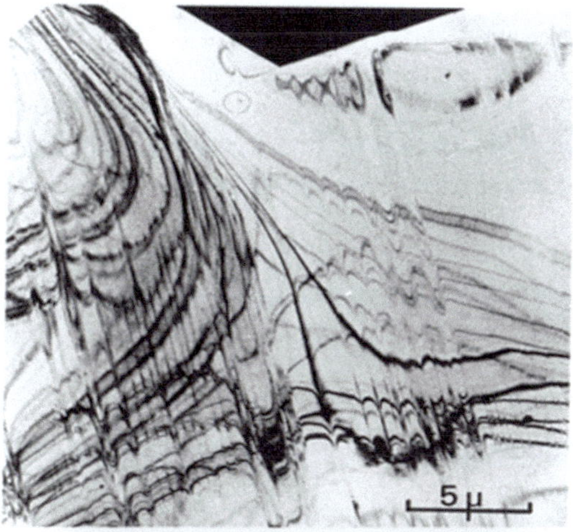

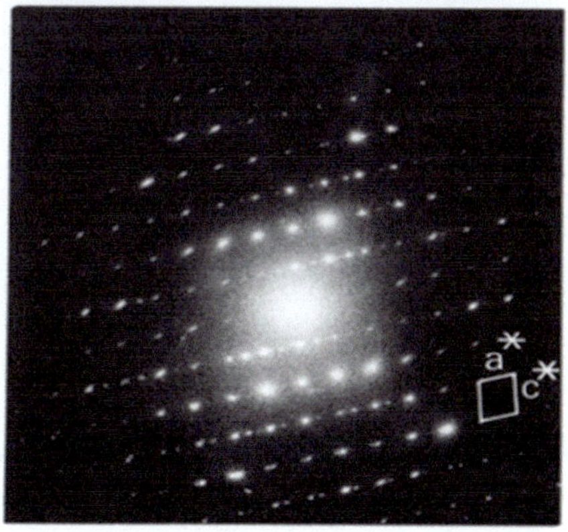

Fig. 9.8 Microtwins in an organic crystal and associated diffraction pattern

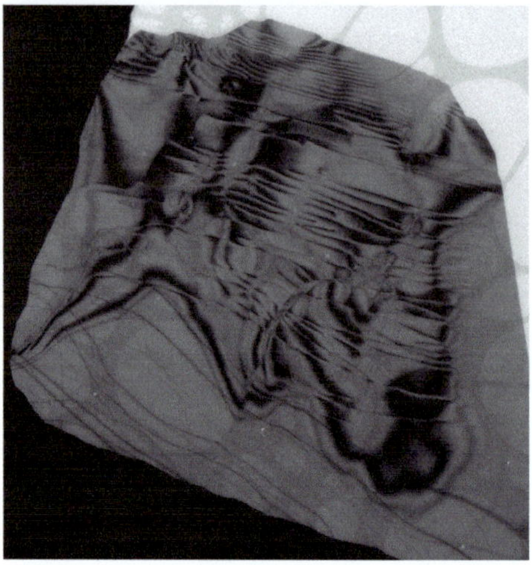

Fig. 9.9 Large number of dislocations in a sample of theophylline as seen by TEM [15]

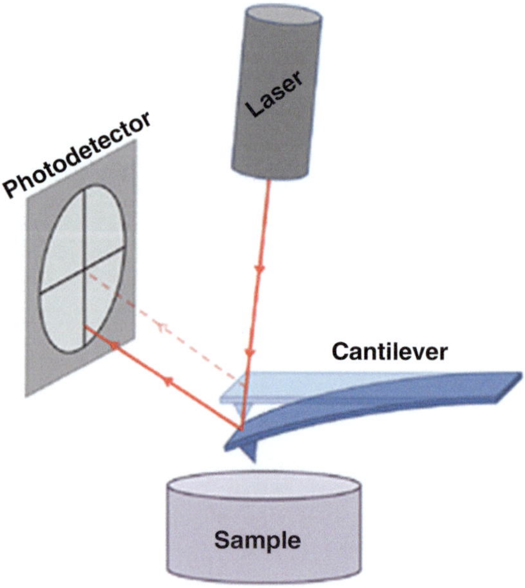

Fig. 9.10 Laser, photodetector and cantilever [17]

were found to exhibit a high degree of mobility on cocrystal surfaces that facilitates recrystallisation events on the surface. This study demonstrated that the information about surface changes that are not apparent from bulk measurements (such as powder X-ray diffraction) can be detected by AFM, and may be important for predicting the stability of active pharmaceutical ingredients (API), as well as at excipient solid-state interfaces.

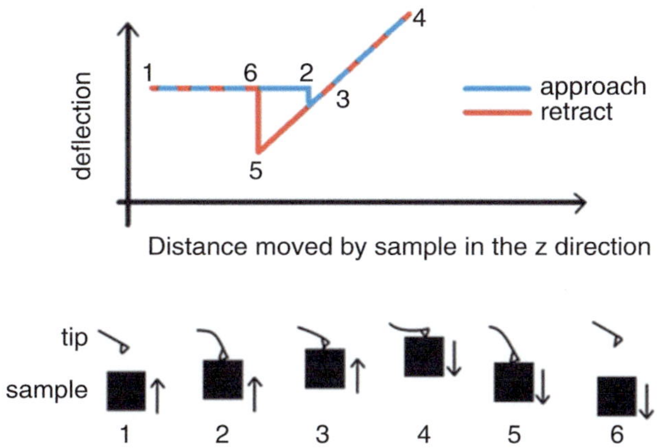

Fig. 9.11 Various stages of tip-surface contact

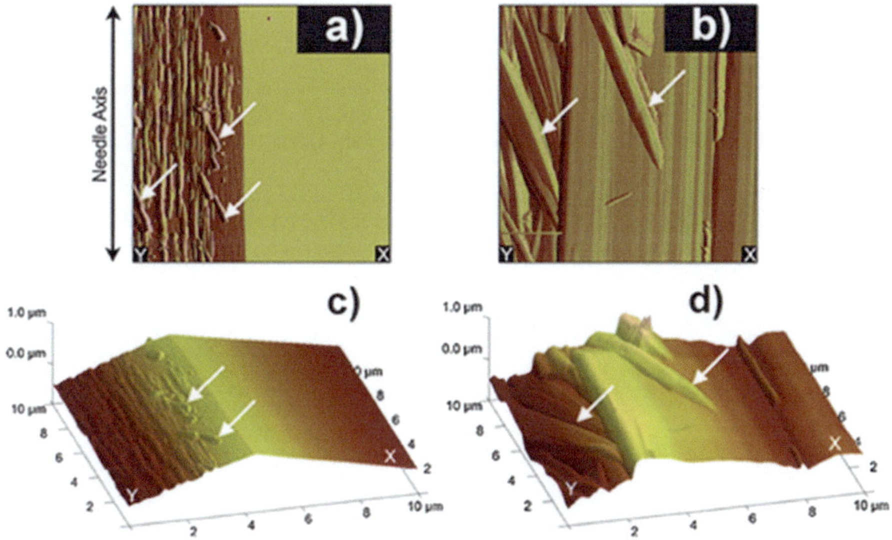

Fig. 9.12 Changes in surface structure of a crystal as a result of exposure to high humidity [19]

References

1. Schmidt GMJ (1971) Photodimerisation in the solid state. Pure Appl Chem 27:647–678
2. Cohen MD, Schmidt GMJ (1964) Topochemistry Part II The photochemistry of trans-cinnamic acid. J Chem Soc:1996–2000
3. Hirschfeld FL, Schmidt GMJ (1964) Topochemical control of solid state polymerisation. J Polym Sci 2:2181–2190

4. Nakanishi H, Hasegawa M, Sasada Y (1972) 4-center type photopolymerisation in crystalline state 5 X-ray crystallographic study of polymerisation of 2,5-distyrylpyrazine. J Polym Sci Part A-2(10):1537–1553
5. Wegner G (1977) Solid state polymerisation reactions. Pure Appl Chem 49:443–454
6. Craig DP, Sarti-Fantoni P (1966) Photodimerisation in crystalline anthracenes. Chem Commun:742–743
7. Tilley RD (1998) Principle and applications of chemical defects. Stanley Thorne Publishers, Cheltenham
8. Byrn SR (1982) Solid-state chemistry of drugs. Academic Press, New York
9. Hull D, Bacon DJ (2011) Introduction to dislocations, 5th edn. Elsevier, Amsterdam
10. Sherwood JN (2004) Fifty years as a crystal gazer: life as an imperfectionist. Cryst Growth Des 4:863–877
11. Thomas JM (1974) Topography and topology in solid-state chemistry. Phil Trans Roy Soc A 277:251–287
12. Thomas JM, Williams JO (1972) In surface and defect properties of solids. R Soc Chem 1:30–143
13. Salzillo T, Zaccheroni S, Valle D, Venuti E, Brillante A (2014) Micro Raman investigation of the photodimerization reaction of 9-cyanoanthracene in the solid state. J Phys Chem C 118:9628–9635
14. Jones W, Thomas JM (1979) Applications of electron microscopy to organic solid-state chemistry. Prog Solid State Chem 12:101–124
15. Eddleston MD, Bithell EG, Jones W (2010) Transmission electron microscopy of pharmaceutical materials. J Pharm Sci 99:4072–4083
16. Eaton P, West P (2010) Atomic force microscopy. Oxford University Press, New York
17. Chow EHH, Bučar D-K, Jones W (2012) New opportunities in crystal engineering – the role of atomic force microscopy in studies of molecular crystals. Chem Commun 48:9210–9226
18. Trask AV, Motherwell WDS, Jones W (2005) Pharmaceutical cocrystallization: engineering a remedy for caffeine hydration. Cryst Growth Des 5:1013–1021
19. Cassidy AMC, Gardner CE, Jones W (2009) Decoupling the effects of surface chemistry and humidity on solid-state hydrolysis of aspirin in the presence of dicalcium phosphate dihydrate. Int J Pharm Res 379:59–66

Chapter 10
Synthonic Engineering Modelling Tools for Product and Process Design

Jonathan Pickering, Robert B. Hammond, Vasuki Ramachandran, Majeed Soufian, and Kevin J. Roberts

Abstract A new software implementation built into the Cambridge Crystallographic Data Centre's (CCDC) Mercury suite is described. VisualHabit is a tool that calculates intermolecular interaction energies for crystal structures using atomistic potentials. Specific types of directed interatomic-interactions, e.g. such as constitute hydrogen bonds, can be quantified in terms of an interaction energy both within the crystal lattice, the intrinsic synthons, and at the terminating surfaces of crystals, thereby mediating their interaction with the surroundings, the extrinsic synthons. Lattice energy is calculated and crystal shape predicted by application of the attachment energy model. The distribution of interaction energy of probe molecules, such as common solvents, with selected crystal surfaces can be interrogated using SystSearch, a systematic search tool, in conjunction with VisualHabit.

Keywords Attachment energy • Crystal chemistry • Crystal engineering • Crystal morphology • Crystal shape • Crystal system • Crystallisation energy • Lattice energy • Intermolecular interactions • Miller/Crystal planes • Slice energy • Synthons

List of Symbols

(hkl)	*Miller plane* – two dimensional surface cut through lattice
a, b, c	unit cell lengths
α	angle between b and c
β	angle between a and c

J. Pickering (✉) • R.B. Hammond • V. Ramachandran • M. Soufian • K.J. Roberts
School of Chemical and Process Engineering, University of Leeds, Leeds LS2 9JT, UK
e-mail: j.h.pickering@leeds.ac.uk; r.b.hammond@leeds.ac.uk; v.ramachandran@leeds.ac.uk; k.j.roberts@leeds.ac.uk

K.J. Roberts et al. (eds.), *Engineering Crystallography: From Molecule to Crystal to Functional Form*, NATO Science for Peace and Security Series A: Chemistry and Biology, DOI 10.1007/978-94-024-1117-1_10

γ	angle between a and b
E_{cr}	lattice energy
E_{sl}	slice energy
E_{att}	attachment energy
τ	shape factor
A	surface area of crystal habit
V	volume of crystal

List of Abbreviations

CSV	comma separated value
SYBYL©	The name of a computational informatics software suite, available from Certara

10.1 Introduction

It has been estimated [1] that 60% of all manufactured chemical products have processing steps involving the solid form and it may be anticipated that the proportion of high value-added, speciality materials which are solids could be significantly greater still. Further improvements in the quality and efficiency of production, such as the Six Sigma program proposed for the pharmaceutical industry [2], may depend on the ability to engineer the properties of crystalline solids, which in turn requires a quick and reliable means of predicting properties, such as the chemical stability, adhesion and dissolution together with mechanical properties. These can be derived both from the chemistry of the constituent molecules and the lattice into which they are packed. This chapter describes the development of a software package that predicts crystal properties based on the interaction energies of the molecules within the crystal lattice.

Molecular crystals can be regarded as ordered supramolecular structures, and any ability to predict the properties of these structures will require knowledge of the spatial arrangements and energy of the intermolecular interactions, which hold the crystal together. In the context of supramolecular structures the concept of supramolecular-synthons was introduced by Desiraju [3], and a refined definition as the "spatial arrangements of intermolecular interactions" was given by Thalladi [4]. Synthonic Engineering (SE), or synthonic modelling, enables the physicochemical properties of crystals to be predicted from analysis of the synthons. This allows molecular-scale design and control of the physicochemical properties.

Synthons consist of isotropic intermolecular van der Waals and electrostatic interactions combined with highly-directional hydrogen bonds. For a pair of molecules in a given configuration the synthon (inter-molecular energy) can be found by summing the atom pair interactions as calculated using an inter-atomic potential

[3, 5, 6]. The approach can be applied to crystals by considering a central molecule and summing the molecular interaction energy with all the molecules within a given convergence radius. If the energy converges it should represent the lattice energy, which is closely equivalent to the sublimation enthalpy of the material.

Synthonic analysis requires knowledge of the basic chemistry of the molecules together with details concerning the crystal lattice and its associated molecular arrangement. Such details can be derived from the crystallising material's crystallographic structure using molecular modelling techniques. The combination of a need to access and manipulate crystallographic data and the provision of a usable tool for the process engineers led to implementation as a new plug-in to the Cambridge Crystallographic Data Centre's (CCDC) existing Mercury product [7–10].

The applications of SE tools include: predicting the morphology, predicting the mediation of crystal growth by additives or impurities, predictions of hydration, prediction of stability of mixtures of crystals and the prediction of physical and chemical properties of the formulated compounds. The general approach is to provide fast in-silico predictions with the aim of reducing the time and cost of experimental studies. The overall roll of Synthonic Engineering in the production process is anticipated to be desigining the crystallisation process of the solid together with facilitating its subsequent formulation see Fig. 10.1.

The SE tools have two main components: VisualHabit and SystSearch which are based on the intrinsic (bulk) synthons and the extrinsic (surface) synthons respectively. VisualHabit was a re-implementation of the program HABIT98 [11, 12] which predicts crystal morphology, whilst SystSearch investigates the surface chemistry of specific crystal faces. It was also developed as a stand-alone program but has now been combined into VisualHabit.

10.2 Calculation of Bulk and Surface Crystal Properties

In this section the theory behind the molecular modelling of crystals is introduced. This covers both the physical science of inter-molecular interactions and the summations used to convert these to lattice energies. The lattice energy consists of synthons surrounding a molecule that has been saturated or is in equilibrium; the tool and VisualHabit investigate these saturated synthons. In order to understand the saturated synthons a brief summary on crystallography is given in the next section.

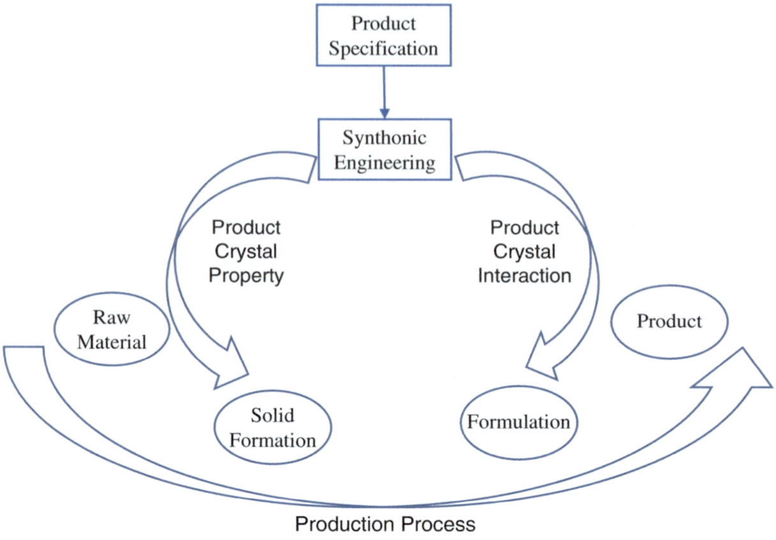

Fig. 10.1 Schematic of the application of synthonic engineering tools to solid formation and product properties. Predictions of crystal properties would assist in both setting up the crystal production process (Solid Formation) and the process of combining crystals into a finished product (Formulation)

10.2.1 Crystallography

A fuller account of the crystallographic background can be found in the Ramachandran's chapter of this volume [13]. A crystal is made up of an infinite number of unit cells. A crystal lattice is an imaginary grid in three dimensions. It is a periodic array of units of atoms or molecules. These structural units can be represented by *lattice* which consists of a periodic array of points and groups of atoms or molecules about every point. A group of atoms or molecules that is being repeated in the lattice is called a motif.

There are fourteen different types of lattices which are known as *Bravais lattices*. When a motif is repeated after a geometry operation (also known as symmetry element) is performed, the object is said to possess symmetry. Examples of a symmetry element are translation, inversion, rotation or mirror [14]. Symmetry elements can occur alone or in combination in a crystal and such a combination is characteristic of that crystal. A set of symmetry elements is known as a point group and there are 32 crystallographic point groups. The fourteen Bravais lattices together with the point groups belong to one of the 7 crystal systems in which the crystals are grouped.

The orientation of the planes that define the external crystal growth morphology are denoted by the Miller indices, expressed as $(h\ k\ l)$, where h, k, and l are the reciprocal of the fractional intercepts on the cell axes, a, b, and c respectively.

These planes are defined using the convention if a plane intercepts an axis the index, with respect to the axis, is the inverse of the coordinate, in unit cell size fraction, of the intercept. If the plane is parallel to an axis the index is zero. For example the plane that is parallel to the a axis, intercepts the b axis at the limit of the unit cell, and intercepts the c axis half way to the limit is denoted by the Miller indices (0, 1, 2).

The spacing between repeated planes is called the d-spacing or inter-planar spacing. The symmetries that form the space group can be used to group the planes into Laue classes. Laue classes correspond to the eleven centrosymmetric point groups formed by adding central symmetry to all the point groups that do not already possess it. All the planes in a given class will share the same d-spacing and surface chemistry. More information on crystallography and crystal growth can also be found in the chapters on crystal science fundamentals [13] and crystallisation [15].

The X-ray diffraction of a crystal gives the inter-planar spacing (d-spacing) between the layers of atoms, molecules or ions. It is reasonable to assume that all the faces in a given form will have the same growth rate, so the growth of a crystal can be considered as the growth of the various forms [7, 8] . The surface area of the form appearing in the final crystal is governed by its morphological importance, which expressed as the relative centre to face distances. The morphology is defined by the list of all the forms visible in the crystal and the habit is the exact proportion of each form (in practice these two terms can be interchangeable). The regular appearance of crystals of the same compound is caused by the differential growth rates of the forms. Unless the faces of two forms meet at right angles the fast growing forms will 'grow out' and disappear beneath the slow growing forms, see Fig. 10.2. Hence the morphological importance is inversely related to the relative growth rates of the forms.

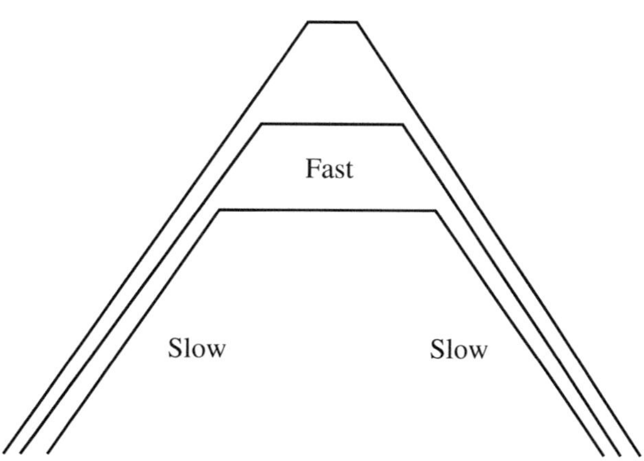

Fig. 10.2 Schematic showing the effect of growth rate on the surface area of a crystal face. The fast growing face is effectively grown out by the slow growing faces. The result is that the habit is dominated by slow growing faces

10.2.2 Lattice Energy Calculations

As a crystal is a stationary arrangement of molecules, atoms or ions, an analysis of the properties of the crystal from first principles can only consider the forces holding these components in place, there is no need to consider motion or kinetic energy. If the inter-body forces within the crystal are considered as a potential field, the lattice energy can be defined as the change in the potential field caused by forming one mole of components from infinite seperations. For a stable crystal this must be. The energies are calculated as a combination of the Lennard-Jones formalism for the van der Waals interactions with a Coulombic term to account for electrostatic interactions. In this the internal energy between two atoms, with atomic charges q_1 and q_2, a distance r apart is given by Eq. 10.1, where A and B are constants and ε_0 is the permittivity of free space.

$$E = -\frac{A}{r^6} + \frac{B}{r^{12}} - \frac{4\pi\varepsilon_0 q_1 q_2}{r} \tag{10.1}$$

The van der Waals component may be replaced with a Buckingham potential, see Eq. 10.2. Also, the attractive component of the van der Waals energy can be modelled using a tenth power in to model hydrogen bonding.

$$E = Ae^{-Br} - \frac{C}{r^6} \tag{10.2}$$

Finally, a term for the electrostatic attraction of the atoms in the molecules must be included. In Hartree atomic units this is given by the following equation.

$$E = -\frac{q_1 q_2}{r} \tag{10.3}$$

The interaction energy for a pair of atoms can be found by summing Eq. 10.4 with one of the first two equations. To find an intermolecular energy we sum the interatomic energies for all the interatomic interactions of the two molecules, and for a lattice energy (E_{cr}) we sum for all the intermolecular interactions between an arbitrary central molecule and all the molecules out to a given radius.

$$E_{cr} = \frac{1}{2} \sum_{k=1}^{N} \sum_{i=1}^{n} \sum_{j=1}^{n} E_{kij} \tag{10.4}$$

The index j runs through the atoms in the central molecule, the index i runs through the atoms in a distant molecule, and the index k runs through all the molecules within the limiting radius. The factor of one half arises as the energy of any pair of molecules is shared between the two molecules. The process can be repeated for all the molecules in a unit cell and averaged to give an energy per mol. A fuller account can be found in R Hammond's chapter of this book [16].

10.2.3 Prediction of Crystal Growth Morphology

Relative to any crystal plane (hkl) the lattice energy (E_{cr}) can be partitioned into a slice energy (E_{slice}) and the attachment energy for a slice (E_{att}), see Eq. 10.5.

$$E_{cr} = E_{slice}(hkl) + E_{att}(hkl) \tag{10.5}$$

The difference between the slice energy and the lattice energy is the attachment energy, originally introduced by Hartman [17–19], which is a measure of the unsaturated synthons on that surface. The morphology is the list of all the forms visible in the crystal and the habit is the exact proportion of each form. The regular appearance of crystals of the same compound is caused by the differential growth rates of the forms. Unless the faces of two forms meet at right angles the fast growing forms will 'grow out' and disappear beneath the slow growing forms, see Fig. 10.2. Hence the morphological importance is inversely related to the rate of growth in the faces of the a given form, which was developed by Bravais Friedel, Donnay and Harker (BFDH) [20–22]. In the energy-based method, the growth rate is assumed to be proportional to the attachment energy of a slice of material one d-spacing thick on to the face. Larger attachment energies give rise to faster grow and hence smaller morphological importance. This underlying approach, as used in HABIT [11, 23–26] and was extended to allow the calculation of the polarity of molecular urea [27] and investigate the effect of disruptive and blocking additives [28, 29]. For further details see the chapters of this volume by Rosbottom [12] and Hammond [16].

10.2.4 Systematic Search

SystSearch uses the same atomistic approach, as described in the previous section, to model and predict the interaction, or binding, of a probe molecule with a given face of a crystal, which, for example, can be applied to understand the effect of a solvent or an additive with a crystal surface [24]. In brief, a probe molecule in a fixed conformation is selected and the atoms are typed using the same potential as was used for the lattice energy calculation. The probe is now located above a simulated face of the crystal and the interaction energy of the probe and face is calculated in the same way as for calculating the lattice energy. The calculations are repeated in a three-dimensional grid over the surface. At each node in the grid several calculations are carried out as the probe is rotated through a grid of Euler angles. The process is described in detail in the chapter by Hammond in this volume [16].

10.3 Development of the Software with Case Study

Since the aim of the software is to apply molecular modelling techniques to crystallographic data, with the intention of predicting the bulk and surface properties of crystals, including their equilibrium habits, it made sense to build the software into an existing crystallographic data application. Such an application would provide the ability to read files and extract and visualize the crystallographic data. As a result it this software has been developed as a plug-in to Cambridge Crystallographic Data Centre (CCDC)'s Materials Mercury programme.

Materials Mercury is designed as an access, visualisation and support tool for the Cambridge Structural database (CSD) and can also read a range of crystallographic information file formats. It provides tools for visualising and investigating the crystal system. It also allows for crystal habit plane prediction based on the (BFDH) method [21].

The tool is comprised of two separate components: VisualHabit, which calculates the lattice energy of a crystal system; and SystSearch, which calculates the interaction energy between a probe molecule and a crystal face. VisualHabit was a re-implementation, in C++, of the original FORTRAN77 (F77) program HABIT [11], and has been integrated with Materials Mercury in the form of a plug-in.

SystSearch [30, 31] was developed as FORTRAN95 (F95) program, which was modified and built as a dynamic link library, callable from the C++ plugin. This allowed inputs and outputs to be passed between VisuaHabit and SystSearch as C++ data structures, thereby avoiding the use of multiple input and output files.

10.3.1 VisualHabit

VisualHabit calculated lattice energies summing all the interaction energies between a central molecule and all the other molecules within a fixed radius of the central molecule. For any pair of molecules, the molecule-molecule interaction energy was found by using an interatomic potential and coulombic equation to calculate interaction energy for every atomic pair. For crystal systems with Z' greater than one, the process was repeated for every molecule in the asymmetric unit and averaged to produce an energy per mole. To achieve this one unit-cell's worth of molecules had to be generated, the atoms typed for the user's chosen potential (Table 10.1 gives the list of potential available in the programme) and the atomic charges found for the atoms. The coordinates of the atoms and the molecular bonding could be provided by Mercury. Some of the potentials require typing of the atoms beyond the element, for example carbon can be aliphatic or aromatic, or hydrogen can be ordinary or hydrogen bonding. Mercury provides functions for determining the SYBYL atom type [35], for detecting hydrogen bonding, and, with some extensions these provided the atom typing. Charges were found using the

Mercury implementation of Gasteiger charges [37, 38]. In addition, it is possible to input other charges from a text file.

Table 10.1 List of the available inter-atomic potentials, including whether they use a coulombic term, and include a separate hydrogen bonding calculation

Potential	Coulombic	Type	Hydrogen bonding
Dreiding II [32]	Yes	6–12	10–12
Momany [33]	Yes	6–12	10–12
Williams [6]	Yes	6-exp	None
Gavezzotti [34]	No	6-exp	None
Tripos 5.2 [35]	Yes	6–12	None
Govers 1 [5, 36]	Yes	6-exp	None
Govers 2 [5, 36]	Yes	6-exp	None

This method of calculation has the disadvantage of being vulnerable to the non-convergence of the electrostatic component, which can occur in strongly charged systems such as salts. The problem is caused by a series of reciprocals being conditionally convergent. Having calculated the lattice energy, the program attempts to build an energy-based crystal habit. This is calculated by iterating through a list of crystallographic forms and calculating the attachment energy for a slice of crystal with a thickness of one d-spacing. The attachment energies are used in place of the d-spacing when calculating a crystal habit.

The software is run from a pop-up dialog (Fig. 10.3), and the only input required from the user is to select a potential and set the limiting radius for the calculation. If the user selects a potential that does not include all the atoms in the crystal system, an error is signalled. When complete, the results appear as an item in the combo box at the bottom of the dialog. The user may make many runs using various potential, and/or radii, (Fig. 10.3a) and browse the results (Figs. 10.3b–e). The predicted morphology appears in the Mercury main window (Fig. 10.3d). For each lattice energy calculation the user is provided with three means of viewing the data, in all cases the data can be saved to comma-separated text files (CSV), which can be read by most spreadsheets and maths packages.

As a case study example of the use of the packages, one of the common analgesics, aspirin (acetylsalicylic acid, $C_6H_8O_4$), was considered. The molecule is shown in Fig. 10.4 and it has a carboxylate ester and a carboxylic acid group. Aspirin belongs to $P2_1/c$ space group and the cell parameters are listed in Table 10.2.

By selecting the BFDH morphology within the drop down menu 'Calculate', the BFDH morphology was displayed. Within the *Morphology* tab the '*Visual Habit*' tab was chosen. From the drop down menu the Momany potential was selected and the lattice energy calculated. The atom types and charges assigned by the program are listed as shown in Table 10.3. These can be exported as a CSV file.

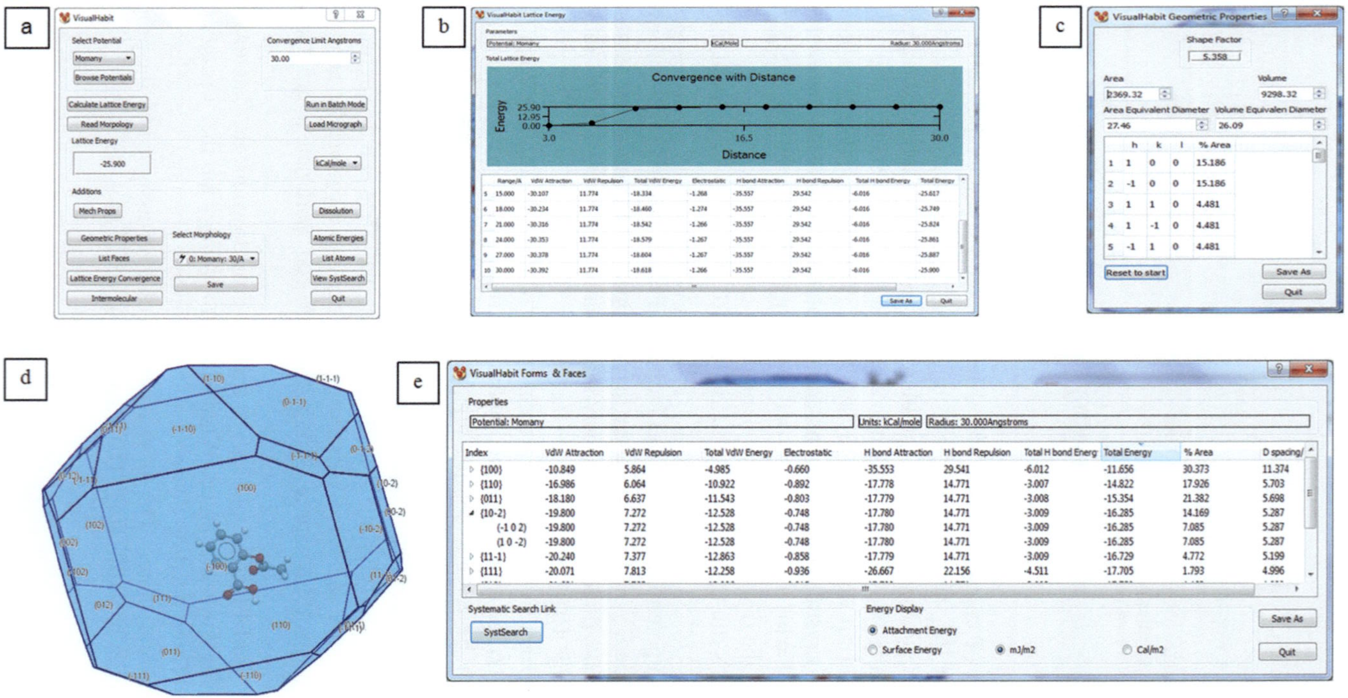

Fig. 10.3 VisualHabit dialogs: (**a**) the main dialog showing the results of a lattice energy calculation using the Momany potential and convergence radius of 30 Å, the lattice energy is shown in the box on the *left*; (**b**) the convergence of the lattice energy with distance; (**c**) the geometrical properties results; (**d**) the predicted morphology; (**e**) the attachment energy results of the forms in the morphology

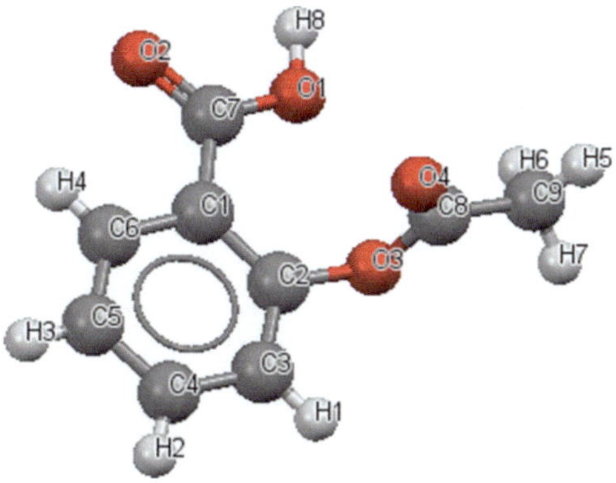

Fig. 10.4 Single molecule of aspirin shown in the style of a *ball and stick*. Colour code: *Grey*-carbon, *red*-oxygen and *white* – hydrogen

Table 10.2 The cell parameters for aspirin polymorph I, (CCDC ref. code: ACSALA01)

a	b	c	α	β	γ
11.430/Å	6.591/Å	11.395/Å	90.00°	95.68°	90.00°

Table 10.3 Gasteiger charges and Momany atom types for the atoms in a single molecule of aspirin

Atom	Charge/AU$_{(Hartree)}$	Momany type
C1	0.087	C aromatic
C2	0.143	C aromatic
C3	−0.018	C aromatic
C4	−0.058	C aromatic
C5	−0.061	C aromatic
C6	−0.045	C aromatic
C7	0.271	C carbonyl
C8	0.24	C carbonyl
C9	0.02	C aliphatic
O1	−0.487	O hydroxyl
O2	−0.15	O none
O3	−0.436	O none
O4	−0.156	O none
H1	0.066	H aromatic
H2	0.062	H aromatic
H2	0.062	H aromatic
H4	0.063	H aromatic
H5	0.034	H aliphatic
H6	0.034	H aliphatic
H7	0.034	H aliphatic
H8	0.295	H hydroxyl

10.3.1.1 Convergence of Lattice Energy

The user can view the convergence of the lattice energy in ten equally spaced radii out to the limit. This can be used to check the extent of energy convergence. The energy is broken down into van der Waals attractive, van der Waals repulsive, hydrogen bonding attractive, hydrogen bonding repulsive and electrostatic. The convergence plot against the radial cut of distance (in angstroms) was displayed and the data exported and plotted in Excel (Fig. 10.5).

10.3.1.2 Visualize Interactions

The user is allowed to select a molecule in the unit cell and view it with a visualization of the energy interactions with its neighbours. This is provided by rendering the chosen molecule, but in the location of each neighbour within the limiting radius a cube, scaled in proportion to the inter-molecular energy between the central molecule and the molecule occupying the location is displayed. The result gives an indication of the primary interactions building up the crystal, see Fig. 10.6 which shows the case study example of aspirin molecule which is at the centre of a sphere of radius 30 A and the interaction energies represented as cubes. The size of the cube is directly proportional to the strength of the interaction and it is seen that there is one strong interaction which could be as a result of the hydrogen bonds between the carboxylic acid groups.

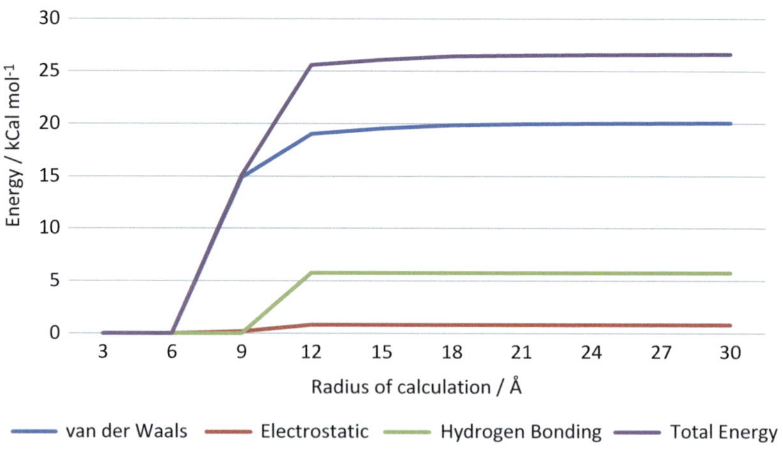

Fig. 10.5 The convergence of the lattice energy with summation radius; the figures are broken down into van der Waals, electrostatic and hydrogen bonding components

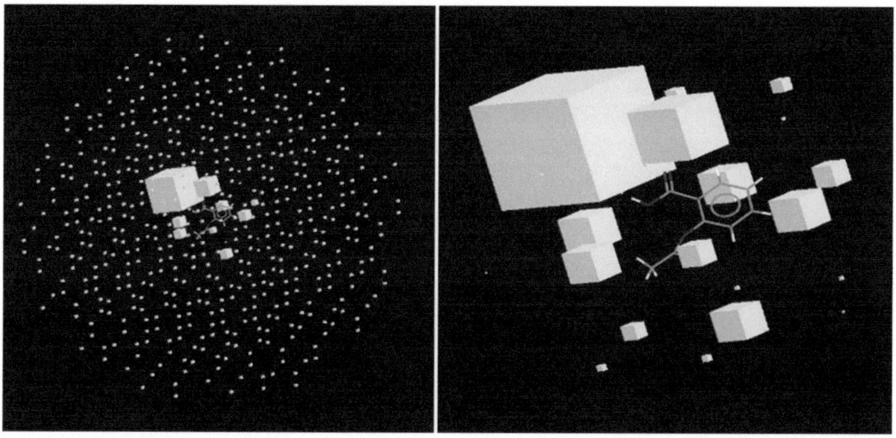

Fig. 10.6 A visualization of the interaction energies of aspirin within a cut-off radius of 30: (*left*) cubes located at molecular sites scaled to represent the total interaction energies between the molecule and the central molecule. All the molecular locations in the convergence radius indicate that these few interactions dominate the lattice energy. The image is zoomed in (*right*) to show only the interactions round the molecule, showing the dominance of the hydrogen bond on the carboxylic group

10.3.1.3 Forms and Faces

The user can view the attachment energies for each of the forms in the habit and each of the faces in the form. The percentage of the habit's surface area accounted for by the form/face is also included, as is the d-spacing. This table includes the faces, percentage surface area of the shape, and breakdown of attachment energy in kcal/mol. It is observed from Fig. 10.7 and Table 10.4 that {1 0 0} form has the lowest attachment energy (absolute value) and thus replicating slowest growth rate which is reflected in the highest percentage surface area. It is also noted that the attachment energy of this form is highly hydrogen bonded unlike the attachment energies of the other forms which dominate in the van der Waals component.

The dominant faces observed on the crystal morphology were listed and exported as given in Table 10.4. Each form opens up a tree of faces and the corresponding breakdown of energy in kcal/mol was displayed.

10.3.1.4 Geometrical Data

The user is presented with the shape factor τ for the crystal habit [39], the surface area (A) divided by the square of the cube root of the volume (V) as given below.

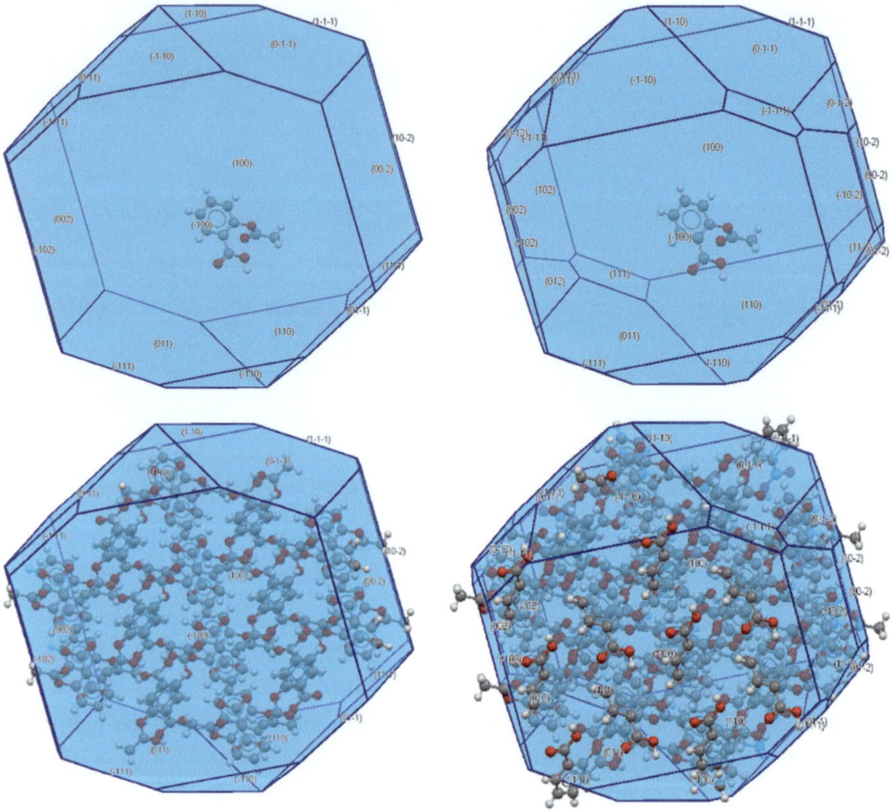

Fig. 10.7 Predicted crystal habits of Aspirin using, (*top left*) BFDH, (*top right*) VisualHabit attachment energy method. The *bottom* two images show the morphologies packed with molecules allowing the chemistry of the faces to be interpreted

Table 10.4 The attachment energies of the form in the energy based crystal habit prediction for aspirin polymorph I

Form	van der Walls	Electrostatic	H bond	Total Energy	Area %	d spacing
{100}	−4.99	−0.66	−6.01	−11.66	27.41	11.37
$\{10\bar{2}\}$	−13.02	−0.52	0.00	−13.54	18.10	5.29
$\{10\bar{1}\}$	−14.00	−0.70	0.00	−14.70	14.12	5.20
{110}	−10.92	−0.89	−3.01	−14.82	12.70	5.70
{111}	−11.85	−0.82	−3.01	−15.67	11.18	5.00
{011}	−14.86	−0.62	0.00	−15.48	8.60	5.70
{002}	−13.40	−0.81	−3.01	−17.22	4.02	5.67
{012}	−12.96	−0.75	−3.01	−16.72	3.87	4.30

$$\tau = \frac{A}{V^{2/3}} \tag{10.6}$$

This is a constant for any crystal habit and provides a scale independent way of converting surface area to volume. The user is also provided with the surface area, the volume, the area equivalent diameter and the volume equivalent diameter of the crystal habit. These last four items are adjustable and changing any one will alter the other three.

In the case of aspirin the shape factor was found to be 5.3, which is approximately half way between the values for a cube (6.0) and a sphere (4.84). From the data in Table 10.4 it can be seen that two forms $\{100\}$ and $\{10\bar{2}\}$ account for over 50% of the surface area. The area equivalent diameter was 26.5 and the volume equivalent diameter was 25.3, with no specific units associated with these numbers as they are derived from the geometry of the model crystal which has no scale. This allows data size data in the form of volume or area equivalent diameters, such as may come from a laser diffraction particle sizer, to be converted to a surface area and volume.

10.3.2 SystSearch

SystSearch is a contraction of systematic search, a program which has been developed to investigate the effect of binding or interaction of a probe molecule on a crystal plane in the crystal system [30], and outputs a grid of energies representing a scan of the probe molecule across one reticular area of the unit cell on the surface defined by the plane. The intention is to model the interaction energy field between the probe molecule and the surface. An early version was used to validate the software and investigate the use of surface interaction fields to support the prediction of solvent mediated morphology and to understand the cohesive/adhesive balance between API and excipients [40].

The algorithm converts the unit cell to an orthogonal coordinate system with the X axis perpendicular to the specified plane. The symmetry operators are then applied to the asymmetric unit (ASU) to build a packed unit cell, which will be one d-spacing thick. Additional unit cells are added around and below the newly constructed cell to account for edge effects, resulting in the construction of a slab. Figures 10.8 shows the dialogs where the input parameters are selected, and where the progress of calculation is displayed. The probe molecule is then moved on a grid, defined by the user, over the surface of the original, central, unit cell. At each grid point the probe molecule is rotated about the principal axes, in an angular grid specified by the user. It is possible for the probe to 'crash' into the molecules resulting in a large positive energy. To prevent these being recorded a high energy cut-off is provided, which has to be set by the user. Any energy that is below the cut-off is stored and returned as part of the results. In addition, an offset to the

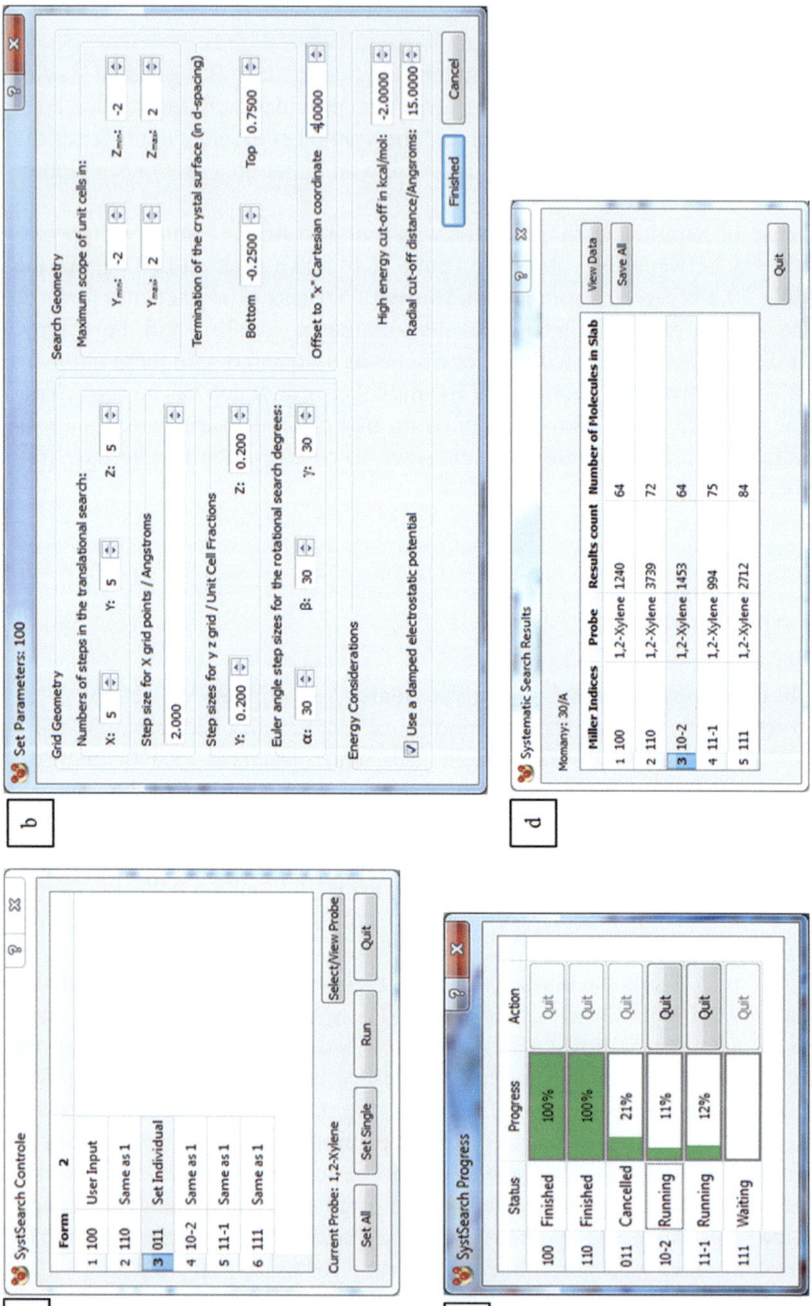

Fig. 10.8 SystSearch dialogs. (**a**) Selection of form or face for running systematic search calculations which can be used to set the same parameters for all forms or different individual parameters; (**b**) Initial dialog allowing the user to select a probe molecule and set the input parameters. (**c**) The progress dialog visible during the run; SystSearch will attempt to run as many forms in parallel as there are free cores on the processor. The user has the ability to quit any individual run. (**d**) The main results dialog listing forms, solvents and the number of results that passed the energy filter for each run

search grid, which moves the entire grid up or down, is provided allowing the user to move the search so as to minimize the number of collisions.

The results can be viewed by selecting the VisualHabit results and clicking a "SystSearch" button, which reveals all systematic search results derived from that VisualHabit result. Once a run of SystSearch is selected, the results may be viewed in a separate dialog.

As a case study example, SystSearch was carried out on aspirin crystal surfaces and 2-butanol as the probe. In the dialog the first three statistical moments of the results are listed (Fig. 10.9a), and the results themselves are tabulated (Fig. 10.9b), with the energies broken down into van der Waals and hydrogen bonding, attractive and repulsive, plus electrostatic, as in VisualHabit. The entire energy field can be viewed in which case the lowest energy at each spatial location which returned an energy is displayed. The display places a cube at each location and colour codes them according to the energy as shown in Fig. 10.9c. Alternatively, is possible to select a single result and observe a three dimensional view of the slab and molecule (Fig. 10.9d).

As a case study example, SystSearch was run on the $\{10\bar{2}\}$ form of aspirin, which has the lowest attachment energy with 2-butanol as the probe. As was described above the probe was scanned over the surface in a spatial and angular grid, with the interaction energy with the surface calculated at each node in the grid. This data can be used to give an indication of the degree to which the solvent will interact with the surface, potentially altering its growth or dissolution rates. The data had a mean of -2.38 kCal mol^{-1}, a variance of 5.79 kCal mol^{-1} and a skewedness of -14.53 kCal mol^{-1}, a histogram of the results is shown in Fig. 10.10. The figure shows the results clustered round the high energy cut off limit, with only a very shallow low energy tail, which explains the large skewedness.

The data was visualized by placing a grid of cubes over the surface located at the special nodes that had any results passing the low energy filter. Each node held a list of orientation energy pairs that were found at that location. All the cubes were coloured according to lowest energy found at each location, ranging from red for the highest through green, to blue for the lowest (most negative) energy. Two views of the results are shown in Fig. 10.11. Finally the most favourable position found for the probe with respect to binding energy is shown in Fig. 10.12.

10.4 Concluding Remarks

This chapter discusses the development of the SE tool, Habit (via Visualhabit and SystSearch) for quantitative understanding of the intrinsic and extrinsic synthons. The close integration of crystallographic, chemical and molecular modelling software, within a graphical user interface, allows relatively quick and intuitively simple analysis of the bonding of the crystal. It can also generate models of the interaction energy field on the crystal surfaces.

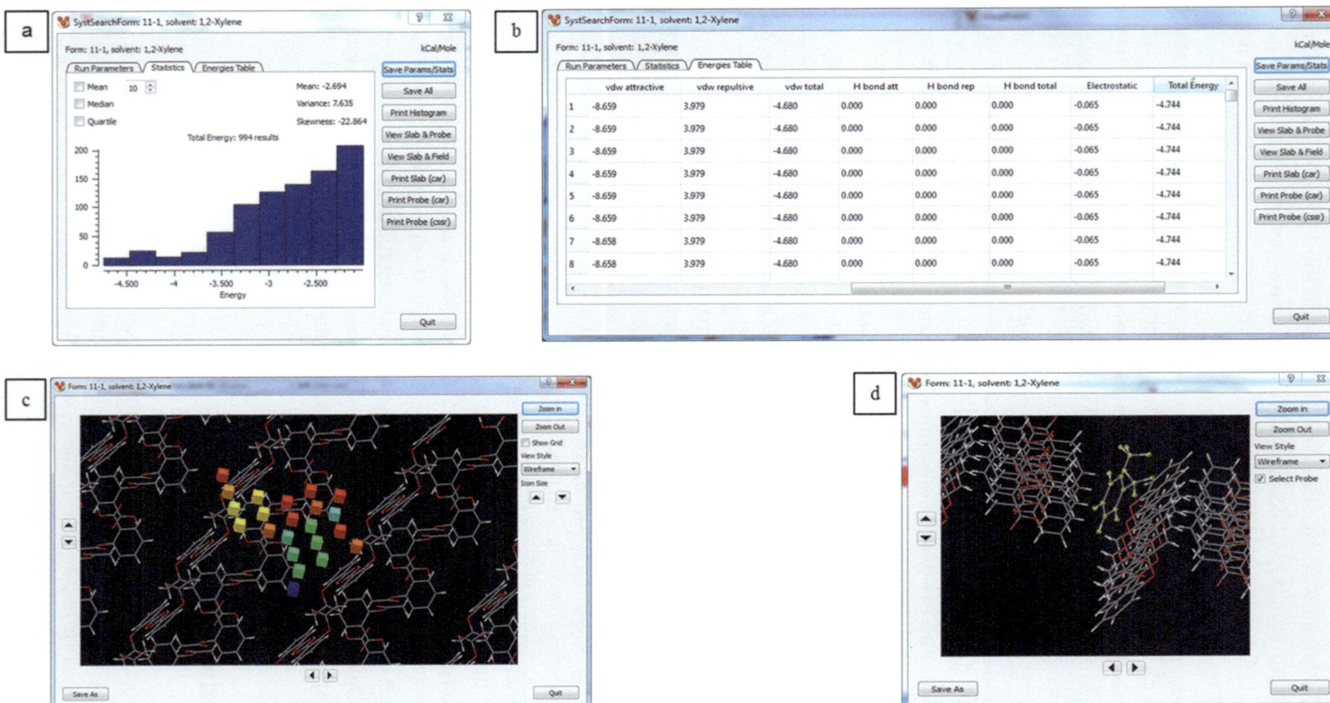

Fig. 10.9 Screenshot of the results produced by SystSearch. (**a**) The dialog showing the complete results set for a run of SystSearch. The statistics tab is selected showing the mean variance and skewness, together with a histogram of the results binned by energy. (**b**) The energies tab is selected showing the table of interaction energies is visible, the other two tabs reveal the parameters used for the run and the statistics of the energies in the table. (**c**) The dialog showing the energy field over the slab, each cube is centred on a point in the search. The cubes colouring is a function of the value of the lowest energy found in the angular search carried out at that point, with red being the highest energy and blue the lowest. (**d**) This is the dialog showing the probe location and orientation for a single energy value, the 1, 2-xylene molecule (*highlighted*) can be seen within a 'valley' of ibuprofen molecules

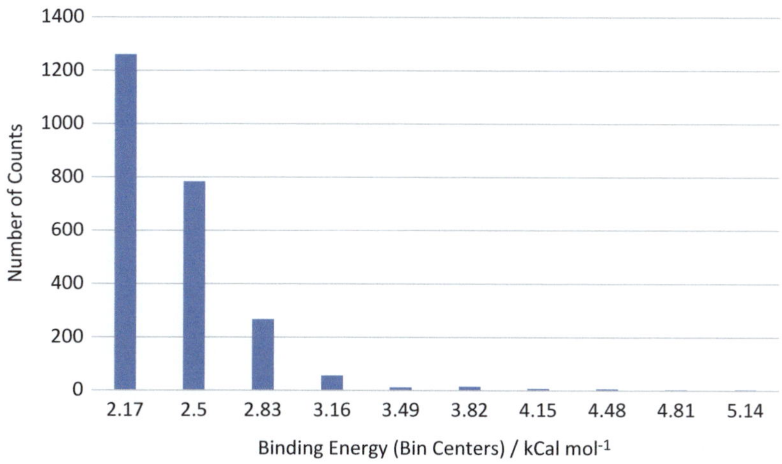

Fig. 10.10 A histogram of the interaction energies found by using SystSearch to scan 2-butanol over the {102} face of aspirin. The results show a very *shallow tailed* distribution with only a handful of strongly bonding sites

Fig. 10.11 Two views of the energy field formed by 2-butanol over the {102} face of asprin. Each *cube* represents a location in the spatial grid at which at least one energy result from the angular search passed the low energy filter. The colouring of the boxes is driven by the value of the lowest (most negative) energy found at each location, with *blue* representing the lowest (most strongly binding) and the *red* the highest (least strongly binding) energy. It can be seen that the best locations appear to be at the edge of the "valleys"

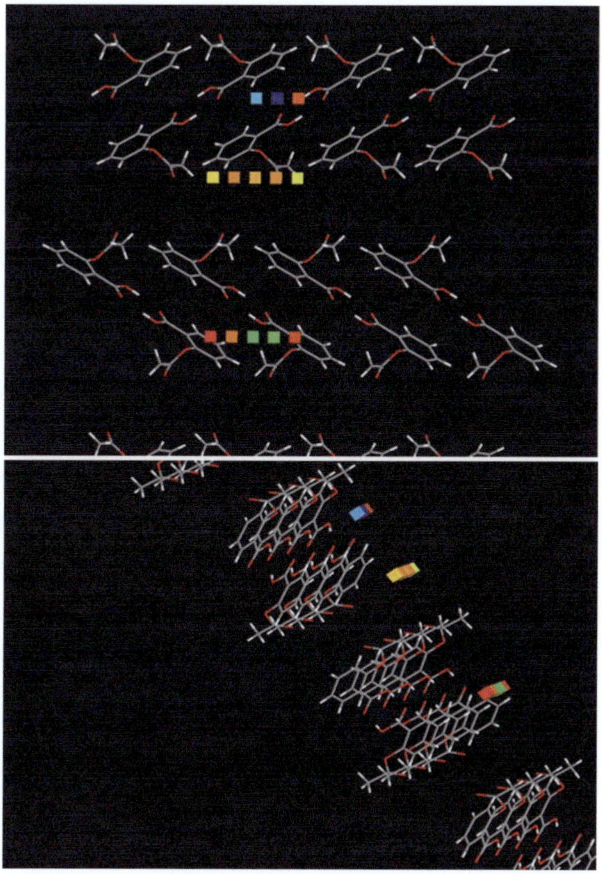

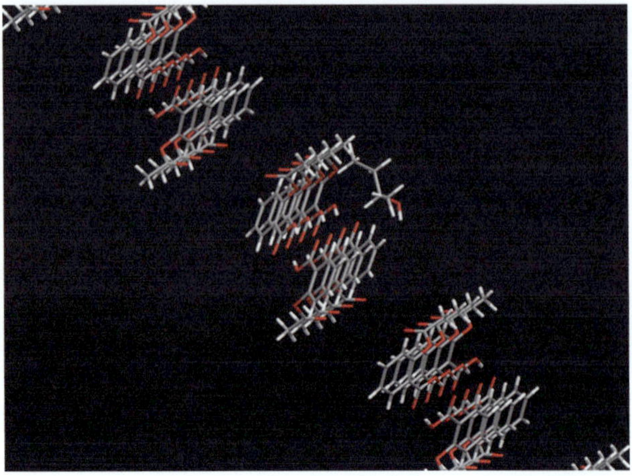

Fig. 10.12 The location and orientation of the 2-butanol molecule with the lowest (most negative) interaction energy with the {102} face of asprin (Note the location of the molecule at the edge of one of the "valleys" in the surface)

References

1. Charpentier JC (2007) Among the trends for a modern chemical engineering: CAPE an efficient tool for process intensification and product design and engineering. In: Plesu V, Agachi PS (eds) 17th European Symposium on Computer Aided Process Engineering, pp 11–18
2. Crom S (2006) Right first time in pharmaceuticals: Six Sigma for continuous and breakthrough improvement. Chim Oggi 24(3):20–21
3. Desiraju GR (1995) Supramolecular synthons in crystal engineering – a new organic-synthesis. Angew Chem Int Ed Engl 34(21):2311–2327
4. Thalladi VR, Goud BS, Hoy VJ, Allen FH, Howard JAK, Desiraju GR (1996) Supramolecular synthons in crystal engineering. Structure simplification, synthon robustness and supramolecular retrosynthesis. Chem Commun 3:401–402
5. Kitaygorodsky A (1961) Interaction curve of non-bonded carbon and hydrogen atoms and its application. Tetrahedron 14(3–4):230
6. Williams DE (1966) Nonbonded potential parameters derived from crystalline aromatic hydrocarbons. J Chem Phys 45(10):3770
7. Macrae CF, Bruno IJ, Chisholm JA, Edgington PR, McCabe P, Pidcock E, Rodriguez-Monge L, Taylor R, van de Streek J, Wood PA (2008) Mercury CSD 2.0 – new features for the visualization and investigation of crystal structures. J Appl Crystallogr 41:466–470
8. Macrae CF, Edgington PR, McCabe P, Pidcock P, Shields GP, Taylor R, Towler M, van de Streek J (2006) Mercury: visualization and analysis of crystal structures. J Appl Crystallogr 39:453–457
9. Bruno IJ, Cole JC, Edgington PR, Kessler MK, Macrae CF, McCabe P, Pearson J, Taylor R (2002) New software for searching the Cambridge structural database and visualising crystal structures. Acta Cryst B58:389–397
10. Taylor R, Macrae CF (2001) Rules governing the crystal packing of mono- and di-alcohols. Acta Cryst B57:815–827
11. Clydesdale G, Roberts KJ, Docherty R (1996) HABIT95 – a program for predicting the morphology of molecular crystals as a function of the growth environment. J Cryst Growth 166(1–4):78–83
12. Rosbottom I, Roberts KJ (2017) Crystal growth and morphology of molecular crystals, chapter 7. In: Roberts KJ, Docherty R, Tamura R (eds) Engineering crystallography: from molecule to crystal to functional form. Springer Advanced Study Institute (ASI) Series, 2017, in press

13. Ramachandran V, Halfpenny P, Roberts KJ (2017) Crystal science fundamentals, chapter 1. In: Roberts KJ, Docherty R, Tamura R (eds) Engineering crystallography: from molecule to crystal to functional form. Springer Advanced Study Institute (ASI) Series, 2017, in press
14. Authoer (1971) Crystallography and crystal chemistry. Holt Rinchart and Winston Inc., New York
15. Camacho Corzo D, Ma CY, Ramachandran V, Roberts KJ (2017) Crystallisation route map, chapter 11. In: Roberts KJ, Docherty R, Tamura R (eds) Engineering crystallography: from molecule to crystal to functional form. Springer Advanced Study Institute (ASI) Series, 2017, in press
16. Hammond RB (2017) Modelling route map: from molecule through the solution state to crystals, Chapter 6. In: Roberts KJ, Docherty R, Tamura R (eds) Engineering crystallography: from molecule to crystal to functional form. Springer Advanced Study Institute (ASI) Series, 2017, in press
17. Hartman P, Perdok WG (1955) On the relations between structure and morphology of crystals. 1. Acta Crystallogr 8(1):49–52
18. Hartman P, Perdok WG (1955) On the relations between structure and morphology of crystals. 2. Acta Crystallogr 8(9):521–524
19. Hartman P, Perdok WG (1955) On the relations between structure and morphology of crystals. 3. Acta Crystallogr 8(9):525–529
20. Bravais A (1866) Du cristal considere comme un simple assemblage de points. Etude Cristallographiques
21. Donnay JDH, Harker D (1937) A new law of crystal morphology extending the law of bravais. Am Mineral 22(5):446–467
22. Friedel G (1907) Studies on the law of Bravais. Bull Soc Fr Mineral 30:326–455
23. Clydesdale G, Docherty R, Roberts KJ (1991) A predictive approach to modeling the morphology of organic-crystals based on crystal-structure using the atom-atom method. Cryst Growth, Pts 1 and 2
24. Bennema P, Meekes H (2004) Two centuries of morphology of crystals: integration of principles of mathematical crystallography, statistical mechanics of surface models and chemistry. In: Liu XY, De Yoreo JJ (eds) Nanoscale structure and assembly at solid-fluid interfaces. Kluwer Academic Publisher, Boston
25. Roberts KJ, Docherty R, Bennema P, Jetten L (1993) The importance of considering growth-induced conformational change in predicting the morphology of benzophenone. J Phys D Appl Phys 26(8B):B7–B21
26. Authoer (1997) Theoretical aspects and computer modelling of the molecular solid state. Wiley, Chichester
27. Docherty R, Roberts KJ, Saunders V, Black S, Davey RJ (1993) Theoretical-analysis of the polar morphology and absolute polarity of crystalline UREA. Faraday Discuss 95:11–25
28. Clydesdale G, Roberts KJ, Docherty R (1994) Modeling the morphology of molecular-crystals in the presence of disruptive tailor-made additives. J Cryst Growth 135(1–2):331–340
29. Clydesdale G, Roberts KJ, Lewtas K, Docherty R (1994) Modeling the morphology of molecular-crystals in the presence of blocking tailor-made additives. J Cryst Growth 141 (3–4):443–450
30. Hammond RB, Pencheva K, Ramachandran V, Roberts KJ (2007) Application of grid-based molecular methods for modeling solvent-dependent crystal growth morphology: aspirin crystallized from aqueous ethanolic solution. Cryst Growth Des 7(9):1571–1574
31. Hammond RB, Jeck S, Ma CY, Pencheva K, Roberts KJ, Auffret T (2009) An examination of binding motifs associated with inter-particle interactions between facetted nano-crystals of acetylsalicylic acid and ascorbic acid through the application of molecular grid-based search methods. J Pharm Sci 98(12):4589–4602
32. Mayo SL, Olafson BD, Goddard WA (1990) Dreiding – a generic force-field for molecular simulations. J Phys Chem 94(26):8897–8909

33. Momany FA, Carruthe LM, RF MG, Scheraga HA (1974) Intermolecular potentials from crystal data .3. Determination of empirical potentials and application to packing configurations and lattice energies in crystals of hydrocarbons, carboxylic-acids, amines, and amides. J Phys Chem 78(16):1595–1620
34. Gavezzotti A (1998) The crystal packing of organic molecules: challenge and fascination below 1000 Da. Crystallogr Rev 7(1):5–121
35. Clark M, Cramer RD, Vanopdenbosch N (1989) Validation of the general-purpose tripos 5.2 force-field. J Comput Chem 10(8):982–1012
36. Govers HAJ (1978) Atom-atom approximation and lattice energies of 2,2′-bis-1,3-dithiole (ttf), 7,7,8,8-tetracyanoquinodimethane (tcnq) and their 1-1 complex (ttf-tcnq). Acta Crystallogr A 34(Nov):960–965
37. Gasteiger J, Marsili M (1978) New model for calculating atomic charges in molecules. Tetrahedron Lett 34:3181–3184
38. Gasteiger J, Marsili M (1980) Iterative partial equalization of orbital electronegativity – a rapid access to atomic charges. Tetrahedron 36(22):3219–3228
39. Mandar VD, Carstensen JT (1995) Effect of change in shape factor of a single crystal on its dissolution behaviour. Pharm Res 13(1):155–162
40. Ramachandran V, Murnane D, Hammond RB, Pickering J, Roberts KJ, Soufian M, Forbes B, Jaffari S, Martin GP, Collins E and others (2015) Formulation pre-screening of inhalation powders using computational atom-atom systematic search method. Mol Pharm, 12(1): 18–33

Part II
Formation

Chapter 11
Crystallisation Route Map

Diana M. Camacho Corzo, Cai Y. Ma, Vasuki Ramachandran, Tariq Mahmud, and Kevin J. Roberts

Abstract A route map for the assessment of crystallisation processes is presented. A theoretical background on solubility, meta-stable zone width, nucleation and crystal growth kinetics is presented with practical examples. The concepts of crystallisation hydrodynamics and the application of population balances and computational fluid dynamics for modelling crystallisation processes and their scaling up are also covered.

Keywords Solubility • Supersaturation • Critical undercooling • Metastable Zone Width (MSZW) • Nucleation kinetics • Induction times • Isothermal method • Polythermal method • Crystal growth kinetics • CFD • Morphological population balance • Process scale-up

List of Symbols

(hkl)	*Miller plane* - 2D surface cut through lattice
a_c	dimensionless molecular latent heat of crystallisation
a_1 and a_2	free parameters in PN model
A_m	a constant from literature [35, 36]
A_o and B_o	system related constants
b	bulk compartment in SFM
b_o	dimensionless thermodynamic parameter
$B_a(\mathbf{x}, \mathbf{y}, t)$ and $D_a(\mathbf{x}, \mathbf{y}, t)$	birth and death terms of n for agglomeration
$B_d(\mathbf{x}, \mathbf{y}, t)$ and $D_d(\mathbf{x}, \mathbf{y}, t)$	birth and death terms of n for breakage
$B_0(\mathbf{x}, \mathbf{y}, t)$	nucleation term
C	solution concentration

D.M. Camacho Corzo (✉) • C.Y. Ma • V. Ramachandran • T. Mahmud • K.J. Roberts
School of Chemical and Process Engineering, University of Leeds, Leeds LS2 9JT, UK
e-mail: D.M.CamachoCorzo@leeds.ac.uk; C.Y.Ma@leeds.ac.uk; v.ramachandran@leeds.ac.uk; T.Mahmud@leeds.ac.uk; k.j.roberts@leeds.ac.uk

K.J. Roberts et al. (eds.), *Engineering Crystallography: From Molecule to Crystal to Functional Form*, NATO Science for Peace and Security Series A: Chemistry and Biology, DOI 10.1007/978-94-024-1117-1_11

C^*	equilibrium concentration
C_{max}	maximum supersaturated concentration
C_0	concentration of nuclei at the IN point
D	characteristic dimension
d	dimensionality of crystal growth
ds	diameter of a stirrer
f_1, f_2	feed compartments in SFM
g_i	face growth rate in *i* direction
G_i	Growth rate
J	nucleation rate
k_j	empirical parameter of nucleation rate
k_n	nucleus shape factor
k_v	crystal shape factor
K_G	growth rate constant
K_J	nucleation rate constant
m_1	order of nucleation
n	number population density function of crystals
n_0	initial number population density function of crystals
n_g and m_g	growth exponents
N	number of internal variables for a crystal
N_s	stirrer speed
N_{det}	number of crystals at detection point
q	cooling rate
q_0	free parameters in PN model
Q	flow rate of a stream
r	nucleus radius
R	gas constant
R_i	rate of growth of a crystal face
Re	Reynolds number
r_c	nucleus critical radius
S	supersaturation
T	temperature
T/T_0	scale up ratio
T_0	diameter of the laboratory reactor
T_c	crystallisation temperature
T_{diss}	dissolution temperature
T_e	solution equilibrium temperature
T_f	fusion temperature of pure solute
t	time
t_{micro}, t_{meso}	time constants of micro-mixing and meso-mixing
t_{sat}	saturation temperature
$u_{1,2}, u_{1,3}$ & $u_{2,3}$	flow rates
V	velocity
v_0	volume occupied by a solute molecule in the crystal

$\boldsymbol{v}$	velocity vector
x	mole fraction
x_1, x_2, x_3	normal distances
$\boldsymbol{x}$	internal variable vector with N components
$\boldsymbol{y}$	external variable vector such as spatial coordinates (y_1, y_2, y_3)
α	surface entropy factor
α_{det}	relative volume of crystals at detection point
Y	activity coefficient
γ_e	interfacial tension and/or surface energy
ρ	density
ΔC_{max}	maximum concentration difference
ΔH_{diss}	enthalpy of dissolution
ΔH_f	enthalpy of fusion of pure solute
ΔS_{diss}	entropy of dissolution
ΔS_f	entropy of fusion of pure solute
ΔT_c	critical undercooling for crystallisation
Δt_{max}	metastable zone width
ε	specific power input
ζ	anisotropy factor
k	Boltzmann constant
λ	molecular latent heat of crystallisation
μ	fluid dynamic viscosity
u_0	relative critical undercooling at the IN point
u_c	relative critical undercooling
σ	relative supersaturation
τ	induction time
τ_r	residence time
ω	molecular volume
∇	gradient operator for the $\boldsymbol{y}$ coordinates

11.1 Introduction

Crystallisation is a key process used in the manufacture of drugs, pharmaceuticals and fine chemicals which enables e.g. the separation and purification of particulate materials. In some cases crystallisation needs to be avoided, e.g. fuels operating in cold weather or within drugs stabilised in highly concentrated liquid-formulations. Compared with other techniques, such as distillation crystallization it is a more energy-efficient process involving lower temperatures, which are more appropriate for processing heat-sensitive chemical products. The crystallisation process is driven by supersaturation and this affects solid-liquid separation and product purification.

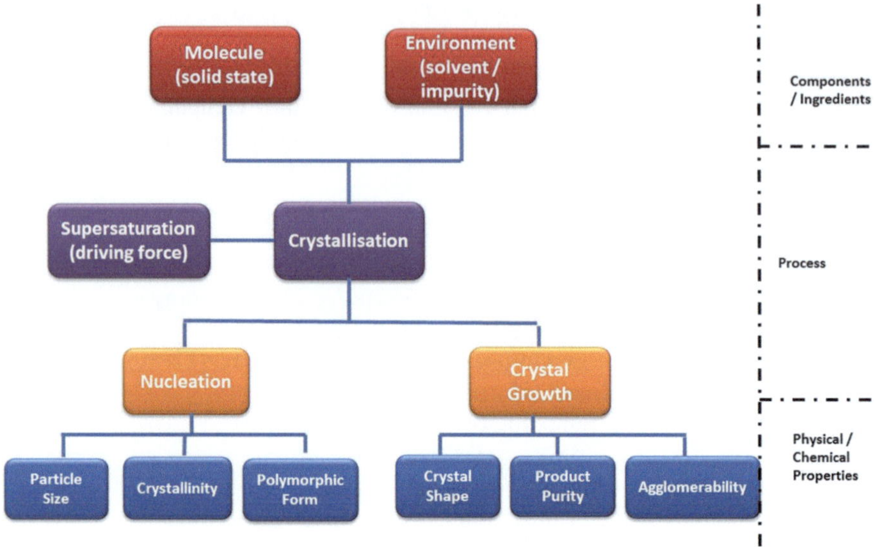

Fig. 11.1 A schematic showing the role played by the fundamental parameters of crystallisation (nucleation and growth) in directing the physical properties of the resulting solid forms. Prediction of the outcomes would allow greater control of the final product

The fundamental concepts related to a crystallisation process involve two key stages viz.: nucleation, which is three dimensional (3D) (assembly of molecular clusters on nm size scale) and growth which is two dimensional (2D) (on all atomically smooth particle surfaces) as shown in the schematic in Fig. 11.1. Nucleation will have an impact on the particles' size, crystallinity and the polymorphic form obtained, whilst crystal growth will have a direct impact on the shape of the final crystals, their purity and agglomerability.

11.2 Solubility

11.2.1 Solubility and the van't Hoff Plot

When excess solid is mixed with a solvent at a constant temperature, the solid will dissolve until equilibrium is established. The resulting solution is said to be saturated and the composition of the solution is the equilibrium solubility. The most common ways of expressing solubility are: mass of solute per unit volume of solution and mole fraction (x) (moles of solute per moles of solution).

If the solute-solute interactions in the solid are the same as the solute-solvent interactions the only enthalpy change involved is the enthalpy of fusion. Based on this, the ideal solubility can be predicted using the Hildebrand equation:

$$Ln\, x = \frac{\Delta H_f}{R}\left[\frac{1}{T_f} - \frac{1}{T}\right] \tag{11.1}$$

where

ΔH_f = enthalpy of fusion of pure solute
T_f = fusion temperature of pure solute
R = gas constant
T = temperature

Since $\Delta H_f = T_f\, \Delta S_f$ this can also be rewritten as:

$$Ln\, x = \frac{-\Delta H_f}{RT} + \frac{\Delta S_f}{R} \tag{11.2}$$

where ΔS_f = entropy of fusion of pure solute.

In a non-ideal solution the Hildebrand equation neglects the enthalpy and entropy of mixing. This effect can be included by replacing ΔH_f by ΔH_{diss} (the enthalpy of dissolution) and ΔS_f by ΔS_{diss} (the entropy of dissolution) yielding the van't Hoff equation [1, 2]:

$$Ln\, x = \frac{-\Delta H_{diss}}{RT} + \frac{\Delta S_{diss}}{R} \tag{11.3}$$

A plot of $\ln x$ versus $\frac{1}{T}$ is expected to yield a straight line with a gradient of $\frac{-\Delta H_{diss}}{R}$ and an intercept of $\frac{\Delta S_{diss}}{R}$.

11.2.2 Understanding Solution Structure

A plot of the solubility and the solution's ideality in the same plane of coordinates can give an indication of the strength of the solution's molecular interactions. The difference between the two lines will give the activity coefficient at saturation (Y). Y = 1 ideal solution, Y > 1 less than ideal and Y < 1 more than ideal. Figure 11.2, after [3], shows the ideal and experimental solubility for succinic acid in water and isopropanol. In water solubility is greater than ideal solubility, i.e. solute-solvent interactions are enhanced. In isopropanol solubility is lower than ideal, i.e. solute – solute interactions are enhanced [1, 2].

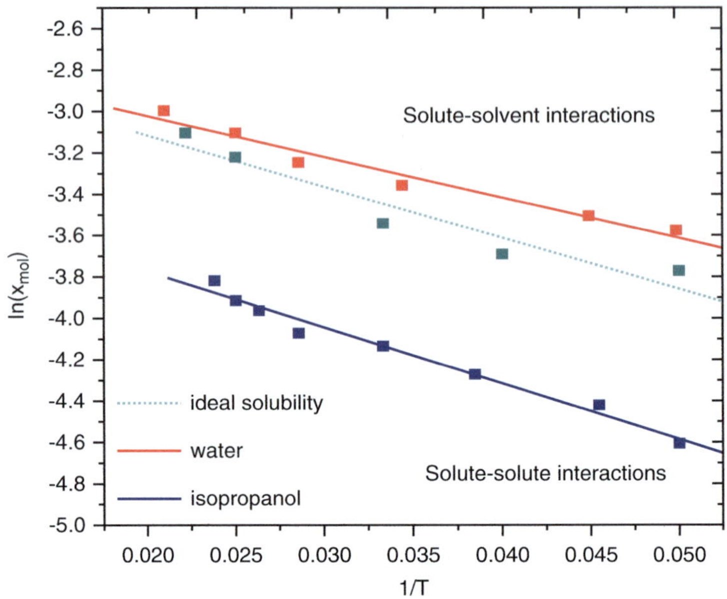

Fig. 11.2 Ideal and experimental solubility for succinic acid in water and isopropanol [3]

11.3 Supersaturation and Metastable Zone Width

11.3.1 Supersaturation

The extent to which a solution exceeds equilibrium solubility can be expressed either by supersaturation ratio S or relative supersaturation σ as given below [4].

$$S = \frac{C}{C^*} \tag{11.4}$$

$$\sigma = \frac{C - C^*}{C^*} = S - 1 \tag{11.5}$$

where C is the solution concentration and C^* is the equilibrium concentration.

11.4 Metastable Zone Width (MSZW)

The temperature-concentration diagram can be divided into three regions (stable, metastable and labile) [4] as shown in Fig. 11.3. As cooling rate increases, so does supersaturation and nucleation rate but the nucleation cluster size and particle size

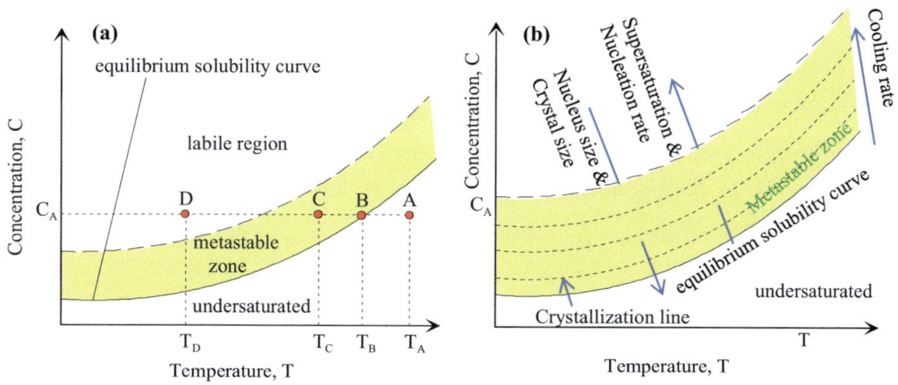

Fig. 11.3 (**a**) Schematic of solubility-supersolubility; (**b**) Meta stable zone width is wider and nuclei size decreases as the cooling rate increases

decrease. In extreme cases at very high supersaturations when the cluster size is the same as the molecular size, an amorphous phase instead of crystalline solid will be obtained.

In the stable region, the solution is undersaturated and there is no possibility of crystallisation and this region is represented by a point A at temperature T_A. When the temperature is reduced to T_B, the solution is at equilibrium (represented by point B on the schematic). T_B is the dissolution temperature where the solute particles are completely dissolved and the solution cannot take any more solid particles. The region between the equilibrium curve and the broken line is called the metastable zone.

In this region, represented by point C at temperature T_c, the solution is saturated and can remain in this state for long periods without spontaneous crystallisation if not disturbed.

The metastable zone width (MSZW) can be defined as the difference between the maximum solution concentration in the supersaturated state before crystallisation takes place C_{max} and the solution concentration at equilibrium C^*

$$\Delta C_{max} = C_{max} - C^* \tag{11.6}$$

MSZW can be defined in terms of undercooling as the difference between dissolution temperature T_{diss} and crystallisation temperature T_c.

$$\Delta T_c = T_{diss} - T_c \tag{11.7}$$

The relationship between supersaturation and undercooling is given by:

$$\Delta C_{max} = \left(\frac{dC^*}{dT}\right)\Delta T_c \tag{11.8}$$

The third region above the metastable zone is the labile region (represented by the point D at temperature, T_D) and in this region precipitation is almost instantaneous.

11.5 Nucleation Kinetics

11.5.1 Classical Nucleation Theory (CNT)

Nucleation refers to the step in which individual solute molecules (or atoms) that are dispersed in the solvent structure will begin to cluster together. Some of these clusters may grow sufficiently large to form stable nuclei and subsequently form crystals. Nucleation is best understood by examining the free energy changes associated with nucleus formation. The volume effect on the free energy of a nucleus is associated with the decrease in free energy per molecule, when the molecule is transferred from the supersaturated solution to the solid phase, and has an r^3 dependence (Fig. 11.4). The surface of the new solid phase has an energy associated with it and this results in an increase of free energy per unit surface area of the cluster which has an r^2 dependence. The sum of these two contribution yields the total free energy. As the total free energy has to decrease for the spontaneous process to take place, large nuclei will be favoured during the process due to their lower free energy. A nucleus that achieves a size greater than the critical radius r_c will grow into a crystal [4].

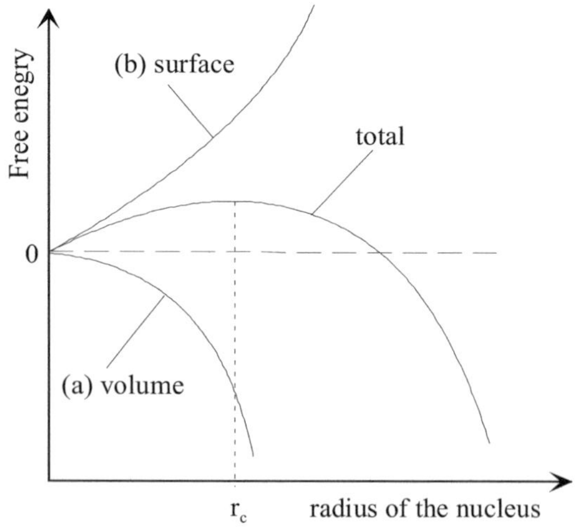

Fig. 11.4 Change in free energy as a function of radius

The critical radius for the case of a spherical 3D nucleus is given by:

$$r_c = \frac{2\omega\gamma_e}{kT \ln(S)} \tag{11.9}$$

where:

γ_e = surface energy
ω = molecular volume
k =Boltzmann constant
T = Temperature
S = supersaturation

r_c decreases with increasing S so nucleation becomes easier and the nucleation rate, J, is given by:

$$J = K_J \exp\left[\frac{-16\pi\gamma_e^3\omega^2}{3k^3T^3(\ln S)^2}\right] \tag{11.10}$$

where K_J is the nucleation rate constant.

11.5.2 Methods to Study Nucleation: Polythermal Method

Nucleation can be studied by measuring the turbidity of a crystallising solution by either polythermal or isothermal methods. As nuclei form and grow in an originally clear solution, the optical transmittance of the medium decreases. The polythermal method is based on the determination of the MSZW and the effect exerted on it by the rate at which supersaturation is created. The isothermal method is based on determination of induction time (τ),i.e. the time taken for crystallisation to be detected at constant temperature and the influence of the supersaturation on this time.

In polythermal method, as shown in Fig. 11.5, the difference between the dissolution temperature T_{diss} and the crystallisation temperature T_c is measured as a function of cooling rate (q). Polythermal experimental data can be analysed using different approaches:

11.5.2.1 Polythermal Method

Nyvlt Approach
Nyvlt approach [5, 6] uses an empirical expression for nucleation rate:

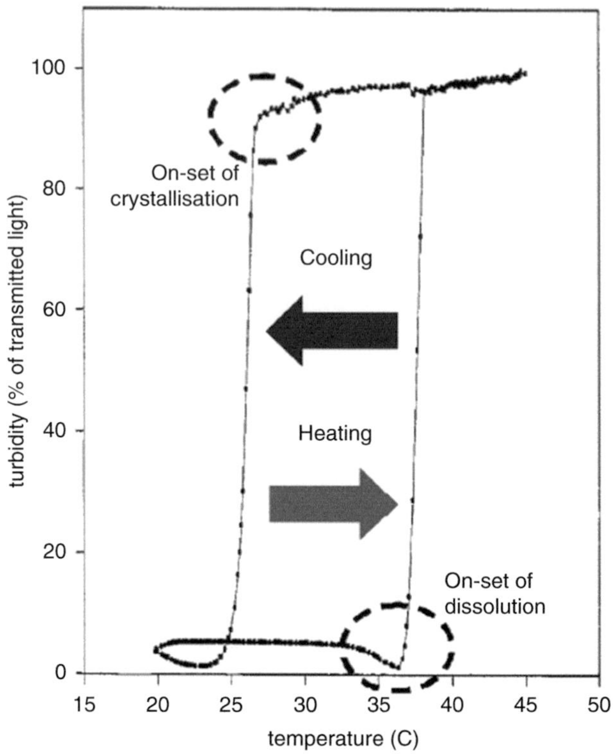

Fig. 11.5 Transmittance versus temperature for a turbidity probe in a crystallising solution (Reproduced by consent of J Crystal Growth from Ref. [7])

$$J = k_j(\Delta C_{max})^{m_1} \tag{11.11}$$

where k_j is an empirical parameter of nucleation rate and ΔC_{max} is related to ΔT_c by Eq. 11.12:

$$\Delta C_{max} = \left(\frac{dC^*}{dT}\right)\Delta T_c \tag{11.12}$$

$$\ln(q) = (m_1 - 1)\log\frac{dC^*}{dT} + \ln\ k_j + m_1 \ln \Delta T_c \tag{11.13}$$

The slope of a linear fit of experimental polythermal data in $\ln q$ versus $\ln \Delta T_c$ coordinates will deliver the order of nucleation m_1.

KBHR Approach

The KBHR approach [8–11] uses the expression of CNT defined in terms of relative critical undercooling u_c given by:

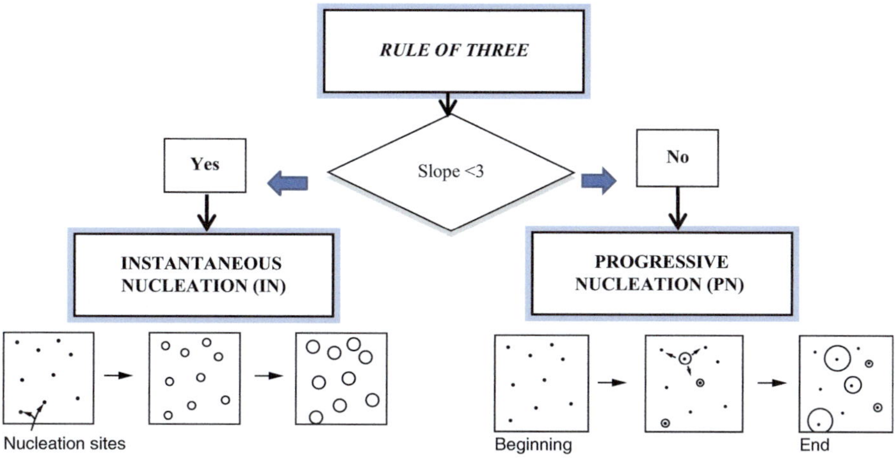

Fig. 11.6 Assessment of nucleation mechanism from polythermal data using the KBHR model (Reproduced by consent of CrystEngComm from Ref. [10])

$$u_c = \frac{\Delta T_c}{T_e} = \frac{T_e - T_c}{T_e} \tag{11.14}$$

Where T_e is the solution equilibrium temperature.

The slope of a linear fit of experimental polythermal data according to the expression $\ln\ q = \ln\ q_0 + (slope) \ln u_c$ will deliver the mechanism ruling the crystallisation process as defined by the rule of three (Fig. 11.6).

Further fitting of the data to specific models related to each case of nucleation allows determining interfacial tensions and nucleation rates in the case of progressive nucleation (PN) or the concentration of nuclei C_0 at the nucleation point in the case of instantaneous nucleation (IN).

Progressive Nucleation [9] Where new crystal nuclei are continuously formed in the presence of the already growing ones.

$$\ln q = \ln q_0 + a_1 \ln u_c - \frac{a_2}{(1-u_c)} {u_c}^2 \tag{11.15}$$

Here the three free parameters a_1, a_2 and q_0 are given by

$$a_1 = 3 \tag{11.16}$$

$$a_2 = b_o \tag{11.17}$$

$$q_0 = \frac{V K_J T_e}{N_{det}\, 2 b_0} \tag{11.18}$$

where N_{det} is the number of crystals at the detection point, V is the volume of the solution and b_o is defined by

$$b_o = \frac{k_n v_o^2 \gamma_e^3}{kT_e \lambda^2} \tag{11.19}$$

In this expression k_n is the nucleus shape factor, v_0 volume occupied by a solute molecule in the crystal and λ is the molecular latent heat of crystallisation.

Instantaneous Nucleation [8] Where all nuclei emerge at once at the beginning of the crystallisation process to subsequently grow and develop into crystal.

$$\ln q = \ln q_0 + \left(\frac{1}{m_g}\right) \ln\left[u_c{}^{(n_g+1)m_g} - u_0{}^{(n_g+1)m_g}\right] \tag{11.20}$$

with q_0 given by

$$q_0 = \left[\frac{k_v C_0}{\left(n_g + 1\right)^d \alpha_{det}}\right]^{\frac{1}{m_g d}} a_c{}^{n_g} K_G T_e \tag{11.21}$$

where u_0 is the relative critical undercooling at the IN point, n_g and $m_g$are the growth exponents, d is the dimensionality of crystal growth, α_{det} is the relative volume of crystals at the detection point, K_G is the growth rate constant, k_v is the crystal shape factor and a_c is defined by

$$a_c = \frac{\lambda}{kT_e} \tag{11.22}$$

Examples of the polythermal method applied to the case of α-para-aminobenzoic acid (α-PABA) crystallising from ethanol [12] and methyl stearate crystallising from kerosene [10] are presented below. Using the KBHR approach, the analysis of the crystallisation process for α-PABA solutions showed that crystals are formed by means of IN (Fig. 11.7). The concentration of PABA nuclei C_0 at the nucleation point, for different solution concentrations, was found to be in the order of 6.6×10^8–1.3×10^9 m^{-3}.

In contrast, the example of methyl stearate crystallising from kerosene (Fig. 11.8) reveals a PN nucleation mechanism.

In this case the application of the methodology to different solution concentrations allowed obtaining interfacial tensions (γ_e), critical radius (r_c) and nucleation rates (J) in the range of 1.21–1.91 mJ m^2, 0.7–0.9 nm and 5.1×10^{16}–7.9×10^{16} nuclei ml^{-1} s^{-1}, respectively.

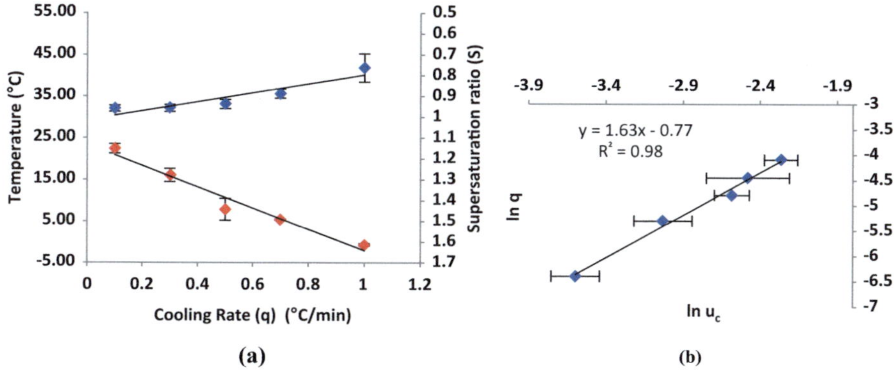

Fig. 11.7 Example of application KBHR approach to the analysis of polythermal data collected for α-PABA crystallising from ethanol with concentration (170 g/Kg). (**a**) Crystallisation and dissolution temperatures as a function of cooling rate (**b**) Plot of ln q vs ln u_c (Reproduced by consent of Faraday Discussions from Ref. [12])

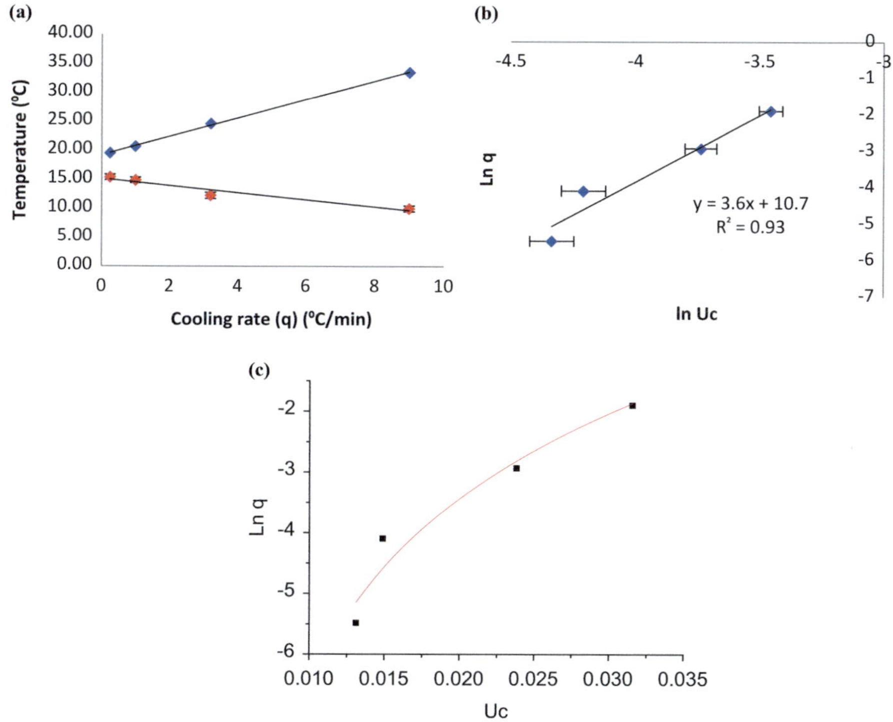

Fig. 11.8 Example of application KBHR approach to the analysis of polythermal data collected for methyl stearate crystallising from kerosene with concentration (250 gr/l) (**a**) Crystallisation and dissolution temperatures as a function of cooling rate (**b**) Plot of ln q vs ln u_c (**c**) Plot of ln q vs u_c (Reproduced by consent of CrystEngComm from Ref. [10])

11.5.3 Methods to Study Nucleation: Isothermal Method

In the isothermal method as shown in Fig. 11.9, a solution is crash cooled to different temperatures within the MSZW and the induction time τ is monitored by the change in the solution tubidity, from the time at which the solution reached the predetermined temperature to that of the crystallisation onset, which corresponds to the time at which the light transmittance decreases [11, 13].

The interfacial tension γ_e can be calculated from the slope of the line of a plot of experimental data in ln (τ) versus $T^{-3}(\ln S)^{-2}$ according to the expression below

$$\ln \tau = \left[\frac{16\pi {\gamma_e}^3 \omega^2}{3k^3T^3(\ln S)^2}\right] - \ln A_o \tag{11.23}$$

The plots obtained by application of isothermal method for n-eicosane crystallising from n-dodecane solvent in the presence of different impurities [14] are given in Fig. 11.10. The corresponding interfacial tensions obtained according to Eq. 11.23 are given in Table 11.1.

11.6 Crystal Growth Kinetics

The second step in a crystallisation process is the growth of stable nuclei into crystals. This process can occur through different mechanisms in each of the crystal faces.

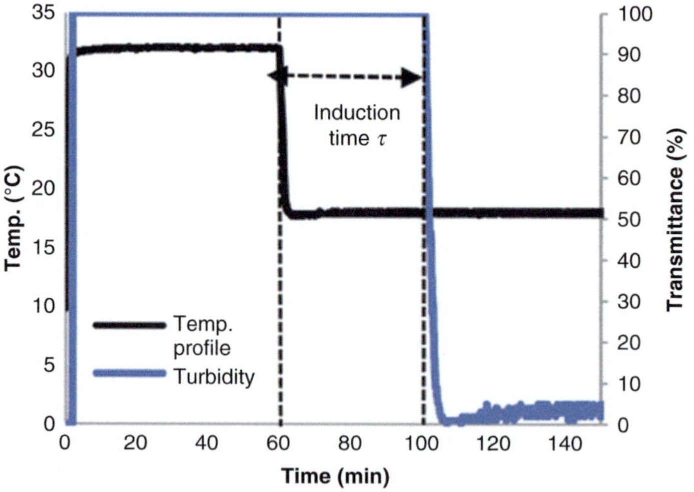

Fig. 11.9 Turbidity profile of a solution crystallising at a fixed temperature

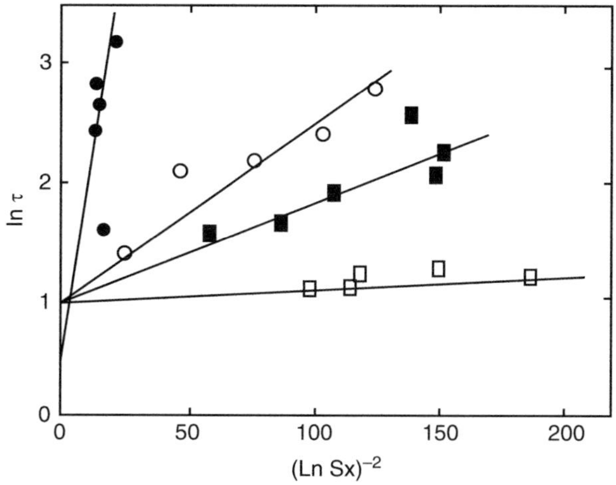

Fig. 11.10 Plot of ln (τ) versus (ln S)$^{-2}$ for n-eicosane (C20) crystallising from n-dodecane in the presence of different impurities. (•) C20/C21, (○) C20/C22, (■) C20/C18 and (□) C20/C19 (Reproduced by consent of J. Crystal growth from Ref. [14])

Table 11.1 Interfacial tensions obtained by the application of the isothermal method for n-eicosane (C20) crystallising from n-dodecane in the presence of different impurities [14]

Solute composition	Interfacial tension γ_e $\left(\frac{mJ}{m^2}\right)$
C20/C18	7.75
C20/C19	8.44
C20//C21	17.13
C20/C22	11.70

11.6.1 Birth and Spread (B&S) Model

This involves the formation of a stable cluster of molecules on a flat face (Fig. 11.11), i.e. nucleation is required. As for 3D nucleation, the 2D nucleus must reach a critical radius to become stable [4, 15].

The B&S model follows an exponential tendency with growth mediated by 2D nucleation:

$$R_i = A_o \sigma^{5/6} \exp\left(\frac{B_o}{\sigma}\right) \tag{11.24}$$

where

R_i = rate of growth of a crystal face (m/s)
σ = relative supersaturation
A_o and B_o = system related constants

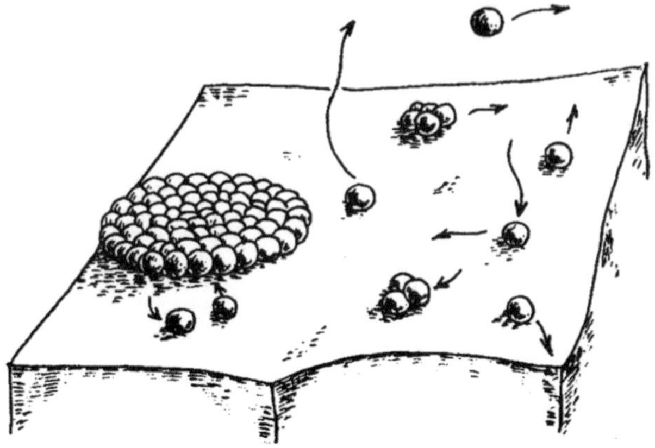

Fig. 11.11 Schematic of B & S crystal growth process

11.6.2 Burton Cabrera Frank (BCF) Model

Some crystals contain imperfections known as dislocations. Screw dislocations produce half a step where they emerge on a crystal face. As molecules attach to the step, it winds up into a spiral on the crystal face [4, 15] (Fig. 11.12).

The BCF model follows a parabolic tendency with growth being mediated by the presence of screw dislocations on the crystal surface.

$$R_i = A_o\sigma^2 \tanh\left(\frac{B_o}{\sigma}\right) \tag{11.25}$$

where

R_i = rate of growth of a crystal face (m/s)
σ = relative supersaturation
A_o and B_o = system related constants

11.6.3 Rough Interface (RIG) Model and the Jackson α Factor

If a crystal surface is rough at the molecular level every growth unit which impinges on the surface can be expected to be incorporated [17, 18].

This model follows a linear tendency as the growth occurs on a molecular roughened surface:

$$R_i = A_o\sigma \tag{11.26}$$

where

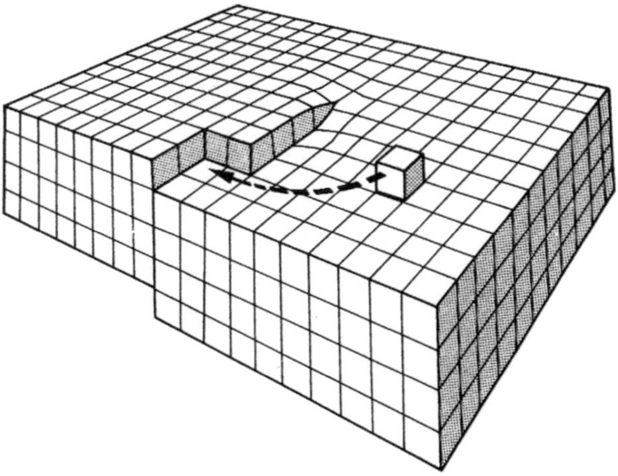

Fig. 11.12 Schematic of a face growth mediated by screw dislocations (Reproduced by consent of CrystEngComm from Ref. [16])

R_i = rate of growth of a crystal face (m/s)
σ = relative supersaturation
A_o = system related constant

A useful parameter in assuring surface roughness is the Jackson α factor [19].

The α factor also referred to as the surface entropy factor, describes the roughness of a crystal face [20, 21]. It is the product of the anisotropy factor ζ and the entropy change upon crystallisation:

$$\alpha = \frac{\zeta \Delta H_f}{RT} \tag{11.27}$$

where ζ = anisotropy factor.

The anisotropy factor is related to the number of nearest neighbour site at the interface and in the bulk of the crystal. The entropy change is influenced in part by the shape and complexity of the crystallising species.

$\alpha < 2$ rough interface leading to a continuous growth mechanism.

$\alpha > 5$ molecularly flat interface upon which 2D nucleation will be unfavourable resulting in the dominance of a spiral growth mechanism.

$2 < \alpha < 5$ two-dimensional nucleation is expected to dominate.

11.6.4 Methods to Measure Crystal Growth Rates

Measurements of growth rates for specific crystal faces (*hkl*) can be carried out using microscopy coupled with a growth cell set up [22]. The growth rates of the individual faces R_i can be obtained by following the increase with time of the normal distance from the centre of the crystal to the faces. Figure 11.13 shows the methodology used to measure the growth of individual faces of Ibuprofen crystals

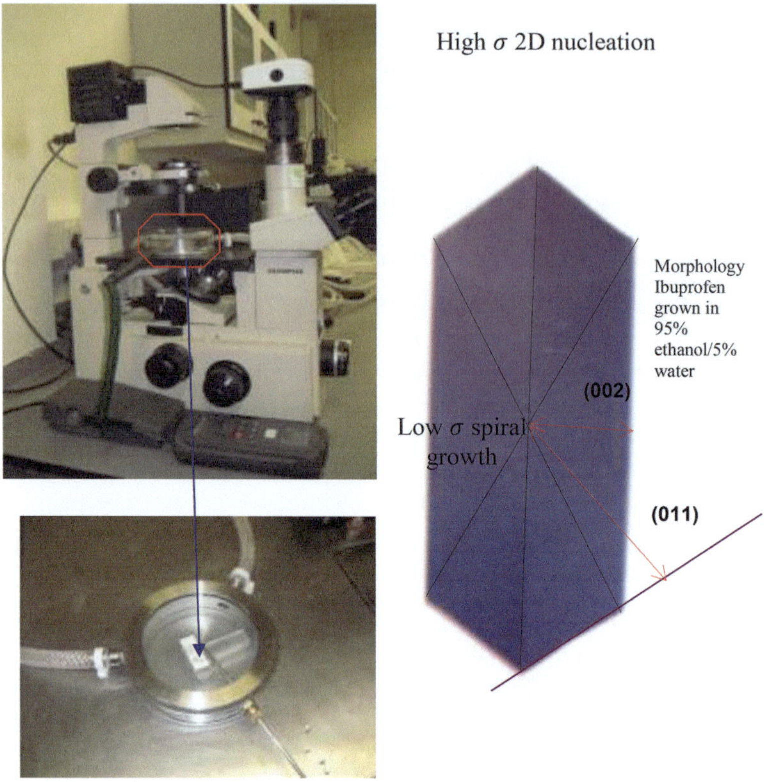

Fig. 11.13 Experimental set up and methodology for the measurement of growth rates of the (011) and (002) faces of Ibuprofen crystals: Olympus IMT-2 inverted optical polarising microscope integrated with Lumenera Infinity 3.3 megapixel CCD camera; enlarged picture of the crystal growth cell and example of measurement of normal distances from the centre of the crystal to the faces (Reproduced by consent of CrystEngComm from Ref. [22])

and Fig. 11.14 an example of the corresponding growth rate values obtained experimentally.

The mechanism of crystal growth of a specific crystal face can be assessed by fitting of measured growth rate data as a function of supersaturation to the models described. Figure 11.15 shows an example of this assessment for Ibuprofen crystals growing from different solvents.

11.7 Crystallisation Process Engineering

This section will provide a practical example of crystallisation process design which includes:

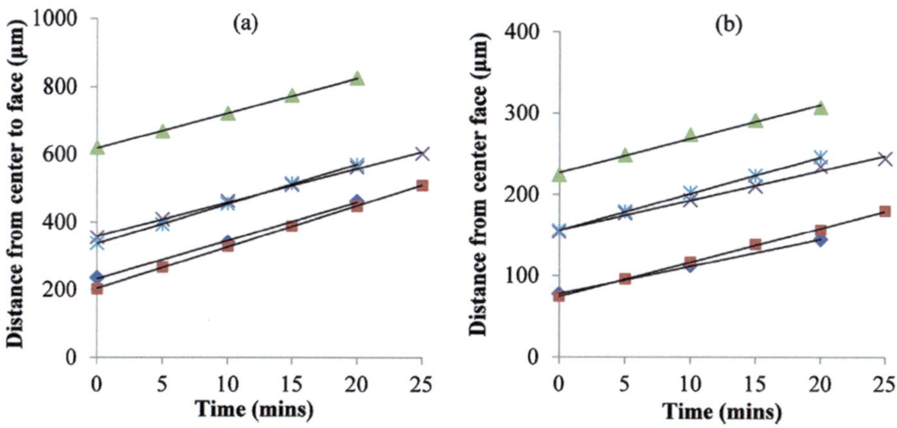

Fig. 11.14 Example of normal distances from the center of the crystals to the {011} and {001} faces as a function of time. Each *line* represents the growth rate of an individual crystal over time: (**a**) the {011} face in ethanol at $\sigma = 0.66$ and (**b**) the {001} face in ethanol at $\sigma = 0.66$ (Reproduced by consent of CrystEngComm from Ref. [22])

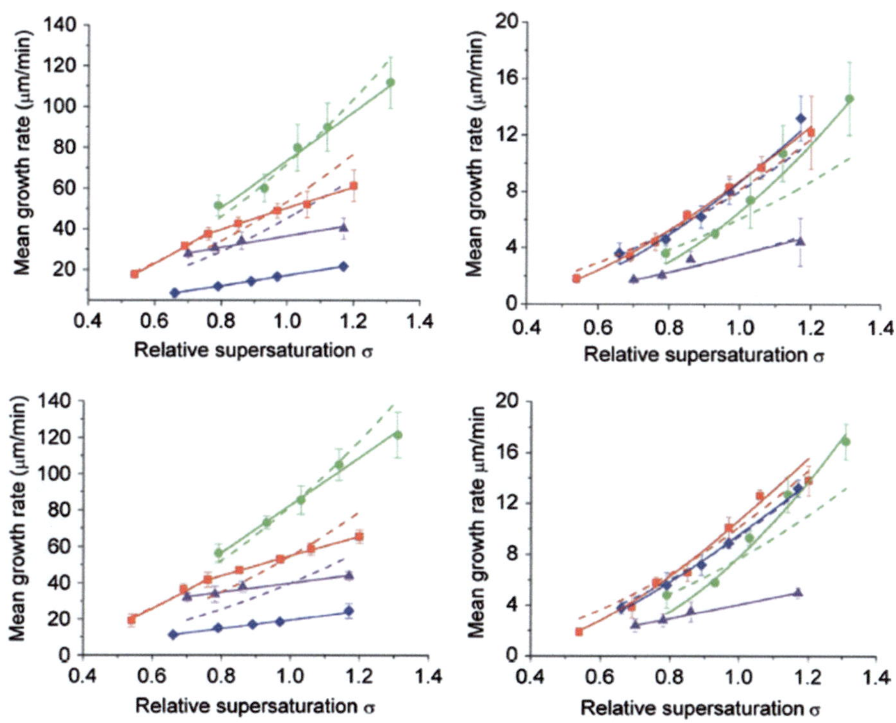

Fig. 11.15 Growth rate versus relative supersaturation of ibuprofen crystals growing from ethanol (*blue*), ethyl acetate (*red*), acetonitrile (*green*) and toluene (*purple*) together with fitted B&S (*solid lines*) and BCF (*dotted lines*) mechanism models for the (011) (*left*) and (001) (*right*) faces and for both 0.5 ml (*top*) and 15 ml (*bottom*) scale sizes (Reproduced by consent of CrystEngComm from Ref. [22])

- Hydrodynamics and mixing in a crystalliser using computational fluid dynamics (CFD) methods,
- Methodology of multi-zonal models based on the detail CFD simulation results of mixing and hydrodynamics in crystallisers,
- Morphological population balance (MPB) modelling approach for capturing crystal size and shape distributions and their evolution during crystallisation processes,
- Process scale-up studies linking to crystallisers' size scales/configurations, operating conditions such as agitation speed, and the internals (impeller types/materials . . .).

11.7.1 Hydrodynamics of a Batch Crystallisation Process

Hydrodynamics in a crystalliser can be very complex because the configuration of a crystalliser may be composed of moving components (such as impellers for better mixing), stagnant zones (such as the use of baffles) and the different locations of components input and product output, hence leading to non-homogeneous distributions of flow pattern, temperature, concentration, particles, etc. in the crystalliser (see for example [23–28]).

To accurately characterise the hydrodynamics in a crystalliser, both experimental measurements and numerical simulations can be used. With the advances in computational fluid dynamics (CFD) and powerful high performance supercomputers, CFD modelling methods have been widely used to characterise and capture the important flow and mixing features in a crystalliser. Such simulations can be validated with the help of experimental measurements (such as Laser Doppler Anemometry, LDA, which measures fluid velocity properties throughout a vessel as a function of agitation conditions and reactor scale size, etc.). The Reynolds number ($Re = \rho\, Ns\, D^2/\mu$), defined as the ratio of inertia forces to viscous forces and consequently quantifying the relative importance of these two types of forces for given flow conditions, in a crystalliser, is one of the most important parameters to affect the mixing and flow profiles. The Re is used to characterise different flow regimes such as laminar, transition and turbulent flow according to the Re values. Generally speaking, higher Re will produce better mixing. For batch crystallisation processes, higher Re usually is generated by higher impeller speed, hence requiring higher power input and also leading to high possibility of crystal breakage. The Re can also be used to investigate the scale-up of crystallisation processes.

To simulate a crystallisation process, a multi-phase and multi-component system is required to be solved using multi-phase CFD for velocity, temperature, concentration distributions, coupled with population balance (PB) modelling for crystal size/shape distributions. The coupling can be at different levels. Traditionally, for a batch crystalliser with an impeller, a well-mixed condition is assumed, therefore a PB model can be applied to obtain the crystal size/shape distributions. However, this can cause big errors in the PB simulation as the actual conditions in a crystalliser are not homogeneous. The high level of coupling is to fully couple CFD with PB through each CFD mesh cell, i.e. treating each mesh cell as a well-mixed small reactor and applying PB to obtain size/shape distribution in this cell with the dynamic exchange of flow features information with its neighbouring cells. The fully coupling method will significantly increase the required computational time because the total mesh cells for the CFD simulation of a crystalliser can be millions in order to capture the flow/mixing features, hence the same number (millions) of PB equations needs to be solved for each time step over the crystallisation period.

Turbulence exists in almost all flows of practical engineering interests including crystallisation processes, and is inherently three-dimensional and time dependent. Due to its extreme complexity, turbulence has been recognised as the major unsolved problem of classic physics [29]. To address this issue, various turbulence models have been developed including zero–/one-equation, two-equation ($k-\varepsilon$, $k-\omega$ and the variations), and Reynolds-stress models, etc. The most commonly used $k-\varepsilon$ model can be employed to represent the Reynolds-stress terms with the two equations being derived from the Navier-Stokes equation with some closure assumptions to allow simplification of the Reynolds-stress terms [29, 30].

The continuity, momentum and enthalpy conservation equations based on time-averaged quantities, together with turbulence models, are derived from Reynolds averaging of the corresponding instantaneous equations and numerically solved to obtain hydrodynamic and heat transfer profiles in crystallisers. The mixing behaviours in the crystallisers can be simulated by introducing an inert tracer into the reactors. The solution of the Reynolds-averaged species transport equations produces the spatial and temporal distributions of the tracer concentrations. For multiphase flow, two-fluid / multi-fluid method can be used to treat each phase as an inter-penetrated fluid. Each phase will have the corresponding conservation equations with a weighing factor to be its mass fraction. For further details, please refer to literature such as [31–34] and textbooks.

For a batch crystalliser with 3-blade retreat curve impeller (typically used in glass lined reactors), the reactor configuration is shown in Fig. 11.16 [35].

With the proper mesh cells, initial and boundary conditions and solution schemes, the CFD simulation of the crystalliser with water and 100 rpm impeller speed was validated by the measured velocity using LDA in a crystalliser with the same configurations (Fig. 11.17).

11.7.2 Zonal Models for Crystallisers

The compromised method is to use multi-zonal model or Villermaux's segregated feed model [36] by which CFD simulations are performed first, and then different zones in the crystalliser are identified according to the flow/mixing features. These zones will be treated as well-mixed reactors to couple with PB modelling. This will dramatically reduce the number of PB equations to be solved, hence saving

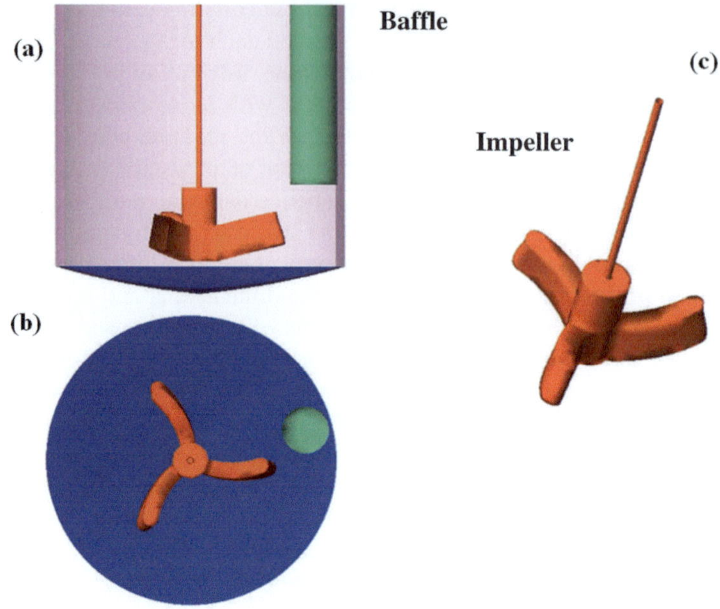

Fig. 11.16 (**a**) A baffled batch crystalliser with 3-blade retreat curve impeller; (**b**) *top-view* of the reactor; (**c**) the impeller (Reproduced by consent of Ind. Eng. Chem. Res. from Ref. [35])

computational time. This method is particularly useful when involving large scale industrial crystallisers and investigating process scale-up.

The flow and turbulence characteristics obtained from the CFD simulation can be used to identify different mixing regions. As shown in Fig. 11.18, for a zonal coupling method, the crystalliser can be divided into four zones such as impeller zone (1), bottom zone (2), top zone (3) and wall zone (4).

A tracer was introduced into the crystalliser to simulate the turbulent mixing via solving a species transport concentration equation to obtain the tracer distributions over time [33] as shown in Fig. 11.19.

The Segregated Feed Model (SFM) [32, 36] is a compartmental mixing model. The combination of SFM with CFD and population balance can simulate the effect of various operating conditions and reactor configurations on the nucleation rate and crystal size distribution.

Villermaux [36] proposed the SFM model based on physically meaningful mixing parameters involving:

- diffusive micro-mixing time;
- convective meso-mixing time.

The SFM is particularly suitable for modelling mixing effects as it combines advantages of both

- compartmental model;
- physical model.

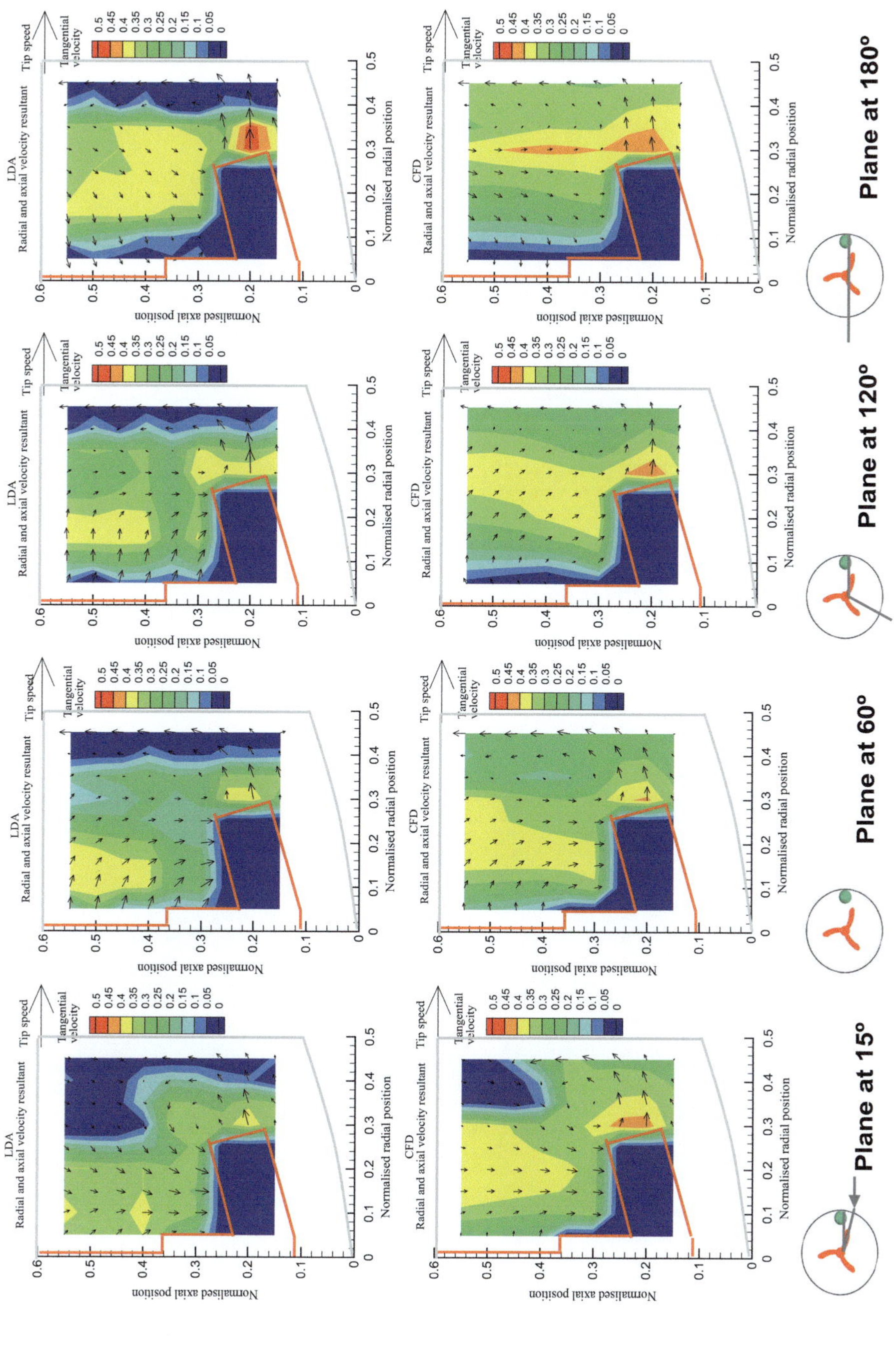

Fig. 11.17 Flow patterns at various vertical planes: (*Top*) LDA measurements; (*Bottom*) CFD simulations (Reproduced by consent of Ind. Eng. Chem. Res. from Ref. [35])

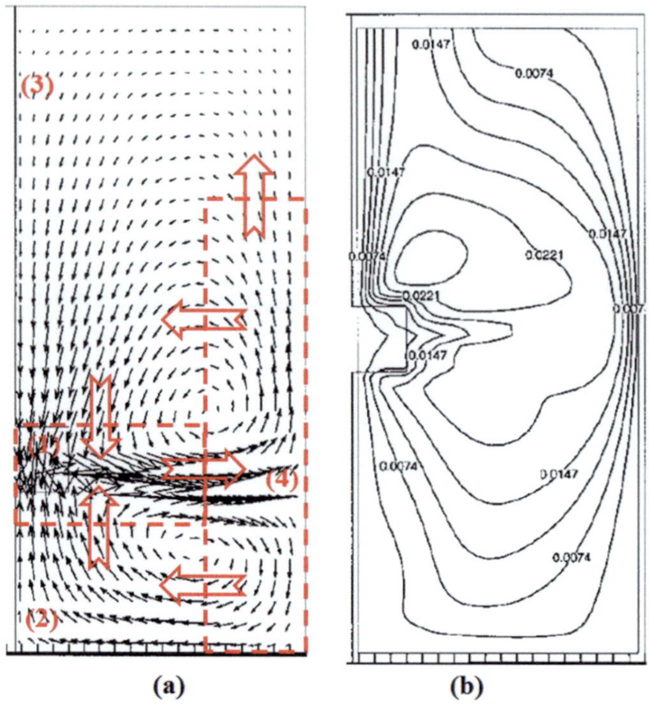

Fig. 11.18 (**a**) Velocity vectors; (**b**) eddy viscosity contour (Reproduced with permission from Ref. [37])

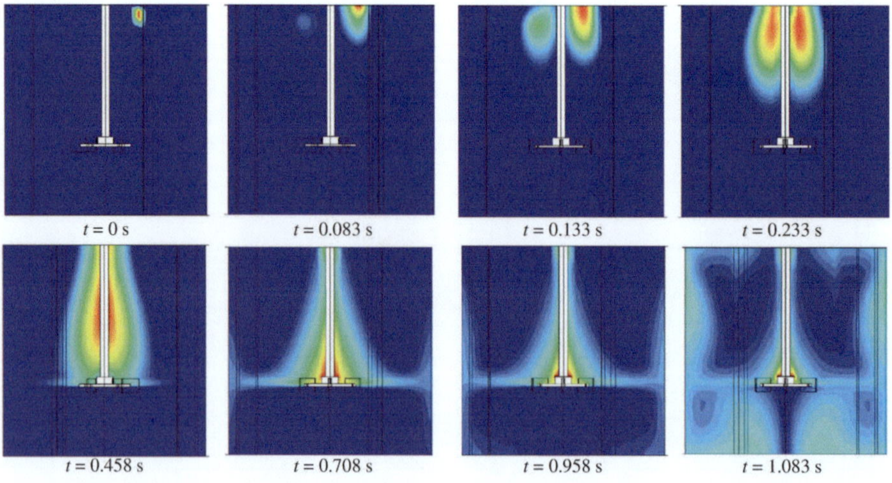

Fig. 11.19 Tracer concentration distributions in the vertical planes at 0 ° (tracer injection plane) and 180 ° angular positions, which reveal the overall mixing process in the tank (Reproduced by consent of Chem. Eng. Process. from Ref. [33])

SFM divides the reactor into three zones:

- two feed zones f_1 and f_2;
- bulk b.

The feed zones exchange mass with each other and also with the bulk zone. The process depicted by flow rates $u_{1,2}$, $u_{1,3}$ and $u_{2,3}$, respectively is shown in Fig. 11.20a [32]. The time constants characteristic can be used to identify the micro-mixing and meso-mixing [38, 39]:

$$t_{micro} = 17.3 \times \left(\frac{v}{\varepsilon_{loc}}\right)^{1/2} \qquad t_{meso} = A_m \frac{\varepsilon_{ave}}{\varepsilon_{loc}} \frac{Q^{1/3}}{N_s^{4/3}\, d_s} \tag{11.28}$$

where ε is the specific power input; A_m is a constant from literature [38, 39]; d_s is the diameter of a stirrer; Q is the flowrate of a stream.

As shown in Fig. 11.20b [32], the scale-up methodology using SFM involves the following steps:

- Carry out laboratory scale measurements;
- Model hydrodynamics via computational fluid dynamics (CFD);
- Use population balance model for particle properties (number/size distribution);
- Link two models via segmented feed model (SFM);
- Predict precipitation performance as a function of scale size.

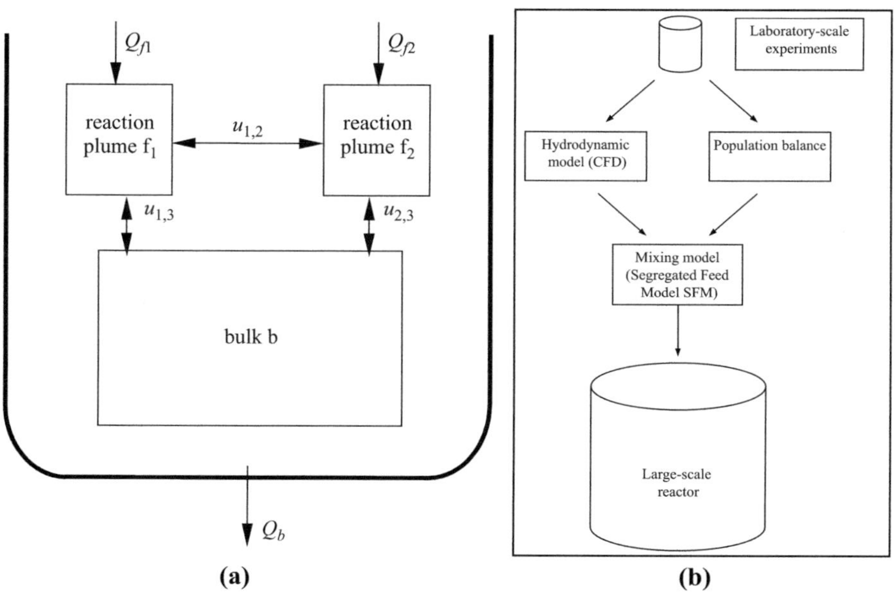

Fig. 11.20 (**a**) Schematic of segregated feed model, (**b**) scale-up methodology with SFM approach (Reproduced by consent of Ind. Eng. Chem. Res. from Ref. [32])

11.7.3 Morphological Population Balance Models

A population balance (PB) model generally accounts for the convective processes that involve both the motion of particles in a system through their defined domains and their birth-and-death processes that can both terminate existing particles and produce new particles. The generic mathematical formulation for multidimensional PB modeling can be given by the following equation:

$$\begin{aligned}&\frac{\partial n(\boldsymbol{x},\boldsymbol{y},t)}{\partial t}+\nabla\cdot[n(\boldsymbol{x},\boldsymbol{y},t)\boldsymbol{v}]+\sum_{i=1}^{N}\frac{\partial}{\partial x_i}[n(\boldsymbol{x},\boldsymbol{y},t)G_i(\boldsymbol{x},\boldsymbol{y},t)]\\&+\frac{n(\boldsymbol{x},\boldsymbol{y},t)-n_0}{\tau_r}=B_a(\boldsymbol{x},\boldsymbol{y},t)-D_a(\boldsymbol{x},\boldsymbol{y},t)+B_d(\boldsymbol{x},\boldsymbol{y},t)-D_d(\boldsymbol{x},\boldsymbol{y},t)\\&+B_0(\boldsymbol{x},\boldsymbol{y},t)\end{aligned}\tag{11.29}$$

where N is the number of internal variables for a crystal, $\boldsymbol{x}$ is the internal variable vector with n components, which can be parameters related to crystal size, shape, and other properties; $\boldsymbol{y}$ is the external variable vector such as spatial coordinates (y_1, y_2, y_3); n is the number population density function of crystals in the internal variable range (x_i, $x_i + dx_i$, $i = 1, N$) and in the differential volume of $dy_1dy_2dy_3$; ∇ is the gradient operator for the $\boldsymbol{y}$ coordinates. On the left-hand side, the 1st term is the accumulation term of n; the 2nd term denotes the convection of n in the $\boldsymbol{y}$ space with $\boldsymbol{v}$ being the velocity vector; the 3rd term is the convection of n due to particle growth in the $\boldsymbol{x}$ space with G_i being the growth rate; the 4th term is the net change of n during residence time, τ_r, due to the inlet and outlet flows of continuous crystallization processes with n_0 being the initial number population density function of crystals. The terms on the right-hand side of Eq. 11.29, $B(\boldsymbol{x}, \boldsymbol{y}, t)$ and $D(\boldsymbol{x}, \boldsymbol{y}, t)$, represent the birth and death terms of n for agglomeration and breakage, and the third term $B_0(\boldsymbol{x}, \boldsymbol{y}, t)$ for nucleation. Indices a, d, and 0 relate to agglomeration, breakage and nucleation.

In a well-mixed crystalliser, Eq. 11.32 becomes the PB equation proposed by Randolph and Larson [40]:

$$\begin{aligned}&\frac{\partial n(\boldsymbol{x},t)}{\partial t}+\sum_{i=1}^{N}\frac{\partial}{\partial x_i}[n(\boldsymbol{x},t)G_i(\boldsymbol{x},t)]+\frac{n(\boldsymbol{x},t)-n_0}{\tau_r}\\&=B_a(\boldsymbol{x},t)-D_a(\boldsymbol{x},t)+B_d(\boldsymbol{x},t)-D_d(\boldsymbol{x},t)+B_0(\boldsymbol{x},t)\end{aligned}\tag{11.30}$$

Although population balance (PB) modeling for crystallization processes is for all crystals in a crystallizer, crystal shape was often ignored with an over-simplified crystal-size definition, i.e., the volume equivalent diameter of spheres or simplified as length and width for some needle-like crystals [41]. A morphological population balance (MPB) model is able to incorporate any complicated crystal structures/shapes into PB modeling, therefore, can simulate the size-related dimensional evolution of crystals for each identified independent crystal face (for further details, see [42, 43]). From the predicted growth of different faces at different times during crystallization process, many important crystal properties, such as shape and growth rate can be evaluated and used for real-time monitoring, control and manipulation of crystal morphology.

The MPB models define the multiple dimensions of PB number density as the distances of each face to its geometric centre, hence being able to fully re-construct the shape of any one crystal at any time and also taking into account different growth kinetics for different crystal faces. Taking a potash alum crystal as an example, a 3D MPB model, the corresponding three parameters, x_1, x_2, x_3, shown in Fig. 11.21, can be formed to model its morphological changes in a well-mixed batch crystallizer with breakage, agglomeration and nucleation being ignored. The MPB equation of Eq. 11.30 can, thus, be written as follows:

$$\frac{\partial n(x,y,z,t)}{\partial t}+\frac{\partial}{\partial x}[n(x_1,x_2,x_3,t)g_1(x_1,t)]+\frac{\partial}{\partial y}[n(x_1,x_2,x_3,t)g_2(x_2,t)]$$
$$+\frac{\partial}{\partial z}[n(x_1,x_2,x_3,t)g_3(x_3,t)]=0 \tag{11.31}$$

The crystal shape evolution as shown in Fig. 11.22 demonstrates that the {100} and {110} faces will eventually disappear with the crystal expected to exhibit pure octahedral diamond-like morphology at steady state under the current simulation conditions, which has been observed in literature.

For the crystallisation of potash alum crystals, the growth rate dispersion (GRD) may play a role particularly on the faster growing {100} faces, especially for larger crystals (~cm). Growth sector boundaries predicted from the MPB simulations with and without GRD and measured by experiment are plotted in Fig. 11.23. The faces {100} and {110} clearly tend to disappear completely (Fig. 11.23) if GRD effect is not included in the simulations. However, when the GRD is included, all three habit

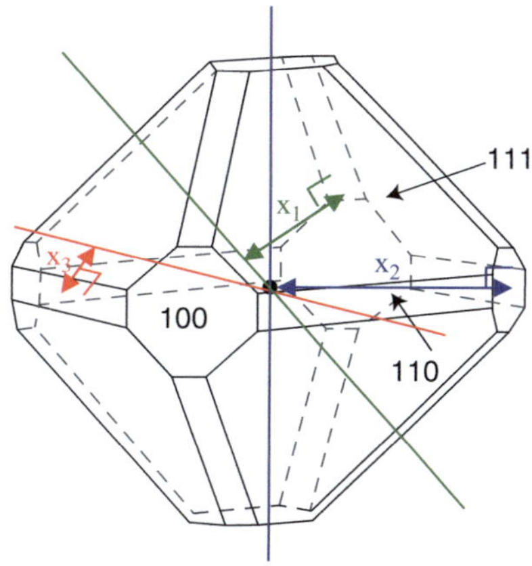

Fig. 11.21 Morphology of potash alum crystal and schematic diagram of the three size characteristic parameters (x_1, x_2, x_3) to be used in a MPB model for each independent crystal face in a potash alum system (Reproduced by consent of AIChE J. from Ref. [42])

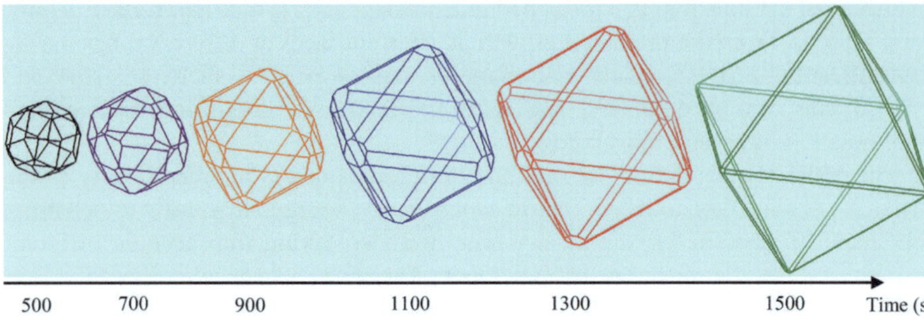

Fig. 11.22 Crystal shape evolution during the crystallisation process of potash alum (Reproduced by consent of AIChE J. from Ref. [42])

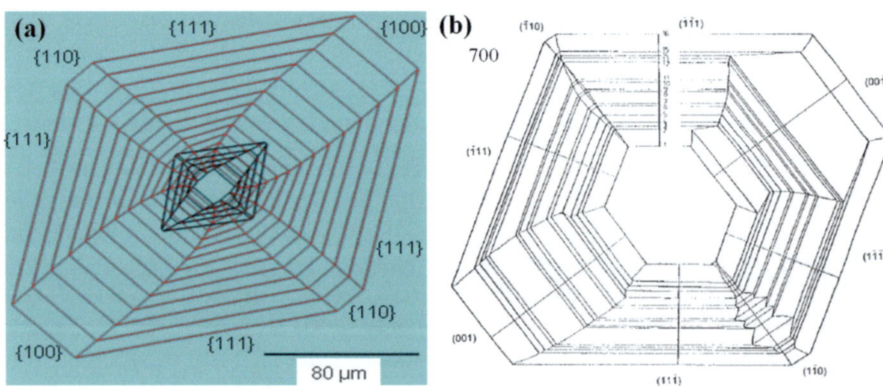

Fig. 11.23 Growth sector boundaries of a typical potash alum crystal (**a**) predicted by the MPB model with (coloured in *red*) and without (coloured in *black*) including GRD effect (Reproduced by consent of AIChE J. from Ref. [43]) and (**b**) measured by experiments (Reproduced by consent of Chem. Eng. Technol. from Ref. [44])

faces show continuous growth at variable speeds, which is close to that found in experiments.

By comparing the predicted growth sector boundaries with GRD effect (Fig. 11.23a) with the experimentally measured boundaries (Fig. 11.23b), the results are in qualitative agreement.

11.7.4 Crystallisation Process Scale-Up

The PB modelling method is a scalable technique which can be directly applied to large and small scale systems. However, as shown in Eq. 11.29, the crystal size/shape distributions are affected by the mechanisms/kinetics of nucleation, growth, agglomeration, breakage, etc. which are directly related to the flow pattern and mixing, mass and heat transfer in a reactor. The two-way interactions between CFD and MPB modelling can be achieved via either zonal coupling or mesh-to-mesh fully CFD-MPB coupling. The coupled CFD-MPB modelling method will have great potential for the scale-up of crystallisation processes and further research is needed. As we understand, the physical phenomena in a crystalliser play very important roles in the process scale-up, which include mass transfer, heat transfer, physical properties of solute and solvent, etc. The process scale-up can be performed by simple geometrical similarity which may not achieve the corresponding mixing, particle quality, if flow characteristics in the two reactors are not similar. For dynamic and kinematic similarities, the ratio of forces in a process, and the velocity at similar locations (such as impeller tip) can be kept constant during scale-up. Furthermore, the crystallisers' internals such as impeller types and materials can also affect the performance of the crystallisation processes during scale-up.

Liang et al. [28, 45] investigated the effects of reactor internals [28] and reactant mixing [45] on the measured MSZW associated with the batch crystallisation of L-glutamic acid from supersaturated aqueous solutions. The cooling crystallisation experiments were carried out at three reactor scales (450 mL, 2 L and 20 L) agitated at various stirring speeds using an industry-standard retreat curve impeller with a single beaver-tail baffle. Nucleation kinetic parameters at 450 mL were evaluated using a method proposed by Nyvlt et al. [5]. It was found that increased mixing generally is capable of enhancing the nucleation rate, but with a further increase of the stirrer speed beyond a critical value, aeration was observed and this may contribute to the reduced nucleation. The measured MSZWs are mostly found to decrease with increasing stirring speed, with enhanced nucleation also being observed as the reactor scale increased (Fig. 11.24); albeit hindered nucleation was found at higher stirrer speeds in the 450 mL reactor experiments (Fig. 11.25).

From Fig. 11.24, a linear relationship (Eq. 11.32) between MSZW and the stirrer speed was obtained by least-squares linear regression analysis.

$$\Delta t_{max} = -0.0464 N_s + 35.109 \tag{11.32}$$

Where Δt_{max} is the maximum possible supercooling of the system and N_s is the stirrer speed. This revealed that the stirrer speed is a significant parameter of the current system. However, for a general model application it needs to correlate the nucleation process to the reactor hydrodynamics as characterised by the Reynolds number. Through combining the influence of reactor hydrodynamics and scale on mass transfer during the crystallization process, a general correlation was postulated, including the method for nucleation kinetics from Nyvlt et al. [5] and the definition of the maximum possible supersaturation Δc_{max} as a function of Δt_{max}. It was fitted with experimental data (as shown in Fig. 11.26) using multiple nonlinear regression analysis to yield Eq. 11.33:

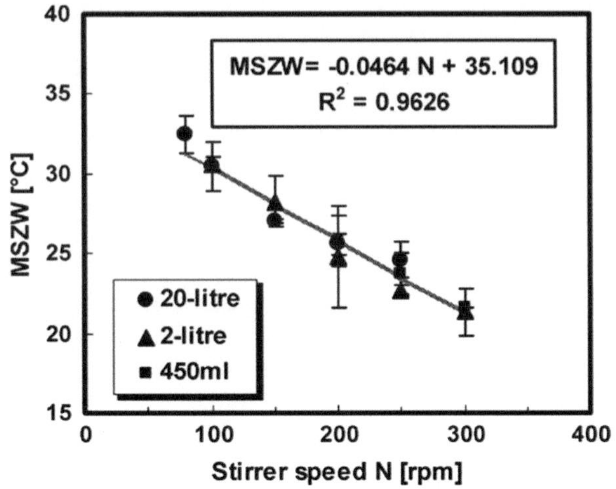

Fig. 11.24 MSZW as a function of the stirrer speed in 450-mL, 2-L, and 20-L reactors at a cooling rate of 0.2 °C/min (Reproduced by consent of Ind. Eng. Chem. Res. from Ref. [45])

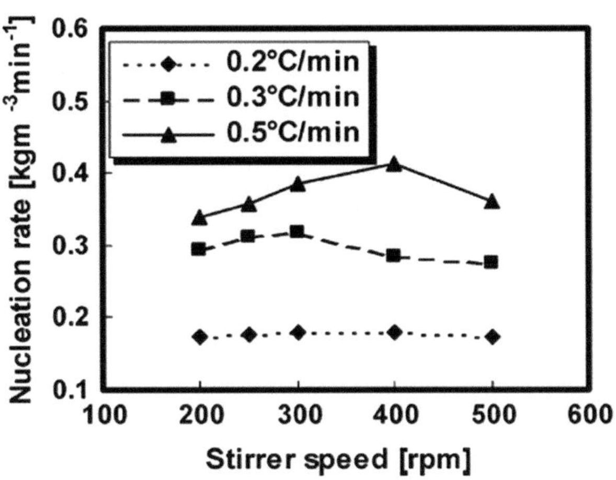

Fig. 11.25 Calculated nucleation rate as a function of the stirrer speed in a 450-mL reactor (Reproduced by consent of Ind. Eng. Chem. Res. from [45])

$$J \propto \left(\frac{\Delta t_{max}}{t_{sat}}\right)^{1.54} = 18.95\, Re^{-0.5} \left[\frac{T}{T_0}\right] \tag{11.33}$$

where t_{sat} is the saturated temperature, *Re* is the Reynolds number, T_0 is the diameter of the laboratory reactor and T/T_0 is the scale-up ratio. It is suggested that the correlation provides a good estimate of MSZW in an agitated vessel for the system examined. It is also concluded that mixing affects the surface-induced

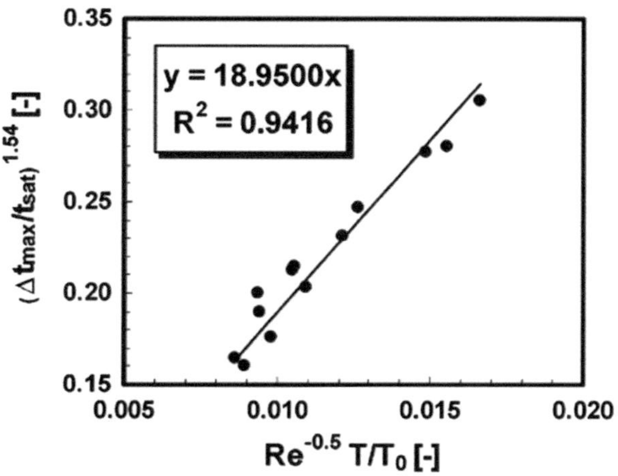

Fig. 11.26 Fitting experimental data with Eq. 11.33 showing a good agreement of the experimental data and the proposed model (Reproduced by consent of Ind. Eng. Chem. Res. from [45])

heterogeneous nucleation process by thinning the boundary layer at the stirrer blade surface.

Liang et al. [28] reported that primary nucleation can be affected by stirrer material. In the study, batch cooling crystallizations of L-glutamic acid aqueous solutions were carried out in a 450 mL reactor using stirrers with identical geometry but made from different materials: stainless steel and Perspex. It was found that there existed a high degree of crystal encrustment on the surface of the stirrers and impeller shaft with much denser crystal attachments on the blades of the stainless steel stirrer than on the Perspex one. This strongly indicated that the nucleation process initially started on the surface of the stirrer rather than in the preferentially cooled regions adjacent to the reactor wall as would be more conventionally accepted. By using measured MSZW for nucleation kinetics analysis, both stirrers showed similar MSZW profiles as a function of stirring rate, though nucleation was found to be much easier with the stainless steel stirrer. Nucleation order obtained from the experiments performed with the stainless steel stirrer were found to be greater than those with the Perspex stirrer, i.e., consistent with a much lower energy barrier for nucleation in the case of the stainless steel stirrer. However, the data also show that the nucleation rate constant for experiments carried out using the Perspex stirrer were much higher than those when using the stainless steel impeller for the same agitation rate. Figure 11.27 shows the measured advancing contact angle and free energy ratio of L-glutamic acid aqueous solutions on both Perspex and stainless steel flat plates.

The encrustment observations together with experimental measurements (Fig. 11.27) of contact angle for different stirrer materials, from which the free energy ratio was calculated, confirm that the energy needed to form critical nuclei on the stainless steel surface would be much lower than on Perspex. Surface roughness is also believed to play an important role. Overall, these observations

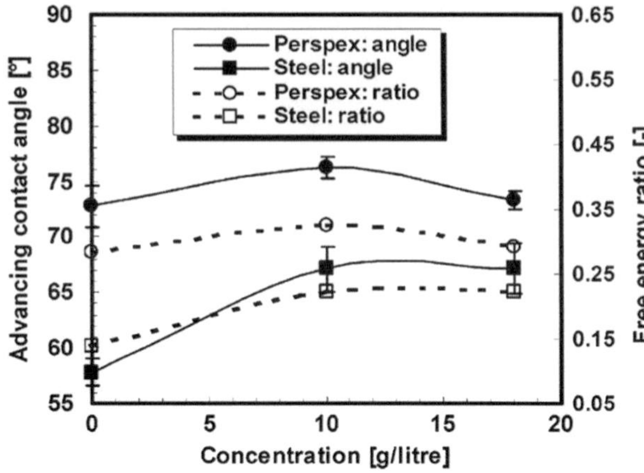

Fig. 11.27 Dependence of advancing contact angle and free energy ratio on surface types and concentration of aqueous L-glutamic acid solutions (Reproduced by consent of Crystal Growth & Design from [28])

are consistent with the stirrer surface providing preferred nucleation sites in L-glutamic acid crystallization with both the stirrer material and its surface roughness being important factors dictating the nature of the primary nucleation process. It also appears that nucleation occurs first on the surface of the stirrer, where the strongest turbulent kinetic energy is present. Hence, it is reasonable to conclude that these newly formed nuclei grow continuously to critical nuclei as freshly supersaturated solution is transported to the region due to better micromixing compared to that in the rest of the bulk. These stable nuclei may then be washed away by the strong fluid shear force and quickly dispersed into other parts of the bulk in the crystallizer and then the overall nucleation event is triggered. Therefore, this study [28] reveals a heterogeneous nucleation mechanism involving a surface-induced process on the stirrer surface with the surface properties and its material of construction playing an important role by the overall crystallization process.

Li et al. [46] used CFD simulations to investigate the scale up effects for three geometrically similar laboratory scale vessels of 0.5, 2 and 20 L with retreat curve impellers and cylindrical baffles, which mimic reactors widely used in the pharmaceutical and fine chemical industries. CFD results have then been validated using LDA measurements and empirical power consumption literature data. The comparisons of power number, discharge flow number, secondary circulation flow number and pumping efficiency at three different scales suggest that the scale up with the selected laboratory vessels has little effect on the macro mixing performance for optimisation of the configuration and operating conditions of an industrial scale reactor. Further details about the experimental and modelling investigations of process scale-up can be found in the literature, such as [28, 32, 45–49].

11.8 Concluding Remarks

The objective of this crystallisation route map is to lay the foundation for the crystallisation processes, which covers the solubility and solution ideality for crystallisation processes, the supersaturation, MSZW and its impact on product form through the use of supersaturation to control nucleation and growth processes, the nucleation and its kinetic characterisation, the crystal growth and its measurement together with characterising the growth mechanisms, and the hydrodynamics of crystallisation processes, population balance modelling, and crystallisation scale-up.

References

1. Prausnitz JM (1969) Molecular thermodynamics of fluid-phase equilibria. Prentice-Hall Inc., Englewood Cliffs
2. Dickerson RE (1969) Molecular thermodynamics. W. A. Benjamin, New York
3. Davey RJ, Mullin JW, Whiting MJL (1982) Habit modification of succinic acid crystals grown from different solvents. J Cryst Growth 58:304–312
4. Mullin JW (2001) Crystallization, 4th edn. Butterworth-Heinemann, Oxford
5. Nyvlt J (1968) Kinetics of nucleation in solutions. J Cryst Growth 4:377–383
6. Nyvlt J, Rychly R, Gottfried J, Wurzelova J (1970) Metastable Zone Width of some aqueous solutions. J Cryst Growth 6:151–162
7. van Gelder RNMR, Roberts KJ, Chambers J, Instone T (1996) Nucleation of single and mixed straight chain surfactants from dilute aqueous solutions. J Cryst Growth 166:189–194
8. Kashchiev D, Borissova A, Hammond RB, Roberts KJ (2010) Dependence of the critical undercooling for crystallization on the cooling rate. J Phys Chem B 114:5441–5446
9. Kashchiev D, Borissova A, Hammond RB, Roberts KJ (2010) Effect of cooling rate on the critical undercooling for crystallization. J Cryst Growth 312:698–704
10. Camacho D, Borissova A, Hammond R, Kashchiev D, Roberts K, Lewtas K, More I (2014) Nucleation mechanism and kinetics from the analysis of polythermal crystallisation data: methyl stearate from kerosene solutions. CrystEngComm 16:974–991
11. Kashchiev D (2000) Nucleation: basic theory with applications. Butterworth-Heinemann, Oxford
12. Toroz D, Rosbottom I, Turner T, Camacho DM, Hammond RB, Roberts KJ (2015) Towards an understanding of the nucleation of alpha-para amino benzoic acid from ethanolic solutions: a multi-scale approach. Faraday Discuss 179:79–114
13. Sangwal K (2007) Additives and crystallization processes: from fundamentals to applications. Wiley, Chichester
14. Roberts KJ, Sherwood JN, Stewart A (1990) The nucleation of n-eicosane crystals from solutions in n-dodecane in the presence of homologous impurities. J Cryst Growth 102:419–426
15. Boistelle R, Astier JP (1988) Crystallization mechanisms in solution. J Cryst Growth 90:14–30
16. Frank FC (1949) The influence of dislocations on crystal growth. Faraday Discuss 5:48–54
17. Elwell D, Scheel HJ (1975) Crystal growth from high temperature solutions. Cryst Res Technol 11:K28–K29
18. Weeks JD, Gilmer GH (1979) Dynamics of crystal growth. Adv Chem Phys 40:157–227
19. Jackson KA (1958) Mechanisms of growth. In: Metals ASf (ed) Liquid Metals and Solidification. American Society for Metals, Cleveland

20. Jetten LAMJ, Human HJ, Bennema P, van der Eerden JP (1984) On the observation of the roughening transition of organic crystals, growing from solution. J Cryst Growth 68:503–516
21. Human HJ, Van der Eerden JP, Jetten LAMJ, Odekerken JGM (1981) On the roughening transition of biphenyl: transition of faceted to non-faceted growth of biphenyl for growth from different organic solvents and the melt. J Cryst Growth 51:589–600
22. Nguyen TTH, Hammond RB, Roberts KJ, Marziano I, Nichols G (2014) Precision measurement of the growth rate and mechanism of Ibuprofen {001} and {011} as a function of crystallisation environment. CrystEngComm 16:4568–4586
23. Gron H, Borissova A, Roberts KJ (2003) In-process ATR-FTIR spectroscopy for closed-loop supersaturation control of a batch crystallizer producing monosodium glutamate crystals of defined size. Ind Eng Chem Res 42:198–206
24. Gron H, Mougin P, Thomas A, White G, Wilkinson D, Hammond RB, Lai XJ, Roberts KJ (2003) Dynamic in-process examination of particle size and crystallographic form under defined conditions of reactant supersaturation as associated with the batch crystallization of monosodium glutamate from aqueous solution. Ind Eng Chem Res 42:4888–4898
25. Haque JN, Mahmud T, Roberts KJ, Rhodes D (2006) Modeling turbulent flows with free-surface in unbaffled agitated vessels. Ind Eng Chem Res 45:2881–2891
26. Haque JN, Mahmud T, Roberts KJ, Liang JK, White G, Wilkinson D, Rhodes D (2011) Free-surface turbulent flow induced by a Rushton turbine in an unbaffled dish-bottom stirred tank reactor: Ldv measurements and Cfd simulations. Can J Chem Eng 89:745–753
27. Mahmud T, Haque JN, Roberts KJ, Rhodes D, Wilkinson D (2009) Measurements and modelling of free-surface turbulent flows induced by a magnetic stirrer in an unbaffled stirred tank reactor. Chem Eng Sci 64:4197–4209
28. Liang K, White G, Wilkinson D, Ford LJ, Roberts KJ, Wood WML (2004) An examination into the effect of stirrer material and agitation rate on the nucleation of L-glutamic acid batch crystallised from slow-cooled supersaturated solutions. Cryst Growth Des 4:1039–1044
29. Wilcox DC (2006) Turbulence modeling for CFD. DCW Industries, Inc, La Cañada
30. Launder BE, Reece GJ, Rodi W (1975) Peogress in the development of a Reynolds-stress turbulence closure. J Fluid Mech 68:537–566
31. Rane CV, Ganguli AA, Kalekudithi E, Patil RN, Joshi JB, Ramkrishna D (2014) CFD simulation and comparison of industrial crystallizers. Can J Chem Eng 92:2138–2156
32. Zauner R, Jones AG (2000) Scale-up of continuous and semibatch precipitation processes. Ind Eng Chem Res 39:2392–2403
33. Javed KH, Mahmud T, Zhu JM (2006) Numerical simulation of turbulent batch mixing in a vessel agitated by a Rushton turbine. Chem Eng Process 45:99–112
34. Ma CY, Liu JJ, Zhang Y, Wang XZ (2015) Simulation for scale-up of a confined jet mixer for continuous hydrothermal flow synthesis of nanomaterials. J Supercrit Fluids 98:211–221
35. Li MZ, White G, Wilkinson D, Roberts KJ (2004) LDA measurements and CFD modeling of a stirred vessel with a retreat curve impeller. Ind Eng Chem Res 43:6534–6547
36. Villermaux J (1989) A simple model for partial segregation in a semibatch reactor AIChE Annual Meeting, San Francisco, pp paper 114a
37. Jones RM, Rouge B, Harvey AD III, Acharya S (2001) Two-equation turbulence modeling for impeller stirred tanks. Trans ASME J Fluids Eng 123:640–648
38. Batdyga J, Bourne JR, Hearn SJ (1997) Interaction between chemical reactions and mixing on various scales. Chem Eng Sci 52:457–466
39. Batdyga J, Podgorska W, Pohorecki R (1995) Mixing-precipitation model with application to double feed semibatch precipitation. Chem Eng Sci 50:1281–1300
40. Randolph AD, Larson MA (1962) Transient and steady state size distributions in continuous mixed suspension crystallisers. AICHE J 8:639–645
41. Ma CY, Wang XZ, Roberts KJ (2007) Multi-dimensional population balance modeling of the growth of rod-like L-glutamic acid crystals using growth rates estimated from in-process imaging. Adv Powder Technol 18:707–723

42. Ma CY, Wang XZ, Roberts KJ (2008) Morphological population balance for modeling crystal growth in face directions. AICHE J 54:209–222
43. Ma CY, Wang XZ (2008) Crystal growth rate dispersion modelling using morphological popultion balance. AICHE J 54:2321–2334
44. Lacmann R, Herden A, Mayer C (1999) Kinetics of nucleation and crystal growth. Chem Eng Technol 22:279–289
45. Liang KP, White G, Wilkinson D, Ford LJ, Roberts KJ, Wood WML (2004) Examination of the process scale dependence of L-glutamic acid batch crystallized from supersaturated aqueous solutions in relation to reactor hydrodynamics. Ind Eng Chem Res 43:1227–1234
46. Li MZ, White G, Wilkinson D, Roberts KJ (2005) Scale up study of retreat curve impeller stirred tanks using LDA measurements and CFD simulation. Chem Eng J 108:81–90
47. Khan S, Ma CY, Mahmud T, Penchev RLY, Roberts KJ, Morris J, Ozkan L, White G, Grieve B, Hall A, Buser P, Gibson N, Keller P, Shuttleworth P, Price CJ (2011) In-process monitoring and control of supersaturation in seeded batch cooling crystallisation of L-glutamic acid: from laboratory to industrial pilot plant. Org Process Res Dev 15:540–555
48. Li R, Penchev R, Ramachandran V, Roberts KJ, Wang XZ, Tweedie R, Prior A, Gerritsen J, Hugen F (2008) Particle shape characterisation via image analysis: From laboratory studies to in-process measurements using an in-Situ Particle Viewer (ISPV) system. Org Process Res Dev 12:837–849
49. Zauner R, Jones G (2002) On the influence of mixing on crystal precipitation processes - application of the segregated feed model. Chem Eng Sci 57:821–831

Chapter 12
Phase Diagrams for Process Design

Gerard Coquerel

Abstract The knowledge of stable and metastable heterogeneous equilibria is paramount for the rational management of separation/purification processes based on crystallization.

Starting from a collection of binary and ternary phase diagrams the design of crystallization processes is illustrated by careful details on different pathways corresponding to crystallizations induced by: (i) cooling (ii) antisolvent (iii) evaporation. By taking into account the stable and metastable heterogeneous equilibria, the access to different final or transient states are detailed. The fine analyses of the pathways inside the relevant phase diagrams also lead to assess the robustness of the processes including their risks of lack of control. In addition, the thermodynamic optimum yield can be reckoned and thus, the tuning of the kinetic parameters can be framed within sensible limits.

Keywords Phase diagrams • Design of crystallization processes • Optimization • Risk assessment

List of Symbols and Abbreviations

<P>	symbolizes crystals of compound or component P
Mixed solvate:	at least two different solvent molecules are incorporated in the same crystallographic site.
Heterosolvate:	at least two different solvent molecules are incorporated in two different crystallographic sites.
Efflorescence:	release of solvent molecules in the gas phase by either: a solvate, a mixed solvate or a heterosolvate.
Deliquescence:	uptake of solvent from the gas phase up to dissolution of the crystals (often applied to moisture uptake and

G. Coquerel (✉)
University of Rouen Normandy, Rouen, France
e-mail: gerard.coquerel@univ-rouen.fr

K.J. Roberts et al. (eds.), *Engineering Crystallography: From Molecule to Crystal to Functional Form*, NATO Science for Peace and Security Series A: Chemistry and Biology, DOI 10.1007/978-94-024-1117-1_12

	aqueous solution). This phenomenon can be followed by the recrystallization of a solvate (hydrate in case of moisture uptake).
u.s.s.:	Undersaturated solution i.e. monophasic domain with a single liquid phase.
Non-congruent soluble compound:	In a ternary system, the line connecting the point representative of the solid and that of the solvent does not intersect the stable solubility curve of the compound. In order to crystallize that compound in solution without any trace of any other solid, an excess of one component has to be added in the system. A compound can have a non-congruent solubility at one temperature and a congruent solubility at another temperature.
Binary eutectic:	Three phase invariant which involves the simultaneous and reversible crystallization and melting of two solids from (to give) a liquid. The temperature and the composition of the three phases are fixed.
Ternary eutectic:	four phase invariant which involves the simultaneous and reversible crystallization and melting of three solids from (to give) a liquid. The temperature and the composition of the four phases are fixed.
Binary peritectic (non-congruent fusion):	Three phase invariant involving two solid phases and one liquid, all of them of specific compositions. At a defined temperature, it corresponds to the reversible decomposition of an intermediate compound on heating.
Binary monotectic (stable miscibility gap at high temperature):	Three phase invariant involving two liquids and one solid, all of them of specific compositions. At a defined temperature, it corresponds to the reversible decomposition of a liquid into a solid phase and another liquid on cooling.
RH:	relative humidity. This is a ratio (thus non unit) valid at a fixed temperature and defined as: the vapor pressure of water in the system over the saturated vapor pressure.

12.1 Introduction

When temperature and or composition are changed in a system the Gibbs function evolves; a stable state can move from homogeneity (a single phase has a Gibbs function below all the other) to heterogeneity (at least two Gibbs functions are necessary to describe the stable state of the system).

This happen for instance for a crystallization induced by cooling. Let us suppose water and sodium chloride: starting from a single phase at high temperature, the system will evolve towards a biphasic system at low temperature. The solute which was initially completely solvated is, at the end of the process, in two different compartments of the system: the crystallized particles and the saturated solution. Thus, a part remains as solvated entities the other part as a solid phase. We will not try to describe the mechanism of nucleation and crystal growth we will try to understand how the system evolves and how the physico-chemist can globally 'conduct' the evolution of the system.

A phase diagram is a tool which gives the interesting: segments, planes and hyperplanes tangent to the minima of the Gibbs functions in competition at a certain pressure and temperature. Figure 12.1 below is an illustration of this property. It shows the 3D representation of the Gibbs functions of the two pure enantiomers G<R> and G<S> the racemic compound G<RS> – those functions can be considered as almost vertical segments (i.e. no solid solution), there is also the Gibbs function of the liquid. The tangent planes between G<R>, G<S> and G<RS> and the G(liquid) are projected along the vertical axis and delineate the ternary isothermal phase diagram between the solvent, R and S. Six domains can be distinguished. (i) a single monophasic domain. It corresponds to the under-saturated solution: domain 1 (ii) Three biphasic domains: two symmetrical domains of pure solid enantiomers: <S> with its saturated solution (domain 2) and<R> with its saturated solution (domain 6) and the racemic compound <RS> plus its saturated solution (domain 4) (iii) a pair of symmetrical triphasic domains: <S> + <RS> + doubly saturated solution (domain 3) and <R> + <RS> + doubly saturated solution (domain 5).

Phase diagrams describe in a geometrical way (i.e. very concise) the limits and the nature of the heterogeneities of a system in stable or metastable thermodynamic equilibrium. Phase diagrams contain crucial data when separation processes (i.e. purification) are at stake. The following questions that can be directly addressed are:

How many phases co-exist in equilibrium (stable or metastable equilibrium)?
What are the compositions of the phases in equilibrium?
What is the mass of every phase? (Application of the lever rule)
At any moment in a process, how far the system is from equilibrium?

For a given crystallization process, phase diagrams give a geometrical representation of the pathway of the solution point and the nature of the solid phase(s) which should crystallize. At every temperature for a crystallization conducted by cooling

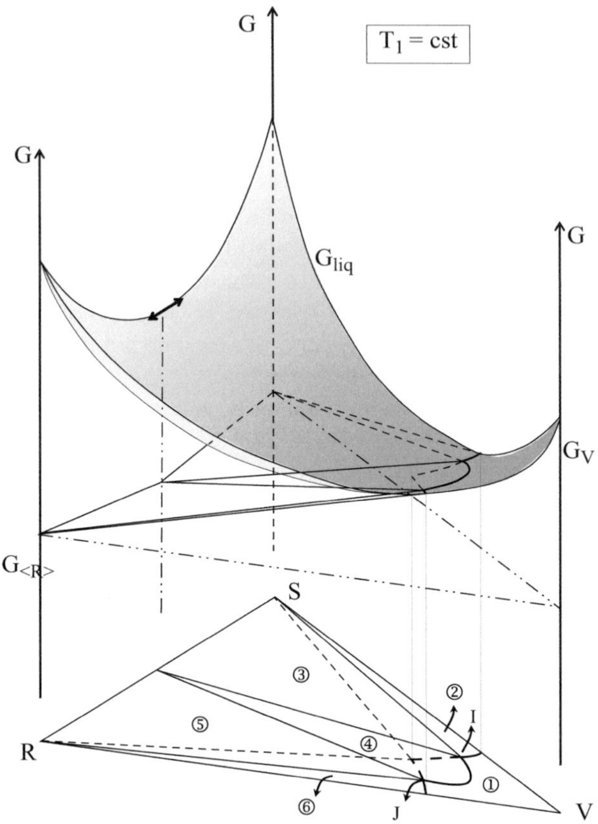

Fig. 12.1 Isothermal ternary phase diagram of a couple of enantiomers S and R with a solvent V. A stable racemic compound is form at that temperature. This phase diagram is obtained by using the vertical projection of the G versus composition diagram at that temperature. The domains 1,2,3,4,5 and 6 correspond to several co-existing phases. 1: the undersaturated solution; 2:<S>plus its saturated solution. 3:< S> + <RS>and the doubly saturated solution I, 4:< RS>plus saturated solution, 5:<RS> + <R> plus the doubly saturated solution J, 6: <R> plus its saturated solution. Dashed lines represent the metastable equilibria of the conglomerate

and at every composition for crystallization conducted by anti-solvent addition or for crystallization conducted by evaporation it is possible to compute the mass balance in the system and thus to know quantitatively the mass of crystals which should exist in the system if the system has reached a stable or a metastable equilibrium.

Several books and reviews treat the subject of phase diagrams [1 – 9] but indeed, the connection between phase diagrams and separation processes has not been systematically treated so far.

Among organic components there is a huge number of possible systems (virtually infinite). For a given solute the number of possible counter ions for any basic or acidic molecule is considerable. On top of that the number of organic solvents exceeds several hundreds and more than often mixtures of organic solvents are used! Organic components have also common features such as:

- The possibilities to crystallize as different: polymorphs, stoichiometric or non-stoichiometric solvates, co-crystals, host-guest inclusions. Possible disordered phases or partially disordered phases.
- One can see a number of miscibility gaps in the liquid state.

- The difficulty to crystallize countless compounds is well known.
- It is common to isolate efflorescent solvates or by contrast, organic solids can exhibit a high hygroscopic character.
- A large proportion of organic compounds are chemically fragile which brings limit to the phase diagram that can actually be explored. They can also lead to: isomerization, racemization, epimerisation, etc.

12.2 Example of Utility of Phase Diagrams in the Understanding of Polymorphism vs. Miscibility in the Solid State

Figure 12.2 illustrates the possible effects of a second component on the polymorphism of a first one.

In case a, the temperature of transition (τ) of component M is not modified by the presence of the second component A. G<M1> and G<M2> versus composition at all temperature are quasi vertical functions. When the system is in equilibrium the two polymorphs can only co-exist at a single temperature Tτ.

In case b, component B can enter in the crystal lattice of <M1> and <M2> and this lower the Gibbs free energy. It results two functions G<M1> and G<M2> versus composition which are in competition. Even when the system is in equilibrium, the two polymorphs can now co-exist in a temperature range depending on the overall composition. The temperature of transition drops down to the metatectic invariant below which only the low temperature form is stable.

In case c, component C can also enter in the crystal lattice of <M1> and <M2> but the low temperature form has more affinity with the second component than the high temperature form. Thus, this is the opposite situation as that depicted in case b. When the system is in equilibrium, G<M1> and G<M2> functions intersect showing that the two polymorphs can co-exist at different temperature -composition values. The temperature of transition increases up to a peritectic invariant above which the high temperature form only is stable.

It should be noticed that hundreds of ppm of a second component can be enough to change the temperature of transition by more than 10 °C. For instance Urea has a transition at 56 °C which is dropped down to 32 °C by 300 ppm of water. In many books and publications there is a confusion between the temperature of that metatectic and the genuine temperature of transition of pure Urea which is 24 °C higher than that of the metatectic.

At the industrial scale, there are basically three ways that can be used to induce a crystallization:

- By cooling
- By evaporation and
- By anti-solvent addition

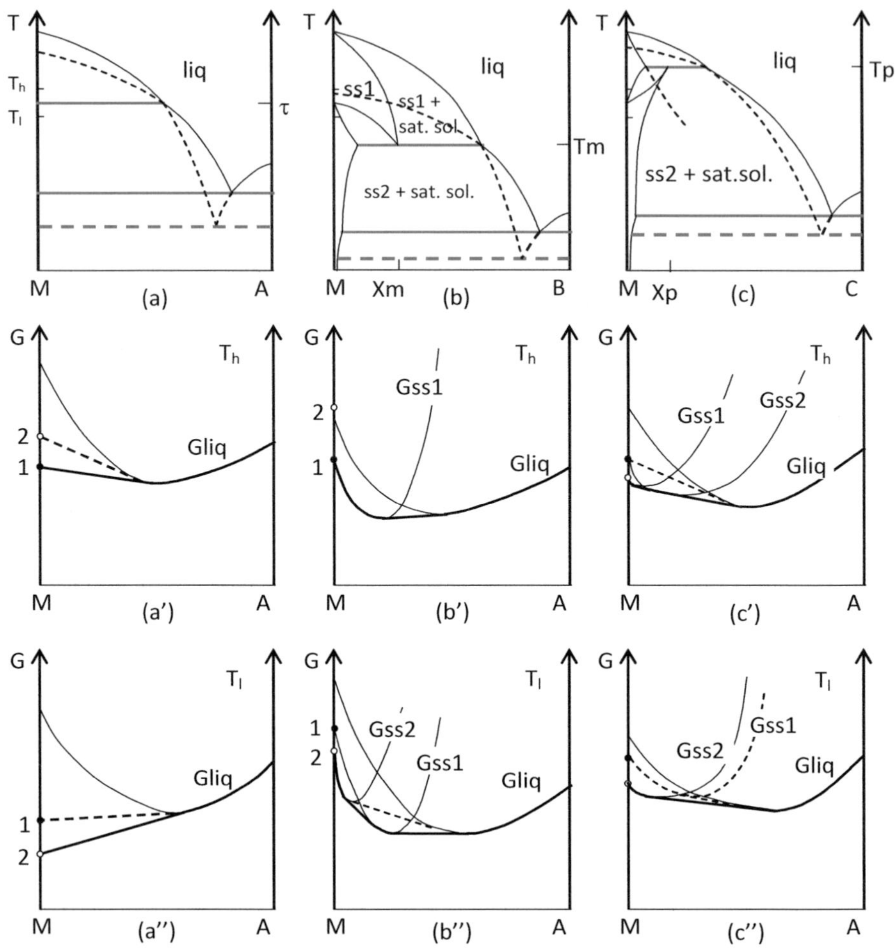

Fig. 12.2 Schematic representation of the possible effects of three components A, B and C on the polymorphism of M. A does not change the temperature of the enantiotropic transition τ because A does not enter in the two crystals lattices of M. By contrast, component B can enter in the two crystal lattice of M with a higher affinity with the high temperature form. It results in the formation of a metatectic invariant i.e. the temperature of transition is lowered as the composition in B increases. Component C can also enter in the two crystal lattices of M but with a higher affinity for the low temperature form. The temperature of transition is now increased with a higher composition in C up to the peritectic transition. The **a'**, **a''**, **b'**, **b''**, **c'**, **c''** diagrams above illustrate the shape of the G versus composition functions at two different temperatures (Adapted from reference 7)

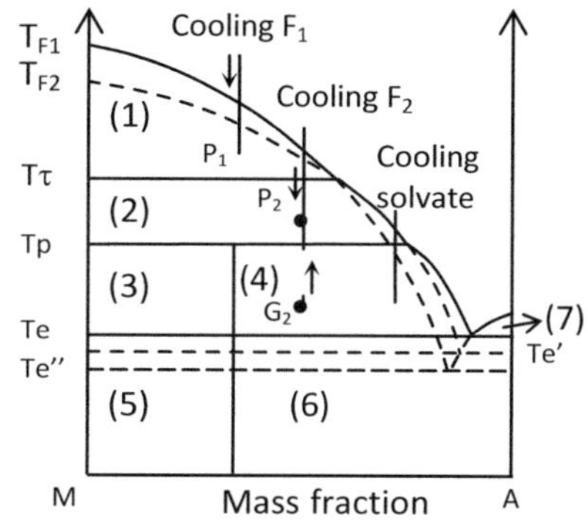

Fig. 12.3 Binary system between a solute **M** and a solvent **A**. M has two polymorphs enantiotropically elated: <M1> (High temperature form and <M2> (low temperature form) with a transition at Tτ. below Tp the stoichiometric solvate <M−Ak> is the stable solid phase

12.3 Crystallization Induced by Cooling

Figure 12.3 depicts the stable and metastable equilibria involving a solute M with two polymorphs <M1> the high temperature variety (melting point TF1) and <M2> the low temperature variety (melting point TF2). In addition to these two solid phases a solvate <M−A_K> is stable up to Tp –the peritectic temperature. A stands for a solvent.

This binary system contains three stable invariants (horizontal full line): at low temperature –below the melting point of the solvent, there is a eutectic at Te. At Tτ the temperature of reversible polymorphic transition between <M1> and <M2> (there is no solid solution here). At Tp the solvate reversibly decomposes into a doubly saturated solution and the solute <M2>.

There are also two metastable eutectic invariants: one, at T = Te′, corresponds to the three phase invariant: doubly saturated solution and the low temperature form <M2>. The other one at the lowest temperature Te″, corresponds to the most metastable eutectic between the high temperature form <M1>, a doubly saturated solution and <A>

Design of polythermic crystallization processes to access selectively to: <M1> or, <M2> or <M-Ak>.

When <M1> is desired, its crystallization can be implemented by following the 'cooling F1' pathway. Starting from a homogeneous solution, the system is cooled down and maintained at the temperature of point P1 (i.e. the point representative of the whole system: P1 is the lowest point of the vertical segment). In order to save time (there might be a concern in chemical stability of the component for long standing at relatively high temperature. It is possible to seed with the desired high temperature form (<M1>) at a temperature which corresponds to the intersection between the vertical segment and the stable solubility curve. Subsequent filtration and drying should be carried out at a temperature above Tτ.

When <M2> is desired, the 'cooling F2'pathway can be used. Starting from a homogeneous solution, the temperature is rapidly dropped below the intersection of the metastable solubility curve of <M2> and the vertical segment 'cooling F2′. It is beneficial to seed the system by inoculating pure seeds of <M2> or a suspension containing pure <M2> [10]. In case the spontaneous primary nucleation of <M1> is fast it might be better to shift the overall concentration towards a lower concentration. In order to secure the process, filtration and drying should be operated at a temperature above T_P and below T_τ.

It is also possible to start from a suspension of the solvate: <M-A_K> plus its saturated solution represented by point G2. By heating the system up to P2 (i.e. dotted line pathway) the pure <M2> phase will be obtained as soon as the invariant at Tp will be crossed. The benefit of that process comes from the reshaping of the particles inside the saturated solution. Therefore, the particles have a better shape, less internal defects, i.e. a better crystallinity and lower specific surface area compared to <M2> obtained by desolvation via a solid – gas process. Indeed it sounds strange but, it is perfectly possible (and sometime highly beneficial) to dehydrate a hydrate in aqueous solution above the peritectic transition.

When <M-A_K> is desired, the 'cooling solvate' pathway can be used. Starting from a homogeneous solution, a fast cooling is implemented down to Tp where seeding with fine crystals of pure <M-A_K> can be operated. If <M2> is fast to nucleate spontaneously it might be more secure to diminish the concentration of the solution so that the vertical segment 'cooling solvate' will intersect the stable solubility curve of the desired solute. The process will be even more robust by adjusting the cooling program so that the system will be kept close enough to equilibrium so that neither <M2> nor <M1> could appear. Of course, the filtration and drying need to be operated at a temperature below Tp.

12.4 Design of a Selective Crystallization Process by Using Auto-seeding

The auto-seeding (or self-seeding) has been initially proposed for preferential crystallization [11]. It consists in adjusting the composition and the temperature of the system so that a single crystallized phase is present in equilibrium -as numerous small particles- with their saturated solution. The selective crystallization starts from this suspension in thermodynamic equilibrium. The cooling program is thus adjusted so that the crystallization remains selective even if, it finishes in a domain where other components or compounds should also crystallize. This fundamental principle can be used for a lot of systems of different order binary, ternary, quaternary systems, etc.

For example, it is possible to start from the low temperature form of the solute M and to adjust the composition and temperature so that the overall synthetic mixture is represented by point P1 (Fig. 12.3). A slurrying at that temperature would induce the conversion of the low temperature form 2 into the high temperature form 1. When the conversion is completed the suspension is exclusively composed of crystal of form 1 and the saturated solution. It is possible to cool down the system below Tp so that only secondary nucleation and crystal growth of form 1 is ensured.

How far the system can stand its metastable character is a matter of a case by case basis. Of course the suspension must be filtered off as fast as possible and the wet solid has to be dried also rapidly (preferably at a temperature above the polymorphic transition).

By using the same principle, it is also possible to create *in-situ* the solid particles of the solvate (thermodynamic equilibrium) below Tp. Then a cooling program has to be designed so that only the hydrate can growth. Indeed 100% pure seeds is not the warranty that only the stable phase can grow. If the cooling rate is excessive a heterogeneous primary nucleation of the metastable phase onto the stable phase can happen [12].

12.5 Design of a Process Which Avoids the Problem of a Metastable Miscibility Gap

Let us consider the binary system displayed on Fig. 12.4. From I to F, if the cooling rate is fast and no seeding has been performed, according to the Oswald rule of stages [13], it is very likely that a miscibility gap will appear first with two co-exiting liquids with different compositions in solute: 2 and 2′. By an annealing at TF the crystals of <A> should appear and then an irreversible evolution of the system towards equilibrium will happen: <A> crystal and a saturated solution 3′ should then be reached as the final evolution of the system. This two-step process is usually undesired because: two types of particles are obtained: (i) those resulting from phase 2 are likely to be hard spheres (if the system was stirred); the initial highly concentrated liquid phase could contain a lot of impurities: the chemical purity of that type of particles is likely to be questionable (ii) sheaf-like aggregates resulting from the crystallization of phase 2′ which is far less supersaturated than phase 2. The remedy to that ill-defined crystallization process is to seed at $T < T_H$ and to cool down the system at a pace that crystal growth and secondary nucleation can stand [14 – 17].

12.6 Polythermic Crystallization Process in a Ternary System

Let us consider now Fig. 12.5 which shows a perspective view of the polythermic ternary system A, B and C. At high temperature, there are three surfaces corresponding to the solubilities of the three pure components (no solid solution, no intermediate compound). Those three surfaces intersect two by two, forming three monovariant valleys which gather at low temperature to a single point: the ternary eutectic point ε. Those monovariant lines connect the three binary eutectic points E1, E2 and E3 to the common ternary eutectic point ε. (This is why those

Fig. 12.4 Design of a crystallization process in order to avoid a miscibility gap in the liquid phase

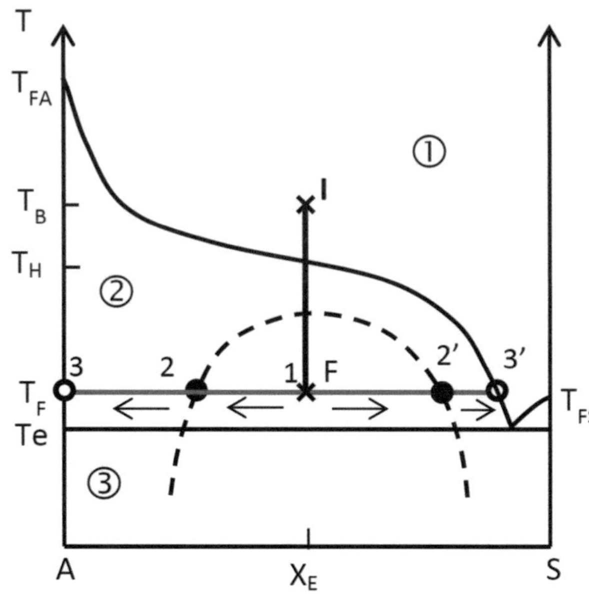

curves are sometimes called eutectic valleys). If we consider the mixture of composition M at $T>T_S$ the system is monophasic, i.e. it is simply constituted of an under saturated liquid. The bold line indicates the trajectory followed by the liquid phase on cooling if the system remains in equilibrium at whatever T: $T_\varepsilon<T<T_S$. From T_B to T_S, the system is monophasic, just a single liquid can exist. If seeding of pure particle <A> is performed at T_S and the cooling rate is adapted to the crystal growth and secondary nucleation capabilities of <A> in this complex medium, the solution point glides on the solubility surface of <A>. The trajectory of the liquid point belongs to the isoplethal section TAK; as long as <A> is the only solid which crystallizes. When the solution point reaches D (point D is the intersection between the TAK isoplethal section and the monovariant valley E_1-ε), <A> continues to crystallize but now concomitantly with <B> crystallizing as well. As a matter of consequence, the solution point gets out of the Isoplethal section TAK and follows the monovariant valley E_1- ε. When T_ε is reached, the liquid has the composition of point ε and it will completely crystallize at temperature T_ε. <A>, <B> and <C> are crystallizing simultaneously in a specific ratio. In other words below T_ε there are three solid phases and no more any liquid.

In many separation processes by means of crystallization <A> only is desired [9], the cooling should be stopped at T_D for a robust process. Preferential crystallization is an exception [11] because below point D it is possible to keep on selectively crystallizing <A>. It means that the solution point continues to evolve on the solubility surface of <A> (it is now metastable) on a trajectory which is still in the TAK isoplethal section.

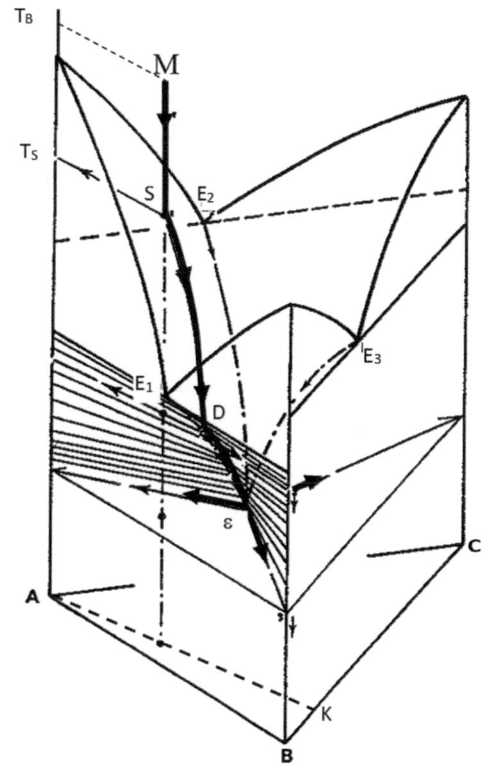

Fig. 12.5 Perspective view of a ternary system showing solid-liquid equilibria. A, B C components do not form any intermediate compound nor any solid solution and no polymorphism. The trajectory of the solution point is shown for an overall composition M. The MSD part of that trajectory belongs to the TAK isoplethal section

12.7 Crystallization Induced by Cooling or Evaporation

Cooling crystallization: Starting from a homogeneous solution represented by point I the cooling ends at point III. It means that the process implies a vertical pathway inside the diagram. As soon as the solution reaches the solubility curve, the solution starts to be supersaturated. In this state seeding is possible and beneficial in order to control the crystal growth and the secondary nucleation by imposing close to equilibrium conditions to the system. If the seeding has been implemented in the form which is stable (for instance in case of polymorphism of <A>), the cooling rate has to be adapted so that the secondary nucleation remains under control [12]. It might happen that the stable phase has a slow crystallization rate so that the cooling rate has to be below a threshold above which the metastable polymorph is able to hetero nucleate on the stable form (Fig. 12.6).

If no seeding has been implemented below the shaded zone (Ostwald zone) a spontaneous crystallization is likely to take place. This limit is not strictly representable in a binary system because it is simply related to kinetics. Nevertheless, it is roughly parallel to the solubility curve. When the process involves a monophasic supersaturated solution at a temperature below the Ostwald zone, erratic results

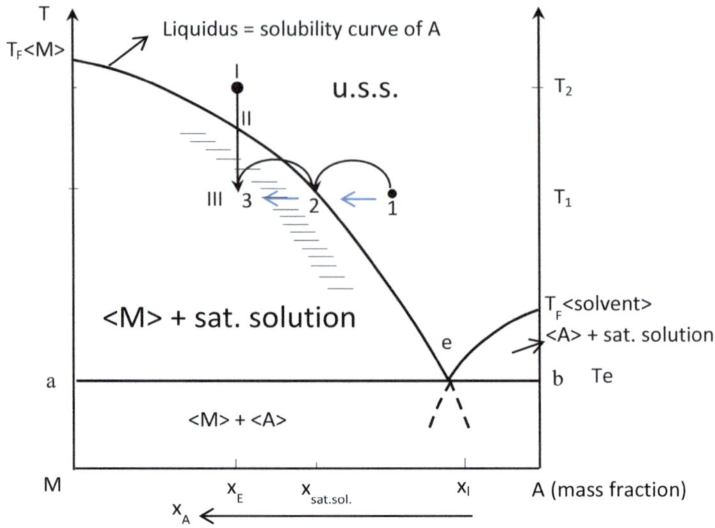

Fig. 12.6 shows two ways to design a crystallization process: cooling and evaporation. U.S.S. stands for under saturated solution

might appear with new polymorphs, difficulty to filtrate, poor crystallinity, inappropriate Crystal Size Distribution (CSD), etc.

12.8 Evaporative Crystallization [18]

Starting from point 1 (Fig. 12.6) it is possible at T_1 to evaporate the solvent and thus to create the supersaturation necessary for the crystallization. We will consider here an isothermal process, thus, the solution point pathway is horizontal. When the system is represented by point 1 there is a single phase which is an undersaturated solution. Evaporation of the solvent shifts the point representative of the system towards the other component. Point 2 is reached when a sufficient amount of solvent has been removed: saturation is then obtained. By continuing evaporation point 3 is reached, the solution becomes supersaturated, because the concentration at that point exceeds the Ostwald zone and hence a spontaneous nucleation should occur. By contrast, if seeding by particles of <M> is performed when the solution point reached point 2 and the evaporation is carried out at a pace compatible with the crystal growth and secondary nucleation of <M>; the crystallization process is thoroughly under controlled.

The evaporative crystallization can be conducted to ensure a selective crystallization and it is compatible with auto-seeding [19]. The injection of dry nitrogen (or air) in the suspension could be used to ensure the ‘stripping’ of the solvent and the stirring of the suspension. In another mode the evaporation could be ensured by depression over the solution and then the suspension.

12.9 Crystallization Induced by Addition of a Third Component (Mainly Anti-solvent Addition)

Figure 12.7 shows a ternary isotherm at T1; the apex A stands for a solute which presents a contrasted solubility between: S1 (the antisolvent) in which the solubility at T1 is poor and S2 the 'good' solvent. Starting from the under saturated solution represented by point B, the addition of S1 creates supersaturation as soon as the point representative of the system enter in the biphasic domain. At that point it is useful to seed with small particles of <A>. On further slow addition of S1 it is possible to conduct the crystallization rather smoothly. If a swift antisolvent addition is performed and no seeding is performed the solution point is likely to enter inside the zone limited by the metastable solubility of a metastable variety. According to the Ostwald rule of stages [13] it is highly probable that the metastable <A'> form appears spontaneously first. This method can be recommended when the experimenter aims at screening the polymorphs of a given solute.

The process looks simple nevertheless, in addition to the problem of an easy access to a metastable form, the process is easily uncontrolled and may lead to: (i) particles with a poor crystallinity raising some concern about chemical stability, caking, etc. (ii) particles with inappropriate size (extremely fine, difficult to filtrate) and/or with extreme habitus elongated fibers (sometime bent crystals) or very fine sheet like crystals (filtration could be unmanageable even at laboratory scale).

Among other difficulties of crystallization process induced by antisolvent addition, a frequent apparition of a metastable miscibility gap can be encountered. This

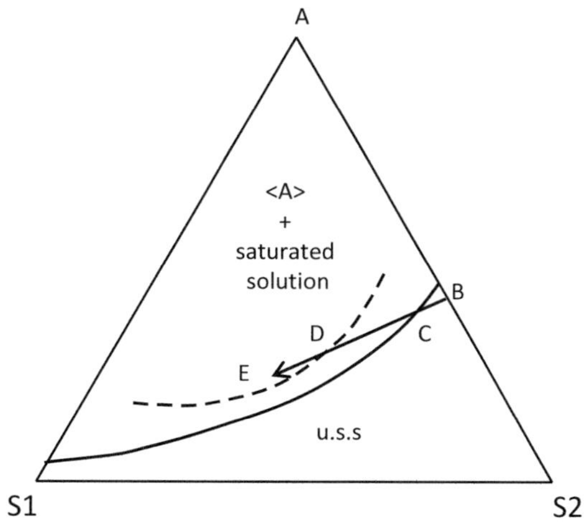

Fig. 12.7 Crystallization of <A>induced by slow antisolvent addition (S1). If no seeding is performed -between C and D and the addition of S1 is fast <A'>another polymorph of <A> is likely to appear first. Full line and dashed line represent respectively the solubility curve of <A> the stable variety and the solubility of <A'> a metastable polymorph

is usually detrimental to a homogeneous crystal size distribution and a substantial part of the expected purification effect is often lost when this transient phenomenon appears.

Figure 12.8 (1) (2) (3) depict different probable cases that can be observed when a DMSO solvate is not desirable as a final product (e.g. pharmaceutical compound, it is worth noting that this DMSO-solvate could be a beneficial intermediate for purification purpose) and water an antisolvent. Starting from a suspension of DMSO solvate in equilibrium with its saturation solution one can observe.

In Fig. 12.8 (1): addition of a sufficient quantity of water results in the crystallization of pure <M>. It is worth noting that the attributes of the final particles will depend on the operating condition (stirring mode, stirring rate, presence or not of surfactant, etc.) and also on the initial concentration of the suspension represented by B or B′.

In Fig. 12.8 (2): addition of a sufficient quantity of water induces the crystallization of the hydrate. Nevertheless, if the process is run fast it is possible to crystallize <M> as a metastable phase. The figure also shows two starting suspension; B′ is more concentrated than B and is even more likely to give <M> instead of the hydrate on fast addition of the antisolvent because the supersaturation will be enhanced. The more the system departs from equilibrium the more likely the Ostwald rule of stages applies.

In Fig. 12.8 (3): more possibilities can be observed. If equilibrium is assumed all along the process of antisolvent addition, the solids or mixtures of crystallized phases that could be observed are: DMSO solvate, DMSO solvate and heterosolvate (or mixed solvate), heterosolvate (or mixed solvate), heterosolvate (or mixed solvate) and hydrate and then solely the hydrate. If the nucleation barriers of the solvates are high and the process is run quickly, it might be possible to form <M> directly from the DMSO solvate but this solid phase is metastable.

Dehydration (like any desolvation) can be smoothly performed by addition of a third crystallized component in a suspension of the hydrate in equilibrium with saturated solution. Indeed, Fig. 12.9 illustrates this possibility. At a given temperature, the isothermal phases diagram is composed of six domains:

- the undersaturated solution,
- the di-hydrate plus its saturated solution,
- a doubly saturated solution plus <M$-2H_2O$> and <M>.,
- <M> plus its saturated solution,
- a doubly saturated solution and <C> and <M>.,
- <C> and its saturated solution.

The successive additions of small portions of component <C> shift the overall composition of the system on the isoplethal section BC. Starting from the suspension of the dehydrate: point B in domain 2, on successive additions of <C> the overall composition moves to the three-phasic domain 3 and then the biphasic domain 4 where there is the end point: E. If we assume that the thermodynamic equilibrium is reached: the solid particles are exclusively composed of crystallized

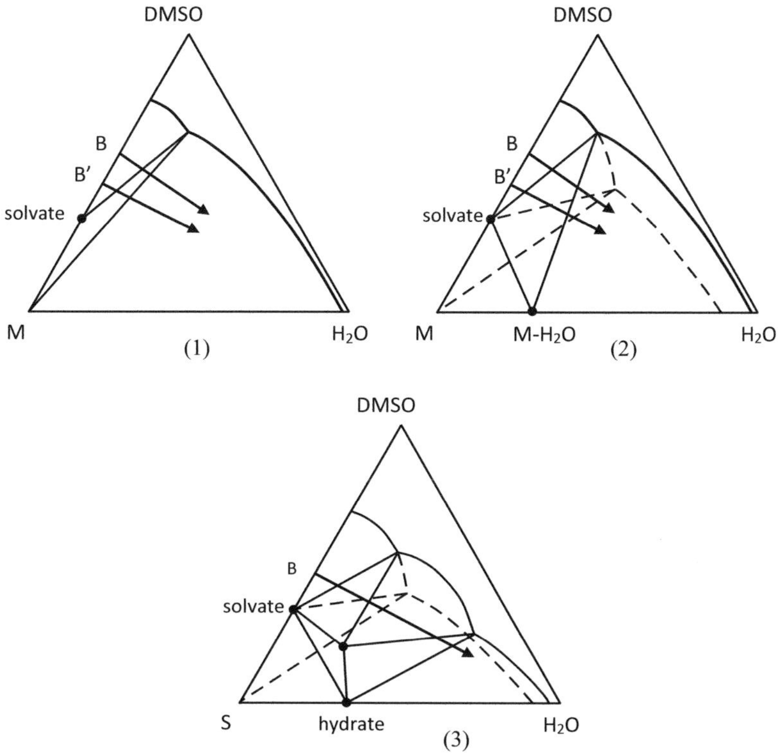

Fig. 12.8 Antisolvent addition on suspension illustrated by three common cases: (1) desolvation of the DMSO solvate by addition of water with two possible initial concentrations of the suspension (2) Switch from a DMSO solvate to a hydrate by means of water addition (3) two possible phases can crystallize in sequence on water addition in the DMSO suspension: starting from the DMSO solvate a heterosolvate (or mixed solvate) can first crystallize and on further addition a hydrate will be the single solid phase in equilibrium with its saturated solution

<M> and the liquid phase is its saturated solution. The net result of this addition is a smooth dehydration in water at a temperature which could be far below the peritectic transition (and possibly a temperature of chemical degradation). The final particle can be well-shaped crystals with several attributes different from the same phase <M> but obtained by means of a thermal dehydration (specific surface area, crystallinity, crystal size distribution, chemical impurity profile, etc).

If we assume that the ternary isotherm (Fig. 12.9) is represented in mass fraction and the stable thermodynamic equilibrium is obtained, the mass of <M> that can be retrieved from this operation is given by:

$$m(< M >) = m_{TOTAL} \cdot SE/SM \tag{12.1}$$

m_{TOTAL} stands for the total mass of the system at point E, that is the mass of the initial suspension plus the mass of <C> added.

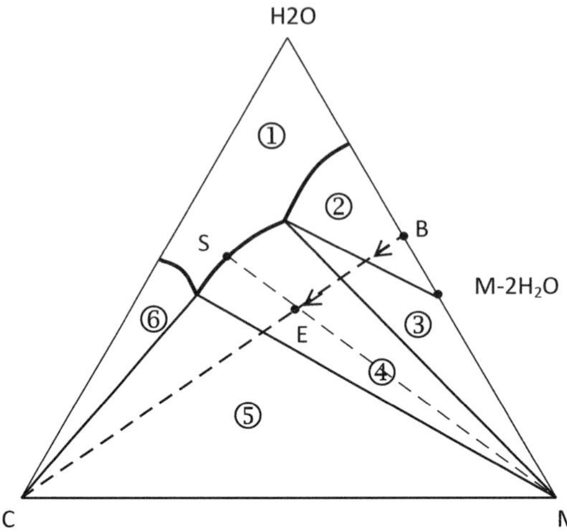

Fig. 12.9 Isothermal ternary diagram represented in mass fraction with: water (solvent), component M forming a dihydrate and another component C more soluble than <M> and <M-$2H_2O$> in water at that temperature. Domain 1 corresponds to the undersaturated solution. Domain 2 corresponds to a bisaphic domain <M−$2H_2O$> plus its saturated solution, domain 3 is a three phase domain composed of: <M−$2H_2O$>, <M> and a doubly saturated solution, Domain 4, which includess the point representative of the end of the process (E), is a biphasic domain composed of: <M> plus its saturated solution. Domain 5 is composed of three phases: <M>; <C> and a doubly saturated liquid, domain 6 corresponds to <C> plus its saturated solution. Starting from point B i.e. a suspension of the dihydrate, addition of component C can be represented by the pathway from B to E. At point E the solid has been smoothly dehydrated in water to yield <M> in equilibrium with its saturated solution

12.10 Crystallization Based on Solid – Gas Interactions

One common way to purify components is based on sublimation that is to say on selective crystallization from the gas phase. There are other crystallization processes which involve the gas phase such as formation of clarthrates or uptake of moisture resulting in crystallization of hydrates or the opposite process: efflorescence [20]. We will just contemplate a couple of examples, with a stoichiometric hydrate only [21], to illustrate those facts which can deeply change the properties of the solid phase. Let us consider Fig. 12.10a at constant temperature (T1), the phase diagram teaches the value of Relative Humidity (RH) that <M> can stand without up-taking moisture up to deliquescence (RH = 60%). Deliquescence means that a hydrate should be formed but for kinetics and/or other reason(s) the hydrate does not appear and the system tends to a solution. In Fig. 12.10b at the same temperature a trihydrate <M−$3H_2O$>has appeared. It has significantly extended the domain of stability of a solid phase towards high RH, because up to 80% RH, <M−$3H_2O$> behaves properly without: stickiness, chemical degradation and the usual trail of undesired properties. Down to 15% RH this hydrate is stable. At that lower limit there is an invariant <M−$3H_2O$>⇔ <M> + doubly saturated vapor (15%RH); it

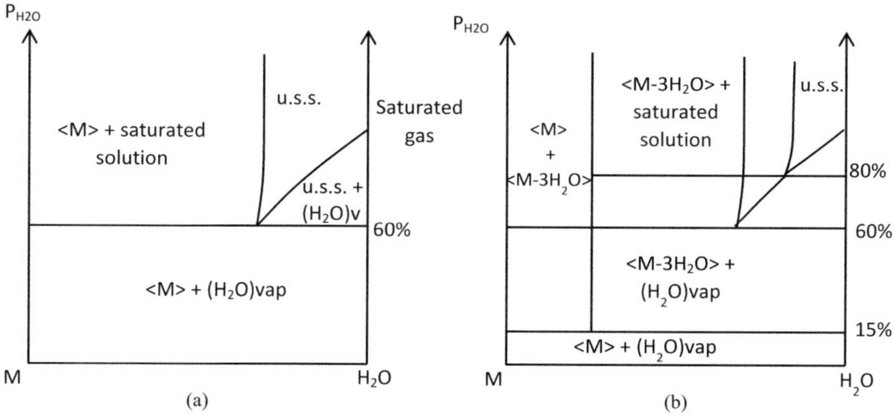

Fig. 12.10 (**a**) Isothermal binary P_{H2O} –composition phase diagram. Component M becomes deliquescent for pH_2O >60%. (**b**) Same diagram as 10(a) with the apparition (sometime difficult to observe) of a trihydrate. This solvate is deliquescent for RH $\geq$ 60% and efflorescent below RH $\leq$ 15%

results in the efflorescence of the trihydrate which converts into the anhydrous phase.

In order to limit the hygroscopic character of a solid, it is thus recommended to find the hydrate which should exist and which should enlarge the domain of stability of a solid phase. Numerous RH cycling around the invariant value of deliquescence has proved to be a useful method. Conversely, hygroscopic solids usually exhibit a high solubility in water and the crystallization of those hydrates can sometimes be unmanageable. One solution to that problem can be to saturate a hydrophobic solvent with water such as Toluene (or Methylcyclohexane) to ensure a facile crystallization of the hydrate [22].

The solubility curves are exaggeratedly slanted to differentiate them from a stoichiometric compound.

12.11 Conclusion

This chapter has provided a short illustration of the usefulness of:

- phase diagrams in the design of: separation/purification processes based on crystallization,
- the management of polymorphism,
- formation of solvates,
- desolvation.

Phase diagrams give a full and concise description of the stable and metastable heterogeneous equilibria. This precious information could be used to rationally design a crystallization process (e.g. resolution of a chiral molecule by the Pasteurian method [9], crystallization by using antisolvent addition, formation of co-crystal with a non-congruent solubility [23]). Moreover, it can be used to calculate the ideal yield that could be reached even if the crystallization has been conducted far from equilibrium [24]. It also gives the opportunity to assess, at any moment, the gap between the (stable and metastable) equilibria and the actual situation of a system, i.e. the strength of the driving force and the robustness of the process (when the process is run too far from equilibrium there is a risk of a lack of control, e.g. a new solid phase can appear or a transient miscibility gap might also happen). When industrial production needs to be optimized this is an inescapable tool to ensure a rational approach. The author acknowledges that the construction of phase diagrams can be time consuming but, the time and money spent in defining the lines of stable and metastable equilibria are clearly rewarded when the kinetics parameters have to be experimentally tuned within reasonable limits.

Acknowledgements Dr M-N Delauney is thanked for her important contribution in the illustrations.

References

1. Ricci JE (1966) The phase rule and heterogeneous equilibrium. Dover Publications Inc., New York
2. Findlay A (1904) The phase rule and its applications. Longmans, Green and Co, London
3. Zernike J (1955) Chemical phase theory. A comprehensive treatise on the deduction, the application and the limitations of the phase rule.; N.V. Uitgevers-Maatschappij E. Kluwer: Dventer-Antwerp-Djakarta
4. Hillert M (1998) Phase equilibria, phase diagrams, phase transformations. Their thermodynamic basis. Cambridge University Press, Cambridge
5. Coquerel G (2015) J Pharm Pharmacol 67:869–878
6. Coquerel G (2014) Chem Soc Rev 43:2286–2300
7. Coquerel G (2006) Chem Eng Technol 29(2):182–186
8. Coquerel G (2000) Enantiomer 5:481
9. Viedma C, Coquerel G., Cintas P (2015) Crystallization of chiral molecules. In: T. Nishinaga (ed) Handbook of crystal growth, volume IA: fundamentals: thermodynamics and kinetics, Elsevier, Amsterdam ISBN: 9780444563699:951–1002
10. Coquerel G (2006) Chem Eng Process 45:857–862
11. Coquerel G (2007) Preferential crystallization in topics in current chemistry novel optical resolution technologies. (ed) Springer GmbH, ISSN: 0340-1022: 1–50
12. Courvoisier L, Mignot L, Petit MN, Coquerel G (2003) Org Process Res Dev 7:1007–1016
13. Ostwald WZ (1897) Phys Chem 119:227
14. Veesler S, Lafferrère L, Garcia E, Hoff C (2003) Org Process Res Dev 7:983
15. Bonnett PE, Carpenter KJ, Dawson S, Davey RJJ (2003) Chem Commun 21(6):698–699
16. Codan L, Bäbler MU, Mazzotti M (2010) Cryst Growth Des 10:4005
17. Brandel C, Gbabode G, Cartigny Y, Martin C, Gouhier G, Petit S, Coquerel G (2014) Chem Mater 26(14):4151
18. Mullin J W (2001) Crystallization, 4th edn. Butterworth Heinemann, ISBN 0 7506 4833 3
19. Mahieux J, Sanselme M, Harthong S, Melan C, Aronica C, Guy L, Coquerel G (2013) Cryst Growth Des 13(8):3621–3631

20. Fours B, Cartigny Y, Petit, S Coquerel, G (2015) Faraday Discuss 179:475-488
21. Authelin J-R (2005) Int J Pharm 303:37–53
22. Couvrat N, Cartigny Y, Linol J, Vaysse-Ludot L, Langlois P, Coquerel G (2015) Chem Eng Technol 38(6):999–1005
23. Coquerel G (2012) 'Co-crystals and beyond' ch 13 in 'Pharmaceutical salts and co-crystal' RSC Publishing, Editors: Johan Wouters and Luc Quéré ISBN: 978-1-84973-158-4:300–317
24. Prigogine I (1961) Introduction to thermodynamics of irreversible processes. Interscience, New York

Chapter 13
Seeding in Crystallisation

Jose V. Parambil and Jerry Y.Y. Heng

Abstract Crystal seeding is the process of adding homogeneous or heterogeneous crystals to a crystallising solution to nucleate and/or grow more crystals. Seeding has emerged as one of the most critical steps in optimising the crystallisation process (O'Sullivan B, Smith B, Baramidze G, Recent advances for seeding a crystallization process. Mettler Toledo Auto-Chem, Columbia, 2012). An aptly designed seeding technique would ensure product reproducibility between batches or over time. This is achieved primarily by controlling the crystal size distribution and polymorphism of the crystals that are formed. In this chapter, aspects of crystal nucleation, the importance of seeding and crystallisation methods employed will be discussed.

Keywords Seeding • Secondary nucleation • Crystallisation • Crystal growth

13.1 Crystal Nucleation and Growth

Nucleation is the birth of a new phase from a bulk phase. It involves the formation of prenucleation aggregates, which, upon reaching a critical size, forms stable nuclei with some degree of molecular order. More molecules then add onto the stable nuclei thereby resulting in the growth of the crystal. However, as pointed out by Bergfors [1], the optimal conditions for crystal nucleation and growth are different. Nucleation is more likely to happen at higher supersaturations while ordered growth of the crystal is favoured at lower supersaturations.

Nucleation can be primary or secondary nucleation. In primary nucleation, the event occurs according to the classical theory of nucleation and the stable nuclei is formed directly from the solution in the absence of a previously formed crystal. However, in secondary nucleation, a new crystal would be formed from a pre-existing crystal. Within primary nucleation, the nucleation could occur in a

J.V. Parambil • J.Y.Y. Heng (✉)
Department of Chemical Engineering, Imperial College London, South Kensington Campus, London SW7 2AZ, UK
e-mail: jerry.heng@imperial.ac.uk

K.J. Roberts et al. (eds.), *Engineering Crystallography: From Molecule to Crystal to Functional Form*, NATO Science for Peace and Security Series A: Chemistry and Biology, DOI 10.1007/978-94-024-1117-1_13

homogeneous media termed as homogeneous nucleation or occur at an interface between two phases, then termed as heterogeneous nucleation. The interface could be contributed by a foreign particle in the media such as dust or at the liquid-air interface or liquid-vessel interface.

Nucleation is a stochastic event that might show 100–200% variation in the induction time, based on supersaturation and volume of the crystallising solution [2]. Many approaches to control crystallisation hence attempt to avoid nucleation by adding seed crystals. This also helps to separate out crystal nucleation and growth which, as mentioned before, has different optimal conditions. Common batch crystallisations also utilise the metastable zone in the crystallisation phase diagram along with seeding to avoid unnecessary nucleation and regulate crystal growth.

13.2 Theory of Nucleation

During nucleation, small molecular clusters from the metastable phase aggregates to form a stable phase. However, if the size of the newly formed cluster is below a threshold value, referred to as the critical radius, the cluster would be unstable and dissolve back into the initial phase. This is due to the surface energy of small crystals that result in a significant thermodynamic barrier. As per classical nucleation theory (CNT), the Gibbs free energy of the nucleus can be written as below:

$$\Delta G_r = -\frac{4}{3}\pi r^3 \Delta G_V + 4\pi r^2 \sigma_{SL} \tag{13.1}$$

Where

ΔG_r – Gibbs free energy (J) of the nucleus with radius r (m);
σ_{SL} – interfacial free energy of the nucleus (J/m^2)

It is to be noted that the CNT assumes that the nucleus is spherical and is formed by the addition of single units of the component system. In the above equation, the first term on the right side of the equation stands for the free energy contribution from the bulk volume of nucleus and the second term denotes the free energy requirement for creating a new interface. As it can be seen from the above relationship, while the bulk free energy of the crystal stabilises the structure, the surface energy provides a barrier. Thus, a stable nucleus will be formed only when the energy gain from the bulk overcomes the surface energy barrier. This is represented in Fig. 13.1.

Based on above relationship, the radius of the critical nucleus (r_c) can be found out and can be written as below:

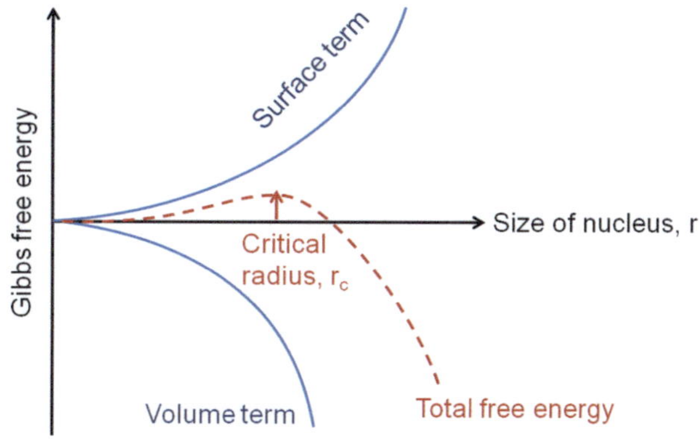

Fig. 13.1 Free energy plot for nucleation

$$r_c = \left|\frac{2\sigma_{SL}}{\Delta G_V}\right| \tag{13.2}$$

Similarly, the critical free energy required for a stable nucleus can then be written as:

$$\Delta G_{crit} = -\frac{4}{3}\pi r_c^2 \sigma_{SL} \tag{13.3}$$

The nucleation rate of crystals in a supersaturated solution can be expressed through an Arrhenius type of equation and can be represented as:

$$J = F \exp\left(-\frac{\Delta G}{kT}\right) \tag{13.4}$$

For homogeneous nucleation, this can be written as:

$$J = F \exp\left(-\frac{16\pi\sigma^3\nu^2}{3k^3T^3(\ln S)^2}\right) \tag{13.5}$$

In case of heterogeneous nucleation, the above relationship would become:

$$J = K_n \Delta c^n \tag{13.6}$$

Although nucleation could theoretically occur as soon as supersaturation is established, there generally occurs a delay between attainment of supersaturation and the detection of the first nucleus from the supersaturated phase. This delay is termed as the induction period (t_i) for nucleation. The induction time is inversely related to the nucleation rate and might exhibit wide variations based on the method of crystallisation. It is to be noted that the CNT does not account for induction time as it assumes ideal steady state conditions which nucleates spontaneously once the supersaturation is established.

13.3 Seeding

As pointed out before, seeding allows to efficiently separate nucleation and growth of crystals during large scale crystallisation processes. This avoids the burden of controlling nucleation in large scale operations and set optimum conditions for crystal growth, thereby controlling crystal size (distribution) and crystal polymorphism. Some of the critical parameters in seeding include seed loading, temperature profile of crystallisation, seed size and shape, seed surface area and porosity, and seed chemistry. Some of these factors are discussed further in the following sections.

13.4 Seed Loading

Seed load denotes the amount of seed that has to be added to the crystallisation solution to achieve the desired result. This could be expressed in terms of mass or surface area. The seed surface area available for crystal growth determines the available area for crystal growth and thereby regulates the supersaturation consumption rate during crystallisation. Generally, for product crystals of smaller size, large numbers of small seed crystals will be used and if the required product crystals are of a larger size, small numbers of larger crystals will be used. These would ensure fast consumption of the supersaturation and quicker crystallisation experiments. However, the desired results in supersaturation profile and product size distribution can be attained only by regulating the cooling profile along with planned seeding. Nonetheless, a general idea on the seed load can be achieved from a ratio of the seed mass to the theoretical crystal yield of the crystalline product [3].

Crystal yield (y) in a general crystallisation experiment can be calculated from the solvent and solute balance equations. Assuming that w_1 and w_2 are the initial and final solvent contents; c_1 and c_2 are the initial and final concentrations; R is the ratio of solvate to non-solvate molecular weights, the solute balance can be written as:

$$w_1c_1 = w_2c_2 + y\frac{1}{R} \tag{13.7}$$

(i.e., initial solute content in solution = final solute content in solution + solute content in crystals)

Considering that E is the ratio of the evaporated solvent to initial solvent, the solvent balance can be written as:

$$w_2 = w_1 - y\frac{R-1}{R} - w_1E \tag{13.8}$$

(i.e., final solvent content = initial solvent content – solvent removed as solvate in crystals – evaporated solvent content)

By substituting solvent balance in solute balance we can get:

$$w_1c_1 = c_2\left(w_1 - y\frac{R-1}{R} - w_1E\right) + y\frac{1}{R} \tag{13.9}$$

Upon rearranging to obtain crystal yield:

$$y = \frac{Rw_1(c_1 - c_2\,(1-E))}{(1 - c_2(R-1))} \tag{13.10}$$

If no undesired nucleation occurs during crystallisation the mass of seeds (m_s) can be estimated using the equation:

$$m_s = y\left(\frac{d_s^3}{d_p^3 - d_s^3}\right) \tag{13.11}$$

where, d_s and d_p are the seed and product particle sizes (assuming narrow mono-disperse particle size for both seeds and particles).

Although the above equation provides the maximum theoretical crystal yield, it does not take into account the kinetics of crystal growth which is also essential in optimising the seed loading and the temperature profile during/after seeding. The temperature profile should follow the growth kinetics and solute consumption within a close gap so as to ensure a moderate level of supersaturation that allows crystal growth and avoids undesired nucleation. Typical industrial seed loading could vary from 0.1 wt% to 5 wt% based on the product requirement [4].

13.5 Temperature Profile

Designing a suitable temperature profile and identifying the point of seeding is a significant aspect in seeding. A general rule of thumb is to add seeds midway between the solubility curve and the metastable zone [4]. Solubility curve and the metastable zone width will have to be established before a suitable temperature profile can be set. For most crystallisations, the temperature has to be maintained in such a way that the supersaturation is always maintained within the metastable zone

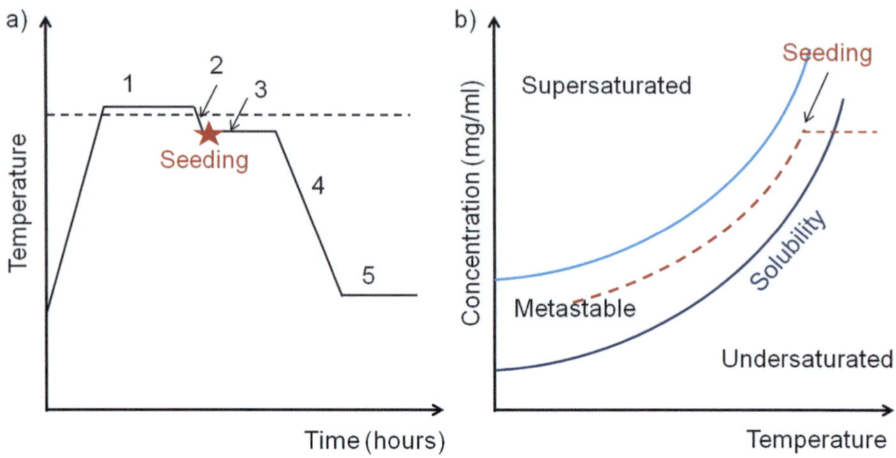

Fig. 13.2 (**a**) Temperature profile and (**b**) an ideal concentration profile to be followed in seeding crystallisation

to avoid inadvertent secondary nucleation while maintaining growth. A general temperature profile that can be used for a seeding crystallisation and the various stages in it are shown in Fig. 13.2.

In the above figure (Fig. 13.2a) step 1 is to warm the solution above the solubility limit, into the undersaturated region (the dotted line in the figure represents the saturation temperature). The solution is then cooled down into the metastable region in step 2 and the seeds are added in step 3. Once the seeds are added, an isothermal condition is generally maintained for some time for the seeds to be dispersed into the solution and to consume the initial supersaturation [4]. This is followed by cooling down (step 4) to achieve the required concentration profile. The rate of cooling could be linear or variable based on the nature of the crystallising solution and the metastable zone width. Once the required final supersaturation is achieved, the temperature could be maintained for some time (step 5) to maximise the yield.

13.6 Other Factors

In addition to the above factors, there are a number of other factors that affect seeded crystallisation. As can be seen from the relationship to find the seed loading, seed size distribution can affect the final product size distribution. Seed processing and handling can also cause problems with seeding. Milled seeds often exhibit fractured edges and faces that might result in multiple crystal growth areas leading to particle agglomeration or other undesired growth effects [4]. Proper mixing is also essential to ensure seed dispersion and to avoid temperature gradient within vessel. Temperature gradients can alter the crystal size distribution, thereby affecting the final product quality [5]. Scale-up of seeded crystallisation is also challenging due to the fact that, although solubility is a scale independent thermodynamic parameter, metastable zone width is scale dependent and also affected by the presence of a small quantity of impurities in the solution.

13.7 Seeding Techniques

Seeds could be added into the supersaturated solution as dry seeds or as slurry in a saturated solution. During the addition of dry seeds care has to be given to ensure that the seeds have been dispersed uniformly in the solution before cooling. Dry powder adhesion is more prone to particle clumping or aggregation if not properly dispersed. This would lead to different particle size distributions if not attended to. Hydrophobic seed particles are more susceptible to agglomeration due to strong non-polar interactions. Wet loading or slurry loading attempts to overcome this difficulty by dispersing the seeds in a small volume of liquid prior to their addition into the mother liquor. The slurry could be prepared in a saturated solution of the same solvent or in a different solvent as in anti-solvent crystallisation. Slurry preparation also helps to heal processed seed particles by allowing fractured faces and edges of the seed particles to be smoothened through growth and Ostwald ripening.

13.8 Feedback Control

Crystal yield, polymorphism, crystal shape, and size distribution of the final particles are some of the key factors that dictate the success of a crystallisation. Based on the product requirements, some of these or related parameters can be monitored during crystallisation and can be used to control the process. Solution concentration, number of particles and particle dimensions are some of the common parameters that can be measured online and used in a feedback control unit to regulate crystallisation [6]. However, use of feedback control strategies would require prior knowledge of the system kinetics which would commonly be established from small scale studies.

A simple feedback control set-up for seeded crystallisation is to set fixed cooling or heating rates and to monitor the particle size (using an FBRM) to follow crystal growth after seeding. In this approach, the cooling profile would be started only when the particle size distribution confirms sufficient dispersion of seeds and further cooling profile(s) will be set based on the particle chord length detected by the system. Once the required particle size (chord length) is achieved, the cooling can be changed or stopped. A similar approach for cooling crystallisation and antisolvent crystallisation has been demonstrated in the literature [4].

Another approach is to avoid any pre-determined cooling rates and allow the feedback control loop to fix the cooling/heating rates based on the data that it measures. Such a methodology was demonstrated in the 'Direct Nucleation Control' technique implemented by Abu Bakar et al. [7]. This approach constantly monitors the number of particles in a controlled volume and generates heating or cooling (or antisolvent addition) rates based on the measured value. This can be used for the automatic detection of metastable zone width as well as for the control of crystal size distribution in the final product.

13.9 Seeding in Continuous Crystallisers

Most of the crystallisation process in industry is conducted in batch mode. However, many designs of continuous crystallisers have been developed and used. A simple example would be the Forced circulation Swenson crystalliser. This crystalliser involves a secondary loop to re-circulate the fine crystals and slurry along with additional feed through a heater to dissolve the fines before returning the mixture into the main crystalliser. Big crystals are continuously removed from the bottom of the crystalliser through gravitational settling.

Novel designs of continuous crystallisers have been developed in recent years. Tubular plug flow crystallisers and continuous oscillatory baffled crystallisers are two of the widely studied design configurations in this context. Uncertainties associated with the nucleation in a continuous process are avoided in these designs by continuous seeding with freshly prepared seed suspension or by re-circulating a part of the product stream. Seed particles of uniform size could be continuously prepared in a separate nucleation unit using impinging jet precipitation, antisolvent, or quench crystallisation [8]. The slurry of seed crystals is then injected into the saturated solution along with the flow. By adjusting the temperature trajectory along the flow path, required supersaturation and corresponding crystal growth rate can be maintained [8].

13.10 Case Studies

13.10.1 Seed Loading and Mean Crystal Size in a Continuous Crystalliser

Seed loading is an important factor that will affect the kinetics of crystal growth and thereby affect the final product quality. The effect of seed loading with respect to the amount of seeds and the number of seed particles on the crystal size distribution of the final crystals in a continuous tubular crystalliser was reported by Eder et al. [8]. The study showed that a narrow unimodal crystal size distribution of the product crystals can be obtained by optimising the seed loading and seed size distribution. For a higher seed loading (seed mass per volume of crystallising solution), it was observed that the increased number of particles consumed the supersaturation much faster and the product size distribution followed very closely to that of the seed particles. Similarly, for a constant seed loading, when particles of smaller size were used (therefore, number of particles change even though the mass per unit volume of the solution does not change), the difference between seed and product crystal sizes reduced considerably with a similar size distribution profile. Thus, a close control on the final crystal size and size distribution in a continuous tubular reactor was demonstrated by altering seed loading and the number of seed particles.

13.10.2 Producing Metastable Polymorphs Through Seeding

Stable polymorphs are commonly preferred for product development as they avoid the risk of product transformation and subsequent quality issues. However, if the metastable polymorphs exhibit superior product attributes such as higher solubility, bioavailability and product handling properties, they are sometimes selected for product development. One such example is that of Abecarnil [9], whose metastable form A is chosen for the final product due to its higher solubility. (The least stable form B is not chosen for development since it transforms quickly into form A; transformation time from form A to the most stable form C was about a day). However, in order to produce pure form A, a suitable seeding technique is to be used. Pure form A seeds have to be first prepared through cooling crystallisation from methanol or by conversion of the least stable form B crystals into form A. Preliminary experiments showed that seed pre-treatment (seeds held in a slurry mixture prior to seed addition to remove any impurities and clean the growth faces) results in a significant improvement in obtaining pure form A crystal. Still, form A seeds cannot be maintained in slurry for a long time due to the risk of conversion to the stable form C. The seed addition has to be performed below the metastable region of form B to avoid its nucleation. Seed loading and cooling rate are closely related; faster cooling rate would demand a higher seed loading to avoid nucleation of form B crystals. On the other hand, a slower cooling rate would result in longer batch time and thereby increase the risk of conversion into form C. Hence, a compromise in the seed loading and cooling profile has to be made. Also, a faster initial cooling rate followed by a slower cooling rate (or a non-linear temperature profile) would be more appropriate as the small seed crystals would initially provide large surface area for crystal growth resulting in faster consumption of supersaturation. Furthermore, once the required final set point or crystal yield is obtained, it is highly important to separate the form A crystals and dry them quickly to avoid any solution-mediated transformation that can occur in the presence of solvent. Thus, by combining seeding with a suitable temperature profile, pure form A crystals of Abecarnil can be produced.

13.10.3 Heterogeneous Seeding

Heterogeneous seeding is a related area in seeding where the seed material is not made of the same crystals that are to be produced. The advantages of heterogeneous seed materials are that they can be prepared independent of the crystallising solution and this allows for the fine tuning of the seed properties. Seed shape, porosity and chemistry can be altered and used to achieve the required crystal nucleation/growth. Heterogeneous seeds are also useful in cases where homogeneous seeds are not readily available or they cause undesirable effects such as cross-nucleation that will affect the final product quality [1]. Oils, precipitants, and crystals of related molecules have been used as heterogeneous seeding surfaces [1]. Seed particles can also be prepared from inorganic materials such as SiO_2, ZnO, TiO_2, or with polymers [10]. Different properties of the heterogeneous seeds such as their molecular arrangement, surface chemistry, pore sizes and surface area are considered to be important in regulation nucleation and growth of crystals [11]. In cases where the heterogeneous seed material and the nucleating crystal exhibit a similarity in their molecular arrangement, they are said to have an epitaxial relationship. In case of seeds with an ordered molecular arrangement, such epitaxial relationships are said to be a major factor in controlling nucleation of new crystals on the seed surface [12].

Crystallisation of macromolecules is a realm where application of heterogeneous seeds would be very promising. High molecular macromolecules are sometimes difficult to crystallise due to their large sizes and high molecular mobility. Porous heterogeneous templates have been proved to be efficient in inducing nucleation of such molecules. Crystallisation of large protein molecules is an essential in deciphering their molecular structure. Hence, heterogeneous seeds have often been used in their crystallisation screening procedure. A grain of sand along with the microseeds of analogous proteins was used for the crystallisation of *E. coli* thioredoxin for its structural determination [13].

Heterogeneous seeds are specifically useful during the crystallisation screening stages of new molecules in pharmaceutical, biopharmaceutical, agro-chemical, and food industries. Heterogeneous seeds with well ordered structures, pore sizes and definite chemistries could be used to induce nucleation of crystals [11]. Such seeds can act as templates that could potentially alter both thermodynamics as well as the kinetics of crystallisation. This would result in crystallisation under conditions which otherwise may not provide the crystals [14].

References

1. Bergfors T (2003) Seeds to Crystals. J Struct Bio 142:66–76
2. Jiang S, ter Horst JH (2010) Crystal nucleation rates from probability distributions of induction times. Cryst Growth Des 11:256–261

3. Doki N, Kubota N, Yokota M, Chianese A (2002) Determination of critical seed loading ratio for the production of crystals of uni-modal size distribution in batch cooling crystallization of potassium alum. J Chem Eng Japan 35:670–676
4. O'Sullivan B, Smith B, Baramidze G (2012) Recent advances for seeding a crystallization process. Mettler Toledo Auto-Chem, Columbia
5. Hojjati H, Rohani S (2005) Cooling and seeding effect on supersaturation and final crystal size distribution (CSD) of ammonium sulphate in a batch crystallizer. Chem Eng Process 44:949–957
6. Barrett P, Smith B, Worlitschek J, Bracken V, O'Sullivan B, O'Grady D (2005) A review oft he use of process analytical technology for the understanding and optimization of production batch crystallization processes. Org. Process Res Dev 9:348–355
7. Abu Bakar MR, Nagy ZK, Saleemi AN, Rielly CD (2009) The impact of direct nucleation control on crystal size distribution in pharmaceutical crystallisation processes. Cryst Growth Des 9:1378–1384
8. Eder RJP, Schmitt EK, Grill J, Radl S, Gruber-Woelfler H, Khinast JG (2011) Seed loading effects on the mean crystal size of acetylsalicyclic acid in a continuous-flow crystallisation device. Cryst Res Technol 46:227–237
9. Beckmann W, Nickisch K, Budde U (1998) Development of a seeding technique for the crystallisation of the metastable A modification of Abecarnil. Org. Process Res Dev 2:298–304
10. Lee AY, Ulman A, Myerson AS (2002) Crystallization of amino acids on self-assembled monolayers of rigid thiols on gold. Langmuir, 18:5886-5898; Diao Y, Myerson AS, Hatton TA, Trout BL (2011) Surface design for controlled crystallization: The role of surface chemistry and nanoscale pores in heterogeneous nucleation. Langmuir 27:5324–5334
11. Shah UV, Williams DR, Heng JYY (2012) Selective crystallisation of proteins using engineered nanonucleants. Cryst Growth Des 12:1362–1369
12. Hiremath R, Varney SW, Swift JA (2004) Oriented crystal growth of 4-Iodo-4′-nitrobiphenyl on polar self-assembled monolayer templates: A case for "Chemical Epitaxy". Chem Mater 16:4948–4954
13. Hendrickson W, Horton J, LeMaster D (1990) Selenomethionyl proteins produced for analysis by multiwavelength anomalous diffraction: a vehicle for direct determination of three dimensional structure. EMBO J 9:1665–1672
14. Shah UV, Allenby MC, Williams DR, Heng JYY (2012) Crystallisation of proteins at ultralow supersaturations using novel three-dimensional nanotemplates. Cryst Growth Des 12:1772–1777

Chapter 14
Preparation, Stabilisation and Advantages of Metastable Polymorphs

Ana Kwokal

Abstract This chapter examines the challenges faced by the pharmaceutical industry in developing metastable polymorphs of an API, and discusses some key advantages with examples. The new perspective of polymorph-directing crystallization using functionalised templates is also discussed. The basic principles and examples of heterogeneous template crystallisation are given.

Keywords Metastable polymorphic forms • Pharmaceutical development • Stability • Crystallisation • Templates

14.1 Introduction

The majority of the active pharmaceutical ingredients (API) are small organic molecules produced as a crystalline powder. The API crystals often show many crystalline forms (polymorphs) but only one of them is thermodynamically stable under defined temperature and pressure and this form is usually selected as the one to be produced. However, there are occasions when development of the metastable form is deemed the most viable option from a manufacturability, chemical purity and/or efficacy point of view. In such cases, appropriate handling and storage conditions would be required and implemented to ensure the kinetic stability throughout the assigned shelf life.

A. Kwokal (✉)
GlaxoSmithKline, R&D, New Frontiers Science Park, Third Avenue, Harlow, UK
e-mail: ana.x.kwokal@gsk.com

K.J. Roberts et al. (eds.), *Engineering Crystallography: From Molecule to Crystal to Functional Form*, NATO Science for Peace and Security Series A: Chemistry and Biology, DOI 10.1007/978-94-024-1117-1_14

14.2 Screening of Solid Forms: Relevance to the Pharmaceutical Industry

The aim of solid form screening in the pharmaceutical industry is to make, by first intent, the thermodynamically stable crystal form of the Active Pharmaceutical Ingredient (API) under given storage and process conditions. However, the thermodynamically stable form does not always behave as desired in terms of chemical stability or physical properties such as flowability, filterability, mechanical and morphological properties. These physical properties can impact on the manufacturability and the ability to deliver a robust product to the patients. In this case screening for a different solid form, which would be metastable, with different crystal structures may offer the opportunity for improved physical properties. The metastable forms also have better solubility, which in principle, enhance bioavailability. Despite the stability risk, some metastable forms show neglectable or infinitely slow kinetics of transformation to the thermodynamically stable form.

The need to find the best possible crystal form having the optimal physical properties comes, not only from a manufacturability point of view, but also from a quality and business perspective.

Whilst early drug development provides an opportunity to identify and select a compound with the optimal physicochemical properties, there are several risks to performing screening very early in the development [1]:

- The level of attrition of compounds is high in the early development phase. Hence, a large proportion of the work may be wasted due to a large number of compounds being abandoned at this stage.
- The work is limited by the availability of API at this stage as the majority of a very limited supply of drug substance is primarily used in toxicological trials.
- The API may be of poor quality often owing to a quite lengthy synthetic route; hence it is likely to be unsuitable for screening for suitable crystalline solid forms as the presence of related impurities is common.
- The API may be produced by non-representative routes or processes hence physical properties may change on scale-up.

Conversely, there are risks associated to screening late in the development process [1]:

- It may be unlikely to have a significant influence on compound selection even though bulk is quite readily available and more representative of the likely marketed quality.
- If solubility is found to be limiting in clinical trials, the introduction of a new salt form will require costly bridging toxicological studies.
- Formulation development will be required to identify a suitable formulation for early clinical trials; hence changes in the API may be difficult.

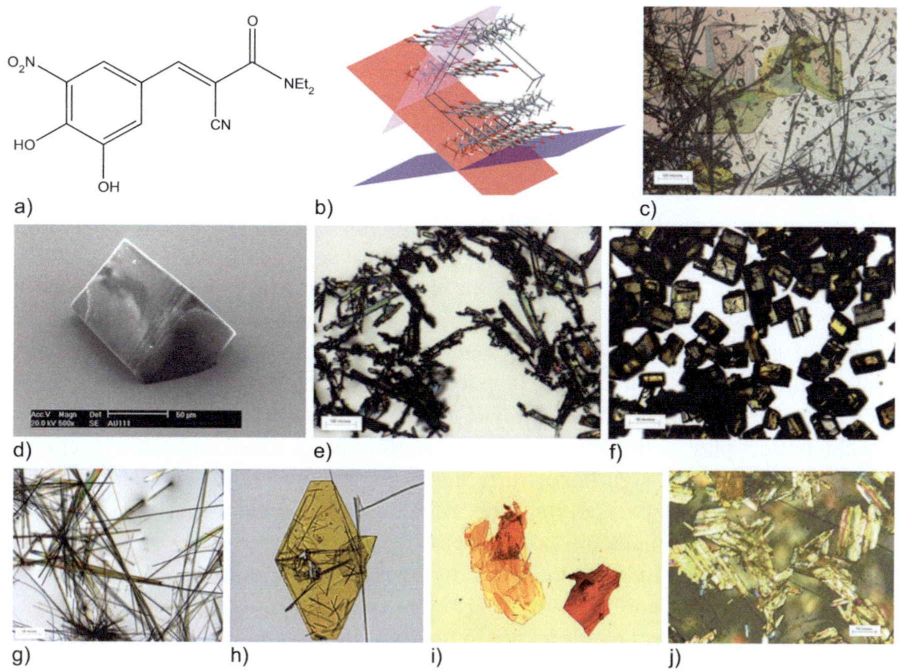

Fig. 14.1 (**a**) Entacapone ((E)-2-cyano-N,N-diethyl-3-(3,4-dihydroxy-5-nitrophenyl)propenamide). (**b**) Inter-molecular packing of form A, based on crystal structure [2]. (**c**) Concomitantly grown forms A, D and δ from 12% aqueous/ 88% acetone (12A88W) solution. (**d**) Form A grown from quiescent crystallisation of 12A88W solution at Au (111) surface. (**e**) Form A harvested after seeded (with form A) crystallisation from 12A88W solution. (**f**) Form A harvested after templated (Au(100) + Entacapone molecular layer) crystallisation from 12A88W. (**g**) Form D harvested after crystallisation from 12A88W solution without seeding. (**h**) Form δ harvested after crystallisation from 12A88W solution, grown concomitantly with form A and D. (**i**) form α grown at negatively polarised Au(100) surface after crystallisation from 12A88W solution. (**j**) Form A harvested after crystallisation from acetone solution [3–5]

Therefore, a prudent approach should be employed when conducting screening of polymorphs. It should be a risk-based and value-adding exercise that should be embedded at an appropriate stage during the drug development cycle.

Figure 14.1 shows a number of polymorphs of an active pharmaceutical ingredient, Entacapone, grown from the same solution under very similar crystallisation conditions, where form A is the most stable form under room conditions.

14.3 Advantages and Disadvantages of Metastable Forms

There are many cases when metastable polymorphs show superior physical and chemical properties over stable forms, for example, improved morphology as shown in Fig. 14.2. However, production and stabilisation of metastable polymorphs is often tricky and seeks extensive understanding and control. The pharmaceutical industry often has to balance the advantages and disadvantages whilst cost benefit analysis is helpful to define the quality and the business perspective.

It is generally acknowledged that the thermodynamically stable polymorph is more chemically stable than the metastable polymorph. This has generally been attributed to higher crystal packing density of the thermodynamically favoured polymorph (i.e. the "density rule"), but recent investigations suggests that other factors, such as optimized orientation of molecules, and non-hydrogen bonds in the crystal lattice play a more important role. Relatively small changes in crystal packing may lead to significant differences in the crystal packing density and chemical reactivity of two polymorphs. An example of polymorphs with this type of behaviour are indomethacin polymorphs [6, 7].

Another such example is donepezil hydrochloride monohydrate, disclosed in patents [EP 296560 and US 4895841]. Donepezil hydrochloride monohydrate exhibits valuable properties for its use in pharmaceutical compositions, particularly in that it is chemically more stable than the anhydrous forms. The monohydrate form is not hygroscopic, and it is compatible with pharmaceutical excipients. However, the problem with crystalline donepezil hydrochloride monohydrate is that it is susceptible to solvent mediated conversion to different polymorphic forms.

Examples where the mechanical properties of metastable forms are superior to the thermodynamic stable forms are described as follows. The mechanical properties of two enantiotropic polymorphs of metoprolol tartrate, metastable form I and the more stable form II (at room temperature) were examined [8]. Tablets formulated with form I possessed lower porosity and yield pressure when compared with tablets of form II. This suggests that the less dense metastable form I may have less

Fig. 14.2 Example of superior morphology of metastable solvate form (*right*) in comparison to most stable anhydrous form (*left*)

strength in the crystal lattice and be easier to deform. Form I also had higher elastic recovery, probably due to higher elasticity and/or lower porosity of the tablets. The tablets of the metastable form I was stronger at low pressures than those of form II, probably due to the higher compressibility of form I. Another example is sulfamerazine. Summers et al. [9] proposed that the presence of slip planes in form I of sulfamerazine was found to be the reason for its higher plasticity than form II, the more stable form at room temperature. This higher plasticity results in greater compressibility and tabletability. Another example is of paracetamol, the better compressible properties for the orthorhombic form II of paracetamol enable direct compression while the most stable monoclinic form I has to be manufactured by an intensive granulation process using more resource [10, 11].

An interesting case is US 8580796 patent, which discloses "Low hygroscopic aripiprazole drug substance and processes for the preparation thereof". The patent claims a new form of aripiprazole that has lower hygroscopicity than the conventional anhydrous form of this material which provides improvement in the physical properties of this material since identification of hygroscopicity in the prior art form confirms that this material can be difficult to handle. The current process has been found to be costly and burdensome and involves special measures to ensure the material's protection from moisture during processing and formulation. The patent also states that exposure to moisture can convert the anhydrous to a hydrous form, which presents several disadvantages, such as: (i) the hydrous forms are less bioavailable and less dissoluble than the anhydrous forms of aripiprazole, (ii) the variation in the amount of hydrated versus anhydrous aripiprazole drug substance from batch to batch could fail to meet specifications set by drug regulatory agencies, (iii) the milling may cause the drug substance, conventional aripiprazole, to adhere to manufacturing equipment which may further result in processing delay, additional operator involvement, increased cost, increased maintenance and lower production yield. This is an example of a technical solution given through improved physical properties.

14.4 Crystallisation of Metastable Forms

Crystallisation of metastable forms of an API is not an easy task due to the fact that such crystallisation is sensitive to subtle process variation, i.e. it can suffer from poor robustness and repeatability, especially at the large production scale.

Organic molecules generally nucleate very poorly and, when cooled from saturated solution, show large metastable zones, which in turns facilitates nucleation to start from high supersaturation. Nucleation and crystal growth from high supersaturation can be very fast, which favours growth of metastable forms. Rapid cooling is one of the most common crystallisation methods but in most of the cases high supersaturation is achieved by fast anti-solvent addition.

Crystallisation of metastable forms, when polymorphs are enantiotropically related, is an easier case in comparison to monotropically related forms. Figure 14.3

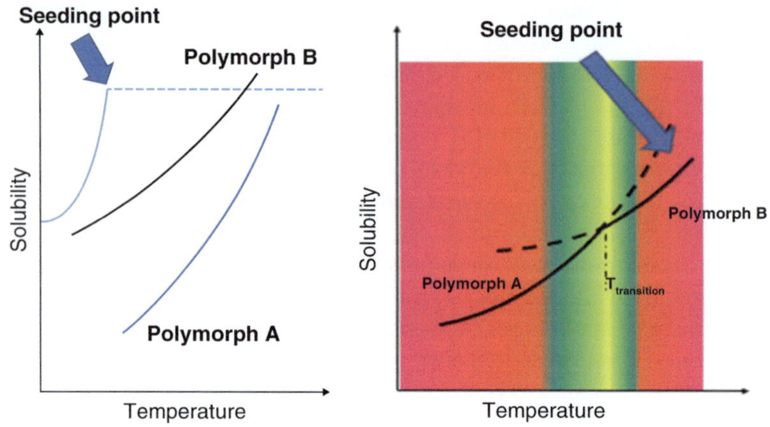

Fig. 14.3 Temperature vs. solubility for two monotropically (*left*) and enatiotropically (*right*) related polymorphs

shows solubility diagram for monotropically and enantiotropically related polymorphs in which the form with lower solubility is the more stable. Nucleation and crystal growth of desired form can be governed within the "safe" zone where the actual form is stable. These crystals can be isolated even from unsafe zone if they are grown to substantial size and when solubility is minimised (for example by anti-solvent).

Washing of the filtration cake is always recommended during isolation of the metastable form to reduce the amount of crystallization solvent, i.e. mother liquor which can recrystallize upon drying into the stable form, or can induce transformation of the metastable form to the stable form under the high temperature of drying. The choice of washing solvent is very important and aims to reduce mother-liquor to a minimum. This is because mother liquors carry impurities and solubilising agents that increase molecular motion at the surface of crystals. It is especially emphasised during drying when crystals are still wet and are exposed to elevated temperature. Washing solvent should be chosen based on its wettability properties and preferential adsorption on the surface of crystals. Once the metastable form is successfully isolated it should be kept under dry conditions (if not a hydrate) and at relatively low temperature.

The particle size distribution of crystals produced from high supersaturation are small and often suffer from poor crystallinity (high disorder), poor flow and are prone to aggregation. Cooling-heating cycles within the metastable zone of crystallisation helps to increase particle size (dissolve undesired small crystals) and crystallinity, but for metastable forms this procedure can be very risky in terms of transformation of form. In fact, metastable forms should be isolated from mother liquor very quickly, since any further mixing in their slurry can lead to form transformation. In addition, isolation and filtration is often performed under low temperature.

Seeding crystallisation is one of the most common techniques employed for crystallisation of organic molecules and especially in the case of metastable polymorphs. The use of seed crystals to initiate crystallization always helps to ensure crystallization of the desired polymorph. Figure 14.3 shows an example of recommended points for seeding towards crystallization of form B (metastable form at room temperature); however setting the right seeding point presents significant problems when crystallization is governed among various other polymorphs. The common opinion is that solvent nature determines the crystallizing polymorph, but it is also known that it is possible, in principle, to get almost any polymorph from any solvent, if the appropriate seed is used. Therefore the correlation between crystal forms and solvent is overestimated, but there is still no full understanding of what drives the formation of forms. Recent studies have shown a strong correlation between surface properties and the crystal structure resulting from surface growth.

Studies have shown that crystallization behaviour can be strongly influenced by the surface of crystallisation vessels, i.e. by material used in construction of the vessels [12, 13]. Indeed, it is a common observation in industrial crystallization practice that the first crystallization in a freshly cleaned reactor can be quite different in nature with respect to subsequent crystallizations (see e.g. Ref. [3]). Such observations are often related to surface effects, e.g. the formation of an ad-layer of adsorbed molecules, which impacts the crystallization process by promoting heteronucleation at the ad-layer surface. It is also well-known that such adsorbed molecules can sometimes persist in being bound to surfaces even after rinsing by solvents and/or heating the solution above its solubility limit. It is also known that the nucleation energy barrier is always lower at the surface than the one in the bulk.

14.5 Nucleation Directed by Heterogeneous Surface Templates

From a phenomenological perspective, nucleation in crystallisation systems is mostly heterogeneous (happening on heterogeneous surfaces) with the process being initiated at preferred local sites where the molecular scale functionality of an available substrate provides a suitable structural environment for its promotion. In practical process systems, such substrates include vessel/piping walls [14], impellers, baffles and probes used for analytical instrumentation [15]. Structural matching between the hetero-nucleation substrate and the surface functionality [16–19] of the crystallising product and/or the porosity/confinement provided by the micro/nano scale geometry and porosity of the substrate [20–22] are important factors in this respect. Templating the crystallisation process is becoming more applied within the crystallisation research community. Examples of this include crystallisation of an API (entacapone) on a single crystal gold surface Au(111) functionalised with a self-adsorbed molecular layer (SALE) that was used as an alternative seed. This study has demonstrated that under quiescent conditions, a bare Au(111) surface can direct the formation of form A of entacapone at its surface with the concomitant crystallisation of entacapone form D in the bulk solution. Only form D normally crystallises spontaneously under the same conditions. In the agitated solutions form D crystallises in 100% of the cases, but when surface template SALE of 5 mm^2 was introduced in a 500 ml vessel, formation of only form A was achieved [3]. This example shows how small amounts of a surface introduced into crystallisation system can change crystallisation outcome in the bulk solution. In addition when gold was polarised the metastable form appeared at edges of the surface [4].

A very interesting study of template governed crystallization was explored with carbamazepine [23]. Cooling crystallization of carbamazepine in glass vials functionalised with different silane molecules, resulted in preferential nucleation of metastable form II or stable form III within a definite range of supersaturation in contrast to the two crystal forms nucleated concomitantly on a control substrate under similar solution conditions [23]. Ward and co-workers have shown the template-directed crystallization of selected organic crystal polymorphs on the surfaces of single crystals of another crystal. Ward was the first to introduce the concept of organic crystal polymorphism by surface control and opened up possibilities for the existence of "a library of organic seeds, which can be used to either control polymorphism or to search for unknown polymorphs" [19, 20, 24, 25].

There are several interesting studies highlighting cases in which different polymorphs were grown on different templates under the same crystallisation conditions. Dressler et al. [26] have shown that the crystallization of a L-glutamic acid on the SAMs resulted in the stabilization of the metastable α-form of L-glutamic acid, while the crystallization onto the SAMs led to the preferential crystal growth with the orientation along the (111) crystal direction. Cox et al. [27] have selectively grown the anhydrous and monohydrate forms of a theophylline on the functionalised SAMs prepared with a hydrophobic (dodecanthiol) and a hydrophilic 11-mercaptoundecanol, 11-mercaptoundecanoic acid and a 16-mercaptohexadecanoic acid. The substrates with a hydrophilic SAM were found to be a selective nucleator of an anhydrous theophyline, while a monohydrate was grown on a hydrophobic substrate.

The study of Hiremath et al. [17] (Fig. 14.4) crystal polymorphism on 1,3-bis (m-nitrophenyl) urea (MNPU) by the use of the SAM's of the substituted 4′-X-mercaptobiphenyls, shows how changing a single functional group can change the inter-molecular interactions and the dipole moment of the orientation, which consequently change the crystal polymorph formed within the crystallisation system. The crystallisation from the ethanolic solution of the MNPU produced a mixture of forms, but in the presence of the SAM, the surfaces under a variety of the solution conditions, consistently crystallised the material with high phase purity.

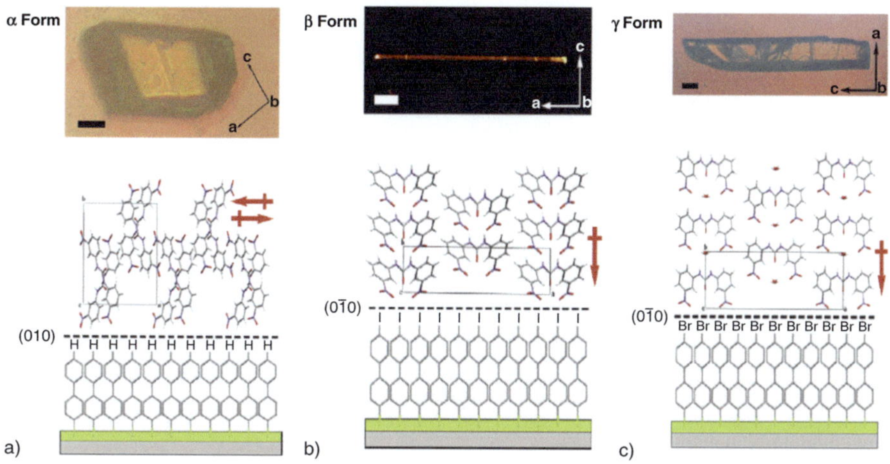

Fig. 14.4 Selective nucleation and growth of the α-MNPU on (**a**) 4-mercaptobiphenyl; (**b**) 4-iodo- ercaptobiphenyl; (**c**) 4-bromo-mercaptobiphenyl SAM templates, Ref. [17]

14.6 Direction of the Polymorphic Form Using an Electrochemical Tuneable Surface Template

Electrochemical techniques are potentially a very powerful method for building nanostructures on solid surfaces which can act as templates for crystallization. The arrangement of surface structures can be changed through the application of very low applied electrochemical potentials. For example, the charge density at an electrode surface can be fine-tuned by an external electric field which can change the configuration and strength of surface adsorption and even result in the formation of different surface structures. The ability to direct the surface crystallisation of different polymorphs of entacapone by tuning the electrochemical potential of metal Au(100) templates has recently been demonstrated [4].

This study of Entacapone crystallization on an Au(100) template under negative charge provided clear evidence for the ability of electrochemical potential to control the surface nucleation process by switching the template towards directing nucleation of the polymorphic forms. The fact that the Entacapone metastable form crystallised on the template surface suggests that this interfacial system is highly selective with a supramolecular structure of the first adsorbed layers that mimics the appropriate intermolecular packing of the directed polymorphic form (Fig. 14.5).

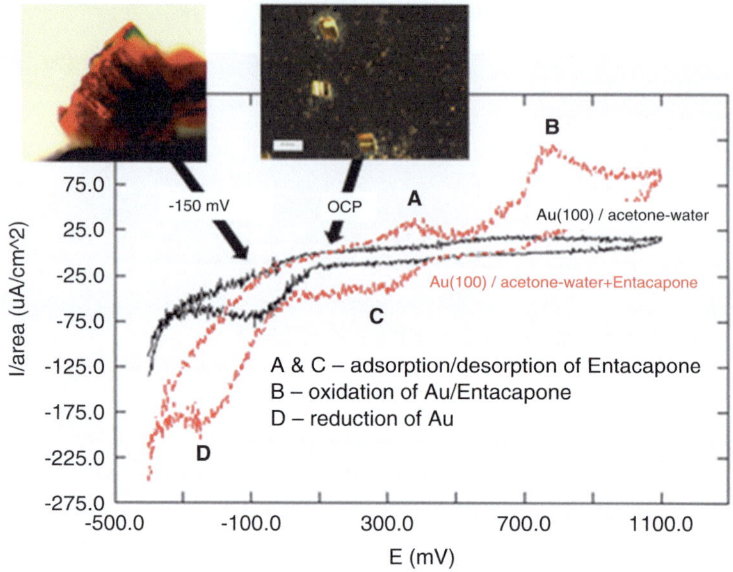

Fig. 14.5 At open circuit potentials (OCP) Entacapone crystallises onto the surface of a gold nucleation template in its stable polymorphic form A whilst at negative polarization (−150 mV) this behavior is suppressed and the polymorphic form is switched to the formation of the metastable form α which nucleate at the template's edges

14.7 Uncontrolled Seeding

Despite all the efforts of controlling crystallization, a full understanding of surface nucleation has not been reached and undesired nucleation can occur through the solid material present in the crystallizer. Tiny particles from the atmosphere as well as the solid impurities, can in principle, act as a seed. Indeed, the atmosphere can become contaminated with the dust of the crystallized material, which could afterwards act as the unintentional seeds in subsequent process. The effect of the presence of such seeds could provide an unwelcome element of surprise. Such types of the uncontrolled seeding frequently occur in the laboratory or in industry, and are probably the reason for many anecdotal accounts related to perverse behaviour in crystallization systems.

14.8 Thermodynamic and Kinetic Stability

Based on thermodynamic relation $\Delta G = \Delta H - T\,\Delta S$ the lowest free energy form is the most thermodynamically stable form. In relation to solubility, the most stable form is the least soluble form. Thermodynamic stability of hydrates/solvates depends upon the relative humidity, i.e. activity of water/solvents and the environmental temperature. The prerequisite for setting safe storage conditions is the knowledge of phase diagram including temperature and water activity parameters.

There is only one thermodynamically stable form in given conditions of temperature, pressure and water activity/humidity (for the case of hydrates) or solvent activity (in the case of solvates). However, some forms have very slow kinetics of transformation to stable form.

The transformation kinetics is largely affected by atmospheric/storage conditions like humidity and temperature. In most of the cases dry conditions and low temperature decrease the kinetics of transformation, except for the hydrates where low humidity might result in dehydration and structure collapse or polymorphic transformation.

Spontaneous transformation of metastable forms is difficult to predict and may occur with extreme variation, from a few hours to even several years. Such a huge difference depends on the activation energy of transformation, but can be affected by the presence of seeds of the stable forms, impurity level, residual solvent level, crystallinity, particle size, or/and surface area, etc. An understanding of the mechanism and prediction of the kinetics of transformation is highly important in terms of assuring product quality.

Stability prediction for the metastable polymorphs rely on accelerated stability testing under stressed conditions of temperature and humidity. Such data can then be extrapolated to predict stability under room temperature conditions. However, it is very challenging to predict long-term stability based on such short-term data mindful that it has been produced under non-representative high temperature conditions. The mechanism of solid-state reactions and therefore their associated reactivity can differ from those taking place at the lower temperatures.

The kinetics of solid-form transformations is usually monitored using techniques such as XRPD, Raman, Mid-IR, NIR, ssNMR, fluorescence, etc. [28–30] and the data (reaction progression against time) are then subsequently modelled. Models currently used in solid-state kinetic studies [31] are classified according to their mechanistic assumptions (such as nucleation, geometrical contraction, diffusion, and reaction order theory) or on the graphical shape of their isothermal curves (acceleratory, deceleratory, linear, or sigmoidal models). Acceleratory models are those in which the reaction rate is increasing as the reaction proceeds; similarly, deceleratory reaction rates decrease with reaction progress while the rate remains constant for linear models, and sigmoidal models show a bell-shaped relationship between rate and reaction progress. For the non-isothermal case it is important to note that the reaction progress-temperature plots are not as distinctive in their shape profiles when compared to the isothermal case.

Isothermal microcalorimetry offers the advantages of having a continuous measurement of the heat flow evolved by chemical or physical processes under conditions of constant temperature and is an important experimental tool in the study of kinetics of solid-form transformation. It allows the determination of not only kinetic parameters but also thermodynamic parameters (n, the order of the reaction; k, the rate constant; and H, the reaction enthalpy change) from power–time curves [32].

The aspect of chemical impurities is also very important to understand but can be very challenging, since the effect of impurities on solid-form stability is not yet fully understood. Impurities can either increase the rates of transformation or they can stabilize metastable forms [30, 33].

The practical approach to mitigate risk of metastable form transformation is usually case-dependent, but some general rules are valid; the amount of the stable form impurities should be minimized as well as amorphous phase; humidity protective packaging should be chosen; level of residual solvents controlled.

14.9 Summary

A general view on importance of crystalline polymorphs in the pharmaceutical industry was given with emphasis on metastable forms from scientific as well as business perspectives. Examples of metastable forms were discussed in view of their opportunities to provide improved chemical stability and physical properties of API over the thermodynamic stable forms. It was emphasised that, although thermodynamically unstable metastable forms can be kinetically stable or can be stabilised by proper handling and storage and can therefore in the right situations be a viable option for development.

The research also reveals the possibility of switching nucleation from uncontrolled nucleation sites toward a more directed crystallization process using well-ordered and defined surface templates, However, despite the significant effort given to control and understanding of polymorphs, crystallisation of metastable forms still presents a significant risk especially at the large scale.

References

1. Storey RA, Ymen I (eds) (2011) Solid state characterisation of pharmaceuticals. Wiley-Blackwell, London
2. Leppanen J, Wegelius E, Nevalainen T et al (2001) Structural studies of acyl esters of entacapone. J Mol Struct 562:129–135
3. Kwokal A, Nguyen TH, Roberts KJ (2009) Polymorph-directing seeding of entacapone crystallization in aqueous/acetone solution using a self-assembled molecular layer on Au (100). Cryst Grow Des 4:4324–4334
4. Kwokal A, Roberts KJ (2014) Direction of the polymorphic form of entacapone using an electrochemical tuneable surface template. CrystEngComm 16:3487–3493
5. Kwokal A, Čavužić D, Roberts KJ (2013) Surface adsorbed templates for directing the crystal growth of entacapone as monitored using process analytical techniques. Cryst Grow Des 13:5324–5334
6. Singhal D, Curatolo W (2004) Drug polymorphism and dosage form design: a practical perspective. Adv Drug Deliv Rev 56:335–347
7. Chen X, Morris KR, Griesser UJ et al (2002) Reactivity differences of indomethacin solid forms with ammonia gas. J Am Chem Soc 124:15012–15019

8. Guillory JK, Poust RI (2002) Chemical kinetics and drug stability. In: Banker GS, Rhodes CT (eds) Modern pharmaceutics, drugs and the pharmaceutical sciences. Marcel Dekker, New York, pp 111–138
9. Summers MP, Enever RP, Carless JE (1976) The influence of crystal form on the radial stress transmission characteristics of pharmaceutical materials. J Pharm Pharmacol 28:89–99
10. Di Martino P, Guyot-Hermann AM, Conflan P et al (1996) A new pure paracetamol for direct compression: The orthorhombic form. Int J Pharm 128:1–8
11. Joiris E, Di Martino P, Berneron C et al (1998) Compression behavior of orthorhombic paracetamol. Pharm Res 15:1122–1130
12. Liang K, White G, Wilkinson D et al (2004) An examination into the effect of stirrer material and agitation rate on the nucleation of l-glutamic acid batch crystallized from supersaturated aqueous solutions. Cryst Growth Des 2:1039–1044
13. Liang K, White G, Wilkinson D et al (2004) Examination of the process scale dependence of l-glutamic acid batch crystallized from supersaturated aqueous solutions in relation to reactor hydrodynamics. Ind Eng Chem Res 43:1227–1234
14. Jones F, Jones P, De Marco R et al (2008) Understanding barium sulfate precipitation onto stainless steel. Appl Surf Sci 254:3459–3468
15. Borissova A, Khan S, Mahmud T et al (2009) In situ measurement of solution concentration during the batch cooling crystallization of l-glutamic acid using ATR-FTIR spectroscopy coupled with chemometrics. Cryst Growth Des 9:692–706
16. Chadwick K, Chen J, Myerson AS et al (2012) Toward the rational design of crystalline surfaces for heteroepitaxy: role of molecular functionality. Cryst Growth Des 12:1159–1166
17. Hiremath R, Basile JA, Varney SW et al (2005) Controlling molecular crystal polymorphism with self-assembled monolayer templates. J Am Chem Soc 127:18321–18327
18. Singh A, Lee IS, Myerson AS (2009) Concomitant crystallization of ROY on patterned substrates: Using a high throughput method to improve the chances of crystallization of different polymorphs. Cryst Growth Des 9:1182–1185
19. Ward MD (2001) Bulk crystals to surfaces: combining X-ray diffraction and atomic force microscopy to probe the structure and formation of crystal interfaces. Chem Rev 101:1697–1725
20. Hamilton BD, Weissbuch I, Lahav M et al (2009) Manipulating crystal orientation in nanoscale cylindrical pores by stereochemical inhibition. J Am Chem Soc 131:2588–2596
21. Diao Y, Harada T, Myerson AS et al (2011) The role of nanopore shape in surface-induced crystallization. Nat Mater 10:867–871
22. Nicholson CE, Chen C, Mendis B et al (2011) Stable polymorphs crystallized directly under thermodynamic control in three-dimensional nanoconfinement: a generic methodology. Cryst Growth Des 11:363–366
23. Parambil JV, Poornachary SK, Tan RBH et al (2014) Template-induced polymorphic selectivity: the effects of surface chemistry and solute concentration on carbamazepine crystallisation. Cryst Eng Comm 16:4927–4930
24. Ward MD (2005) Snapshots of crystal growth. Science 308:1566–1567
25. Mitchell CA, Yu L, Ward MD (2001) Selective nucleation and discovery of organic polymorphs through epitaxy with organic single crystal substrates. J Am Chem Soc 123:10830–10839
26. Dressler DH, Mastai Y (2007) Controlling polymorphism by crystallization on self-assembled multilayers. Cryst Growth Des 7:847–850
27. Cox JR, Dabros M, Shaffe JA et al (2007) Selective crystal growth of the anhydrous and monohydrate forms of theophylline on self-assembled monolayers. Angew Chem Int Edit 46:1988–1991
28. Frenette M, Cosa C, Friščić T (2013) Characterisation of organic solid forms and real-time in situ monitoring of their transformations using solid-state fluorescence. Cryst Eng Comm 15:5100–5106

29. Jiang S, Jansens PJ, ter Horst JH (2010) Mechanism and kinetics of the polymorphic transformation of o-aminobenzoic acid. Cryst Growth Des 10:2123–2128
30. Mukuta T, Lee AY, Kawakami T et al (2005) Influence of impurities on the solution-mediated phase transformation of an active pharmaceutical ingredient. Cryst Growth Des 5:1429–1436
31. Khawam A, Flanagan DR (2006) Solid-state kinetic models: basics and mathematical fundamentals. J Phys Chem B 110:17315–17328
32. Urakam K, Beezer AE (2003) A kinetic and thermodynamic study of seratrodast polymorphic transition by isothermal microcalorimetry. Int J Pharm 257:265–271
33. Laird T (2004) Polymorphism and impurities. Org Process Res Dev 8:301–302

Chapter 15
Crystallisation Control by Process Analytical Technology

Ana Kwokal

Abstract This chapter addresses the basics of Process Analytics Technology (PAT) for crystallisation monitoring and offers examples of the different instrumentation with particular focus on Attenuated Total Reflectance Ultraviolet and Visible spectroscopy (ATR-UVvis), which is one of the main PAT techniques for crystallisation monitoring and control.

Keywords Process Analytical Technology (PAT) • Crystallisation • ATR-UVvis • Raman spectroscopy • FBRM

15.1 Introduction

The aim of Process Analytical Technology (PAT) in the pharmaceutical industry is to assure the production of high quality products through identification and control of critical production parameters in real time. While a traditional control strategy approach included definition of set points and operating ranges for process parameters, the new production approach includes risk management and use of scientific knowledge more extensively to identify and understand process parameters and their impact on critical quality attributes (CQAs) with establishment of design space(s). PAT involves a scientifically-based and multidisciplinary approach that includes analytical and physical chemists, chemical engineers, statisticians, mechanical engineers and programmers.

The benefits of manufacturing based on PAT are [1]:

- process understanding and reducing process failures,
- continuous monitoring and feedback control,
- reducing cycle time to improve manufacturing efficiency,
- identifying the critical production point,
- scientifically based risk assessment.

A. Kwokal (✉)
GlaxoSmithKline, R&D, New Frontiers Science Park, Third Avenue, Harlow, UK
e-mail: ana.x.kwokal@gsk.com

K.J. Roberts et al. (eds.), *Engineering Crystallography: From Molecule to Crystal to Functional Form*, NATO Science for Peace and Security Series A: Chemistry and Biology, DOI 10.1007/978-94-024-1117-1_15

The locations as to where the process measurements need to be made can be summarized:

- off-line: analysis laboratory at separate site;
- at-line: analysis laboratory at manufacturing site;
- on-line: measurements on a diverted sample stream which may be returned to the process after measurement;
- in-line: integrated real-time measurement without sampling, but the process stream may be disturbed (for example, a probe may be inserted);
- non-invasive: integrated real-time measurement where the sensor is not in contact with the material and the process stream is not disturbed [2].

A comprehensive review "Assessment of Recent Process Analytical Technology (PAT) Trends: A Multi-author Review" published in 2015 is recommended for further reading [3]. This review summarizes evolving PAT-based process monitoring and control, sensor development, PAT applications in industrial manufacturing, and the PAT paradigm from a regulatory and science perspective." [3]

15.2 PAT for Crystallization Monitoring

It has been well-recognized that crystallization process parameters can have great impact on the crystallization product quality attributes such as purity, morphology, size, polymorphic form, solubility and dissolution rate. Each of these quality attributes can either independently or collectively impact dissolution, bioavailability, efficacy, and in some cases safety of the drug product.

Crystallization is potentially one of the best processes for PAT implementation due to the stochastic nature of crystallization and the associated need for control of parameters in real time. The non-linearity in batch production provides an additional problem in controlling the crystallisation, thus it is of great importance to improving knowledge of process parameters. Therefore, crystallisation process should ideally be optimised through PAT, so that crystalline product can be produced with predefined properties such as particle size distribution, purity, yield, polymorph and morphology. Optimal crystallization processes require constant crystal growth through the process. It can be achieved through maintaining constant supersaturation, which is very difficult to obtain in practice because the supersaturation level in the real process is changed by the crystal growth process. An additional problem is metastable zone width, which could change due to the presence of impurities.

A range of techniques has been developed for crystallization monitoring, i.e. for monitoring supersaturation, crystal size and shape. The sensors could be grouped according to phase monitoring (liquid or solid):

- solvent concentration measurement (ATR-UV/VIS [4], ATR-FTIR [5, 6], Raman spectroscopy [7],

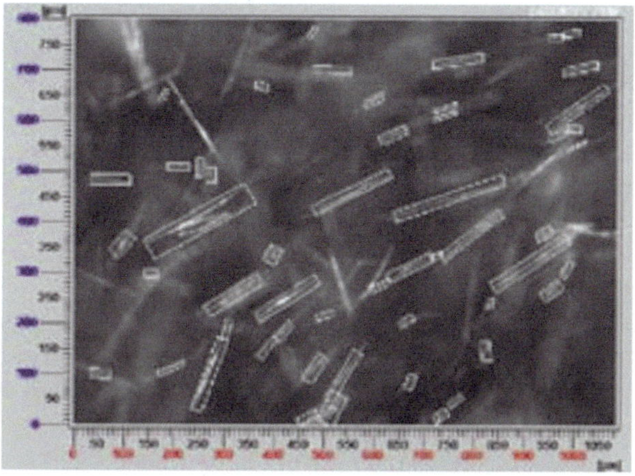

Fig. 15.1 Lasentec PVM, In-process video microscopy [8]

- crystallization on-set measurement (turbidity via UV-VIS spectroscopy, conductivity, Focused Beam Reflectance Measurement, (FBRM) [8]),
- crystal phase determination (on-line X-ray diffraction [9–11], Raman [7], NIR [12]),
- particle size and/or morphology (FBRM, video microscopy Particle Velocimetry Measurement (PVM) as shown in Fig. 15.1, [13] ultrasonic spectroscopy [14, 15]).

The most common technique for supersaturation (concentration) measurement is ATR-FTIR spectroscopy because of its capability to measure dissolved liquid-phase concentrations in the presence of solids. The quantitative approach is based on the univariate approach, establishing a relationship between concentration and individual peak heights, or via chemometric techniques, such as partial least squares (PLS) or principal component regression (PCR). The spectral intensities are dependent on temperature, which makes calibration time-consuming and challenging. However, an alternative method is to track the height of an individual peak, characteristic of the solute of interest, in the absorption spectrum. In this approach the peak height, as measured under supersaturated conditions, is compared to the peak height as obtained from the corresponding saturated solution. This provides the supersaturation value at any temperature point in terms of the differences between these peak heights [16].

Raman spectroscopy is very often used due to its high spectral resolution and ability to measure both solute and solvent simultaneously as well as the ability to measure solute in aqueous media. However, Raman spectra in some cases suffer from inadequate sensitivity and/or fluorescence.

ATR-UVvis spectroscopy is a simple and robust technique for monitoring supersaturation. The other advantages include the wide availability of the UV resistant optical fibres, ATR-probes and diode array detectors, and it is relatively inexpensive.

Near infra-red (NIR) spectroscopy has also been employed [12] for solute concentration monitoring but if compared to ATR-FTIR, the latter shows greater chemical selectivity and enables univariate calibration while NIR spectra are often not very well resolved.

The BedeMonitor™ [17] has been developed for real time X-ray diffraction which does not require a synchrotron X-ray source. This commercially available instrument utilizes a high intensity X-ray source and a curved linear detector to collect a wide 2-θ range simultaneously. On-line measurement of a crystallization process is achieved by recirculation of a sample of the crystallization slurry through a specially designed and thermally regulated cell, through to the X-ray measurement zone and back to the crystallizing vessel. This kind of online PXRD system was shown to be very sensitive to phase impurity and detection was achieved up to 0.2 wt%.

FBRM have been traditionally used for the direct monitoring of particle size. The technique gives a distribution of chord length of particles and a measure of particle count. An alternative to FBRM is Attenuated Ultrasonic Spectroscopy (AUS), which is able to give crystal size distribution and solid concentration simultaneously [14], but has an issue with measurement at high solid fractions.

In addition there are a number of low-cost sensors for crystallization process monitoring, e.g. single frequency ultrasound, dielectric constant, conductivity and imaging using endoscopy which are described in [3].

15.3 Crystallization Control Using PAT – Feedback Control

The most important developing aspect within PAT-based crystallization development is the ability to control and not just monitor critical parameters. The two main parameters of crystallisation, nucleation and crystal growth, are mostly driven by supersaturation. The crystallization control is based on real-time measurement of supersaturation and/or crystal size and the feedback control is achieved through regulation of temperature or anti-solvent addition rate that, in turn, control supersaturation and consequently crystal size. This important discovery [18] in the 1990s showed that operation along a constant supersaturation close to the solubility curve gives the ability to grow large and uniformly distributed crystals.

Feedback (closed-loop) control or supersaturation control (SSC) is based on crystallization direction between the solubility curve and the metastable limit, remaining out of the uncontrolled secondary nucleation zone, even when the process is subject to large disturbances. In this example the metastable limits are determined by FBRM and the solubility curve by ATR-FTIR spectroscopy (Fig. 15.2).

The supersaturation closed-loop control system using in-process ATR-FTIR spectroscopy coupled with the PLS chemometric model was used at the 20 L

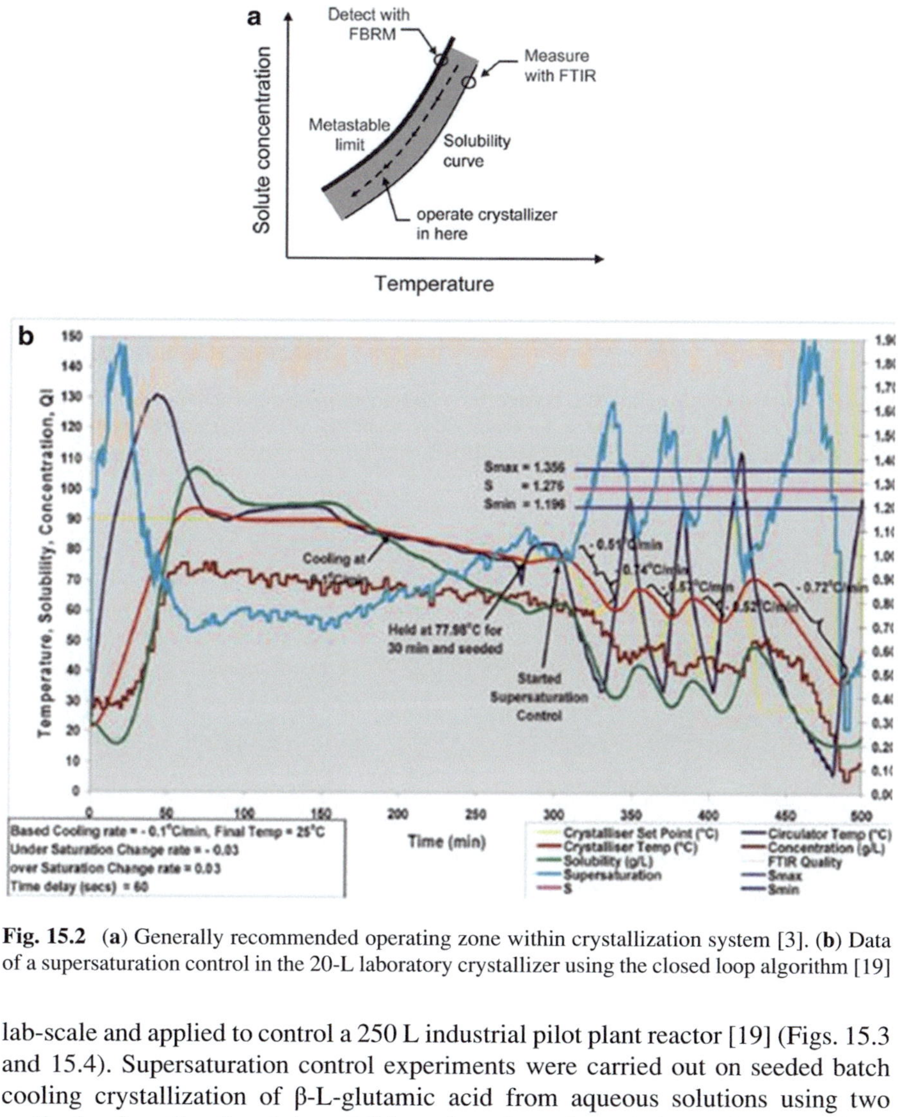

Fig. 15.2 (**a**) Generally recommended operating zone within crystallization system [3]. (**b**) Data of a supersaturation control in the 20-L laboratory crystallizer using the closed loop algorithm [19]

lab-scale and applied to control a 250 L industrial pilot plant reactor [19] (Figs. 15.3 and 15.4). Supersaturation control experiments were carried out on seeded batch cooling crystallization of β-L-glutamic acid from aqueous solutions using two methods of seeding involving addition of seeds to the solution and generation of seeds within the solution. The authors [19] have demonstrated that it is possible to control supersaturation at different levels associated with the crystallization of a challenging compound in an industrial-scale crystallizer overcoming a range of practical challenges based on optimised values of the control parameters at a laboratory scale.

In addition to SSC [20–22] the Direct Nucleation Control (DNC) approach has gained increasing popularity [23–26]. DNC is based on control of the number of particle counts per second provided by FBRM measurement, where the number of particles is in correlation with the particle size distribution. If a smaller number

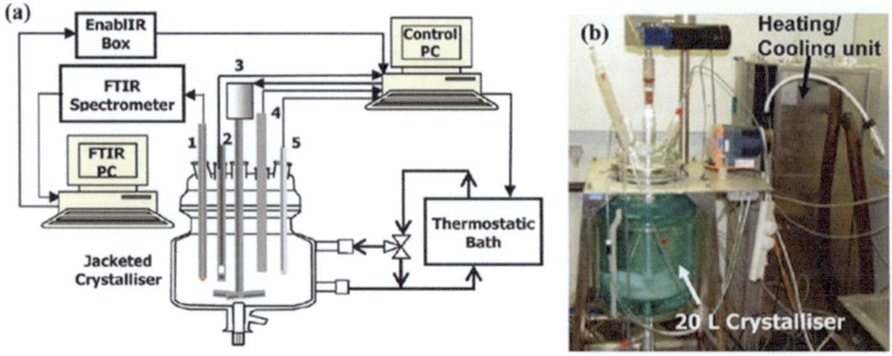

Fig. 15.3 (**a**) Schematic of the 20-L crystallizer with associated instrumentations (*1* – ATR probe, *2* – turbidity probe, *3* – stirrer motor, *4* – temperature probe, *5* – pH probe) and (**b**) photograph of the crystallizer and the heating/cooling unit [19]

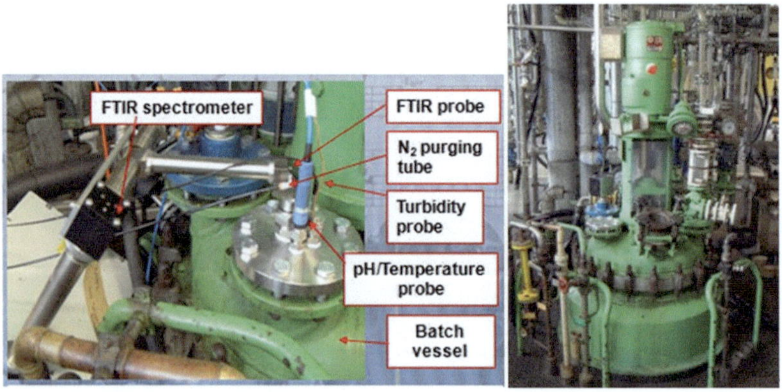

Fig. 15.4 Photograph of the Laboratory 250-L pilot-plant experimental facilities at Syngenta, Munchwilen, Switzerland [19]

of particles is maintained the resulting mean size of the product will be larger and vice versa. The DNC technique automatically manipulates the supersaturation or creates dissolution to control the number of particles around the desired value, as shown in Fig. 15.5 [26].

PAT tools are also often used simultaneously to monitor crystallization processes. The concept is based on the use of composite sensor or composite PAT array (CSA or CPA) for monitoring multiple process parameters and quality indicators at the same time [27, 28]. The information provided by the CSA allows the implementation of robust crystallization control, such as supersaturation control, direct nucleation control, active polymorphic feedback control and a combination of these approaches.

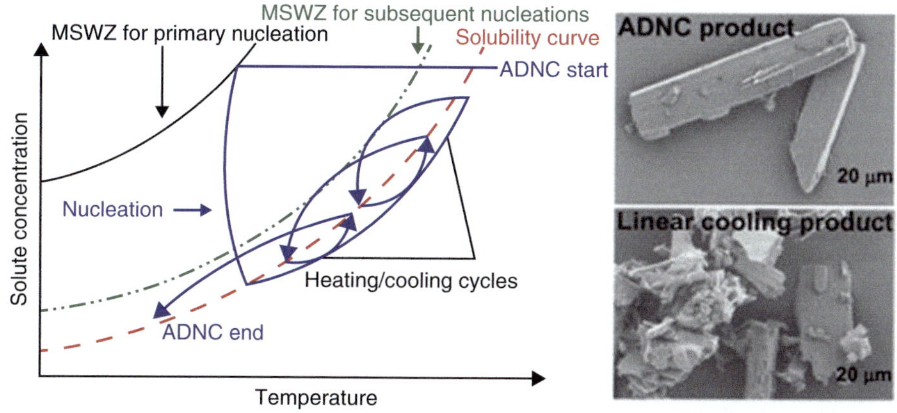

Fig. 15.5 Schematic representation of the Automated DNC operating procedure in the crystallization phase diagram. [26]

15.4 ATR-UV/VIS Spectroscopy

UVvis spectroscopy is based on ultraviolet and visible radiation that interacts with matter which causes electronic transitions (promotion of electrons from the ground state to a high energy state). The ultraviolet region falls in the range between 190–380 nm and the visible region fall between 380–750 nm. While the measuring concentration range of the majority of UVvis instruments falls into the region of ppm, ATR-UVvis spectroscopy enables measurements of much higher concentration typically several g/L which covers most crystallization conditions. The principle of ATR-UVvis spectroscopy is based on the UVvis light beam passing from a material with high refractive index (ATR crystal) to a material with lower refractive index (solution). At an incident angle light is introduced to the probe towards the interface between the probe boundary and the sample. When the angle of incidence is greater than the critical angle, total reflection will occur when the light hits the probe boundary. In practice, the light partially enters the solution in the form of an evanescent wave where it is absorbed partially by the sample. The reflected light is therefore attenuated by the solution, causing a measurable reduction in the output signal dependent upon the absorbance of the solution and hence giving spectral information about physical properties of the sample.

The ATR probe for UV-VIS spectroscopy uses a sapphire ATR crystal (Fig. 15.6), which is directly immersed into the solution to be measured. This type of ATR probe utilises three reflections to bring the light back to the output fibre without using a mirror. The usual optical path length is about 1–2 μm but this depends on the refractive index of the medium.

Analysis of the measurements is based on the well-known Lambert-Beer law which relates the amount of light absorbed by the sample to the quantity of chemical substance.

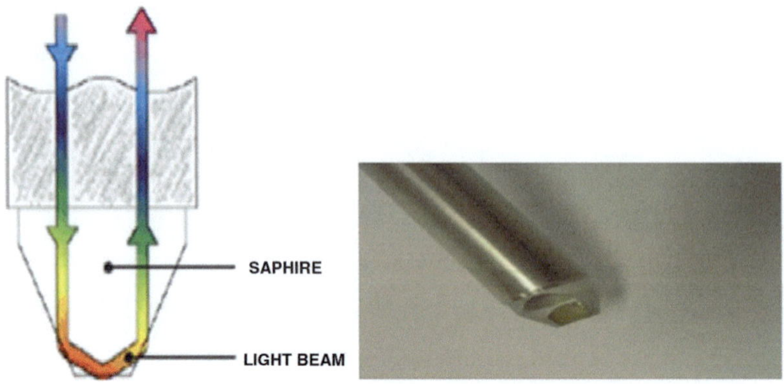

Fig. 15.6 ATR-UV/VIS sapphire crystal probe, Hellma®

$$A = \varepsilon\, b\, c \tag{15.1}$$

Where A is absorption, ε is the molar absorption coefficient and b is path length. The relation between transmittance and absorbance is given by Eq. 15.2.

$$A = \log(1/T) \tag{15.2}$$

Transmittance due to ATR is given by Eq. 15.3.

$$T = (1 - a)z \tag{15.3}$$

where z is the number of reflections and a is the absorption parameter.

The effective path length is a function of wavelength, refractive indices, angle of incidence and state of polarisation. In this respect, it is very important to emphasise that changes of temperature and concentration can significantly influence the refractive index which consequently influences the effective path length and unfortunately the absorption itself. So the proper calibration before any concentration measurement should be obtained.

There are cases when spectra involve overlapping absorption bands from two or more components. In such a case a sophisticated technique should be employed such as background subtraction, first and second derivatives, as well as multi segment background fit and subtraction [29]. For correlation of spectral response to analyte concentration when spectral response in not straightforward chemometric techniques such as PCA, PLS, and multiple linear regression (MLR) are used.

The use of *in-line* UV spectroscopy was recently developed because of the advance in high quality UV grade optical fibres and affordable array detectors. The main advantage of using the ATR technique is that measurements can be performed directly without dilution of the sample even in the presence of solid particles. This is possible through the very short path length obtained without an accompanying risk of clogging [29].

There are several advantages of using an ATR-UV probe for measuring supersaturation levels over other techniques:

- Measurements are relatively insensitive to particles (including bubbles) which enable measurements of solution concentration in high content slurry.
- Fibre optics are not very sensitive to bending and therefore enabling the system flexibility needed for the connection of the instrument to the reactor to be easily facilitated.
- ATR probes are suitable for situations involving the direct measurement of strongly absorbing solutions, where standard transmission probes cannot be used.

However, there are several limitation of ATR-UVvis:

- Solutes should have high UV absorption with respect to the solvent.
- Fibre optic transmission is limited to the range of 220 nm to 600 nm and hence care must be taken that the absorption ranges available meet requirements.

15.5 Simultaneous Use of ATR UVvis and FBRM for Crystal Growth Determination

The following example presents ATR-UV-vis spectroscopy and FBRM as a process tools for accessing the crystal growth kinetics for crystallization performed with and without presence of a surface nucleation template [30].

Figure 15.7a shows a 3D plot of the UV-vis spectra measured as a function of temperature recorded during cooling crystallization. Fig. 15.7b gives the UV-vis absorbance peak maxima as a function of temperature together with its solution concentration. The solution concentration was calculated according to a multilinear calibration model ($c = a1 \times A + a2 \times T + a3$). Examination of the data reveals that the UVvis absorbance maximum increases slightly as temperature decreases but starts to rapidly decrease following crystallization indicating the onset of crystallization (Fig. 15.7c).

Figure 15.8 shows supersaturation (ATR-UVvis) and FBRM data measured during isothermal crystallization in the presence of the template which is overlaid with the data obtained from the crystallization experiments without the template. An examination of the supersaturation versus time curve (Fig. 15.8a) demonstrates that in the absence of the template the onset of crystallization begins at about 50 min at the crystallization temperature of 8 °C, after exhibiting a rapid increase in supersaturation during the cooling period and a subsequent supersaturation plateau. The chord length versus time curve was found to be in rough coincidence in their respective onset points of crystallization with supersaturation and increased relatively slowly within the first 130 min of crystallization and later progressed rather

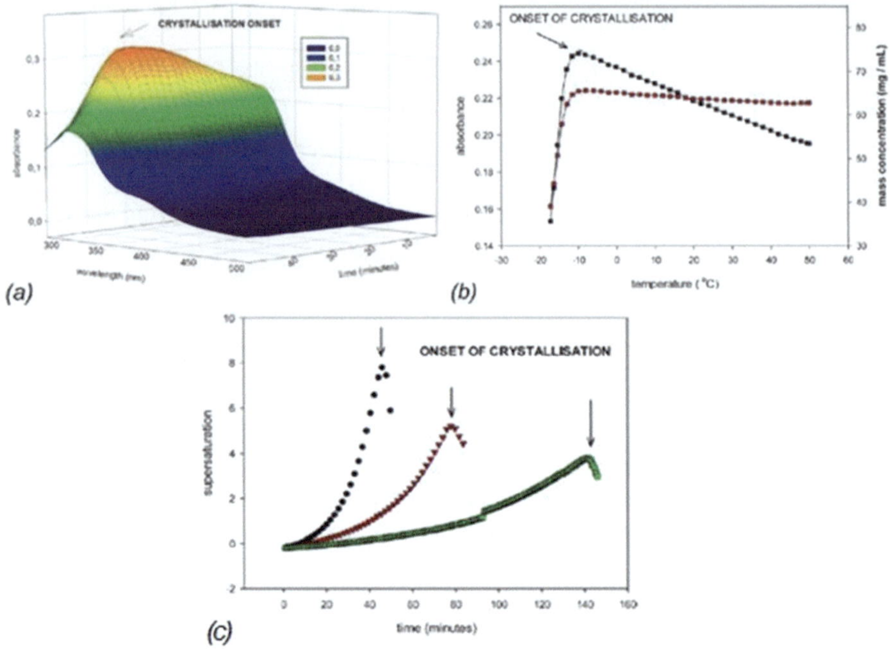

Fig. 15.7 ATR-UVvis spectral data and its analysis: (**a**) 3D plot showing the in-process absorbance vs. wavelength vs. time. (**b**) Plot of the ATR-UVvis peak maxima as a function of temperature as measured (*black*) and recalculated (*gray*) to mass concentration according to eq. (S1). (**c**) Solution supersaturation vs. time as a function of cooling rate for 1 °C/min, 0.5 °C/min, and 0.25 °C/min (from *left* to *right*)

fast until reaching the constant value. However, in the presence of the template, the crystallization process was found to begin significantly earlier (at approximately 40 min). An examination of the chord length measurements revealed that the growth process proceeded more quickly, reaching a constant value within 150 min, at which point the chord length began to retard. Overall, the kinetics of the crystal growth process was found to differ significantly between these two experiments, with the reduction in the crystallization onset points and the increased growth velocities confirming the ability of the template to induce nucleation at a lower supersaturation and to subsequently promote growth. An examination of the initial stages of growth (Fig. 15.8b) close to its onset highlights the process in more detail and demonstrates the ability of the template to enhance the crystal growth rate and to facilitate a shorter induction time and a faster de-supersaturation rate when compared to crystal growth in the absence of the template.

Figure 15.9 shows micrographs taken of the crystals harvested from the crystallization experiments with and without the presence of the template. The crystals obtained in the presence of the template were found to have a narrower size distribution and to be of a much higher quality, notably exhibiting a thickened plate-like morphology with a well-defined crystal shape and smooth edges. By

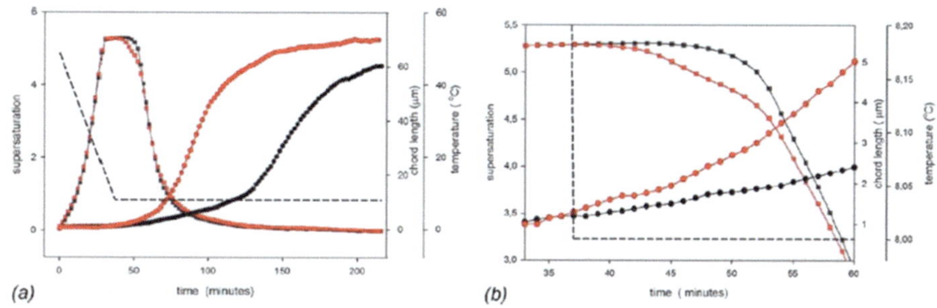

Fig. 15.8 In-process data with and without template: (**a**) supersaturation (*squares*), chord length (*circles*), and temperature (*dashed line*) profile for 70 mg/mL cooling crystallization in the 500 mL reactor in the absence (*black*) and presence (*grey*) of the template; (**b**) enlarged area highlighting the early stage of crystallization

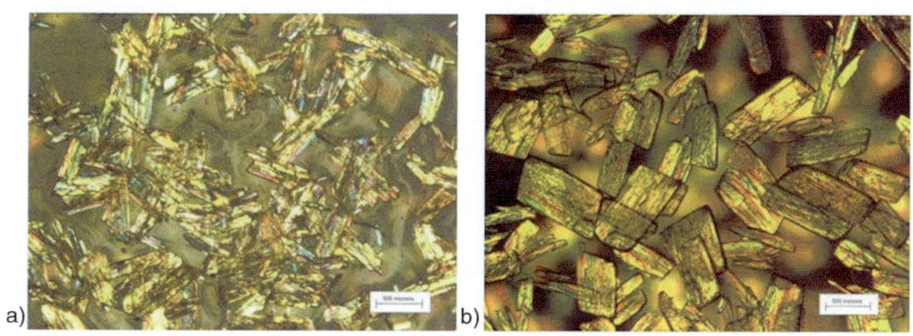

Fig. 15.9 Photomicrograph of crystal harvested after cooling crystallization of 70 mg/mL entacapone/acetone in the first crystallization run without (**a**) and with template (**b**)

contrast, the crystals prepared from crystallization without the template were found to have a wider crystal size distribution and to be of a much lower crystal quality often displaying poorly defined morphology with rough edges and broken surfaces. However, the mean crystal size from both experiments was found to be rather similar, with a median diameter of about 400 μm.

The crystal growth rate parameters (Table 15.1) were calculated from the derivative of the mean chord length with respect to time and supersaturation from UVvis data according to $G = k_G\,(z - \sigma)^b$ where k_G (μm/min) is the rate constant, b (dimensionless) is the order, and z is a correction factor.

Table 15.1 Kinetic parameters derived from the growth rate vs. supersaturation measurements

	Without template		With template	
	K_b /μm min^{-1}	b	K_b /μm min^{-1}	b
Initial rate	0.96	1.28	NA	
Dead zone	0.26	0.02	NA	
Fast growth	0.71	0.63	0.83	0.44

15.6 Summary

The applications of PAT in crystallization process monitoring and control have been discussed. The basic principle, advantages and disadvantages of some of the typical techniques such as ATR-UV-vis, FBRM, and NIR have been presented. In summary, PAT provides the opportunity for better process understanding, faster process optimization and better prediction of the impact of scale-up. It assures the ability to control and correct processes in real time, making it an important tool for use in studying many pharmaceutical processes such as crystallization.

References

1. Bakeev KA (2005) Process analytical technology. Blackwell, Oxford
2. Yua LX, Lionbergera RA, Rawa AS, D'Costaa R, Wub H et al (2004) Applications of process analytical technology to crystallization processes. Adv Drug Deliver Rev 56:349–369
3. Simon LL, Pataki H, Marosi G et al (2015) Assessment of Recent process analytical technology (PAT) trends: a Multiauthor review. Org Process Res Dev 19:3–62
4. Florence AJ, Johnston A (2004) Applications of ATR UV/vis spectroscopy in physical form characterisation of pharmaceuticals. Spectrosc Eur 16:24–27
5. Groen H, Borissova A, Roberts KJ (2003) In-process ATR-FTIR spectroscopy for closed-loop supersatu- ¨ ration control of a batch crystallizer producing monosodium glutamate crystals of defined size. Ind Eng Chem Res 42:198–206
6. Liotta V, Sabesan V (2004) Monitoring and feedback control of Supersaturation using ATR-FTIR to produce an active pharmaceutical ingredient of a desired crystal size. Org Proc Res Dev 8:488–494
7. Févotte G (2002) New perspectives for the on-line monitoring of pharmaceutical crystallization processes ´ using in situ infrared spectroscopy. Int J Pharm 241:263–278
8. Lasentec (2007) http://abstracts.aapspharmaceutica.com/expoaaps07/Data/EC/Event/Exhibitors/127/b3d65da0-150f-4250-9842-6c3341f952b0.pdf
9. Dharmayat S, Hammond R, Lai X et al (2008) An examination of the kinetics of the solution-mediated polymorphic phase transformation between α- and β-forms of l-glutamic acid as determined using online powder X-ray diffraction. Cryst Growth Des 8:2205–2216
10. Hammond RB, Lai X, Roberts KJ et al (2004) Application of in-process X-ray powder diffraction for the identification of polymorphic forms during batch crystallization reactions. Cryst Growth Des 4:943–948
11. Dharmayat S, Calderon de Anda JA, Hammond RB et al (2006) Polymorphic transformation of L-glutamic acid monitored using combined on-line video microscopy and X-ray diffraction. J Cryst Growth 294:35–40
12. Févotte G, Calas J, Puel F et al (2004) Applications of NIR spectroscopy to monitoring and analyzing the solid state during industrial crystallization processes. Int J Pharm 273:159–169

13. Kougoulos E, Jones AG, Jenning KH, Wood-Kaczmara MW (2005) Use of focused beam reflectance measurement (FBRM) and process video imaging (PVI) in a modified mixed suspension mixed product removal (MSMPR) cooling crystallizer. J Cryst Growth 273:529–534
14. Mougin P, Thomas A, Wilkinson D et al (2003) On-line monitoring of a crystallisation process. AICHE J 49:373–378
15. Lyall E, Mougin P, Wilkinson D et al (2004) In situ ultrasonic spectroscopy study of the nucleation and growth of copper sulfate Pentahydrate batch crystallized from supersaturated aqueous solutions. Ind Eng Chem Res 43:4947–4956
16. Barrett M, McNamara M, Hao H et al (2010) Supersaturation tracking for the development, optimization and control of crystallization processes. Chem Eng Res Des 88:1108–1119
17. Bede (2007) http://www.bede.co.uk
18. Gutwald T, Mersmann A (1990) Batch cooling crystallization at constant supersaturation: technique and experimental results. Chem Eng Technol 13:229–237
19. Khan S, Ma CY, Mahmud T et al (2011) In-process monitoring and control of Supersaturation in seeded batch cooling crystallisation of l-glutamic acid: from laboratory to industrial pilot plant. Org Process Res Dev 15:540–555
20. Fujiwara M, Nagy ZK, Chew JW et al (2005) First-principles and direct design approaches for the control of pharmaceutical crystallization. J Process Control 15:493–504
21. Nagy ZK, Chew JW, Fujiwara M et al (2008) Comparative performance of concentration and temperature controlled batch crystallizations. J Process Control 18:399–407
22. Nagy ZK, Aamir E (2012) Systematic design of supersaturation controlled crystallization processes for shaping the crystal size distribution using an analytical estimator. Chem Eng Sci 84:656–670
23. Abu Bakar MR, Nagy ZK, Saleemi AN et al (2009) The impact of direct nucleation control on crystal size distribution in pharmaceutical crystallization processes. Cryst Growth Des 9:1378–1384
24. Saleemi AN, Rielly CD, Nagy ZK (2012) Comparative investigation of Supersaturation and Automated direct nucleation control of crystal size distributions using ATR-UV/vis spectroscopy and FBRM. Cryst Growth Des 12:1792–1807
25. Saleemi AN, Rielly C, Nagy ZK (2012) Automated direct nucleation control for *in situ* dynamic fines removal in batch cooling crystallization. Cryst Eng Comm 14:2196–2203
26. Saleemi AN, Steele G, Pedge NI et al (2012) Enhancing crystalline properties of a cardiovascular active pharmaceutical ingredient using a process analytical technology based crystallization feedback control strategy. Int J Pharm 430:56–64
27. Nagy ZK, Braatz RD (2012) Advances and new directions in crystallization control. Annu Rev Chem Biomol Eng 3:55–75
28. Nagy ZK, Fevotte G, Kramer H et al (2013) Recent advances in the monitoring, modelling and control of crystallization systems. Chem Eng Res Des 91:1903–1922
29. Thompson DR, Kougoulosb E, Jonesb AG et al (2005) Solute concentration measurement of an important organic compound using ATR-UV spectroscopy. J Cryst Growth 276:230–236
30. Kwokal A, Čavužić D, Roberts KJ (2013) Surface adsorbed templates for directing the crystal growth of Entacapone as monitored using process analytical techniques. Cryst Grow Des 13:5324–5334

Chapter 16
Methods for Nano-Crystals Preparation

Samir A. Kulkarni and Allan S. Myerson

Abstract Nanocrystals have the potential to address challenges in very different fields like production of computer chips, delivering drugs of low water solubility, cosmetic research and products, biotechnology, catalysis and textile. In this review, we highlight the industrially relevant technologies, milling, high-pressure homogenization and recent advances in the development of these methods for nanocrystal preparation. The paper summarizes the most up-to-date information and developments available for conventional nanocrystal preparation methods such as antisolvent precipitation, supercritical fluid technology, emulsion method, Impinging Jet Crystallization and spray drying. Finally, we also reviewed recent developments in preparing a nanosized particle of drug crystals using patterned self-assembled monolayers, microfluidic devices and nano confinement in nanoporous material. The present article also describes the merits and demerits of these methods, scale up issues and application in drug delivery of nanocrystals prepared by different methods.

Keywords Drug nano-crystals • Nucleation • Pearl milling • High-Pressure homogenization • Controlled flow cavitation • Impinging jet crystallization • Patterned SAM • Controlled pore glass

16.1 Introduction

Nanocrystal preparation has been an area of significant interest, due to the wide variety of potential applications in cosmetic, computers, semiconductors, pharmaceutical and biomedical area. The applications of nanocrystals can be found in many cosmetic products including moisturisers, deodorants, toothpaste, make up and sunscreen [1]. The nanocrystal products are identified as next generation cosmetic delivery agents that can provide enhanced skin hydration, bioavailability,

S.A. Kulkarni • A.S. Myerson (✉)
Novartis-MIT Center for Continuous Manufacturing and Department of Chemical Engineering, Massachusetts Institute of Technology, 77 Massachusetts Avenue, Cambridge, MA 02139, USA
e-mail: myerson@mit.edu

K.J. Roberts et al. (eds.), *Engineering Crystallography: From Molecule to Crystal to Functional Form*, NATO Science for Peace and Security Series A: Chemistry and Biology, DOI 10.1007/978-94-024-1117-1_16

stability [1]. In the early 1980's the research into semiconductor nanocrystals, also known as colloidal quantum dots started [2]. Ekimov and Efros discovered and demonstrated that the optical properties of semiconductor crystals were dominated by their size, and these nanocrystals could absorb light at much lower wavelengths than expected [2]. In the pharmaceutical industry, about 60% of drugs in company pipelines are poorly water-soluble [3]. As per the biopharmaceutical classification scheme many of Active Pharmaceutical Ingredients (API's) are designated class II compound with low solubility and high permeability [4]. For class II compounds the aqueous solubility is the limiting factor regarding bioavailability [4]. There are many ways to solubilize certain poorly soluble drugs for example solid dispersions are one of the most promising strategies to improve the bioavailability of poorly water-soluble drugs. The main disadvantage is the formation of crystalline solid dispersions rather than amorphous solid dispersion, which was more thermodynamically stable and did not release the drug as quickly as amorphous ones. Other methods include grinding or milling techniques, pH adjustment, freeze-drying and cryogenic engineering, supercritical fluid processing and precipitation techniques. The reduction in sizes of crystals increases the surface area, which further increases the dissolution rates and bioavailability. As per the Noyes-Whitney equation the reduction in size leads to an increased surface area, which further increases the dissolution rate. Micro-crystallization is a suitable way to increase dissolution rates by reducing the size, but nanocrystals can increase the dissolution rates more by increasing the surface area further. The nanocrystals could serve as new building blocks for manufacturing of API's with improved aqueous solubility.

In the case of nano-crystallization, the induction of nucleation needs to be abrupt and homogeneous throughout the growth medium. For many years, the kinetics of the burst nucleation and the growth of nanocrystals have been described by LaMer and Dinegar [5, 6]. As per the classical nucleation theory, from an initial supersaturation, a rapid nucleation of particles would initially occur which is followed by the absorption of building units (monomers) onto these nucleated particles. The resulting reduction in supersaturation strongly depletes the local concentration of building unit (monomers), which lead to a narrow size distribution by decreasing the rate of further nucleation [7, 8]. The Gibbs free energy (Fig. 16.1) is dependent upon the temperature, the supersaturation of the solution S, and its molar volume, v [7, 9] The Gibbs free energy is defined in Eq. 16.1.

$$\Delta G = 4\pi r^2 \sigma + \frac{4}{3}\pi r^3 \Delta G_v \qquad (16.1)$$

Where r is the radius of spherical nucleus, σ is the surface tension and ΔG_v is the volume free energy. Nucleation is the competition between the formation of volume and formation of surface area, because of the positive surface free energy and negative volume free energy. According to the classical nucleation theory, the surface free energy destabilizes nuclei and the nucleation cannot take place until the size of the aggregate reaches the critical size, also known as critical radius (r_{crit}) of the nucleus [9].

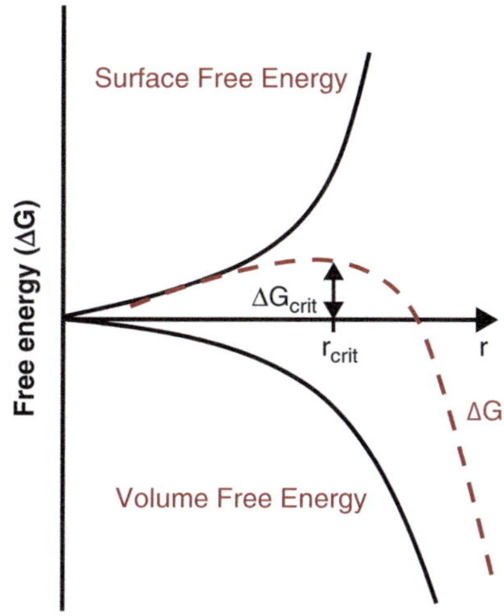

Fig. 16.1 Free energy diagram for nucleation as a function of radius, the figure shows the free energy barrier and the critical radius at the top of the barrier. [7, 9]

$$r_{crit} = \frac{-2\sigma}{\Delta G_v} \tag{16.2}$$

This critical radius corresponds to the minimum size at which a particle can survive in solution, and the particle below the minimum size will dissolve. The same is true for the particle's free energy, where a critical free energy is required to obtain stable particles within solution, Fig. 16.1. Also the critical cluster size at a given supersaturation defines solubility of that size nucleus. The critical free energy is defined in Eq. 16.3 [9].

$$\Delta G_{crit} = \frac{4}{3}\pi\sigma r_{crit}^2 \tag{16.3}$$

Nanocrystal preparation methods can be divided into top-down and bottom-up techniques. [10, 11] The size reduction of relatively large particles into smaller particles are known as top-down technique, whereas bottom-up techniques consist of the growth of small particles from individual molecule [4]. The driving force for the growth of a crystal from individual molecules is supersaturation. Bottom up technique is usually a precipitation or crystallization technique and considered as the oldest technique to prepare nanocrystals [12]. In this Review, we first briefly discuss the methods to produce nanosized particles, which include direct crystallization using high supersaturation (bottom up technique), particle breakage (top-down technique) and crystallization in a constrained environments (bottom up technique). Next, we summarize the merits and demerits of these methods, scale up issues and application in drug delivery of nanocrystals prepared by different methods. We highlight advantages of the Bottom up technique and top-down technique.

16.2 Top-Down Approach: Milling and High-Pressure Homogenization

The top down approach is the most important technique for the production of nanocrystals. Milling and high-pressure homogenization are the two basic top-down techniques for size reduction. Wet milling produces the most nanocrystal products that have reached the market [13]. Wet milling involves mechanical attrition, where particles are wetted by an aqueous solution of surfactants and sheared and grinded by milling balls in a milling container. The particle size is reduced and may reach a few hundred micro-meter, but the conventional milling with modifications can be used for generating nanosized crystals [13]. The preparation process can be carried out in a reproducible manner. The contamination from erosion of metal milling balls or pearls, high energy input, prolonged operation time and decreased crystallinity are the major drawbacks of the technique [13]. The use of polymeric beads may be helpful in minimizing the erosion and contamination.

A new process known as Jet milling to prepare micro-particles without the use of organic solvent was described by Nykamp et al. [14]. The important advantages of solvent free jet milling are the prevention from toxicity due to the absence of organic solvents and very short preparation time [14]. High-pressure homogenization is the process in which two fluid streams of particle suspensions collide under high pressure in a chamber, leading to particle collision and subsequent particle rupture. Nanosized solid particles are produced in Piston-gap homogenizers by forcing a suspension of drug particles with a piston through a thin gap under high pressure [4]. The high shear forces, and turbulent flow fractures the particles and the particle outcome was decided by the power of homogenization, particle hardness, and number of the piston-moving cycles [15]. High-pressure homogenization required high process temperature, high-energy input, required complex equipment and might have possible degradation of the components and may yield less when compared with wet milling and these are the major drawbacks of the technique [15]. The nonexistence of a large-scale production method yielding a product of a quality that is acceptable by the regulatory authorities hampers the introduction of solid nanoparticles to the market [15, 16]. Wet Milling and High-pressure homogenizers are widely used in many industries including the pharmaceutical industry for the production of micro and nano-particles. [15, 16] Hence, Wet milling and High-pressure homogenizer are considered as being industrially the most feasible one for nanocrystal production with no regulatory problem [15, 16].

16.3 Bottom-Up Techniques

16.3.1 Control Flow Cavitation (CFC)

CFC creates the optimum process condition for nanocrystal production after controlling the size, location, density, and intensity of implosion of bubbles in the cavitation zone. The controlled energy released by the implosion of micro bubbles and the ability to control the energy of cavitation, the particles can be brought to desired particle size distributions [17, 18] CFC converts destructive force into constructive with high intensity energy force to produce nano and micro structured materials [17, 18] CFC technology has been exploited in many industries with multiple CFC chamber designs customized for hydrodynamic, chemical, biomedical and cleaning applications [18] The CFC is highly scalable and efficient process with excellent process control and outstanding reproducibility [19].

16.3.2 Spray Drying

Spray Drying is a single-step process for converting solutions, emulsions, suspensions, slurries, and pastes into powders in a continuous manner. It also allows the production of particles with controlled size and morphological aspects [20]. Due to the limited collection efficiency related to cyclone separators, the traditional spray drying process is limited for producing particles of 2–5 μm sizes. A spray dryer with a piezoelectric driven vibrating mesh atomizer and a high-efficiency electrostatic powder collector seems to correct these limitations. A new Nano spray dryer technology B-90, has been developed by Büchi® (Switzerland) and has been applied to perform nano crystallization and drying [21–23]. The other new technology includes the dissolution of drug and a polymeric dispersant system in a suitable solvent. After spray drying the resulting solution the powder containing the drug are produced as either the molecularly dispersant system in the polymer matrix to form a solid solution or dispersed as submicron particles to form solid suspension [24]. The spray drying process is very rapid; can be designed to any capacity, adaptable to a fully automated control system that allows continuous preparation, and also wide ranges of spray dryer designs are available in the market. The spray dryer can be used for heat resistant and heat sensitive products and the feedstock can be as a solution, slurry, gel, suspension or melt form. The current advantage of nanotechnology has increased the stress on existing spray dryer systems to produce nanoparticles with high yield and controlled size distribution.

16.3.3 Supercritical Fluid

Supercritical antisolvent (SAS) processes are lately proposed for the production of micro- and nanosized particles [25, 26]. The solute, solvent, and the supercritical antisolvent are the three important components of SAS. [26]. A very high level of supersaturation was generated due to the high power of supercritical fluids to dissolve the organic solvents as well as by the low solubility of the solute in the SAS. A very fast diffusion and high supersaturation creates precipitation of nanoparticles that are not possible to obtain with antisolvent precipitation or any other techniques [26]. The main advantage is the complete removal of antisolvent by pressure reduction to the gas phase [25]. In case of SAS the solvent power can be obtained by varying pressure and temperature and their diffusivities can be about two orders of magnitude greater than those of liquids [25, 26]. More recently Caputo et al. [26] proposed the use of SAS for the precipitation of sulfathiazole from acetone solution by the use of urea as habit modifier and in the recent past poly (sebacic unhydride) was used by Jarmer et al. as growth inhibitor for griseofulvin using a SAS [26, 27].

16.3.4 Impinging Jet Crystallization

Current advancement to directly produce small particles include impinging jet crystallization in which jets are used to create impinging fluid jet streams and thereby achieve high intensity micro mixing of the fluids prior to nucleation [28]. Two or more jets and two fluids with different solvent composition can be used to micro mix the solvent and antisolvent for initiating the precipitation of solute from solution [29]. Woo et al. [28] studied the control distribution by combining controlled seeding by impinging jet crystallization with a batch crystallizer operating at a controlled constant growth rate. The goal of their study was to propose control strategies to produce crystals with a target crystal size distribution (CSD), that are combinations of optimal control and an impinging jet crystallizer [28] The barium sulphate nanocrystal precipitations are studied experimentally by Schwarzer and Peukert et al. [30] and the mean particle size of the nanoparticles are also predicted accurately by CFD model. The impinging jet crystallization offers very high-energy dissipation rate, efficient micro mixing and tightly controlled conditions to high product quality [29].

16.3.5 Emulsion Method

The method of preparation of nanocrystals by microemulsion is gaining a significant interest in both basic research and in different industrial fields [31]. The

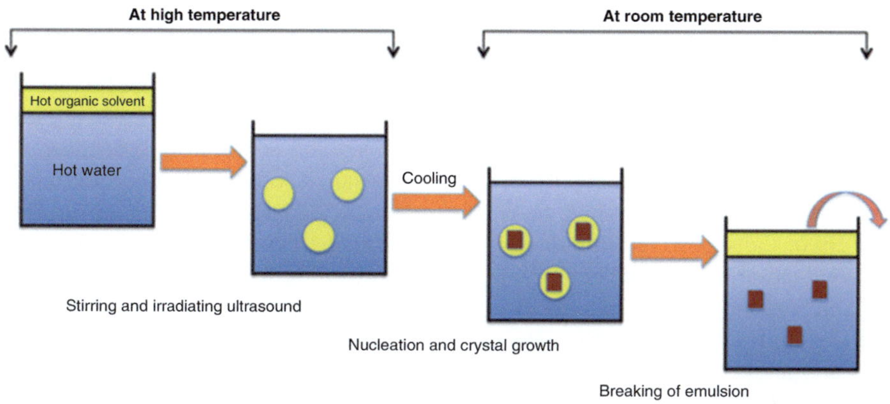

Fig. 16.2 Schematic model of the organic nanocrystal fabrication using the emulsion method [32]

organic nanocrystal fabrication using the emulsion method is a three-step process (Fig. 16.2) [32]. The first step is the preparation of emulsion by the quick addition of solution of compound in the organic phase to the aqueous phase at high temperatures [28]. High stirring speed and irradiating ultrasound was used to produce the stable emulsion [31]. In the second step the solutes are crystallized by gradually cooling the dispersion to low temperatures. In the third step an antifoaming agent was added to break the emulsion and separate the organic solvent. A nanocrystals were obtained as a stable dispersion in an aqueous phase [32]. Ujiiye-Ishii and co-workers [32] described how to prepare perylene nanocrystals using emulsion and Reprecipitation method. Schulman et al. first described the method in 1959 and prepared macroemulsion and microemulsion of hydrocarbons. [33] A significant feature of this method is its ultralow interfacial tension, large interfacial area, and thermodynamic stability of the resulting nanocrystal dispersions in aqueous media [33].

16.3.6 Patterned Micro Well and Patterned Gold Islands

The patterned micro wells are well-defined two-dimensional or confined three-dimensional structures that are mainly used as templates to guide the crystal nucleation and growth [34]. The templates can be self-assembled organic monolayers [34, 35], polymeric matrix [36], silica [37], highly ordered graphite or carbon nanotubes [37, 38]. Recently, significant attention has been directed to design protein nanocrystals with uniform shapes for improving bioavailability and providing alternative release route [39] Wang et al. demonstrated that patterned micro wells provide a platform for controlling the crystallization of protein nanocrystals using industry-standard crystallization conditions [39]. Previously, patterned Self-Assembled Monolayers (SAM) were used to crystallize organic molecules such as glycine [10]. The bifunctional SAMs pattern contains hydrophilic islands surround by hydrophobic regions on which small hemispherical droplets are formed when wetted with polar solvents [10]. The nanocrystals of glycine are formed either by slow cooling, slow evaporation or slow diffusion of an antisolvent [10, 34, 40]. Lee et al. showed that the polymorphic outcome of glycine is primarily influenced by the solvent evaporation rate. [40] The slow evaporation results into α-glycine while fast evaporation favours β-glycine due to creation of high supersaturation [40]. The control of the supersaturation during vapour diffusion experiments results into the crystallization of metastable β-glycine polymorphs [10]. The study shows that the organic vapour diffusion was used to obtain the metastable β-glycine, while slow cooling and slow evaporation are used to prepare the α-glycine and γ-glycine polymorphs [10]. By controlling the initial glycine concentration and the rate of diffusion the nanocrystals with a size between 200 nm to 1.2 μm were prepared based on the island size [10]. Kim et al. also calculated the solubility of glycine crystals in methanol using Ostwald-Freundlich eq. [41] and 100 nm crystals displayed two times more solubility then the equilibrium solubility of glycine [10]. The solubility results calculated from the Ostwald-Freundlich equation for α-glycine and β-glycine polymorphs indicates that β-glycine is more stable than α-glycine due to the effect of surface molecules when the crystal size is under 97 nm [42]. The patterned SAMs technique is very simple, with a large variability of molecules and highest density of SAMs and many functional thiols are commercially available. A major limitation is the removal of crystals from the surface while scraping or ultrasound can be used to remove the crystals from the surface [10].

16.3.7 Microfluidics Devices

A microfluidic system can be used for the continuous production of nanocrystals mainly due to the improved reaction control and performance of mixing, the particle size distribution becomes sharper and the particle size decreases [43, 44]. The microfluidic devices fall into broad categories: capillary- and chip-based systems. In capillary reactors the simple fluidic components can be joined by appropriate lengths of tubing. Chips are precisely tailored and typically fabricated from a plastic, glass or silicon substrate, wet etching or micromachining techniques. Both types of reactor play an important role in nanocrystal synthesis [43, 44]. Single phase or two-phase reactors are two important microfluidic reactors. The single-phase reactors are commonly used reactors, in which miscible streams of reagents are injected into a channel or capillary where they mix and react, making it easy to conduct multistep reactions and produce more complex structures [43]. The undesirable velocity dispersion and fouling on the reactor wall are the two important limitations and can limit the performance of a single phase reactor [43]. In case of two-phase reactors an additional immiscible fluid was injected (which can be a gas or a liquid) into the channel divides and creates a split plug that passes through the reactor at a common speed, eliminating velocity dispersion [44]. The size of the split plugs was proportional to the relative flow rates in the two outlet channels and can be controlled by varying the relative hydrostatic pressures at the two outlets [44]. The microfluidic reactor is still a long way from displacing the conventional reactor for nanocrystal synthesis, which cannot satisfy industrial demand [44].

16.3.8 Nanocrystal Preparation Using Nanoporous Materials

A novel process to generate nanocrystals of active pharmaceutical ingredients within the nanopores of nanoporous materials is the simplest approach for nano-crystal preparation [11]. Many studies have been performed investigating the potential of carrier materials including metal organic frameworks, mesoporous silica, controlled pore glass, porous polycyclohexyl ethylene and polystyrene and nanostructured lipid carriers [11, 45–48].

O'Mahony et al. developed a process with the aim of generating nanocrystalline products within the controlled pore glass (GPG), filling the pores of CPG with API solution [11]. The inhibition time was estimated by the Washburn equation in order to facilitate complete filling of the pores of CPG during the process. The Washburn equation (Eq. 16.4) usually predicts the capillary flow of liquids in porous material to describe the migration of this liquid – air interface with time t within a channel [49].

$$L^2 = \frac{\gamma D t}{4\eta} \tag{16.4}$$

Where t is the time for a liquid of viscosity η and surface tension γ to penetrate a distance L into a fully wettable, porous material whose average pore diameter is D [49] (Fig. 16.3).

The surface area to volume ratio can affect the thermotropic properties of very small crystals, which results in substantial melting point depression due to the limitations on crystal size imposed by the pores [50, 51]. The Gibbs-Thomson equation can be used to describe the melting point depression seen in nanocrystals. The Gibbs-Thomson equation for nanocrystals confined to pores is defined as

$$\Delta T_m = T_m - T_m(d) = \frac{4\gamma_{solid-liquid} M T_m}{d \Delta H_{fus} \rho_{solid}} \cos(\theta) \tag{16.5}$$

where T_m is the bulk melting temperature, T_m *(d)* is the melting temperature of a confined crystal with diameter *d* assumed equal to the pore diameter, *M* is the molecular mass, ρ_{solid} is the density of the solid, $\gamma_{solid-liquid}$ is the surface free energy of the solid-liquid interface, ΔH_{fus} is the molar enthalpy of fusion, and θ is the contact angle between the wall and crystal. The melting point depression for nanosized crystalline material has been reported for particles confined within nanosized pores for a number of compounds [11, 50, 51].

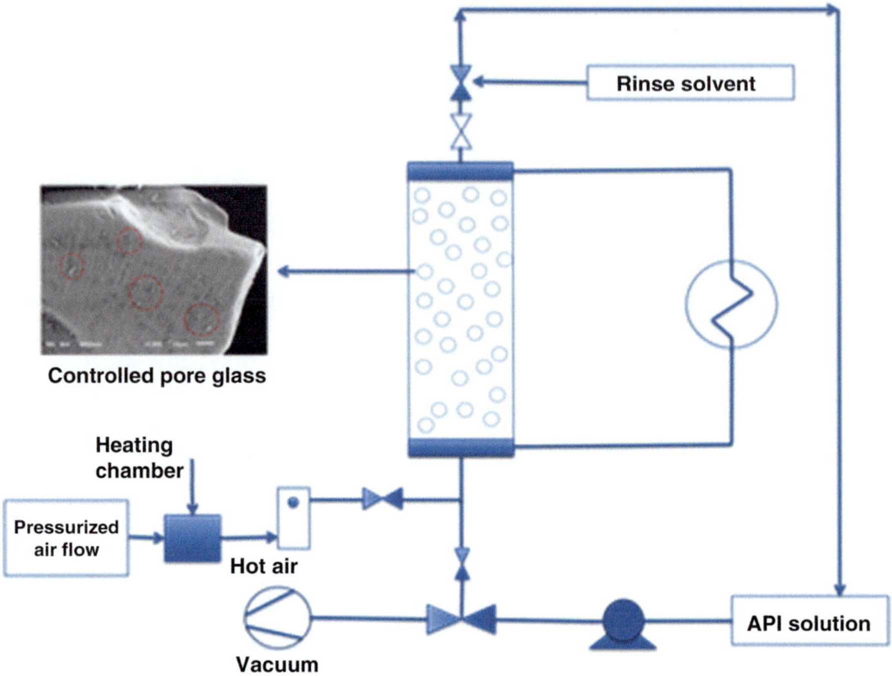

Fig. 16.3 Schematic of process design and setup of column process used for the preparation of nanocrystals within the nanoporous channels of controlled pore glass [11]

The important advantage is the sizes of the nanocrystals are limited by the channel diameter, which in turn limits growth of the crystals along the channel length and the melting behaviour is influenced by the matrix material [47]. CPGs have the advantage of being available in a number of different pore diameters, shapes and size [47].

16.4 Conclusions

Drug nanocrystals are considered as one of the important formulation techniques for poorly water-soluble drugs and can be applied to various routes of administration during the drug delivery of API. The nanocrystal technology offers great benefits that include the increased saturation solubility, and increased dissolution velocity due to increased surface area. We first explained the conventional nanocrystal preparation method and its recent improvements. The Milling and high-pressure homogenization methods are very useful techniques and produced the most nanocrystal products that have reached the market. Supercritical fluid crystallization method, controlled flow cavitation, and impinging jet crystallization methods extend the potential of the re-precipitation method. The key innovations in nanocrystal preparation include microfluidic devices, patterned self-assembled monolayer and generation of nanocrystals within the controlled pore glass. The microfluidic reactor is still a long way from displacing the conventional reactor for nanocrystal synthesis and needs to change before there is a realistic prospect of it. In the future nanocrystal surfaces might be modified to achieve a prolonged or a targeted release that can be the important part of the work for the nanoparticles.

Acknowledgements Novartis International AG is gratefully acknowledged for funding of this research under the Novartis-MIT Center for Continuous Manufacturing. The authors also thank Dr. Marcus O'Mahony, Dr. Siva Rama Krishna Perala and Leia Dwyer for helpful discussions.

References

1. Raj S, Jose S, Sumod US, Sabitha M (2012) Nanotechnology in cosmetics: opportunities and challenges. J Pharm Bioallied Sci 4:186–193
2. Ekimov AI, Efros AL, Onushchenko AA (1985) Quantum size effect in semiconductor microcrystals. Solid State Commun 56:921–924
3. Tran TT, Tran PH, Khanh TN, Van TV, Lee BJ (2013) Solubilization of poorly water-soluble drugs using solid dispersions. Recent Pat Drug Deliv Formul 7:122–133
4. Junghanns J-UAH, Müller RH (2008) Nanocrystal technology, drug delivery and clinical applications. Int J Nanomedicine 3:295–310
5. LaMer VK, Dinegar RH (1950) Theory, production and mechanism of formation of Monodispersed hydrosols. J Am Chem Soc 72:4847–4854
6. Mer VKL (1952) Nucleation in phase transitions. Ind Eng Chem 44:1270–1277
7. Mullin JW (2001) 5 - nucleation. In: Mullin JW (ed) Crystallization, 4th edn. Butterworth-Heinemann, Oxford, pp 181–215
8. Kulkarni SA, Kadam SS, Meekes H, Stankiewicz AI, ter Horst JH (2013) Crystal nucleation kinetics from induction times and metastable zone widths. Cryst Growth Des 13:2435–2440
9. Myerson A (2002) Handbook of industrial crystallization. Butterworth-Heinemann, Boston
10. Kim K, Lee I, Centrone A, Hatton TA, Myerson AS (2009) Formation of nanosized organic molecular crystals on engineered surfaces. J Am Chem Soc 131:18212–18213
11. O'Mahony M, Leung AK, Ferguson S, Trout BL, Myerson AS (2015) A process for the formation of Nanocrystals of active pharmaceutical ingredients with poor aqueous solubility in a nanoporous substrate. Org Process Res Dev 19:1109–1118
12. Shegokar R, Muller RH (2010) Nanocrystals: industrially feasible multifunctional formulation technology for poorly soluble actives. Int J Pharm 399:129–139
13. Peltonen L, Hirvonen J (2010) Pharmaceutical nanocrystals by nanomilling: critical process parameters, particle fracturing and stabilization methods. J Pharm Pharmacol 62:1569–1579
14. Nykamp G, Carstensen U, Müller BW (2002) Jet milling—a new technique for microparticle preparation. Int J Pharm 242:79–86
15. Wise DL (2000) Handbook of pharmaceutical controlled release technology. CRC Press, Boca Raton
16. Souto EB, Wissing SA, Barbosa CM, Müller RH (2004) Development of a controlled release formulation based on SLN and NLC for topical clotrimazole delivery. Int J Pharm 278:71–77
17. Merisko-Liversidge E, Liversidge GG, Cooper ER (2003) Nanosizing: a formulation approach for poorly-water-soluble compounds. Eur J Pharm Sci 18:113–120
18. Salazar J, ller RH, schwitzer JP (2014) Combinative particle size reduction technologies for the production of drug nanocrystals. J Pharm 2014:14.

19. Khadka P, Ro J, Kim H, Kim I, Kim JT, Kim H, Cho JM, Yun G, Lee J (2014) Pharmaceutical particle technologies: an approach to improve drug solubility, dissolution and bioavailability. Asian J Phar Sci 9:304–316
20. Ré M-I (2006) Formulating drug delivery systems by spray drying. Dry Technol 24:433–446
21. Zgoulli S, Grek V, Barre G, Goffinet G, Thonart P, Zinner S (1999) Microencapsulation of erythromycin and clarithromycin using a spray-drying technique. J Microencapsul 16:565–571
22. Wan LSC, Heng PWS, Chia CGH (1992) Spray drying as a process for microencapsulation and the effect of different coating polymers. Drug Dev Ind Pharm 18:997–1011
23. Heng D, Lee SH, Ng WK, Tan RB (2011) The nano spray dryer B-90. Expert Opin Drug Deliv 8:965–972
24. Huang Y, Dai W-G (2014) Fundamental aspects of solid dispersion technology for poorly soluble drugs. Acta Pharm Sin B 4:18–25
25. Reverchon E (1999) Supercritical antisolvent precipitation of micro- and nano-particles. J Supercrit Fluids 15:1–21
26. Caputo G, Adami R, Reverchon E (2008) Supercritical fluid crystallization of adipic acid using urea as habit modifier. Cryst Growth Des 8:2707–2715
27. Jarmer DJ, Lengsfeld CS, Anseth KS, Randolph TW (2005) Supercritical fluid crystallization of griseofulvin: crystal habit modification with a selective growth inhibitor. J Pharm Sci 94:2688–2702
28. Woo XY, Tan RBH, Braatz RD (2011) Precise tailoring of the crystal size distribution by controlled growth and continuous seeding from impinging jet crystallizers. Cryst Eng Comm 13:2006–2014
29. Siddiqui SW, Zhao Y, Kukukova A, Kresta SM (2009) Characteristics of a confined impinging jet reactor: energy dissipation, homogeneous and heterogeneous reaction products, and effect of unequal flow. Ind Eng Chem Res 48:7945–7958
30. Schwarzer H-C, Peukert W (2004) Combined experimental/numerical study on the precipitation of nanoparticles. AICHE J 50:3234–3247
31. Malik MA, Wani MY, Hashim MA (2012) Microemulsion method: a novel route to synthesize organic and inorganic nanomaterials: 1st nano update. Arab J Chem 5:397–417
32. Ujiiye-Ishii K, Kwon E, Kasai H, Nakanishi H, Oikawa H (2008) Methodological features of the emulsion and reprecipitation methods for organic nanocrystal fabrication. Cryst Growth Des 8:369–371
33. Schulman JH, Stoeckenius W, Prince LM (1959) Mechanism of formation and structure of micro emulsions by electron microscopy. J Phys Chem 63:1677–1680
34. Lee AY, Lee IS, Dette SS, Boerner J, Myerson AS (2005) Crystallization on confined engineered surfaces: a method to control crystal size and generate different polymorphs. J Am Chem Soc 127:14982–14983
35. Kulkarni SA, Weber CC, Myerson AS, ter Horst JH (2014) Self-association during heterogeneous nucleation onto well-defined templates. Langmuir 30:12368–12375
36. Shin K, Leach KA, Goldbach JT, Kim DH, Jho JY, Tuominen M, Hawker CJ, Russell TP (2002) A simple route to metal nanodots and nanoporous metal films. Nano Lett 2:933–936
37. Fan S, Chapline MG, Franklin NR, Tombler TW, Cassell AM, Dai H (1999) Self-oriented regular arrays of carbon nanotubes and their field emission properties. Science 283:512–514
38. Han W, Fan S, Li Q, Hu Y (1997) Synthesis of gallium nitride Nanorods through a carbon nanotube-confined reaction. Science 277:1287–1289
39. Wang L, Barton J, Hughes L, Odom TW (2008) Shape-control of protein crystals in patterned Microwells. J Am Chem Soc 130:2142–2143
40. Lee AY, Lee IS, Myerson AS (2006) Factors affecting the polymorphic outcome of glycine crystals constrained on patterned substrates. Chem Eng Technol 29:281–285
41. Mullin JW (2001) 3 – solutions and solubility. In: Mullin JW (ed) Crystallization, 4th edn. Butterworth-Heinemann, Oxford, pp 86–134
42. Yang X, Myerson AS (2015) Nanocrystal formation and polymorphism of glycine. Cryst Eng Comm 17:723–728

43. Phillips TW, Lignos IG, Maceiczyk RM, deMello AJ, deMello JC (2014) Nanocrystal synthesis in microfluidic reactors: where next? Lab Chip 14:3172–3180
44. Song H, Tice JD, Ismagilov RF (2003) A microfluidic system for controlling reaction networks in time. Angew Chem Int Ed 42:768–772
45. Vallet-Regi M, Rámila A, del Real RP, Pérez-Pariente J (2001) A new property of MCM-41: drug delivery system. Chem Mater 13:308–311
46. Vallet-Regi M, Balas F, Arcos D (2007) Mesoporous materials for drug delivery. Angew Chem Int Ed Engl 46:7548–7558
47. Ha JM, Wolf JH, Hillmyer MA, Ward MD (2004) Polymorph selectivity under nanoscopic confinement. J Am Chem Soc 126:3382–3383
48. Tran TH, Ramasamy T, Truong DH, Choi HG, Yong CS, Kim JO (2014) Preparation and characterization of fenofibrate-loaded nanostructured lipid carriers for oral bioavailability enhancement. AAPS Pharm Sci Tech 15:1509–1515
49. Fisher L (1999) Physics takes the biscuit. Nature 397:469–469
50. Ha J-M, Hillmyer MA, Ward MD (2005) Thermotropic properties of organic Nanocrystals embedded in Ultrasmall crystallization chambers. J Phys Chem B 109:1392–1399
51. Jackson CL, McKenna GB (1990) The melting behavior of organic materials confined in porous solids. J Chem Phys 93:9002–9011

Chapter 17
Crystallization Control Approaches and Models

Zoltan K. Nagy

Abstract Detailed review of the crystallization control approaches is provided, focusing on the model free control approaches, including supersaturation control, direct nucleation control and combined techniques. The potential benefits of applying these control techniques are exemplified through a series of example case studies.

Keywords Crystallization control • Supersaturation control • Direct nucleation control

17.1 Introduction

In the majority of industrial crystallization systems, typical feedback control strategies (e.g., PID, cascade) are designed to follow simple heuristic operating policies (e.g., a linear temperature profile or an antisolvent addition or evaporation rate) [1]. The most important driver for the implementation of crystallization control strategies has been the development of the in-situ sensors, which have become available in the last decade of the previous century and enabled real-time precise measurement of the solute concentration (e.g. via ATR-FTIR, ATR-UV/VIS), and the characteristics of the crystal size distribution such as mass, size, shape and morphology. This recent emergence of modern sensor technologies and advances in crystallization modelling and control have enabled more advanced control strategies to be increasingly applied [2–5]. Generally speaking, two main classes of control approaches can be differentiated for crystallization processes: (i) model-based and (ii) model free control techniques (Fig. 17.1). However more recently hybrid control approaches have also been developed that use a combination of model-based and model-free control techniques for crystallization design

Z.K. Nagy (✉)
School of Chemical Engineering, Purdue University, West Lafayette, IN 47906, USA

Department of Chemical Engineering, Loughborough University, Loughborough LE11 3TU, UK
e-mail: zknagy@purdue.edu

K.J. Roberts et al. (eds.), *Engineering Crystallography: From Molecule to Crystal to Functional Form*, NATO Science for Peace and Security Series A: Chemistry and Biology, DOI 10.1007/978-94-024-1117-1_17

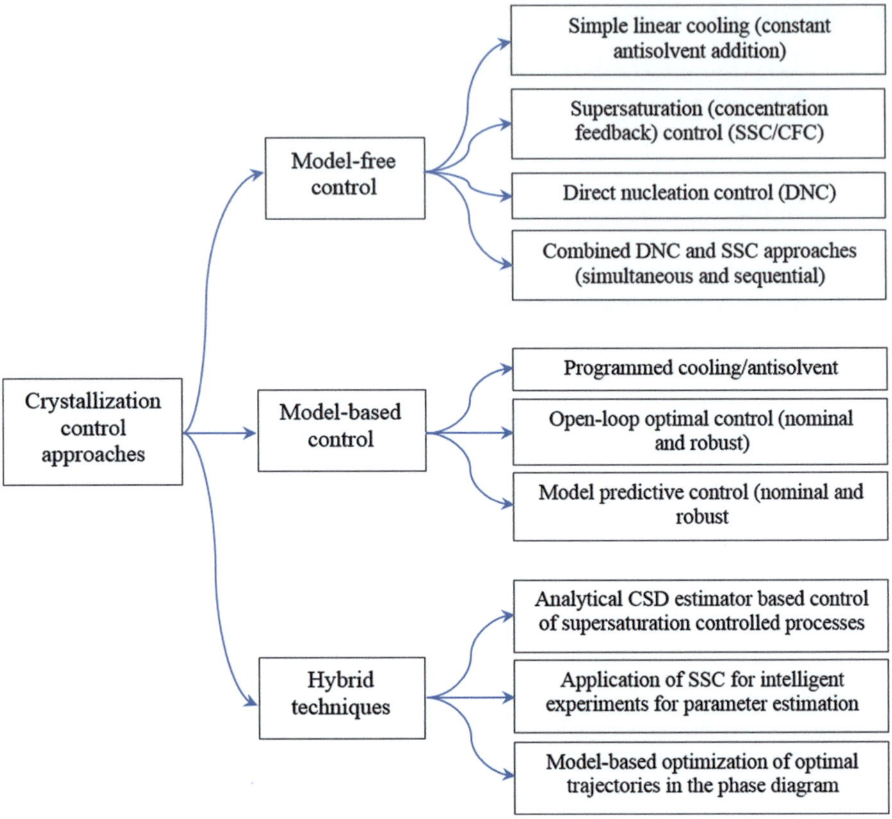

Fig. 17.1 Generic breakdown of crystallization control approaches

and control. For example simplified model-based CSD estimators have been developed and used for crystallization design and control, which are based on the assumption that the crystallization process is controlled at a constant supersaturation level, achieved using model-free (supersaturation) control approaches [6–8]. Additionally, model-free control approaches can be used to selectively trigger crystallization mechanisms by designing intelligent controlled experiments (e.g. growth dominated process versus growth with nucleation) that will provide experimental data that can significantly enhance the estimation of parameters for crystallization models.

17.2 Model-Based Control Approaches

Model based control strategies range back to the early seventies when pioneering work of Mullin [9] and Jones [10] showed the benefits of a programmed off line calculated cooling profile. Since then, especially in the last decades, major advances have been made in model based optimal control of various aspects of the product quality [11]. The model-based optimization is subject to a set of constraints due to equipment limitations (e.g., maximum and minimum temperature values, maximum and minimum cooling rates, maximum volume, limits on antisolvent addition rate). The crystallization must also satisfy a productivity constraint that ensures a desired yield at the end of the batch. Quality constraints can also be included in the optimization [12–14]. The main advantage of the model-based approaches is that in principle they can provide the theoretically optimal recipe or operating conditions and generally require a smaller number of experiments for identification, especially when properly designed, than for statistical experimental design of crystallization conditions. Model-based techniques also have the benefits of increased process understanding, and the possibility of incorporating the effects of non-ideal mixing via computational fluid dynamics [15]. However these advantages also impose several challenges and require the solution of a series of scientific problems.

Jones et al. [11] pointed out in their review a number of problems in the general model based control work:

1. There is a strong emphasis on the optimization of the temperature profile without taking into account the growth characteristics of the nuclei or seed crystals (e.g. growth rate dispersion).
2. There is on overwhelming use of the mean crystal size and coefficient of variation as the main control objective, which can still result in a large number of fines particles.
3. Failure to take into account the uncertainty in the kinetic models, which can lead to very poor performance of the controller when applied in practice.
4. Lack of attention to the operational constraints of the supersaturation during operation, which is defined by the metastable zone for primary or secondary nucleation.

Different model-based control approaches are possible ranging from programmed cooling using a simple tracking controller to enforce a predefined cooling profile in the crystallizer up to multi-variable non-linear optimal control (such as nonlinear model predictive control), which requires a real-time simulation and dynamic optimization of the input trajectories. The main components for a model-based control strategy are:

(i) The process model, which describes the product quality and process performance of the crystallization process with the required accuracy. The model is used either in real-time or off-line to achieve the desired performance or product quality. The model can be a detailed non-linear process model, describing all the details of the crystallization process and mass transfer in the crystallizer; however often a simplified, reduced or linearized model is used to enable real-time optimization of the trajectories. In most cases the process model consists of the population balance equation, material and enthalpy balances and constitutive kinetic models for the crystallization kinetics. The model should give a robust and accurate description of the crystallization process in the current crystallizer and therefore always requires parameter estimation and validation steps before it can be applied in the controller design. In addition to be used for the calculation of the optimal operating profile a process model is required in the observer in the case of real-time model based control for state and parameter estimation.

(ii) A dynamic observer can be used to prevent that the states in the process model drift away from the real states in the process. This is a crucial component of the model-predictive control scheme to achieve robust offset-free control. It is important to note that even if in the case of crystallization processes often full state-feedback can be achieved by using appropriate measurements (measuring concentration, CSD, temperature, etc.), the use of state estimation is still crucial to deal with model-plant mismatch and achieve robust control performance. The states in the model are continuously updated on the basis of the real-time measured process variables and corrected for process disturbances, model/plant mismatch and uncertain initial conditions. The observer is also used to estimate unmeasurable process variables. Different types of observers have been used in crystallization control. In combination with the moment model in many cases the extended (non-linear) Luenberger type of observer is used [4, 16] Other types are available and Mesbah [17] has recently performed a comparison between the different types of observers and concluded that both the extended as well as the unscented Kalman filters can give better performance than the deterministic extended Luenberger observer.

(iii) A dynamic optimizer is used to determine the optimal trajectory to reach the desired state of the process at the lowest possible costs. The possibility to formulate constraints represents an important advantage of the model-based control approaches. The efficiency and robustness of the optimiser plays an important role in the design of the control system. Mesbah [17] compared three different non-linear optimization strategies, i.e. single shooting, multiple shooting, and simultaneous optimization, in a real-time implementation of the model-based control approach. The results indicated that the three optimization strategies perform similarly in terms of optimal process operation. However, the single shooting strategy is computationally more expensive.

Hu [18] studied the optimal control of a batch cooling process, maximizing final-time seed size and minimizing the mass ratio between the nucleated and the grown seed crystal using the cooling profile. Although optimal control gave some improvement of the objective function, hardly any effect was found on the product CSD. Kalbasenka [16] studied the growth behaviour in seeded batch evaporative crystallisation experiments using narrow seed distributions of grinded crystals and reported a strong dispersion in the growth rates and even dissolution of part of these seed crystals, which resulted in a strong broadening of the CSD of the product crystals and in a lack of batch to batch reproducibility. Pre-treatment by ageing of the seed crystals and a high relative supersaturation in the crystalliser at the initial growth phase of the seed crystal were found to be essential to remove the batch to batch variability of the crystal size and to lower the width of the distribution. Aamir and coworkers confirmed the importance of the quality of seed crystals [19]. From a combined model-based and experimental evaluation, they concluded that depending on the pre-treatment of the seed crystals, initial breeding, ripening and agglomeration can drastically influence the performance of the seeds, stressing the importance of the use of well-defined seed preparation techniques. They also proposed a model-based recipe optimization approach that allowed very good tailoring of the product CSD by combining the right amount of sieved seed fractions [19]. Mesbah [4, 20] performed closed-loop model based control studies on a semi industrial batch evaporative crystalliser. In these dynamic optimization studies the growth rate was maximized below a predefined constrained value after the initial phase of the batch by manipulation of the heat input to the crystalliser. The results showed that the model-based growth control approach was able to achieve a substantial increase, i.e. 30%, in the amount of crystals produced in the batch at constant product quality [4, 20, 21]. Although no attempts were reported to improve on the product quality the feasibility of the model-based control approach was clearly demonstrated both based on the reduced (moment model) as well as using the full population balance models.

Nagy [8] continued on this line and designed a simpler controller based on the idea that if the supersaturation is properly controlled at a sufficiently low level using a supersaturation control approach, the seeded batch process will be dominated by growth only. For such a case an analytical CSD estimator has been derived, which directly relates the seed CSD to the product CSD with only two degrees of freedom, i.e. the supersaturation level and the batch time [7, 8]. An off-line optimization can then determine the temperature profile needed to achieve the desired CSD. Experimental evaluation of the method showed a remarkable ability to generate a predefined shape of the product CSD [7]. Zhang [22] continued this line of research and extended it to include both primary and secondary nucleation in unseeded batches using a discretised process model. First of all the reachability of the desired product CSD is determined taking into account model uncertainty, especially for the nucleation parameters [22]. Due to the strong impact of the uncertain nucleation parameters on the crystallization process, leading to infeasible control the authors propose to add an additional control law based on the moments of the distribution, which can be obtained using a state estimator from the concentration measurements. Nagy [5] proposed a robust control scheme consisting of a tracking controller for the temperature and concentration to drive the system to the desired location in the phase diagram and a robust supervising controller to adapt these trajectories in the case of changing process conditions or process disturbances [5]. Several of the proposed techniques have used controlled dissolutions for the elimination of unwanted fine particles [23].

17.2.1 Challenges

The main challenges in the control of crystallisation are mainly related to the large variations in the solution thermodynamics and kinetics, which can be induced by impurities or other components in the crystallizer, the complex nonlinear dynamics with non-ideal mixing, and various disturbances characteristic to these systems, made often more complex due to multiple polymorphic phase transformations and complex shapes (e.g. in the case of large organic molecules). These complex processes are often characterised by different spatial and time scales. Challenges are also related to the increased difficulty in modelling some of the practical objectives (e.g. filterability, purity, tablet stability, dissolution behaviour), and the significant time and engineering effort required for the model development. Additionally, the performance of the model-based approach depends on the model accuracy although the robustness of the model-based control approaches can be improved by incorporating linear or nonlinear robustness analysis into the optimization [5].

17.2.2 Scientific Issues

Current scientific issues are related to the (i) modelling aspects (ii) efficiency of the control and optimisation approaches (iii) availability of reliable sensor technologies. Since generally various mechanisms at different scales are relevant for the properties of the crystalline products it is important to use appropriate multi-scale modelling approaches for the analysis, design and control of crystallization processes. The model-based control approaches often require the solution of one or more population balance equations (PBEs), hence the development of efficient numerical solution approaches is crucial for real-time model predictive control. For shape and polymorphic control, or when dissolution is also considered a system of two- (or multi-) dimensional PBEs must be solved. The incorporation of better mechanistic models in the PBE, which describe nucleation, growth, agglomeration and breakage based on first-principles as well as the effect of hydrodynamics, would significantly enhance the prediction ability and robustness of the model-based control techniques. With these increases in the model complexity the development of efficient numerical approaches for the solution of the model equations as well as for the optimisation problem becomes increasingly important. Efficient model-based estimation approaches which can use the information available from different sensors at different time scales and provide the parameters and consistent initial conditions required for the model prediction must be developed for the feasible real-time application of the crystallisation control approaches.

17.2.3 Emerging Topics

Emerging topics for model-based control include the development of novel control approaches that allow more freedom in designing simultaneously the crystal shape, polymorphic form and CSD. For example the use of growth and nucleation modifying agents as manipulating inputs can provide extra degrees of freedom for better manipulation of the crystal properties. The simultaneous optimisation of different supersaturation generation approaches (e.g. cooling and antisolvent addition) can significantly enhance the efficiency of the crystallisation processes. The application of dynamic seeding has recently proved to be an efficient new control approach to improve the CSD control [8, 24]. Model-based approaches also enable the design and control of crystallization processes as integrated unit operations, taking into account downstream processing units, and considering final product properties as control objectives. These integrated product and process engineering approaches will need to combine advanced sensor technologies using information fusion, as well as model-based and model-free control techniques to design products with tailor-made properties (Fig. 17.2).

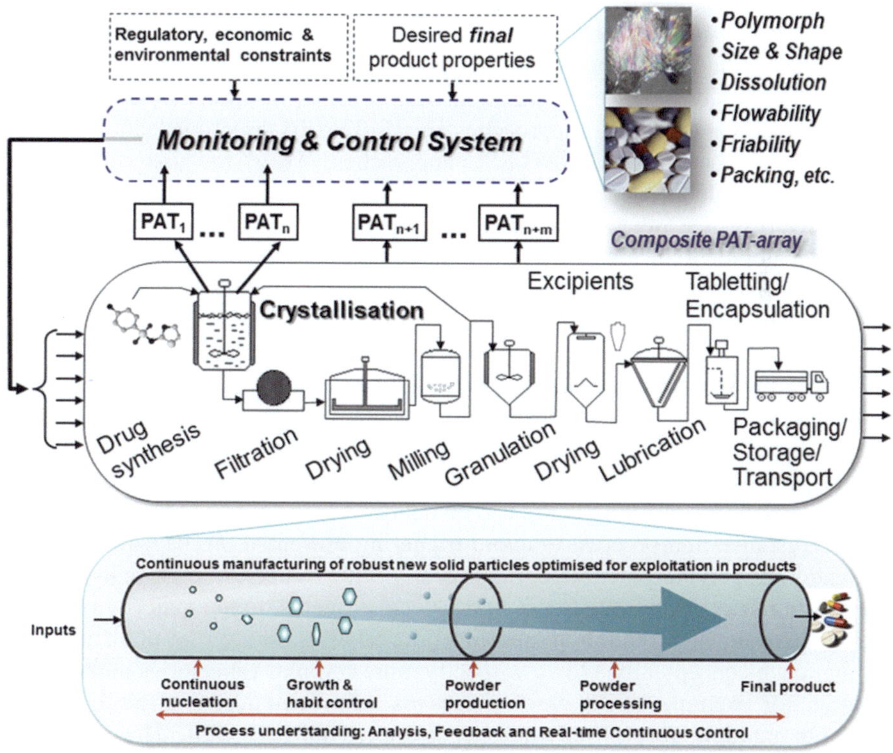

Fig. 17.2 A systems view of the integrated crystallization and product formulation process chain for end-point particle property control. Crystallization is the key unit operation between drug synthesis and the formulation and the process where the key solid properties can be manipulated first

Control of continuous crystallisation processes is of increasing interest in particular in the pharmaceutical industries. The current aim in developing crystallisation control approaches is the implementation of a new generation of integrated and intensified crystallisation systems with drastically improved flexibility, predictability, stability and controllability. New directions of developing control systems for crystallization include the use of hybrid first-principles and data-driven (e.g. statistical, artificial neural networks, response surface methods, etc.) modelling approaches, multi-objective nonlinear model predictive control techniques, as well as the development of robust and intelligent adaptive control methodologies. These approaches will enable the crystallisation as an integrated, intensified and intelligent process, which can react and adapt to changing operating conditions to guarantee the sustainable production of consistent high quality final solid products.

17.3 Model-Free Control Approaches

Model-free control approaches are based on the direct use of PAT-based measurements in feedback control of crystallization processes. Model-free control approaches include: (i) the simple control using linear cooling (or constant antisolvent addition), (ii) supersaturation control (SSC) or often called concentration feedback control (CFC) approaches [3, 25–27] (iii) direct nucleation control [28–31] and (iv) combined DNC-SSC approaches. Whilst the first category of traditional open-loop control approaches are model free, they are often discussed separately from the other model-free techniques, which in fact represent major novel development in crystallization control as opposed to the standard linear cooling or constant antisolvent addition techniques. Supersaturation control is based on the idea to control the crystallization process by following a desired operating curve in the phase diagram. To determine the current state of the crystallization process in the phase diagram it is necessary to measure concentration and temperature (solvent/antisolvent ratio), and the process is controlled by manipulating the temperature (solvent/antisolvent ratio), to follow the desired concentration (supersaturation) trajectory in the phase diagram. The approach provides the advantage that the operating curve can be defined based on detailed understanding of the crystallization process, to avoid triggering unwanted mechanisms, such as nucleation or polymorphic transformation and often can result in close to optimal crystallization performance after only a few experiments. In the majority of applications constant supersaturation (absolute or relative) is used for set-point although the approach conceptually allows to specify more complex trajectories in the phase diagram for example in the case of systems with complicated phase behavior, e.g. with multiple forms (polymorphs, hydrates, solvates etc.). The schematic

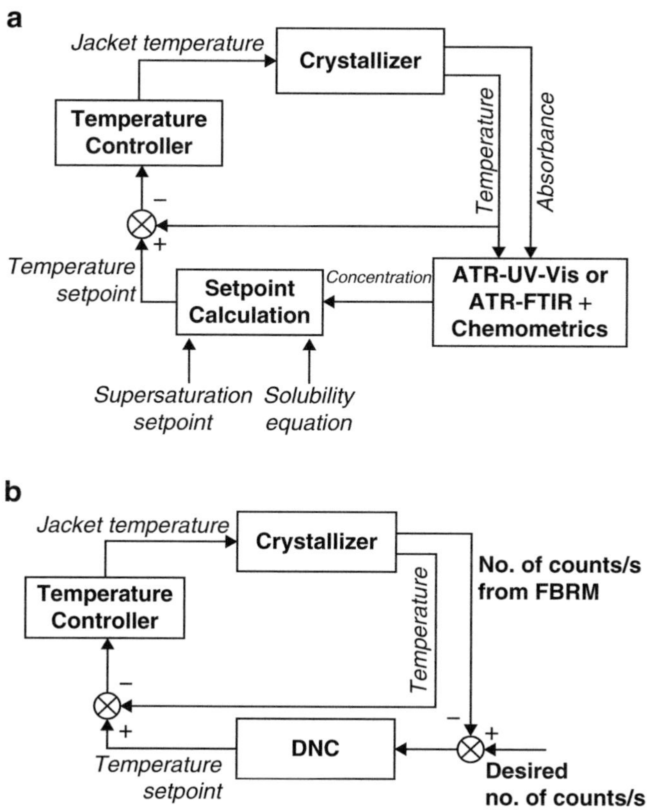

Fig. 17.3 Schematic representation of the two main model-free control approaches: (**a**) supersaturation control (SSC), and (**b**) direct nucleation control (DNC)

representation of the SSC approach is shown in Fig. 17.3a. The SSC approach controls the crystallization process in the phase diagram, thus it is a more direct approach than controlling the process by following temperature trajectories defined in time. Nevertheless, the solid properties are not controlled directly and it is possible for two crystallization batches controlled at the same supersaturation to produce largely different products, due to disturbances, e.g. accidental seeding, attrition, agglomeration or others.

The main characteristic of the direct nucleation control (DNC) approach is that it directly measures and controls properties of the solid particles. The original approach was developed to control the number of particle counts per second provided by FBRM measurement, based on the concept that the number of particles in the system is in correlation with the CSD that can be achieved. If a smaller number of particles is maintained the resulting mean size of the product will be larger and vice versa. The approach detects directly the increase and decrease in the number of particles (or a measure related to this) and increases or decreases the supersaturation or creates dissolution to control the measured number of particles around the desired value, as shown in Fig. 17.3b. The controlled growth and dissolution cycles (GDCs) are achieved by cooling/heating or anti-solvent/solvent addition cycles (or a combination of the two). The GDCs have proved to have numerous benefits in achieving better crystalline product properties. DNC can be used for consistent in situ seed generation, elimination of agglomeration, elimination of fine particles, to provide more uniform and larger particle size, elimination of problems with solvent inclusion, better polymorphic purity. Additionally, the GDCs have an effect of slow surface dissolution and regrowth, which can significantly increase crystal purity by decreasing the amount of foreign molecules that can be incorporated in the crystal lattice. Amongst the many other benefits GDCs can also improve the aspect ratio of needle-shaped particles.

Figure 17.4 shows sample results [31] of using DNC for the design and control of the quality of an active pharmaceutical ingredient (an antiarrhythmic cardiovascular drug). It can be seen that DNC automatically resulted in three GDCs, which have led to the elimination of agglomeration (and solvent inclusion) problem, providing a very rapid crystallization development approach through feedback control. The distinguishing feature of DNC, to provide converging temperature cycles, as opposed to traditional temperature cycling that generally uses same initial and final temperature can also be observed [31]. The DNC yielded high quality crystals with no agglomeration and eliminated the problems related to solvent inclusion (Fig. 17.5).

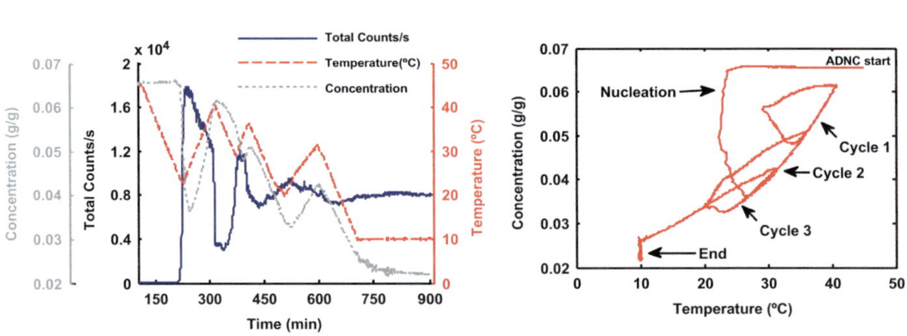

Fig. 17.4 Sample results using direct nucleation control (DNC) as a rapid crystallization design approach for a cardiovascular drug [31]

Although the original DNC approach is based on controlling the number of counts per second resulting from the FBRM other properties can be used in the control algorithm, such as number of counts within certain size ranges or other statistics of the distribution, e.g. standard deviation, as well as combination of statistics. Additionally, other PAT tools can be used in similar feedback control approach to generate GDCs to control different signals provided by the measurement device (e.g. turbidity).

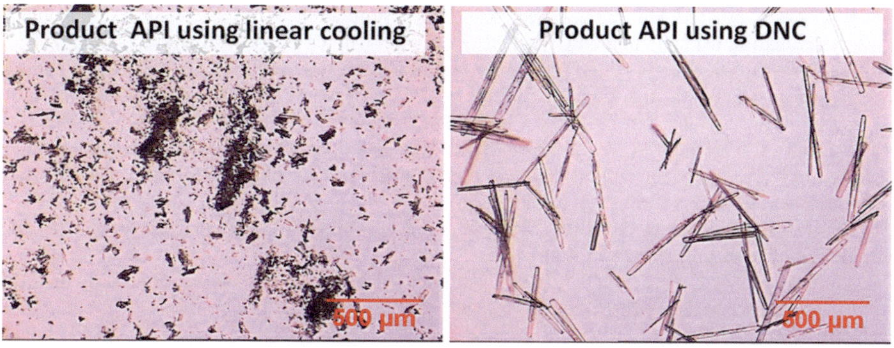

Fig. 17.5 Microscopic pictures of a cardiovascular API product obtained using linear cooling and DNC respectively. DNC produces large, crystals with no agglomeration and eliminated problems with solvent inclusion (Adapted from Ref. [31])

References

1. Rawlings JB, Sink CW, Miller SM (2002) Control of crystallization processes. In: Myerson AS (ed) Handbook of industrial crystallization, 2nd edn. Butterworth-Heinemann, Boston
2. Braatz RD, Fujiwara M, Ma DL, Togkalidou T, Tafti DK (2002) Simulation and new sensor technologies for industrial crystallization: a review. Int J Mod Phys B 16:346–353
3. Fujiwara M, Nagy ZK, Chew JW, Braatz RD (2005) First-principles and direct design approaches for the control of pharmaceutical crystallization, J. Process Control 15:493–504
4. Mesbah A, Huesman AEM, Kramer HJM, Nagy ZK, Van den Hof PMJ (2011) Real-time control of seeded batch crystallization processes. AICHE J 57:1557–1569
5. Nagy ZK (2009) Model based robust control approach for batch crystallization product design. Comput Chem Eng 33:1685–1691
6. Aamir E, Nagy ZK, Rielly CD, Kleinert T, Judat B (2009) Combined quadrature method of moments and method of characteristics approach for efficient solution of population balance models for dynamic modelling and crystal size distribution control of crystallization processes. Ind Eng Chem Res 48(18):8575–8584
7. Aamir E, Rielly CD, Nagy ZK (2012) Experimental evaluation of the targeted direct design of temperature trajectories for growth-dominated crystallization processes using an analytical crystal size distribution estimator. Ind Eng Chem Res 51(51):16677–16687
8. Nagy ZK, Aamir E (2012) Systematic design of supersaturation controlled crystallization processes for shaping the crystal size distribution using an analytical estimator. Chem Eng Sci 84:656–670
9. Mullin JW, Nyvlt J (1971) Programmed cooling of batch crystallizers. Chem Eng Sci 26 (3):369–377
10. Jones AG, Mullin JW (1974) Programmed cooling crystallization of potassium sulphate solutions. Chem Eng Sci 29(1):105–118
11. Nagy ZK, Braatz RD (2012) Advances and new directions in crystallization control. Ann Rev Chem Biomol Eng 3:55–75
12. Worlitschek J, Mazzotti M (2004) Model-based optimization of particle size distribution in batch-cooling crystallization of paracetamol. Cryst Growth Des 4:891–903
13. Corriou JP, Rohani S (2008) A new look at optimal control of a batch crystallizer. AICHE J 54:3188–3206
14. Sarkar D, Rohani S, Jutan A (2006) Multi-objective optimization of seeded batch crystallization processes. Chem Eng Sci 61:5282–5295

15. Woo XY, Tan RBH, Chow PS, Braatz RD (2006) Simulation of mixing effects in antisolvent crystallization using a coupled CFD-PDF-PBE approach. Cryst Growth Des 6:1291–1303
16. Kalbasenka AN, Spierings LCP, Huesman AEM, Kramer HJM (2007) Application of seeding as a process actuator in a model predictive control framework for fed-batch crystallization of ammonium sulphate. Part Part Syst Charact 24(1):40–48
17. Mesbah A, Huesman AEM, Kramer HJM, Van den Hof PMJ (2011a) A comparison of nonlinear observers for output feedback model-based control of seeded batch crystallization processes. J Process Control 21(4):652–666
18. Hu Q, Rohani S, Jutan A (2005) Modelling and optimization of seeded batch crystallizers. Comput Chem Eng 29(4):911–918
19. Aamir E, Rielly CD, Nagy ZK (2010) Evaluation of the effect of seed preparation method on the product crystal size distribution for batch cooling crystallization processes. Cryst Growth Des 10(11):4728–4740
20. Mesbah A, Nagy ZK, Huesman AEM, Kramer HJM, Van den Hof PMJ (2012) Nonlinear model-based control of a semi-industrial batch crystallizer using a population balance modeling framework. IEEE Trans on Control Syst Tech 20:1188–1201
21. Mesbah A, Landlust J, Huesman AEM, Kramer HJM, Jansens PJ, Van den Hof PMJ (2010) A model-based control framework for industrial batch crystallization processes. Chem Eng Res Des 88:1223–1233
22. Zhang K, Nadri M, Xu C-Z (2012) Reachability-based feedback control of crystal size distribution in batch crystallization processes. J Process Control 22(10):1856–1864
23. Nagy ZK, Aamir E, Rielly CD (2011) Internal fines removal using a population balance model based control of crystal size distribution under dissolution, growth and nucleation mechanisms. Cryst Growth Des 11:2205–2219
24. Woo XY, Tan RBH, Braatz RD (2011) Precise tailoring of the crystal size distribution by controlled growth and continuous seeding from impinging jet crystallizers. Cryst Eng Comm 13:2006–2014
25. Gutwald T, Mersmann A (1990) Batch cooling crystallization at constant supersaturation: technique and experimental results. Chem Eng Technol 13:229–237
26. Nagy ZK, Chew JW, Fujiwara M, Braatz RD (2008a) Comparative performance of concentration and temperature controlled crystallizations. J Process Control 18:399–407
27. Groen H, Borissova A, Roberts KJ (2003) In-process ATR-FTIR spectroscopy for closed-loop supersaturation control of a batch crystallizer producing monosodium glutamate crystals of defined size. Ind Eng Chem Res 42:198–206
28. Abu Bakar MR, Nagy ZK, Saleemi AN, Rielly CD (2009) The impact of direct nucleation control on crystal size distribution in pharmaceutical crystallization processes. Cryst Growth Des 9:1378–1384
29. Saleemi AN, Rielly C, Nagy ZK (2012a) Automated direct nucleation control for in situ dynamic fines removal in batch cooling crystallization. Cryst Eng Comm 14:2196–2203
30. Saleemi AN, Rielly CD, Nagy ZK (2012b) Comparative investigation of supersaturation and automated direct nucleation control of crystal size distributions using ATR-UV/Vis spectroscopy and FBRM. Cryst Growth Des 12:1792–1807
31. Saleemi AN, Steele G, Pedge N, Freeman A, Nagy ZK (2012c) Enhancing crystalline properties of a cardiovascular active pharmaceutical ingredient using a process analytical technology based crystallization feedback control strategy. Int J Pharm 430:56–64

Chapter 18
Application of Ultrasound in Crystallization (Sonocrystallization)

Christopher J. Price

Abstract Sonocrystallization is the application of ultrasound to influence crystallization processes. The most widely practiced approach is to initiate controlled nucleation to manage product particle size. This chapter describes the equipment which can be used and how this is applied particularly in pharmaceutical crystallization processes.

Keywords Sonocrystallization • Sononucleation • Ultrasonic crystallization

18.1 Introduction to Ultrasound and Cavitation

The propagation of ultrasound in a liquid is summarised in Fig. 18.1. The sound wave comprises alternating phases of compression and rarefaction the intensity of which depends on the amplitude of the wave expressed in microns of displacement of the vibrations at the source of the sound, which is proportional to the power delivered. Dissolved gases present in the solution will tend to be released as small bubbles during the rarefaction phase and these bubbles shrink during the compression phase and then expand again during subsequent rarefaction phases (stable cavitation). As the amplitude of the sound wave increases, the pressure reduction during the rarefaction phase becomes large enough to allow bubbles of solvent vapour to form in the solution. In aqueous systems this occurs at local intensities exceeding 10 W/cm^2. These bubbles can coalesce and transient cavitation may occur where bubbles increase in size from cycle to cycle until eventually they reach a critical size of about 100–200 μm, at which point they collapse catastrophically. This collapse is accompanied by acoustic and light emission, surface erosion will occur if the bubble nucleates on a surface such as the vessel wall or the agitator. The resulting hot spots are extremely localised but intense. Under the most intense

C.J. Price (✉)
Department of Chemical and Process Engineering, University of Strathclyde, Glasgow G1 1QX, UK
e-mail: chris.price@strath.ac.uk

K.J. Roberts et al. (eds.), *Engineering Crystallography: From Molecule to Crystal to Functional Form*, NATO Science for Peace and Security Series A: Chemistry and Biology, DOI 10.1007/978-94-024-1117-1_18

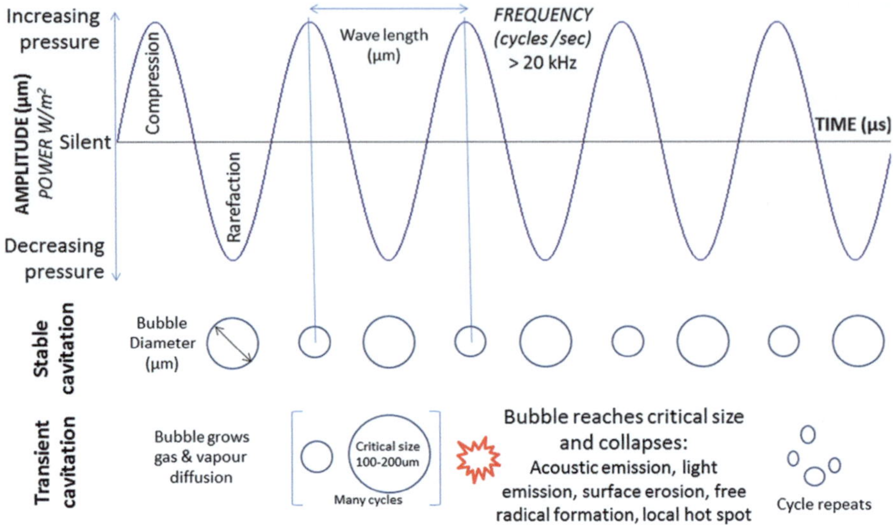

Fig. 18.1 Processes associated with the passage of ultrasound through a liquid

insonation (around 20–30w/cm^2 according to [1] it becomes possible to break carbon-carbon bonds to form radicals.

This catastrophic bubble collapse is shown schematically in Fig. 18.2. During the final compression cycle bubbles adjacent to surfaces collapse unevenly to form a torus shape which eventually breaks up into a series of smaller bubbles. At the point when the bubble opens into a torus a jet of liquid crosses the void of the bubble hitting the solid surface with a high enough velocity to cause surface damage. This is the basis of ultrasonic cleaning and surface erosion leading to pitting [2].

Surface erosion leads to particle shedding which begins to be detectable in aqueous systems when the acoustic intensity exceeds 2 W/cm^2. In the context of pharmaceutical crystallization this would represent a problem for drug substance quality, leading to a specification failure due to the presence of an excessive number of particulate metal impurities. The level of ultrasonic intensity required to produce free radicals is substantially greater at around 20–30 W/cm^2 and would be unlikely to be encountered in manipulation of pharmaceutical crystallization processes. Indeed free radical formation has not been identified as an issue in sonocrystallization of pharmaceuticals. The discipline of sonochemistry suggests that there may be a risk of triggering other undesired chemical transformations with ultrasound, however successful sonocrystallization of highly strained molecules including beta lactam antibiotics which contain a four membered heterocycle has been demonstrated [3]. Local hotspots may also represent a concern when using flammable solvents, however highly energetic materials have been successfully sonocrystallized in flammable organic solvents for example the explosive 3-nitro-

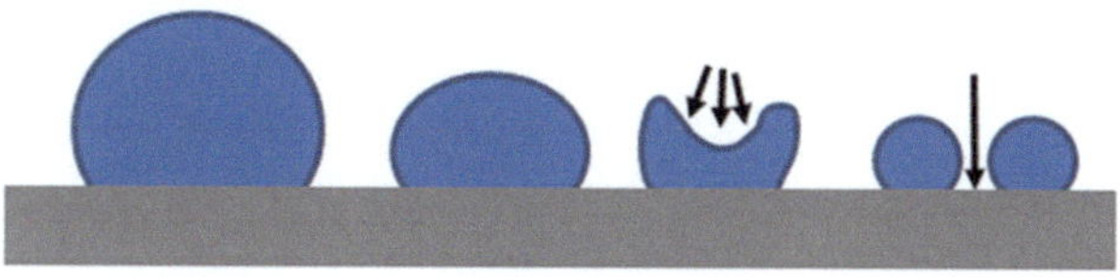

Fig. 18.2 The catastrophic collapse of a cavitation bubble adjacent to a surface

1,2,4-triazol-5-one (NTO) [4] achieving improved stability. Similar findings are reported for another explosive 1,3,5-trinitroperhydro-1,3,5-triazine RDX [5].

The range over which cavitation is possible is limited in aqueous solutions to about 15–20 cm from the acoustic source, due to the shielding effect of cavitation bubbles reducing the intensity of insonation further from the acoustic source. As a result it is necessary to use multiple acoustic sources to insonate large volumes and to adopt tubular designs with a maximum duct diameter of around 40 cm.

The intensity of ultrasound used in different applications ranges from 0.01–1 W/cm^2 in non destructive testing and biomedical imaging, to 1–10 W/cm^2 in biomedical therapy, with slightly higher intensities in industrial processing (e.g. crystallization) and up to 100 W/cm^2 for underwater sonar where transducer array design is complex and proprietary. Biomedical imaging uses higher frequencies in the range 10^6–10^7 Hz to increase resolution, whereas industrial processing uses frequencies of 10^4–10^5 Hz which maximise power delivery (Fig. 18.3).

18.1.1 Transducer Design

A traditional Tonpilz transducer is shown in Fig. 18.4. It comprises a piezo-electric component adjacent to an acoustic resonator. The sound waves can be focussed using a convergent horn tip to increase amplitude at the tip or a divergent horn may be used to reduce amplitude allowing greater total power delivery whilst ensuring the local intensity remains below the threshold for cavitational damage. Piezo ceramic transducers, also shown in Fig. 18.4, offer several advantages including the ceramic's ability to be manufactured in a wide variety of shapes and sizes, its capability to operate efficiently at low driving voltages, and its ability to function at higher temperatures, up to 300 °C. Piezo composites build on these advantages. They comprise arrays of fine piezo ceramic rods bound together with a flexible polymer. Each element in the resulting array of transducers can be addressed individually allowing the acoustic field to be focused. In the context of sonocrystallization the main priority has been on low cost delivery of intense acoustic fields at frequencies below 50 KHz where the Tonpilz transducer is the more cost effective, hence the vast majority of sonocrystallization applications use Tonpilz transducers.

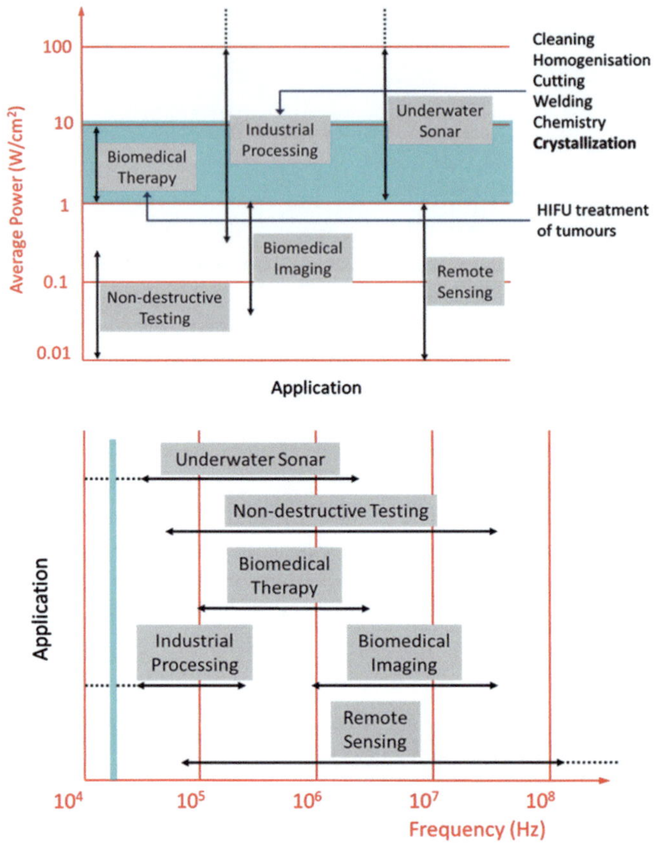

Fig. 18.3 Typical power levels and frequency which define wave length and equipment dimensions, the highlighted region of the figures is that typically employed in pharmaceutical sonocrystallization

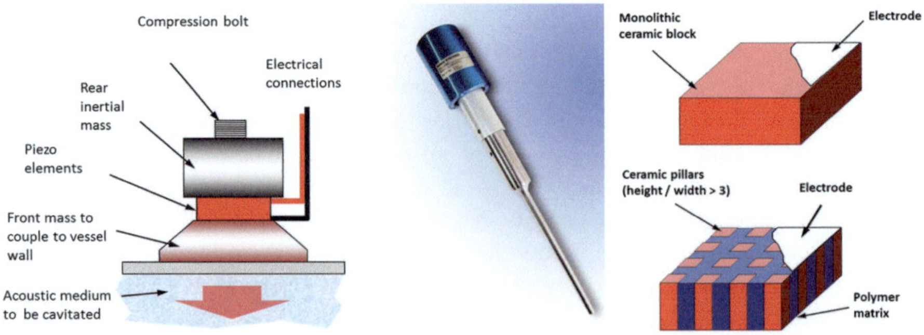

Fig. 18.4 A typical Tonpilz transducer structure (*left*) and as assembled into a typical immersion system with an amplifying acoustic horn (*centre*). A piezo ceramic transducer (*top right*) and a piezo composite transducer array (*bottom right*)

18.1.2 Scale-Up

One of the largest scale applications of sonocrystallization is to precipitate sodium oxalate prior to crystallizing alumina. The oxalate counter ion is a powerful crystal growth inhibitor in the Bayer process used to produce millions of tons of alumina per year. In this application the impurity is sononucleated, crystallized separately and removed by filtration prior to the main alumina crystallization process [6]. The equipment comprises a sequence of tubular sononucleators each with a series Tonpliz transducers bonded radially around the diameter of the tube and stacked longitudinally over its length. These very large transducer arrays are driven asynchronously to achieve a relatively uniform ultrasonic exposure to the material flowing through the duct.

18.2 Sonocrystallization – Historic Context

The first published account of the effect of ultrasound on crystallization is contained in a wide ranging review of ultrasonic phenomena from 1927 [7]. The authors report that crystallization of paraffin waxes can be initiated. The proceedings note that Sir William Bragg remarked that paraffin crystallized in two modifications making an oblique reference to influencing polymorphic form in this first published account.

The first pharmaceutical application is in a patent assigned to Pfizer for the use of ultrasound to reduce the crystal size of procaine penicillin and produce a uniform size distribution [8]. Ultrasound has been used in the preparation of microcrystalline particles of the hormone progesterone as an alternative to size reduction by milling [9]. The authors report data on the effect of insonation intensity on particle size and the width of the size distribution indicating smaller particles were obtained at higher intensities.

A pharmacist at Wyeth Laboratories considered mechanisms by which ultrasound might be effective in producing small uniform crystals and suggested that the beneficial effects of ultrasound on crystallization are linked to cavitation arising from the passage of ultrasound through the solution [10]. The same group [11] suggested that cavitation bubbles act as nucleation sites, and that the localised and transient high pressure might serve to increase the melting point of a solid.

18.3 Mechanisms by Which Ultrasound Affects Crystallization

Ultrasound has the potential to affect almost all of the processes involved in crystallization as illustrated in Fig. 18.5. The effects are interdependent because supersaturation drives both growth and nucleation which together control the size distribution which is further influenced by breakage, agglomeration, encrustation and the polymorphic form of the material crystallizing.

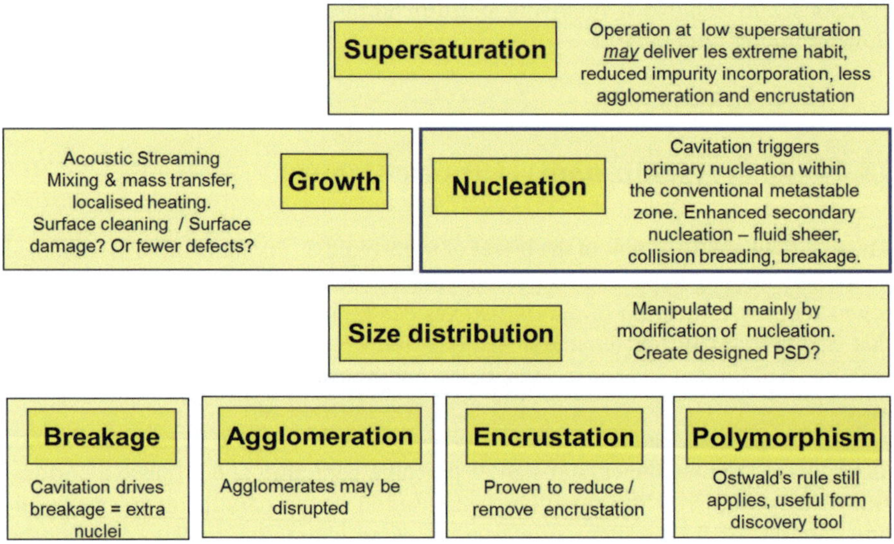

Fig. 18.5 Crystallization processes which may be subject to modification by insonation

18.3.1 Nucleation

The predominant mechanism by which ultrasound is used to influence pharmaceutical crystallization is by inducing nucleation through cavitation. This is an attractive alternative to conventional use of seeds because ultrasound induced primary nucleation (in a particle free solution) occurs at lower supersaturations than conventional spontaneous primary nucleation. If crystals are already present, the secondary nucleation is enhanced. Control of nucleation is the primary way by which the particle size distribution can be controlled.

18.3.2 Growth

The effect of ultrasound on crystal growth is primarily due to enhanced mass transfer at the particle solution interface allowing the solution layer adjacent to the growing crystal from which supersaturation is consumed to be replaced more frequently. At high acoustic intensities it is likely that surface damage may also occur due to collapse of cavitation bubbles which would also generate many secondary nuclei and so reduce the mean particle size.

18.3.3 Supersaturation

Because ultrasound can be used to induce nucleation at modest supersaturations the operating supersaturation level in sonocrystallization processes may be lower than in conventional crystallizations which may deliver less extreme crystal habits and greater crystal purity.

18.3.4 Breakage

Cavitation adjacent to a large crystal is likely to cause breakage, while smaller particles may be less vulnerable. Ultrasonic cavitation has been used successfully as a milling process though this is likely to be less effective than using ultrasound to generate nucleation in a crystallization process. The propensity for crystal breakage is quite material dependent. Images of acoustic cavitation effects on suspended calcite crystals are seen in [12] which shows surface damage and some de-agglomeration but little evidence of breakage. This is also addressed for organic crystals where different breakage mechanisms are investigated [13].

18.3.5 Agglomeration

Ultrasound acts to disrupt agglomerates, resulting in a more uniform distribution of particles. A reduction in the number of agglomerates formed during crystallization may also be linked to the particle size distribution and operating level of supersaturation both of which may also be influenced by insonation.

18.3.6 Encrustation

Ultrasound has been shown to be effective in preventing encrustation forming on vessel walls [14]. Reduction in encrustation propensity may also be linked to operation at lower supersaturations and improved suspension mobility at the vessel wall. For example acoustic streaming may serve to replenish the boundary layer more frequently enhancing the heat and mass transfer environment and reducing the overall supersaturation level at heat exchanger surfaces and so reducing the propensity for encrustation.

18.3.7 Polymorphism

Ostwald's rule, which states that in general the least stable polymorph crystallises first, still applies when inducing primary nucleation. Ultrasound has been used as a tool in polymorph screening especially in hard to nucleate systems. For example ultrasound can be used to influence which polymorphic form in which *p*-amino benzoic acid nucleates [15].

18.4 Example Applications

18.4.1 Sononucleation to Influence Crystal Habit

Figure 18.6 shows the impact of sononucleation on the crystal habit of sorbitol hexa-acetate [16]. The crystallization profile is superimposed on a plot of solubility as a function of temperature. When a crystal free solution with a saturation composition of 40 °C is cooled, spontaneous nucleation occurs at about 33 °C as indicated by the red dot on the figure. Nucleation occurs rapidly as the supersaturation is relatively high at this point and the resulting crystals are small and heavily agglomerated, as shown in the left hand image (a). If the solution is insonated while cooling, nucleation takes place at earlier at about 37 °C and hence at a lower supersaturation so the number of nuclei formed is lower and growth is slower. The resulting product crystals are larger and do not agglomerate as seen in the right hand picture (b). Similar findings for the drug Roxithromycin, where ultrasound was used to reduce the induction time for nucleation, the metastable zone width, and to improve the crystal habit [17].

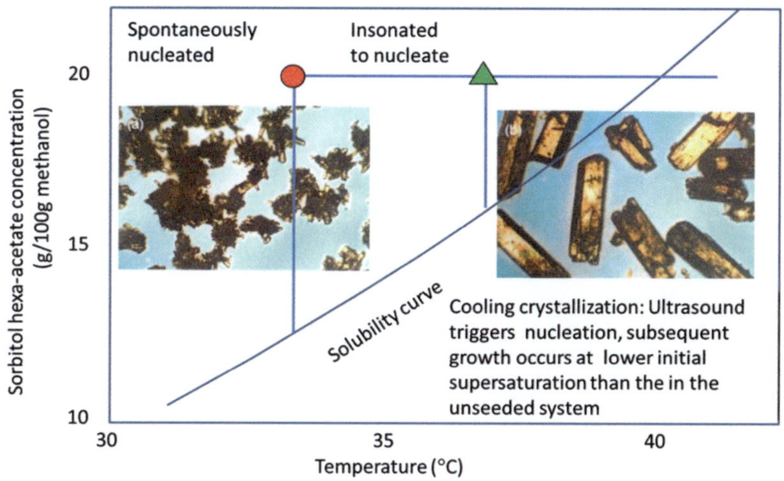

Fig. 18.6 Comparison between conventional un-seeded and sononucleated cooling crystallization of sorbitol hexa-acetate [16]

18.4.2 Seedless Seeding in a Sterile Environment

Sterile crystallization is undertaken in antibiotic manufacturing, but introducing seed crystals potentially threatens the sterility of the batch. However reliance on spontaneous nucleation is undesirable due to the high levels of supersaturation attainable in sterile solutions with very low levels of particulate contamination. When spontaneous primary nucleation occurs the nucleation rate is high and the product particle size is very small leading to subsequent challenges in filtration and drying. Sononucleation offers an attractive solution to this problem. This is exemplified for a beta lactam antibiotic [16].

Figure 18.7 shows the waiting time for nucleation as a function of supersaturation generated by anti-solvent addition for an array of aliquots of the same sterile antibiotic solution. Each vial represents an independent experiment. 50% of the vials were left to nucleate spontaneously and 50% were subjected to 2 sec of insonation. It is apparent that in the absence of insonation, the variance in waiting time and the waiting time itself, is much greater. Insonation reduces the waiting time and narrows the spread of waiting times. The supersaturation level required to ensure that nucleation will occur in every vial is much reduced by insonation. Further benefits arise through control of the particle size distribution.

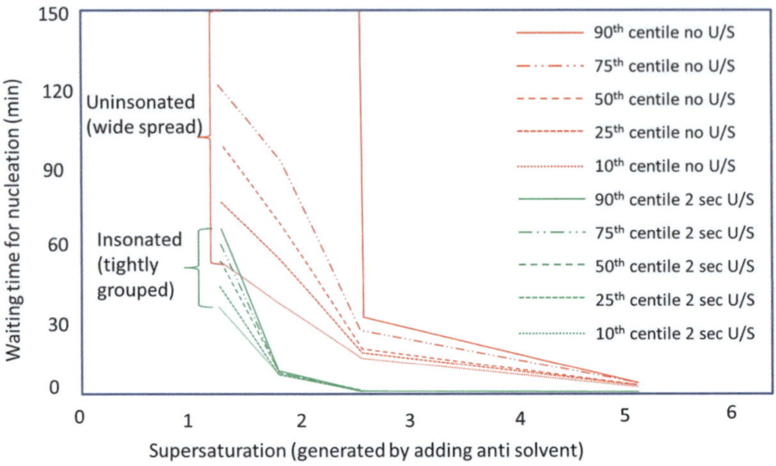

Fig. 18.7 Distribution of induction times of insonated and non-insonated sterile solutions as a function of supersaturation [16]

18.4.3 Sononucleation as a Replacement for Conventional Seeding in Batch Crystallizations

Seeding is routinely used in pharmaceutical batch crystallization processes. Applications include controlling the polymorphic form, in almost all instances this is the most stable polymorph. Seeding is also used to control the size distribution, either by seeding with a large quantity of fine particles so that the majority of growth is on these particles and secondary nucleation is minimised, or seeding with a small quantity of seed at a modest supersaturation such that secondary nucleation delivers product of approximately the target particle size. The drawbacks of seeding are that the seed crystals need to be prepared to a tight specification, and they need to be added to the crystalliser at the correct conditions of temperature and supersaturation level and effectively dispersed throughout the vessel.

Alternatively sononucleation may be employed as illustrated schematically in Fig. 18.8. Reference source not found., in the left hand diagram a side stream is withdrawn from the vessel, optionally cooled slightly to increase supersaturation, insonated and then circulated back into the vessel which acts as a store of nuclei until a defined number of nuclei is generated at which point the cooling phase of crystallization commences. The diagram on the right of Fig. 18.8 shows an alternative approach in which a defined volume of supersaturated solution is driven up into an insonator by pressurising the vessel slightly; the aliquot is then insonated to initiate nucleation and returned to the vessel by reducing the vessel head space pressure. With the addition of appropriate instrumentation such as a Lasentec FBRM this approach has the potential to deliver control over the product crystal size distribution.

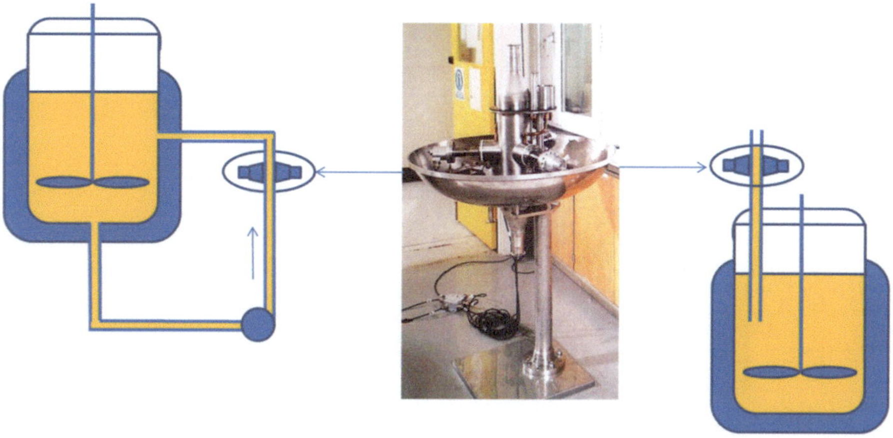

Fig. 18.8 Alternative configurations for batch sononucleation using an intrinsically safe acoustically shielded transducer array

18.5 Application of Experimental Design to Control Particle Size in Continuous Crystallization

Controlling the nucleation rate in continuous crystallization is essential in order to control the particle size distribution. The example shown in Fig. 18.9 is the basis of the sononucleation described in Chapter 19 of this work on continuous pharmaceutical crystallization [18]. The parameters investigated were the ultrasonic power, the acoustic contact area, and the duration of exposure (residence time). The concentration of the feed solution, and the crystallizer operating temperature were fixed. The responses monitored were yield and particle size distribution expressed as L_{50}. From this experimental design it was shown to be possible to control the particle size of this drug substance with great precision within the range of 10–22 μm. Additional experiments verified that it was possible to produce smaller particles under more forcing conditions and under less forcing conditions the process was shown to be vulnerable to the formation of encrustation and agglomeration during prolonged operation (all experiments were run until steady state operation was demonstrated for at least ten residence times).

18.6 Conclusion

Ultrasound offers a useful tool to control industrial crystallization processes. The most widely used application is to induce nucleation in a controlled fashion though other applications are described in the literature.

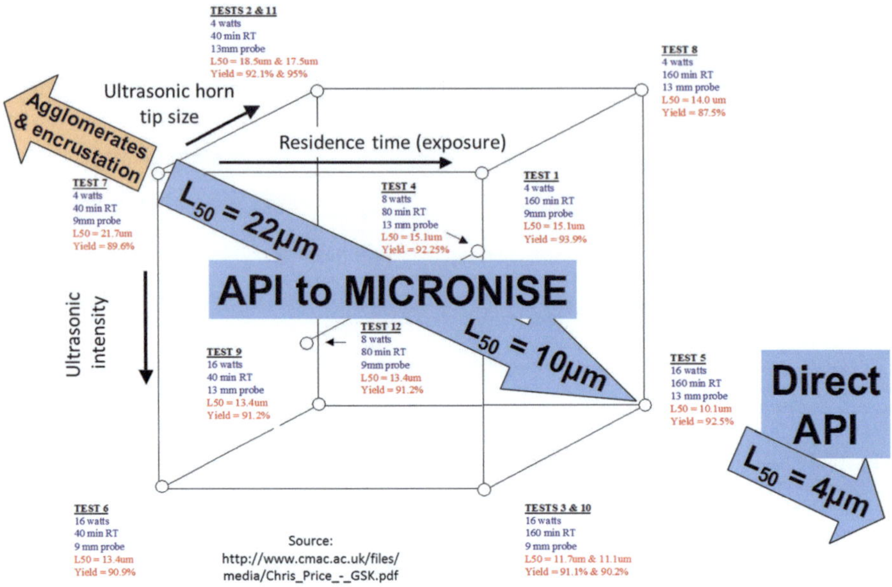

Fig. 18.9 The application of experimental design to evaluate sononucleation as a tool for the continuous crystallization of a novel pharmaceutical product [19]

References

1. Mason TJ, Lorimer JP (2002) Applied sonochemistry, the uses of power ultrasound in chemistry and processing. Wiley-VCH, Weinheim
2. Suslick KS, Price GJ (1999) Application of ultrasound to materials chemistry. Annu Rev Mater Sci 29:295–326
3. Price CJ, Martin PD, McCausland LJ, Ruddick S, Hobden I (1995) Sonocrystallization - designing better crystals for the pharmaceutical industry American institution of chemical engineers 1995 annual meeting p222b Miami Beach 12–17 November
4. Kim YH, Lee K, Koo KK, Shul YG, Haam S (2002) Comparison study of mixing effect on batch cooling crystallization of 3-nitro-1,2,4-triazol-5-one (NTO) using mechanical stirrer and ultrasound irradiation. Cryst Res Technol 37(9):928–944
5. Roberts CW, Hira SM, Mason BP, Strouse GF, Stoltz CA (2011) Controlling RDX explosive crystallite morphology and inclusion content via simple ultrasonic agitation and solvent evaporation. CrystEngComm 13:1074–1076
6. Ruecroft G, Hipkiss D, Fennell M (2005) Improving the Bayer process by power ultrasound induced crystallization of key impurities. The Minerals, Metals and Materials Society Light Metals, New York, pp 163–166
7. Wood RW, Loomis AL (1927) Physical and biological effects of high-frequency sound waves. Philos Mag 4:417–436
8. Umbdenstock RR (1955) Preparation of Procaine Penicillin. US patent 2,727,892 20th Dec 1955
9. Principe JR, Skauen DM (1962) Preparation of microcrystalline progesterone using ultrasound. J Pharm Sci 51:389–390
10. Hem SL (1967) The effect of ultrasonic vibrations on crystallization processes. Ultrasonics 5 (4):202–207

11. Hem SL, Skauen DM, Beal HM (1967) Mechanism of crystallization of hydrocortisone by ultrasonic irradiation. J Pharm Sci 56(2):229–233
12. Wagterveld RL, Boels M, Mayer MJ, Witkamp GJ (2011) Visualization of acoustic cavitation effects on suspended calcite crystals. Ultrason Sonochem 18(1):216–225
13. Zeiger BW, Suslick KS (2011) Sonofragmentation of molecular crystals. J Am Chem Soc 133 (37):14530–14533
14. Anderson HW, Carberry JB, Staunton HF, Sutradhar BC (1994) Crystallization of adipic acid. US 5471001 A 15 Dec 1994
15. Gracin S, Uusi-Penttila M, Rasmuson A (2005) Influence of ultrasound on the nucleation of polymorphs of *p*-amino benzoic acid. Cryst Growth Des 5(5):1787–1794
16. Price CJ (1997) Ultrasound the key to better crystals for the pharmaceutical industry. Pharmaceutical Technology Europe October 78–86
17. Guo Z, Zhang M, Li H, Wang J, Kougoulos E (2004) Effect of ultrasound on anti-solvent crystallization. J Cryst Growth 273:555–563
18. Cross WI, Hannan ML, Johns DM, Lee M-Y, Price CJ (2006) Novel crystalline pharmaceutical product. US 2009/0124585 A1 6 Apr 2006
19. Price CJ (2012) Developing pharmaceutical continuous crystallization processes - knowledge & gaps. Presented at the National Centre for Continuous Manufacturing and Crystallization annual conference September 2012. Available on line: http://www.cmac.ac.uk/files/media/Chris_Price_-_GSK.pdf Accessed 29th October 2015

Chapter 19
Continuous Pharmaceutical Crystallization from Solution

Christopher J. Price

Abstract The pharmaceutical industry has traditionally employed exclusively batch manufacturing for drug substances. The current trend is to explore continuous manufacturing; this chapter describes the process by which a continuous pharmaceutical crystallization process can be developed.

Keywords Continuous crystallization • Pharmaceutical crystallization • Process development • Nucleation • Growth • Purification • Crystal size distribution • Consistent product quality

19.1 Introduction

Batch crystallization dates from antiquity exemplified by the solar evaporation of salt, and crystallization of copper sulphate as illustrated in the 1556 wood cuts of Georgius Agricola [1]. However, continuous crystallization has been around for just over 100 years. The first patent for continuous crystallization was awarded in 1911 for the continuous crystallization of sugar [2]. Although most crystalline commodity chemicals are now produced continuously, in the pharmaceutical and fine chemicals sectors the vast majority of crystallizations are run batch wise.

Both a batch and a continuous crystallization process can be designed to share the same starting and end point conditions and produce the same product albeit with different attributes. However, the processes occurring are distinctly different. For example in a batch cooling crystallization the substrate is dissolved to form an undersaturated solution, filtered to remove insoluble impurities, cooled to generate supersaturation and then is usually seeded to initiate crystallization. A cooling profile is then applied to manage the desupersaturation achieved by crystal growth until equilibrium is approached at the chosen isolation temperature. Every aspect of the crystallization is transient with time making process analysis relatively complex

C.J. Price (✉)
Department of Chemical and Process Engineering, University of Strathclyde,
Glasgow G1 1QX, UK
e-mail: chris.price@strath.ac.uk

K.J. Roberts et al. (eds.), *Engineering Crystallography: From Molecule to Crystal to Functional Form*, NATO Science for Peace and Security Series A: Chemistry and Biology, DOI 10.1007/978-94-024-1117-1_19

and leading to batch to batch variations in product attributes. In the corresponding continuous crystallization, hot undersaturated solution of the same composition and temperature as the starting point of the batch process, is fed into the crystallizer. However, the conditions inside the crystallizer are those of the end point, thus if the input solution cools rapidly as it enters the crystallizer there is a risk of spontaneous nucleation. It is therefore essential that this hot concentrated solution is mixed rapidly into the bulk volume achieving dilution along with cooling. This is one of several design challenges encountered in continuous crystallization development.

Continuous crystallizers are designed to operate at a steady state where the feed rate matches the offtake rate and the quantity of substrate entering matches the quantity leaving, the only difference being that the bulk of the solute enters in solution but leaves as suspended crystals. To achieve this, nucleation and crystal growth must both take place in such a way that the product crystal size distribution remains constant leading to a steady state population of crystals. The underlying principles of modelling particle size distribution in continuous processes are described in [3].

The advantage of a batch process is that a relatively simple general purpose crystallizer vessel may be used. This can be easily adapted for different products, relatively long residence times are possible, and there is a wide body of knowledge and experience to draw from since the number of products prepared this way is large.

The main disadvantages of batch crystallization are that conditions within the crystallizer are continually changing making process monitoring and control complex. In order to minimise batch to batch variation the same crystallization conditions must be replicated time after time. This is especially important to ensure a consistent particle size distribution. Ultimately this might require inter batch cleaning of the crystallizer as small residues of product act as seeds for the next batch and change the particle size distribution of the product.

Traditionally the main advantages of continuous crystallization are associated with the economies of scale. The heating and cooling cycles and materials handling aspects of multiple batch processes, plus the attendant monitoring, render it inefficient when large quantities of product are required. When the process proceeds continuously at steady state, the product and rate of production are more consistent, and monitoring is easier.

One of the main challenges of continuous crystallization is encrustation of the vessel walls. This directly affects heat transfer and has far reaching consequences for product quality. Eventually all continuous crystallization processes have to be halted due to encrustation.

A substantial body of product specific data on the crystallization process is required to design key elements of a continuous crystallization plant whereas a batch process can be adapted to fit into a general purpose vessel. Virtually all continuous crystallization plants are designed around a specific product, making them quite inflexible, in most cases they would need significant reconfiguration for use with other products.

Conventional thinking suggests that a continuous process should be considered for high volume products (at least 250 kg/h) where consistency of product is essential and for which there is a long term market. If the crystallization requires the evaporation of solvent, this is dealt with more easily under continuous operation since evaporation from a batch changes the process conditions and level changes lead to encrustation which is very problematic. The process must be well understood for successful continuous operation.

Industrial experience of continuous crystallization over the last 100 years has confirmed that minimising encrustation is vital, therefore continuous processes tend to operate at lower supersaturations than batch processes to minimise encrustation. A further advantage of operation at low supersaturation is that slow growth favours purer crystals. Many continuous crystallization processes for commodity chemicals also aim to suppress secondary nucleation as far as possible to favour growth on existing crystals. Again this is favoured by operation at low supersaturation in order to produce large crystals which can easily be separated from the impurity rich mother liquors.

Scale up and down is difficult for continuous crystallization processes operating with a relatively small growing crystal surface per unit volume. Successful design of a continuous crystallization process requires a large body of data collected at relatively large scale. Operating at small scale is difficult. As a result most large scale crystallizers are of bespoke usually proprietary designs created by a small number of specialist equipment suppliers. Examples include the GEA group which combines the experience acquired by two major European manufactures prominent in the twentieth century: OSLO and MESSO, while another major player in suspension crystallization is the US based company Swenson which is best known for draft tube baffled crystallizers. Whilst melt crystallization is not widely practiced in the pharmaceutical sector, the situation is similar in that there are a small number of major suppliers; Sulzer who provide melt static, falling film and melt suspension crystallizers; and Armstrong Chemtec providers of scraped surface crystallizers.

19.2 Designing a Classical Continuous Crystallization Process

There are many factors to be considered when designing a continuous crystallization process, these are illustrated in Fig. 19.1. Practical guidance based on large scale industrial experience can be found in [4] and is based on EI DuPont processes.

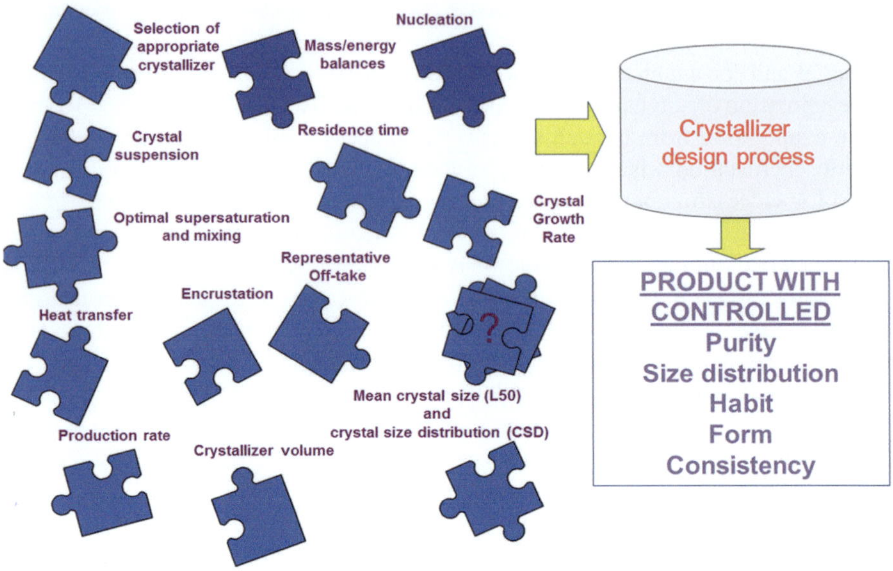

Fig. 19.1 Aspects to consider when designing a continuous crystallizer

19.3 Crystallizer Selection

The type of crystallizer will depend on the method being used to generate supersaturation. Continuous crystallizers are designed and optimised for mode of supersaturation generation and can be classified accordingly. If supersaturation is generated by evaporation, then a boiling surface is required. If it is generated by cooling, a heat exchange mechanism is required, this could be a heat exchange surface or cooling could be achieved by evaporation.

19.4 Optimal Supersaturation and Mixing

Supersaturation is typically generated in two locations; the point where the feed solution is introduced to the bulk suspension and at the supersaturation generating surface eg heat exchange surface, boiling surface or at the interface between the bulk solution and an added reagent such as an anti-solvent. Generating substantial local supersaturation at these points can result in encrustation, high nucleation rates and small crystals. Determining the evaporation rate, heat exchanger sub cooling or addition rate is therefore a key element in the process and along with the agitation rate governs the rapid distribution of feed solution throughout the bulk suspension.

19.5 Crystal Suspension

The agitation rate should be sufficient to keep all the crystals suspended and in motion preventing settling but should not be high enough to cause significant secondary nucleation. Some crystallizer designs deliberately segregate product crystals based on particle size. One example requires crystals to settle against an up flow which is set such that only crystals matching or exceeding a critical size and settling rate can leave the crystallizer, this technique is called classified product removal or elutriation. Another design strategy is to remove fine particles caused by excessive secondary nucleation using a settling zone which allows the majority of suspended fine particles to be removed in a recycle stream, re-dissolved and returned to the crystallizer in the feed solution. This is known as fines destruction and increases the mean product particle size distribution. This can also be seen as a method to reduce the apparent nucleation rate.

19.6 Nucleation, Growth and Size Distribution

Many continuous crystallizers are seeded on start up to avoid spontaneous nucleation at high supersaturation levels as this usually leads to encrustation forming shortly afterwards. The slurry density (mass of crystals per unit volume of suspension), agitation rate, agitator design and degree of supersaturation combine to influence the effective nucleation rate. In fact nucleation in a continuous crystallizer involves a complex interplay of crystal-crystal collisions, crystal-vessel collisions and crystal-agitator impacts. The rate of desupersaturation due to growth depends on the available surface area of crystals per unit volume of suspension and the supersaturation level and the growth kinetics under the prevalent conditions of temperature, impurity loading etc. The impact of a range of supersaturation levels on the growth of indomethacin are described in [5]. The combination of nucleation and growth rates together define the particle size distribution which may also be manipulated by removal of large particles and fines, as described in the text relating to particle suspension. However, excessive manipulation of the crystal size distribution by removal of nuclei and fine particles can leave insufficient crystal surface area for growth to deplete the supersaturation in a reasonable time and presents a significant risk to steady state operation unless the throughput is reduced by increasing the residence time. The underlying message is that the product size is determined by nucleation and growth linked to supersaturation and residence time and may be manipulated by classification and recycle. However, the capacity to modify PSD may be limited as excessive intervention may prevent steady state being reached.

19.7 Residence Time, Production Rate and Crystallizer Volume

Once nucleation and growth are optimised to deliver the target crystal size distribution by setting the crystallizer residence time the throughput or production rate are fixed by selecting the vessel volume.

If the desired crystal size distribution is specified within narrow limits, it is generally not possible to increase the throughput more than a very limited amount because the residence time and hence the product crystal size distribution will be affected. Continuous crystallizers are usually part of a bigger continuous manufacturing process so the scope to adjust throughput may be limited by the upstream and downstream processes or conversely the crystallizer throughput may dictate the flow rate for the entire process train.

19.8 Mass and Energy Balance and Heat Transfer

Mass and energy balances are important in all aspects of process design but a particular concern for continuous crystallization is the effect of sampling during development studies. Sampling may represent a significant process purge which may mask the build-up of impurities in the slurry as fines and mother liquor are recycled into the feed.

If the heat exchange surfaces are insufficient for the slurry volume and feed rate in a cooling crystallizer, encrustation is the likely result, which could render the crystallizer inoperable. It is essential therefore to determine the level of sub cooling a heat exchanger can support before encrustation occurs. Typically this is much less than the metastable zone width expressed as a sub-cooling.

19.9 Minimising Encrustation

Minimising encrustation minimises the down time of the crystallizer reducing the impact on the overall productivity of the process and reducing the quantity of product which is outside of specification limits (associated with start-up and shut-down) requiring reprocessing. Understanding the mechanisms of encrustation formation is an essential prerequisite to its mitigation [6]. Encrustation will definitely occur on surfaces where the local supersaturation exceeds the metastable zone width, however, it can occur at much lower supersaturation levels, which is one of the reasons why most continuous industrial crystallizers operate at rather low supersaturation levels compared with equivalent batch processes. It is important to keep the solution and particles moving to achieve a uniform distribution of supersaturation and particles throughout the crystallizer. The mixing which suspends the particles also serves to minimise encrustation by achieving an even distribution of supersaturation. However, excessively intense local mixing may generate an excessive level of nucleation so a compromise is required, which is typically achieved by appropriate agitator design and positioning. There are a number of well-established encrustation mitigation strategies, for example in evaporative crystallizers encrustation is common just above the boiling surface in the vapour disengagement zone and this is addressed by using a clean in place spay ball to periodically remove encrustation from the vessel walls. A strategy for modelling encrustation in plug flow crystallizers is presented in [7] this is potentially more problematic than in classical stirred vessels. They also present a mitigation strategy based on applying pulses of hot fluid to dissolve material from the vessel walls.

19.10 Representative Off-Take

If the objective is to design a classic mixed suspension mixed product removal crystallizer, the off-take fluid velocity and position must be such that the composition of the offtake stream is representative of the material in the body of the crystallizer. To check that this is the case, a wash-out test may be performed. The crystallizer is fed with saturated solution at the end-point temperature for an extended period during which the off-take should not show any change in particle size distribution even though the suspension density reduces.

There are other design considerations specific to particular continuous crystallizer sub types, this list is not exhaustive. For further information consult an industrial crystallization textbook, for example [4].

19.11 Example Equipment

Figure 19.2 shows the design of a typical Swenson continuous draft tube baffled evaporative cooling crystallizer and the associated phase diagram [8]. The solid line linking points A to E on the phase diagram represents the path in temperature – concentration space taken by an aliquot of solution as it circulates around the crystallizer. The same letters correspond to locations shown on the diagram of the crystallizer. Hot feedstock enters the crystallizer at point A where it mixes with material taken from the upper outlet of the skirt baffle from the body of the vessel. The skirt baffle serves to separate coarse particles which are retained in the body of the crystallizer from the fine crystals which are carried in the mother liquor to be reheated and dissolved before re-entry into the crystallizer. The volume flow of the feed stream is small relative to the fines containing stream. The two streams meet immediately prior to the pump which mixes them intimately and directs the flow into the heat exchanger where the solution becomes slightly undersaturated such that the fine crystals are dissolved. This solution meets the bulk of the circulating

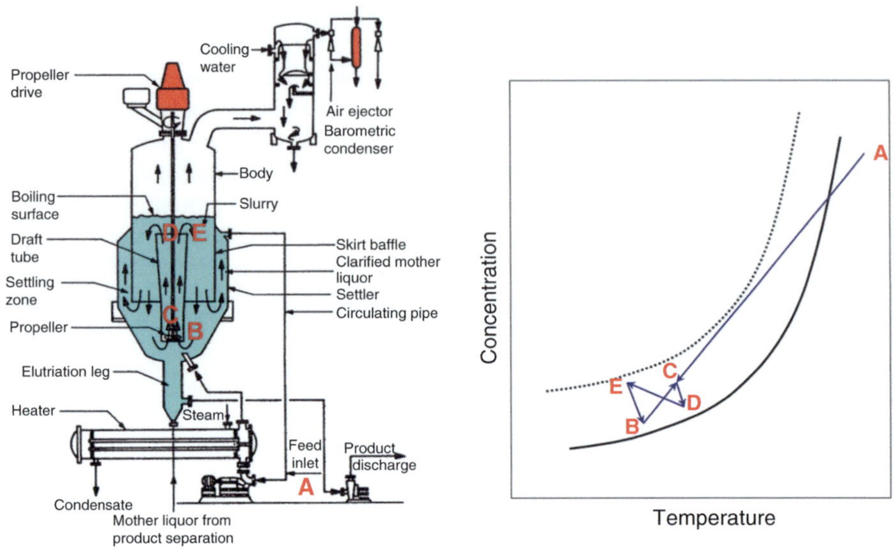

Fig. 19.2 Process composition and temperature trajectory, after [8], as based on a typical Swenson continuous draft tube baffled evaporative cooling crystallizer (*left*) together with the associated phase diagram (*right*)

suspension of composition and temperature B and is directed into the main agitator which drives the circulation of material in the crystallizer and delivers rapid mixing. Again the flow entering the vessel is small relative to the circulation around the vessel such that the hot concentrated stream is diluted in to a much greater volume avoiding a peak in supersaturation. The mixed stream of composition and temperature at C is directed at the boiling surface, and during the transit up the draft tube some crystal growth occurs resulting in the liberation of the heat of crystallization warming the stream slightly to achieve composition and temperature D. This is accompanied by a decrease in concentration as supersaturation is relieved. The crystallizer operates at a reduced pressure to reduce the boiling point, with the pressure reduction being achieved using an air ejector and condenser. The height of the boiling surface above the agitator is sufficiently large to prevent boiling around the agitator as this would result in cavitational damage.

At the boiling surface, solvent evaporates so the solute concentration increases and energy is consumed so the temperature falls (evaporative cooling). This is represented by point E. The suspension continues to be circulated through the annular space around the draft tube and as the solution reaches the base of the crystallizer at B further crystallization occurs so the concentration falls and the temperature increases slightly. A small proportion of the slurry from E becomes trapped behind the skirt baffle where it is recycled as described earlier. The bulk of the volume of the suspension proceeds to point B where it is again mixed with further feed from A; the process continues in this cycle. These changes in the concentration/temperature profile are actually very small; the "bow tie" component of the diagram is exaggerated for clarity. The points B, C D and E are all far below the metastable limit depicted by the dotted line on the phase diagram. Product is discharged from an outlet at the base of the elutriation leg allowing a classified product removal containing only particles above the cut-off size defined by their ability to settle faster than the inlet flow velocity of mother liquor from the downstream product isolation unit.

19.12 Other Classical Designs

The same approach can be used to develop an understanding of other crystallizer designs for example GEA's Oslo and forced circulation crystallizers. The main basis for selection between these competing designs is the ease of attaining the desired particle size distribution which is in turn dominated by the extent to which secondary nucleation can be supressed. The largest crystals are produced by the Oslo fluidised bed designs where there is no agitator in the crystallizer body to cause secondary nucleation, the draft tube baffled design addresses secondary nucleation using the classification and fines dissolution. Forced circulation crystallizers are less effective in supressing nucleation and so produce smaller particles. All evaporative crystallizers share at least one design element in common, the height of boiling surface is sufficiently far above feed points, agitators and pumps to prevent boiling in their vicinity. This is a significant barrier to scale down. The major vendor's pilot units tend to adopt a tall tubular design to overcome this but it is challenging to operate a pilot unit with less than 100 L of suspension. The main application of these crystallizer designs has traditionally been to produce inorganic salts from aqueous solutions where a major consideration is the energy cost of evaporating water. Additional insight can be gathered from [9] which provides an equipment vendors' perspective linking the crystallization processes and principles of operation with the most widely used industrial crystallizer configurations.

19.13 Application to Pharmaceutical Processes

The drivers for continuous crystallization in the pharmaceutical industry differ from those of the commodity chemical industry. The scale of operation is much smaller; the annual demand for most pharmaceuticals is less than 10 tons per year and the market size may be less stable. This can be summarised in a pharmaceutical industry user requirement; A multi-product, scalable, continuous particle formation & purification system delivering consistent API suitable for direct formulation. The material produced needs to be chemically pure and in the required polymorphic form. The crystal size distribution, crystal habit and bulk powder properties should be suitable for direct formulation. The crystallization processes developed on this platform must be suitable for quality by design regulatory filings using in line process monitoring and control. All attributes of the product need to be consistent during scale up or scale down and over the product life cycle. The equipment should be adaptable to most new chemical entities without major equipment redesign or reconfiguration.

Many of the large pharmaceutical companies are actively investigating continuous suspension crystallization with these goals in mind. This is well illustrated by the seven white papers presented at the 2014 International Symposium on Continuous Manufacturing of Pharmaceuticals which addressed implementation, technology & regulatory aspects [10]. There are two main approaches under consideration:- cascades of small continuous stirred tank reactors and oscillatory baffled crystallizers exemplified by those produced by Ni Tec and Cambridge Reactor Design.

19.14 Developing a Continuous Crystallization for a Pharmaceutical Product

19.14.1 Version, Form and Solvent Selection

The process for selecting the solid state version (salt and polymorph) is similar to that used for batch crystallization though crystallizing continuously across a phase boundary or targeting a metastable form poses additional risk over batch processes as the extended duration of operation increases the probability of the most stable form nucleating whereas in a seeded batch process it may be possible to manufacture a metastable form. Solvent selection in pharmaceutical development is constrained by the propensity for drug candidates to form solvates and the level of solvent residues in the final product which must comply with regulatory acceptance criteria for residual solvents set by the US FDA [11]. In continuous crystallization solvent choice is further restricted by the need to link directly to the preceding synthetic steps which may necessitate hybrid approaches to supersaturation generation such as combining cooling and anti-solvent addition to ensure the required yield.

19.14.2 Small Scale Operation

One of the acknowledged challenges of continuous crystallization is operation at small scale. This poses a particular challenge in pharmaceutical development where the available supply of the candidate drug substance and its precursors is likely to be extremely limited in the early phases of development. This necessitates working at small scales perhaps as low as 1 or 2 g/h in order to demonstrate robust continuous operation for sufficiently many residence times to achieve and maintain steady state and to understand the propensity to encrust.

19.14.3 Particle Size Control

Manufacturing size appropriate particles suitable for direct formulation is an attractive opportunity for continuous pharmaceutical manufacture. The particle size requirement is driven by the need to achieve content uniformity in the final formulation – no particle should represent more than 0.1% of the dose for tabletted products which sets an upper limit on the largest particle which may need to be verified by screening. For a typical drug substance this suggests an x_{50} of between 20 and 100 μm. To attain these sizes in a continuous crystallizer requires either a constant feed of seed particles/nuclei or a significant secondary nucleation rate within the crystallizer. High secondary nucleation rates may be achieved relatively easily at high supersaturation but this is not compatible with other objectives including purification and avoidance of encrustation. As a result specific interventions may be needed to manipulate the nucleation rate to attain the desired product crystal size distribution. This is exemplified by a small scale very high sheer system to prepare fine particles of glycine and L-alanine [12]. This could also be achieved using a rotor stator wet mill either in the first crystallizer of a cascade or a side stream feeding back to the first crystallizer. The use of ultrasound is also appropriate [13]; this is presented in detail in Chapter 18.

19.14.4 Purification

Purification is clearly an important objective when crystallizing pharmaceuticals. In pharmaceutical development it is essential to establish the purification capability of the crystallization early and to focus on problematic impurities which are present at levels high enough for the crystallization to struggle to deliver the required purification. Once such impurities are identified it is possible to target them throughout the synthesis, minimising precursors in starting materials, and conducting reactions under conditions which minimise their formation relative to the target substance. This is an area where continuous synthesis may be preferable to batch due to access to a wider range of reaction conditions and tighter process control. In this way it is possible to reduce the impurity burden coming into the crystallizer. Operation at lower supersaturations than typical in batch crystallizations further enhances purification. The final contribution comes from optimising the washing of the product crystals. In this way it is possible in a single continuous crystallization to match or exceed the purification achieved in multiple batch crystallizations. This approach is shown in Fig. 19.3, the upper pair of images shows the HPLC trace of feed solution and product after continuous sonocrystallization. The expansion of the

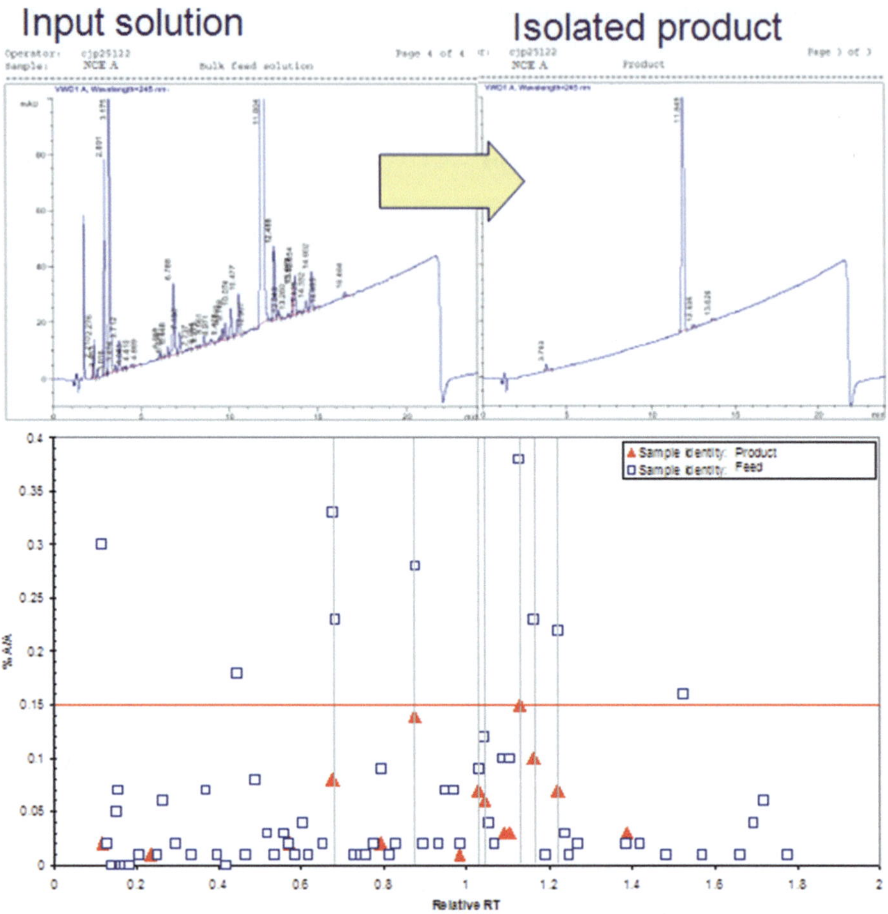

Fig. 19.3 HPLC chromatograms of feed stream and isolated product demonstrating the purification capability of a continuous crystallization process [14]

chromatogram in the lower panel provides further detail. The open squares represent impurities in the feed solution, the closed triangles represent those which remain in the product. The horizontal line set at 0.15% a/a is the target impurity limit, it is apparent that most impurities are readily purged by the crystallization however there are two impurities of concern where the high concentration in the feed solution and the modest level of removal achieved by the crystallization means that they need to be tracked through the synthesis and controlled as described earlier.

Combining these approaches and establishing robust operation without significant encrustation provides a crystallization which can be scaled up and operated continuously for extended periods.

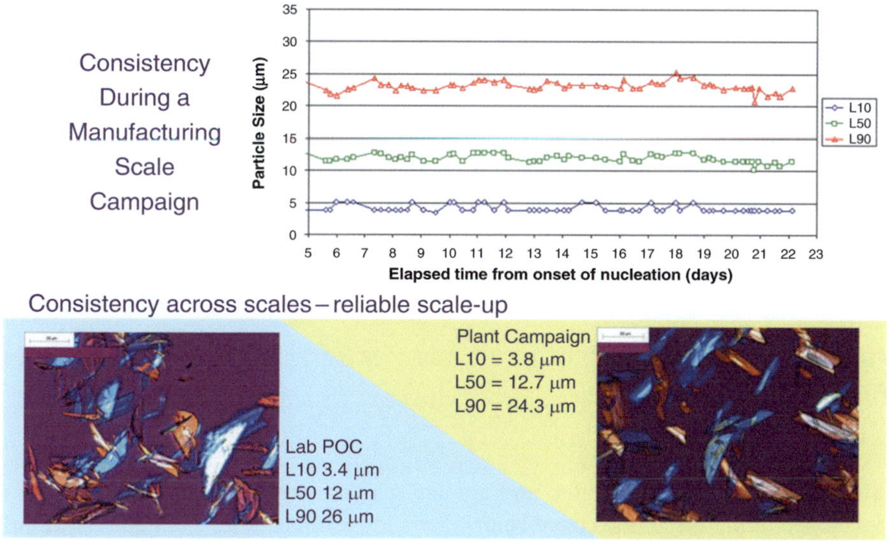

Fig. 19.4 Evidence of consistency of particle size across scales of operation and for extended periods of manufacture [14]

Figure 19.4 shows the consistency of the particle size distribution over extended operations at the manufacturing scale.

This can be articulated as a strategy for the continuous crystallization of pharmaceuticals and fine chemicals:

Enhancing the Nucleation Rate Independently of Growth Rate. This allows control of the particle size distribution to make particles suitable for direct formulation. Combining this with operation at low supersaturations minimises encrustation. This delivers crystals of higher purity and typically with less extreme crystal habits than a conventional batch process.

Operating with a Large Crystal Surface Area of Small Particles. An additional benefit of producing small particles directly is that the process intensity is high compared with traditional monolithic continuous crystallizers allowing a high mass deposition rate at modest supersaturations. Thus the equipment is small compared with conventional continuous crystallizers.

Avoid the Hydrodynamic Constraints Encountered in Conventional Continuous Crystallizer Design. Small organic crystals with sizes appropriate for direct formulation tend not to settle rapidly in their mother liquors. Whilst this precludes the use of settling zones associated with the formation of large particles in classical crystallizers it also frees the design from the associated hydrodynamic constraints. This circumvents compound specific crystallizer design.

This approach described in [13] has been proven for cooling, anti-solvent and reactive continuous crystallizations to deliver chemical and phase pure active pharmaceutical ingredients. The output crystals exhibit consistent particle attributes

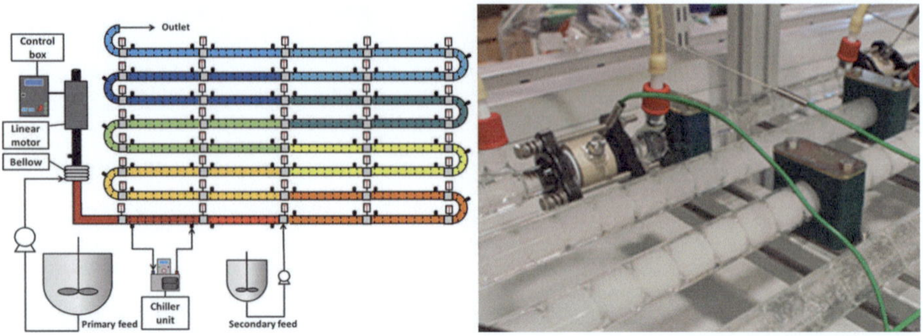

Fig. 19.5 Schematic of CMAC COBC for cooling crystallization (Source: CMAC research program [16])

which can be tightly controlled and are suitable for direct formulation. Processes developed along these lines have been shown to be robust over extended durations and over a range of scales. The approach appears to be applicable to most drug candidates without major equipment redesign or significant reconfiguration. Because the process is continuous it is well suited to on or at line process monitoring and control and so is appropriate for Quality by Design regulatory filing.

An alternative approach to the cascade of continuous stirred tank reactors (CSTRs) described above is the continuous oscillatory baffled crystalliser (COBC) which comprises a series of interconnected small cylindrical chambers contained within a heat exchanger tube for example [15, 16] see Fig. 19.5. Material is pulsed forward and backwards through the chambers, and the resulting oscillatory flow generates turbulence due to the design of the boundary plate which separates adjacent chambers. This turbulence ensures that the particles are suspended uniformly, the solution composition is uniform and temperature gradients are minimised. The residence time is determined by the solution feed rate entering the system. A supersaturation profile can then be established along the length of the crystallizer for example by setting up a cooling profile. In principle the same approaches used to control crystal size distribution in a cascade of CSTRs can be applied in a COBC crystallizer.

Batch ways of working are deep-rooted in the pharmaceutical industry, and this represents a further barrier to implementation of continuous processes. Batch processes continue to be effective in delivering purification and crystalline products which can be successfully formulated. In a risk averse sector the benefits of enhanced consistency smaller equipment footprint and elimination of dry particle size reduction may not be enough to make a compelling case for implementation. For an organisation with an established batch manufacturing capacity, the capital write off may be too large to justify for a single product so a strategic decision is needed. In fact companies with an outsourced supply chain may be best placed to exploit this technology in the near term. The economic aspects are considered further in [17]. Crystallization is a small but vital part of the whole pharmaceutical manufacturing process, which, if continuous and with appropriate manipulation, can deliver consistent particles for direct formulation, reducing complexity in the subsequent formulation process.

References

1. Agricola G (1556) De Re Metallica. Froben, Basel
2. Kestner P (1911) Process of crystallizing sugar US patent 989366 11 April 1911
3. Randolph A, Larson M (1998) Theory of particulate processes analysis and techniques of continuous crystallization, 2nd edn. Academic, San Diego
4. Mersmann A, Rennie FW (2001) Operation of crystallizers. In: Mersmann A (ed) Crystallization technology handbook. CRC Press, Hoboken
5. Patel DD, Anderson BD (2013) Maintenance of supersaturation II: Indomethacin crystal growth kinetics versus degree of supersaturation. J Pharm Sci 102(5):1544–1553
6. Vendel M, Rasmuson Å (1997) Mechanisms of initiation of incrustation. Am Inst Chem Eng J 43(5):1300–1308
7. Majumder A, Nagy ZK (2015) Dynamic modeling of encrust formation and mitigation strategy in a continuous plug flow crystallizer. Crystal Growth Des 15(3):1129–1140
8. Whiting Equipment Canada Inc. (2015) Swenson crystallization equipment. Available on line: http://crystallisation.pbworks.com/f/Swenson+Crystallization+Equipment.pdf. Accessed 17 Oct 2015
9. Samant KD, O'Young L (2006) Understanding crystallization and crystallizers. Chem Eng Prog 102:28–37
10. Baxendale IR, Braatz RD, Hodnett BK, Jensen KF, Johnson MD, Sharratt P, Sherlock J-P, Florence AJ (2014) Achieving continuous manufacturing: technologies and approaches for synthesis, work-up and isolation of drug substance. Presented at The International Symposium on Continuous Manufacturing of Pharmaceuticals MIT 20th 21st May 2014
11. US Food and Drug Administration (2009) Guidance for industry residual solvents in drug products marketed in the United States Office of Pharmaceutical Science in the Center for Drug Evaluation and Research (CDER)
12. Igarashi K, Yamanaka Y, Azuma M, Ooshima H (2012) Control of crystal size distribution using a mL-scale continuous crystallizer equipped with a high speed agitator. J Chem Eng Jpn 45(1):28–33
13. Cross WI, Hannan ML, Johns DM, Lee M-Y, Price CJ (2006) Novel crystalline pharmaceutical product. US 2009/0124585 A1 6 Apr 2006
14. Price CJ (2012) Developing pharmaceutical continuous crystallization processes – knowledge & gaps. Presented at the National Centre for Continuous Manufacturing and Crystallization Annual Conference September 2012. Available on line: http://www.cmac.ac.uk/files/media/Chris_Price_-_GSK.pdf. Accessed 22 Oct 2015
15. Lawton S, Steele G, Shering P, Zhao L, Laird I, Ni X (2009) Continuous crystallization of pharmaceuticals using a continuous oscillatory baffled crystallizer. Org Process Res Dev 13 (6):1357–1363
16. McGlone T, Briggs NEB, Clark CA, Brown CJ, Sefcik J, Florence AJ (2015) Oscillatory flow reactors (OFRs) for continuous manufacturing and crystallization. Org Process Res Dev 19 (9):1186–1202
17. Schaber SD, Gerogiorgis DI, Ramachandran R, Evans JMB, Barton PI, Trout BL (2011) Economic analysis of integrated continuous and batch pharmaceutical manufacturing: a case study. Ind Eng Chem Res 50(17):10083–10092

Chapter 20
Viedma Ripening and Its Role in the Chiral Separation of Optical Isomers

Martin Iggland, Giovanni Maria Maggioni, and Marco Mazzotti

Abstract It has recently been observed that a mixture of two enantiomeric crystals subjected to attrition in a solution containing a racemizing agent undergoes symmetry breaking, with only one of the enantiomers eventually remaining in solid form. This process is believed to occur due to the interplay between racemization in solution, attrition, agglomeration and crystal growth and dissolution caused by the crystal size dependence of solubility. This process is called Viedma ripening and is one of the most fascinating and complex crystallisation processes. In this work, we present a commented review of the most recent literature on Viedma ripening and related topics.

Keywords Chirality • Enantiomers • Deracemisation • Viedma ripening • Modelling

20.1 Introduction

A few years ago, Viedma [1] observed experimentally the transformation of a mixture of both enantiomeric forms of crystalline sodium chlorate to a suspension containing only one enantiomer, achieved simply by stirring an aqueous suspension in the presence of glass beads. Noorduin et al. [2–4] have observed the same behaviour in a system containing the chiral molecule N–(2–methylbenzylidene)–phenylglycine amide (NMPA) and a racemising agent 1,8-Diazabicycloundec-7-ene (DBU). Since these first reports, similar behaviour has been reported for other achiral substances forming chiral crystals [2, 5] as well as for intrinsically chiral substances [5–10].

This process is conceptually different from preferential crystallisation, since it involves starting with a suspension in contact with a saturated solution, rather that the simple transfer of solute from the liquid phase to the solid phase. The

M. Iggland • G.M. Maggioni • M. Mazzotti (✉)
Institute of Process Engineering, Eidgenössische Technische Hochschule Zurich, CH-8092 Zurich, Switzerland
e-mail: marco.mazzotti@ipe.mavt.ethz.ch

K.J. Roberts et al. (eds.), *Engineering Crystallography: From Molecule to Crystal to Functional Form*, NATO Science for Peace and Security Series A: Chemistry and Biology, DOI 10.1007/978-94-024-1117-1_20

requirements to be fulfilled for this deracemisation process to work are that there is a racemisation reaction in solution and that the substance forms a conglomerate. The applicability of Viedma ripening can be extended by using various tricks to obtain a conglomerate. For example, there is the possibility of working with a metastable conglomerate [11] if the substance does not form a stable conglomerate, or the substance can be transformed into a salt which forms a conglomerate [12]. The process has also been demonstrated to work on a system exhibiting epitaxial growth [13]. The name Viedma ripening for this phenomenon was suggested by Noorduin et al. [14], but the process is often also referred to as attrition-enhanced deracemisation.

Recently, it has been shown that the grinding achieved by using glass or metal beads in the conventional Viedma ripening protocol can be replaced by temperature cycles [15], which represents a major practical improvement. The parameters affecting the rate of change of the enantiomeric excess in the solid phase have been investigated experimentally. Increased attrition by addition of glass beads or an increased stirring rate leads to a decrease in the time required to achieve complete chiral conversion [1, 4, 5, 8, 16], as does an increased racemisation rate [4]. It has been reported that the evolution can be directed towards a specific enantiomer using chiral additives [2, 17]. However, it has been found that a small enantiomeric excess, or a size imbalance, is sufficient to direct the system towards the chiral form initially in excess [1, 2, 5, 16].

20.2 The Mechanisms Responsible for Viedma Ripening

The mechanism of this symmetry breaking process has been the subject of intense debate, and there is still uncertainty about the details, as recently discussed in the literature [18]. The mechanism used in the PBE model presented here is based on the propositions of several authors [1, 14, 19, 20], and includes racemisation in solution, a size-dependent solubility, agglomeration and attrition, as shown in Fig. 20.1a. The influence of these mechanisms on particle size and solution concentration is shown schematically in Fig. 20.1b.

Racemisation in solution is often due to the action of a catalyst and is responsible for the conversion of solute molecules of one enantiomer into the other. The case of achiral molecules yielding enantiomeric crystals is equivalent to that of a chiral compound undergoing an infinitely fast racemisation in solution. Attrition leads to a reduction in crystal size and the creation of small crystals having the same chirality as their parent crystal. These small crystals have a higher solubility than larger crystals, and therefore dissolve.

The size-dependence of solubility is a thermodynamic effect, which is also responsible for the process known as Ostwald ripening. Agglomeration of small crystals with larger crystals of the same chirality leads to an increase in size and a decrease in the total number of crystals. Neither attrition nor agglomeration affect the concentration in solution directly, but can cause a crystal to jump between the

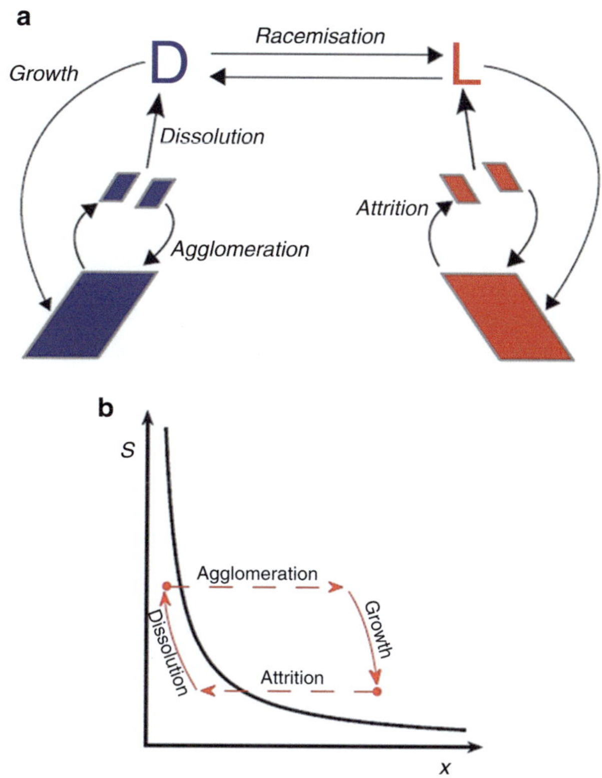

Fig. 20.1 (**a**) Schematic representation of Attrition-enhanced deracemisation. (**b**) The influence of the various phenomena on particle size and solubility

dissolution and the growth regime. Since agglomeration involves two particles, it has a higher order dependence on particle concentration than growth or attrition – in other words, agglomeration is much faster for the population whose crystals are present in a larger number [19]. The relatively simple experimental protocol of this process gives it potential to become a valuable alternative method for chiral resolutions.

The knowledge required to design a robust process is limited: one only needs to be sure that the substance crystallises as a conglomerate and that the substance can be racemised in solution. If these two requirements are fulfilled, Viedma ripening can be employed, leading to a theoretical yield of 100% due to the racemisation reaction. In principle, the solubility of the substance does not need to be known, and the direction of evolution can be controlled by starting with an asymmetry towards the desired enantiomer. Furthermore, the productivity of a process based on Viedma ripening is potentially very high, since the amount of solids which can be obtained in one batch is not limited by the solubility of the substance.

20.3 Modelling Viedma Ripening

However, in order to make control, optimisation and scale-up possible, a deeper understanding of the mechanisms and of the effects of process parameters is required. Mathematical models can aid in this understanding. More importantly, a detailed mathematical model which can successfully reproduce experimental observations is a way [13] to demonstrate that the mechanism is plausible. Several models have been presented earlier in the literature. Uwaha [27] proposed a reaction-type model to explain the evolution towards a homochiral state, considering only the total concentration of crystals of each configuration, monomers in solution and chiral clusters. This model does reproduce some experimental observations, but, due to its assumptions, its ability to describe size-dependent effects is very limited [21, 22]. Noorduin et al. [14] have adapted this model in an attempt to explain experiments carried out without the racemisation reaction. Saito and Hyuga [23] have also presented a reaction-type model, including crystals of various sizes. However, the largest cluster is made up of only 6 monomer units. Wattis [24] used these models, making various simplifications to be able to analytically investigate the model behaviour. Noorduin et al. [20] modelled the chiral conversion process using a Monte-Carlo type model including attrition effects and an exchange of monomers between two crystals, representing Ostwald ripening. However, the critical size as a function of concentration is not considered. All monomer exchange takes place directly and not via the solution, even when there is a racemisation step in between. Agglomeration is not included in the model. Another Monte-Carlo type model was presented by Cartwright et al. [25]. In this model, crystals are not present initially, but are created by primary nucleation. Attrition reduces the size of clusters, by creating new crystals of the same chirality as the parent crystal. Ostwald ripening is assumed to cause the dissolution only of crystals in the smallest size range. Uwaha and Katsuno [26] use a Monte-Carlo model in order to investigate whether Ostwald ripening is essential for attrition-enhanced chiral resolution. In their model, attrition is assumed to limit the size of crystals. Having reached this maximum size, the crystals can break, thus creating fragments of various sizes. Other mechanisms modelled include the reversible attachment of monomer and dimer units to clusters. This model has then been extended to cover an open system, in which some of the largest chiral clusters are taken out of the system and replaced by achiral molecules or by a racemic mixture of clusters [27]. In another paper [28], the same authors consider a different description of Ostwald ripening, incorporating the Gibbs-Thomson effect for the size-dependence of solubility. Saito and Hyuga have used a series of Monte-Carlo type models to simulate the interaction of monomers and clusters on surfaces [29–31], by also considering the action of grinding and the influence of the number of kink sites on the attachment and detachment rates. Hatch et al. [32] have also used a Monte-Carlo simulation to model chiral symmetry breaking, focussing on diffusion on a surface and on the kinetics of the subsequent formation of chiral species.

The assumption that growth and dissolution of particles due to size-dependent solubility leads to the direct exchange of substance molecules between crystals, without passing through the solution, is equivalent to assuming that growth and dissolution do not lead to any change in mass of the crystal population. These assumptions are not always justified when considering Ostwald ripening. Furthermore, several of the models mentioned neglect the concentration dependence of the critical size, which is the basis for size-dependent solubility in the first place. Attrition and agglomeration phenomena are also described in a very simplified manner in the models mentioned above.

Attrition is assumed to limit the size of crystals to an arbitrarily chosen maximum size. Crystals that reach this size are broken into parts. Agglomeration is described by the incorporation of growth clusters, a cluster being a collection of a few, usually two, monomers. The models referenced above are able to handle systems having either chiral or achiral molecules in solution, and are able to qualitatively reproduce the experimental observations. However, they do not accurately reflect the mechanisms which are proposed to be the cause of the observed behaviour. In the most recently published Monte-Carlo model of Viedma ripening [33], published after the original publication of our PBE model [34], these limitations have been overcome. In another recent publication [35], Blackmond and co-workers argue in favour of a PBE model with breakage but without agglomeration, and claim that the model can qualitatively describe the Viedma ripening process. Unfortunately, their model is incorrect, and so is the authors' conclusion that agglomeration is not necessary. In fact, they use a physically incorrect growth rate expression, where the growth rate of a single particle increases not only with supersaturation, as it must, but also with the concentration of crystals, which is physically inconsistent. The interested readers can convince themselves by applying Blackmond's growth model to a simple seeded growth process at constant supersaturation and observe that during growth the initial population of seeds gets distorted in an unphysical way.

20.4 Using Population Balance Equations to Explain Viedma Ripening

Viedma Ripening has been described with a population balance equation (PBE) model incorporating the various mechanisms (growth, dissolution, agglomeration, breakage and racemisation) believed to be the fundamental steps of this deracemisation process. This model has then been used to understand and explain experimental observations. The need for such a model arose from the fact that the experimental observations of Viedma ripening were intriguing, but not thoroughly understood. The previously published models of this process were unsatisfactory in the sense that they included too many simplifying assumptions. Population balance models lend themselves to this analysis, as they represent an established framework describing the various mechanisms involved in crystallisation processes. The major advantage of a rigorous mathematical model is that it enables the investigation of each mechanism separately, something which is not possible to the same extent experimentally.

The key results of such modelling can be summarised as follows:

1. A mixture of two enantiomeric crystals subjected to attrition in a solution containing a racemising agent undergoes symmetry breaking, with only one of the enantiomers eventually remaining in solid form. This process is believed to occur due to the interplay between racemization in solution, attrition, agglomeration and crystal growth and dissolution caused by the crystal size dependence of solubility. A mathematical model of this process has been presented, based on population balances incorporating all these elementary mechanisms [34]. This model is able to explain the experimentally observed behaviour. A sensitivity analysis on several parameters is carried out, leading to the conclusion that racemization, growth and dissolution, and agglomeration are necessary for deracemisation to occur. While deracemisation occurs also in the absence of attrition phenomena, these enhance it by reducing the time needed to achieve complete chiral purity.
2. Ostwald ripening has been incorporated into a detailed population balance model. The model is based on a size-dependent growth rate expression incorporating the Gibbs – Thomson relationship between particle size and solubility, and is solved numerically. The effect of parameters such as average initial particle size, initial width of the particle size distribution, and initial mass as well as solubility are investigated in simulations. This analysis focuses on understanding how the ripening phenomenon can be exploited in a crystallization process. The simulations are compared to the predictions of classical Lifshitz, Slyozov, Wagner (LSW) theory. Using the results of the PBE model for Ostwald ripening, the advantages and disadvantages of the full numerical simulation compared to the LSW model have been assessed [34, 36].
3. The chiral separation process known as Viedma ripening, or attrition- enhanced deracemisation, relies on several interacting mechanisms and is influenced by thermodynamic and kinetic factors as well as by the initial conditions of the experiment, sometimes making the interpretation of experimental results difficult. Thanks to the mathematical model of the process, the large variations in terms of outcome and process time observed in many experiments have been analysed and explained. They have been furthermore discussed in detail in terms of the effect of the initial conditions. It has been shown that the direction of evolution can be predicted by the model and depends on the asymmetries present in the initial particle populations. The process time has been shown to be very sensitive to the initial conditions when the system is not clearly biased toward one enantiomer. This analysis showed that the lack of reproducibility of some Viedma ripening experiments can at least partially be explained by the high sensitivity of the process to the initial conditions and can be prevented by a careful choice of the initial conditions [34, 36–38].

20.5 New Experimental Approaches Towards Deracemisation

High Pressure Homogenisation as a novel approach to carry out attrition- enhanced deracemisation is being developed, in order to combine the effects of mechanical attrition and temperature. The effect of pressure drop and homogeniser temperature on the evolution of deracemisation of the compound N-(2-methylbenzyli- dene)-phenylglycine amide (NMPA) has been studied. Results prove the feasibility of the technique to achieve complete solid state deracemisation, even without a substantial initial enantiomeric excess, and highlight low processing times. The effect of the homogeniser temperature has been demonstrated as a key parameter in HPH processing. These results establish the basis of High Pressure Homogenization as a new effective technique for Viedma ripening and open new possibilities for further development of solid state deracemisation [39].

References

1. Viedma C (2005) Chiral symmetry breaking during crystallization: complete chiral purity induced by nonlinear autocatalysis and recycling. Phys Rev Lett 94(6):065504
2. Noorduin WL, Izumi T, Millemaggi A, Leeman M, Meekes H, Van Enckevort WJP, Kellogg RM, Kaptein B, Vlieg E, Blackmond DG (2008) Emergence of a single solid chiral state from a nearly racemic amino acid derivative. J Am Chem Soc 130(4):1158–1159
3. Viedma C (2007) Chiral symmetry breaking and complete chiral purity by thermodynamic-kinetic feedback near equilibrium: implications for the origin of biochirality. Astrobiology 7 (2):312–319
4. Noorduin WL, Meekes H, van Enckevort WJP, Millemaggi A, Leeman M, Kaptein B, Kellogg RM, Vlieg E (2008) Complete deracemization by attrition-enhanced Ostwald ripening elucidated. Angewandte Chemie-International Edition 47(34):6445–6447
5. Cheung PSM, Gagnon J, Surprenant J, Tao Y, Xu HW, Cuccia LA (2008) Complete asymmetric amplification of ethylenediammonium sulfate using an abrasion/grinding technique. Chem Commun 44(8):987–989
6. Tsogoeva SB, Wei S, Freund M, Mauksch M (2009) Generation of highly enantioenriched crystalline products in reversible asymmetric reactions with racemic or achiral catalysts. Angew Chem 121(3):598–602
7. Wei S, Mauksch M, Tsogoeva SB (2009) Autocatalytic enantiomerisation at the crystal surface in deracemisation of scalemic conglomerates. Chem Eur J 15(39):10255–10262
8. Viedma C, Ortiz JE, Torres TD, Izumi T, Blackmond DG (2008) Evolution of solid phase homochirality for a proteinogenic amino acid. J Am Chem Soc 130(46):15274–15275
9. Van der Meijden MW, Leeman M, Gelens E, Noorduin WL, Meekes H, van Enckevort WJP, Kaptein B, Vlieg E, Kellogg RM (2009) Attrition-enhanced deracemization in the synthesis of clopidogrel – a practical application of a new discovery. Org Process Res Dev 13 (6):1195–1198
10. Noorduin WL, Kaptein B, Meekes H, van Enckevort WJP, Kellogg RM, Vlieg E (2009) Fast attrition-enhanced deracemization of naproxen by a gradual in situ feed. Angew Chem Int Ed 48(25):4581–4583

11. Spix L, Meekes H, Blaauw RH, van Enckevort WJ, Vlieg E (2012) Complete deracemization of proteinogenic glutamic acid using viedma ripening on a metastable conglomerate. Cryst Growth Des 12(11):5796–5799
12. Spix L, Alfring A, Meekes H, van Enckevort WJP, Vlieg E (2014) Formation of a salt enables complete deracemization of a racemic compound through viedma ripening. *Crys Growth Des* 14(0):1744–1748
13. Kaptein B, Noorduin WL, Meekes H, van Enckevort WJP, van Enckevort WJP, Kellogg RM, Vlieg E (2008) Attrition-enhanced deracemization of an amino acid derivative that forms an epitaxial racemic conglomerate. Angew Chem Int Ed 47(38):7226–7229
14. Noorduin WL, van Enckevort WJP, Meekes H, Kaptein B, Kellogg RM, Tully JC, McBride JM, Vlieg E (2010) The driving mechanism behind attrition-enhanced deracemization. Angew Chem Int Ed 49(45):8435–8438
15. Suwannasang K, Flood AE, Rougeot C, Coquerel G (2013) Using programmed heating-cooling cycles with racemization in solution for complete symmetry breaking of a conglomerate forming system. Cryst Growth Des 13(8):3498–3504
16. Noorduin WL, van der Asdonk P, Bode AC, Meekes H, van Enckevort WJP, Vlieg E, Kaptein B, van der Meijden MW, Kellogg RM, Deroover G (2010) Scaling up attrition-enhanced deracemization by use of an industrial bead mill in a route to clopidogrel (Plavix). Org Process Res Dev 14(4):908–911
17. Noorduin WL, van der Asdonk P, Meekes H, van Enckevort WJP, Kaptein B, Leeman M, Kellogg RM, Vlieg E (2009) Complete chiral resolution using additive-induced crystal size bifurcation during grinding. Angew Chem 121(18):3328–3330
18. Skrdla PJ (2011) Kinetics and thermodynamics of efficient chiral symmetry breaking in nearly racemic mixtures of conglomerate crystals. Cryst Growth Des 11(5):1957–1965
19. McBride JM, Tully JC (2008) Physical chemistry: did life grind to a start? Nature 452 (7184):161–162
20. Noorduin WL, Meekes H, Bode AAC, van Enckevort WJP, Kaptein B, Kellogg RM, Vlieg E (2008) Explanation for the emergence of a single chiral solid state during attritionenhanced ostwald ripening: survival of the fittest. Cryst Growth Des 8(5):1675–1681
21. Uwaha M (2004) A model for complete chiral crystallization. J Phys Soc Jpn 73 (10):2601–2603
22. Uwaha M (2008) Simple models for chirality conversion of crystals and molecules by grinding. J Phys Soc Jpn 77(8):083802
23. Saito Y, Hyuga H (2005) Chirality selection in crystallization. J Phys Soc Jpn 74(2):535–537
24. Wattis J (2011) Mathematical models of the homochiralisation of crystals by grinding. Orig of Life Evol Biosph 41:133–173
25. Cartwright JHE, Piro O, Tuval I (2007) Ostwald ripening, chiral crystallization, and the common-ancestor effect. Phys Rev Lett 98(16):165501
26. Uwaha M, Katsuno H (2009) Mechanism of chirality conversion by grinding crystals: Ostwald Ripening vs crystallization of chiral clusters. J Phys Soc Jpn 78(2):023601
27. Uwaha M (2011) Steady chirality conversion by grinding crystals–Supercritical and subcritical bifurcations. J Cryst Growth 318(1):89–92
28. Katsuno H, Uwaha M (2009) Monte Carlo simulation of a cluster model for the chirality conversion of crystals with grinding. J Cryst Growth 311(17):4265–4269
29. Saito Y, Hyuga H (2008) Chiral crystal growth under grinding. *J Phys Soc Japn* 77(11):113001
30. Saito Y, Hyuga H (2010) Crystal chirality selected by mutual antagonism. J Phys Soc Jpn 79 (8):083002
31. Saito Y, Hyuga H (2011) Grinding-induced homochirality in crystal growth. J Cryst Growth 318(1):93–98
32. Hatch HW, Stillinger FH, Debenedetti PG (2010) Chiral symmetry breaking in a microscopic model with asymmetric autocatalysis and inhibition. J Chem Phys 133(22):224502
33. Ricci F, Stillinger FH, Debenedetti PG (2013) A computational investigation of attrition-enhanced chiral symmetry breaking in conglomerate crystals. J Chem Phys 139(17):174503

34. Iggland M, Mazzotti M (2011) A population balance model for chiral resolution via Viedma ripening. Cryst Growth Des 11(10):4611–4622
35. Gherase D, Conroy D, Matar OK, Blackmond DG (2014) Experimental and theoretical study of the emergence of single chirality in attrition-enhanced deracemization. Cryst Growth Des 14(3):928–937
36. Iggland M, MazzottI M (2012) Population balance modeling with size-dependent solubility: Ostwald ripening. Cryst Growth Des 3:1489–1500
37. Iggland M, Mazzotti M (2013) Solid state deracemisation through growth, dissolution and solution-phase racemisation. *Cryst Eng Comm* 15:2319–2328
38. Iggland M, Mazzotti M (2014) Population balance modeling with size-dependent solubility: Ostwald ripening. Cryst Growth Des 14:2488–2493
39. Iggland M, Fernández-Ronco MP, Senn R, Kluge J, Mazzotti M (2014) Complete solid state deracemization by high pressure homogenization. Chem Eng Sci 111:106–111

Chapter 21
Mechanochemistry and Its Role in Novel Crystal Form Discovery

William Jones

Abstract Mechanochemistry has a long history and has recently undergone a surge in interest in many areas from metallurgy to chemical synthesis. It is frequently seen as a Green Chemistry approach since in many instances the volumes of solvent typically used (e.g. in synthetic organic chemistry) can be significantly reduced. The chapter reviews in particular the recent applications of mechanochemistry in the field of pharmaceutical materials. Here it is known to provide a simple way of screening for various polymorphic forms of single and multicomponent crystals. The particular advantage of adding small (catalytic) amounts of a liquid during the process is also considered. To further develop the potential application in pharmaceutical materials attempts to increase the scale of production are presented as well as efforts to follow by in situ methods (both spectroscopic and diffraction) the detailed mechanisms involved.

Keywords Grinding • LAG • Transformations • Reactivity • Defects • Cocrystals

21.1 Introduction

Mechanochemistry has long been used for the synthesis of solids and molecules. One of the first was recorded by Faraday in 1820 who showed it was possible to reduce AgCl to Ag by grinding, in a mortar and pestle, a mixture of AgCl and Zn. Carey Lea (1823–1897) helped develop the approach and the position of mechanochemistry within the framework of chemistry was delineated by Ostwald who noted the equivalency of electrochemistry, thermochemistry, photochemistry and mechanochemistry. A recent extensive review [1] by James et al. has highlighted several of the emergent areas now being actively pursued with regard to the application of mechanical energy in driving various transformations. The authors describe how "mechanochemistry refers to reactions, normally of solids,

W. Jones (✉)
Department of Chemistry, University of Cambridge, Lensfield Road, Cambridge CB2 1EW, UK
e-mail: wj10@cam.ac.uk

K.J. Roberts et al. (eds.), *Engineering Crystallography: From Molecule to Crystal to Functional Form*, NATO Science for Peace and Security Series A: Chemistry and Biology, DOI 10.1007/978-94-024-1117-1_21

induced by the impact of mechanical energy such as grinding in ball mills". The formal definition by IUPAC states "a mechano-chemical reaction is a chemical reaction that is induced by mechanical energy". Note that we now merge the idea of breaking chemical bonds with also the breaking on "non-covalent bonds" in supramolecular chemistry. As a result phase transformations between one polymorph and another can fall under a common definition.

Our interest will be in the application of mechanochemistry in pharmaceutical materials science. We can note that it may be a deliberate action (as in milling or grinding) or it may be an unwanted consequence of compressing at high pressures, e.g. during tablet formation. The amount of energy which can be imparted to a system under mechanical activation can be significant – certainly sufficient to break chemical bonds – but can also be reduced when working with systems such as soft molecular crystals in order to minimise chemical activation while still enabling processes such as the conversion of a crystalline solid to an amorphous product, or the readjustment of molecules within a lattice. Such readjustments may lead to the introduction of lattice imperfections, to polymorphic transformations or to the intimate mixing of two separate crystalline molecular solids to create a crystalline multicomponent (in some cases cocrystal) product indistinct from that which would be obtained by a conventional solution crystallisation, i.e. an intimate mixing without the need for large amounts of solvent. This makes it particularly advantageous with regard to "Green Chemistry" [2]. In terms of commercial interest, the review by James et al. [1] describes a search of patent applications for the terms "mechanochemical" and "mechanochemistry" and reveals a significant increase in usage in the patent literature from 1980 onwards.

21.2 General Areas of Application and Development

The areas in which mechanochemistry has been applied with success include: catalysts, nanoparticles, organic synthesis, MOFs, oxides, biomaterials, supramolecular chemistry, process engineering, dyes and pigments, fluorophores, exfoliation of graphene nanosheets and modification of fullerenes. Numerous further examples are given in the James et al. review.

In synthetic chemistry an important observation is that substances can react by a markedly different pathway in the solid state than in solution leading to new products or to much higher selectivity – Fig. 21.1 [3].

This ability to generate a much broader product-spectrum, as we will demonstrate later, is also particularly important in areas other than chemical transformations, e.g. in crystal polymorph conversions. A good example is shown by anthranilic acid and succinic acid - note that the addition of very small amounts of liquid has a clear impact on the outcome of a solid-to-solid reaction, even though quantities are much less than a conventional solvent-based reaction might require

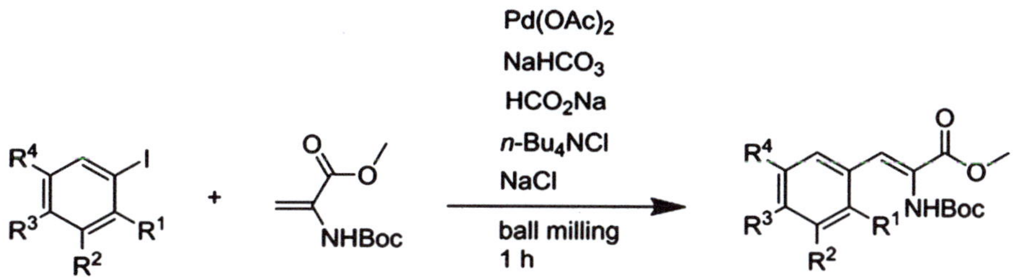

Fig. 21.1 Higher yields for the Heck Pd catalysed coupling reaction using ball milling [3]

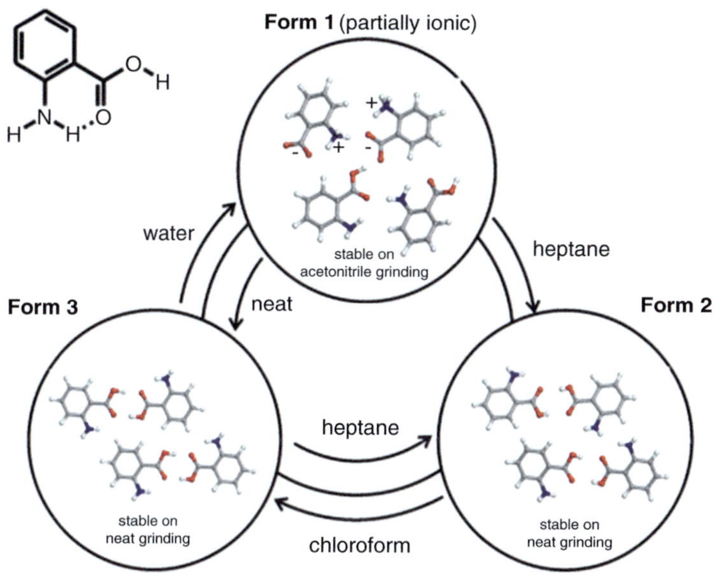

Fig. 21.2 LAG results for anthranilic acid showing variable polymorphic form outcome depending on the liquid added [4]

(Fig. 21.2). This is now called liquid assisted grinding, LAG, and will be considered further below.

Given the variety of materials of interest, ranging from metals and ceramics to pharmaceuticals the processes by which mechanically induced reactions occur can be complex and varied [5–8]. Also to be distinguished is the time-scale over which an event is studied. During the few milliseconds following the impact of two solids at high velocity, significant local consequences will ensue. Hot spots, magma-plasma regions, rapid induction of defects (point, linear and planar) and then propagation of these defects through the crystal are some of the concepts. The distribution of imparted energy will also be different in those solids with high rigidity and few pathways for energy dissipation (so that the effects at the specific point of impact will be great) than in those systems where structural movement

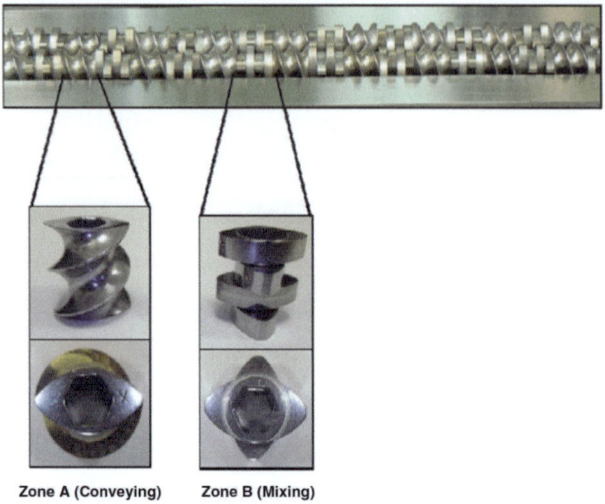

Fig. 21.3 Screw extrusion technology used for making continuous production of cocrystal product. The method is based on the frequently used technique of hot melt extrusion [9]

(e.g. slippage of crystal planes) may allow the energy (and its consequences) to be distributed over a much larger distance. During this early stage, the making and breaking of chemical bonds would seem possible. Following the initial short-term consequences, there will also be the longer term events associated with the final energy dissipation and relaxation of the system and possible subsequent additional transformations.

A key factor for potential commercial exploitation in many instances would seem to be difficulties associated with scale-up. In the laboratory commercial ball-milling equipment is used based on a few grams of material. The progression from gram-scale, lab-based synthesis to kilogram-scale or higher outputs is an issue that has been addressed in many industrial applications and more recently in the area of pharmaceuticals, where twin extrusion approaches have been successfully employed (Fig. 21.3).

Importantly the generation of a new polymorph (e.g. in the case of molecular crystals) at the gram level can allow, through subsequent seeding of a supersaturated solution, production via a conventional solvent-based crystallisation route.

It is also necessary to recognise the need for methods of structurally characterising the products of mechanochemical processes. When applied to any chemical conversions which may occur, molecular identity can be established by conventional analytical methods (e.g. solution NMR and mass spectrometry). For solid-solid transformations, where the identity of crystalline intermediates or final states is required, major improvements in crystal structure determination from powder X-ray diffraction data, Fig. 21.4, (frequently coupled with solid-state NMR) has been pivotal – especially when done *in situ* using synchrotron facilities. [10] In addition, the structural analysis of individual crystallites that result from mechanochemical processes is possible through utilisation of the electron diffraction facility of the transmission electron microscope (Fig. 21.5).

Fig. 21.4 In situ studies of cocrystallization [10]

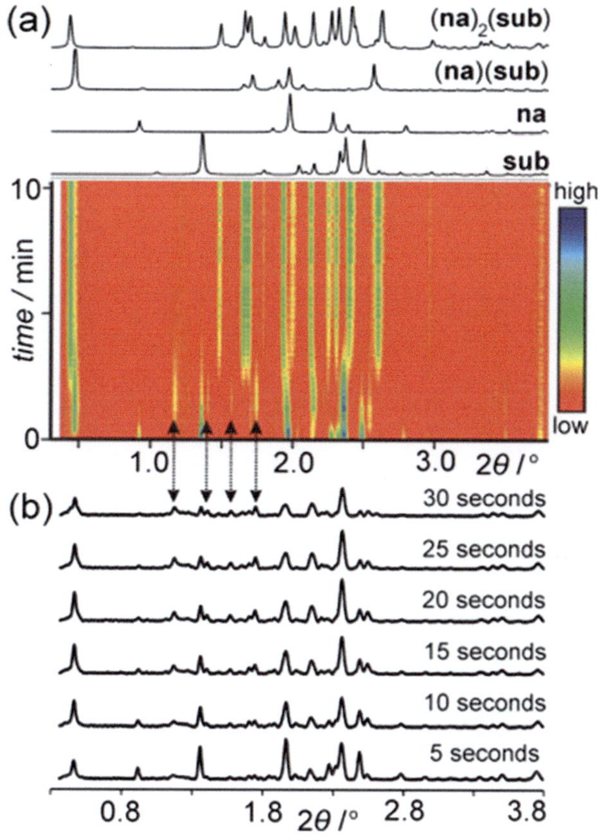

Particle size reduction is an important step in many industrial processes and includes compression (e.g. rollers) or impact (e.g. impact mills and ball mills). All methods involve transfer of varying amounts of energy to the particle(s). As a result and along with size reduction other consequences can ensue including the introduction of defects (such as dislocations, point defects and stacking faults) but in particular amorphisation or polymorphic transformations. Frequently the amorphisation of the material can be a desirable outcome especially as a means of increasing the transient solubility of the material or eliminating compression issues with the crystalline equivalent – see Fig. 21.6.

This attribute is frequently used in the pharmaceutical industry and the complex nature of the resulting solid has been extensively studied. A frequently used particle-size reduction method within the pharmaceutical industry is jet-milling whereby particle-particle collisions occur within a carrier gas (typically air). This can be a very high energy process.

An additional reason for interest in mechanochemical processes comes from the fact that the majority of drugs are formulated as tablets, a process which itself requires the application of significant applied pressures. Tablets are seen as a

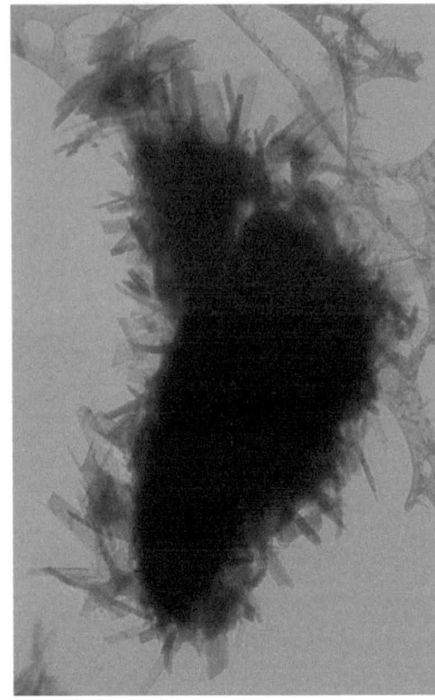

Fig. 21.5 TEM image of a milled material [11]

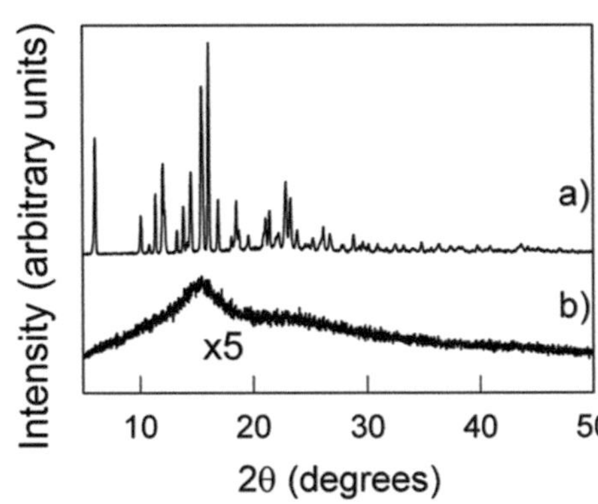

Fig. 21.6 Before and after milling X-ray powder patterns for trehelose [12]

reliable and general way of delivering medicines with approximately 80% drugs delivered via tablets. Equally important is that the tablet contains other components (excipients) to enhance performance, e.g. lubricants, dissolving agents and so on. As a result under pressure there is a risk of surface-surface reactions between the components leading to enhanced drug instability [13].

An approach for directing the polymorphic outcome of a grinding experiment in which a small amount of liquid was added to the material was described above for the cases of anthranilic acid and succinic acid. For anthranilic acid interconversion between the three known polymorphic forms occurred depending upon the choice of the added liquid. In the case of succinic acid while the β-polymorph appeared stable to neat grinding addition of a few drops of a low polarity liquid (e.g. hexane) caused conversion to the α-polymorph. More polar liquids (e.g. water, acetonitrile or methanol) appeared not to induce the transformation. [See reference 8 for review.]

More recent interest has focussed on the formation of pharmaceutical cocrystals by milling or grinding. An early example of a cocrystal obtained by grinding is that of quinhydrone (Wohler published in 1884). Numerous examples of host-complex formation by solid-solid mixing were reported by Toda and co-workers [14]. Work by Etter and colleagues [15] also demonstrated how solid-solid grinding could be used to produce hydrogen bonded cocrystals of adenine and thymine derivatives. They were able to demonstrate a marked selectivity amongst various base pairs. Hollingsworth et al. [16] used grinding as a means of creating cocrystals of dinitriles and urea as part of wide study of inclusion crystals in cases where solution growth was not possible. Pedireddi et al. [17] demonstrated the use of the role of a third component within the cocrystal in order for solid-solid reaction to take place and Kuroda and her colleagues [18] studied the addition of a third component on the optical properties of resulting cocrystals.

An early indication of the potential to form cocrystals containing drug molecules was provided by Caira et al. [19] where grinding was used to easily prepare a series of products containing sulfonamide with crystal structures similar to those obtained from solution. They also reported detailed solid state kinetic data based on X-ray measurements. The work also demonstrated a significant selectivity to product identity from ground mixtures of potential reacting components. In 2004 Almarsson and Zaworotko [20] in 2004 summarised the crystal engineering aspect and potential of such pharmaceutical cocrystals as a means of rationally designing new solid forms of drug molecules with enhanced properties and Trask reported on the potential intellectual property aspects [21].

The outcome of mechanochemical formation of cocrystals compared to other approaches subsequently followed. Trask et al. [22] reported three possible outcomes in the case of cocrystals with caffeine: (i) solution growth and milling produced identical products, (ii) a different stoichiometric product could be obtained by milling compared to solution and (iii) different polymorphic forms were possible.

In terms of property improvement with pharmaceutical cocrystals various examples are known including: stability against hydration of an anhydrate form, developing forms with improved compression properties (for tablet formation) and significantly improving the bioavailability of poorly soluble drugs. In non-pharmaceutical applications benefits of cocrystallisation have been shown for fluorescence properties and explosives.

A challenge, however, is selecting possible coformers which might cocrystallise with the API rather than remain as a separate and distinct crystalline phase. Whilst various strategies have been developed to predict the likelihood of cocrystal formation experimental screening for success remains vital (Fig. 21.7).

While solution growth remains the ideal approach - especially with regard to scale-up in later stages of development - significant differences in solubility

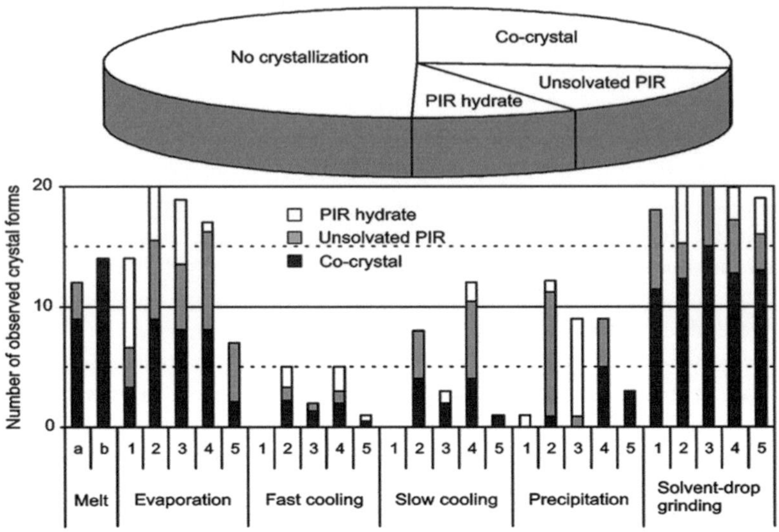

Fig. 21.7 Summary of the outcome of various cocrystallisation approaches for the drug piroxicam [23]

between the drug and coformer often presents difficulties. Mechanochemistry has emerged as a powerful tool in the experimental screening area. It has also emerged as a powerful way of searching for different polymorphic forms of the cocrystal as well as various stoichiometric compositions e.g. 1:1, 1:2 and 2:1. In the case of LAG initial attention focused as a means of accelerating cocrystallisation kinetics although it was subsequently shown to provide a means of increasing product diversity. The advantage of LAG was interpreted by Chiarelli et al. [24], Childs et al. [25] and others on the basis of solution phase chemistry. Numerous recent publications have demonstrated that LAG is likely to be the most efficient way of screening for cocrystal formation (Figs. 21.7, 21.8, 21.9).

Further insight was given by recent work of Bucar et al. [28] for the case of caffeine and benzoic acid. No cocrystals for this pair of molecules were known (despite a strong expectation based on studies of analogous systems) and despite extensive efforts over many years. A systematic study using a large range of cocrystallization methods was used including neat grinding, liquid assisted grinding, melt crystallization and various solution methods. All failed to produce the elusive cocrystal although theoretical work had clearly confirmed a cocrystal was thermodynamically feasible. Grinding experiments, accompanied by the addition of a few seed crystals of an isostructural fluorobenzoic acid cocrystal, however, readily generated the elusive cocrystal with the inference being that, as in solution crystallisation, frequently a high barrier to nucleation of a particular form can exist.

To better understand the processes occurring during milling several recent efforts have concerned developing methods to follow in situ the course of the reaction. Noteworthy has been the use of synchrotron X-ray diffraction. In a

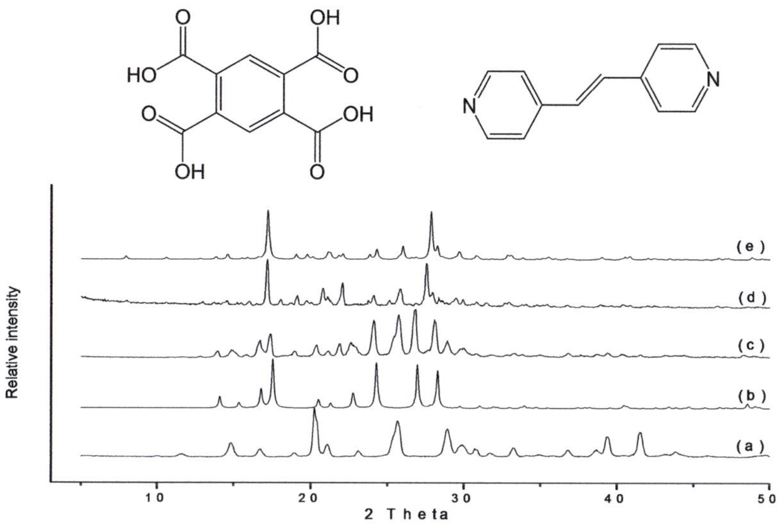

Fig. 21.8 Increased kinetics using LAG [26].

Fig. 21.9 Less amorphous form using LAG [27]

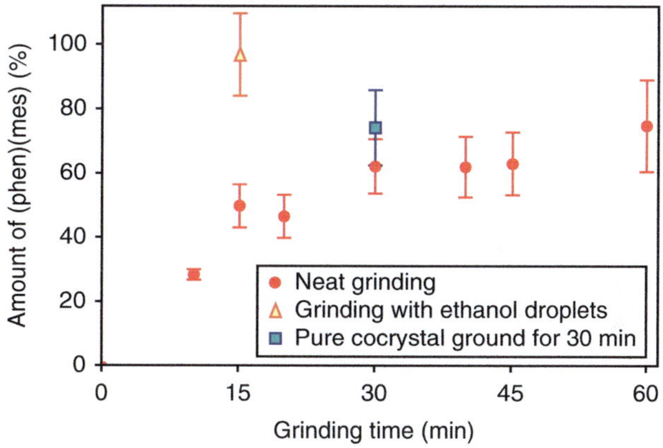

study of the milling of a mixture of carbamazepine and succinic acid Halasz et al. [10] showed that within the jar, and in the absence of any added liquid, the outcome was the gradual disappearance of reflections of the two components and apparent amorphisation, Fig. 21.4 above. A similar experiment, but now with the addition of 50 μL of acetonitrile, showed that within 10 s of milling reflections associated with a triclinic (cbz)(sac) cocrystal emerged. The reaction was complete within 4 min. Such rapidity was to be compared with other solid-solid grinding systems (e.g. the reaction of ZnO with imidazoles) where reaction timescales were much longer. Frenette et al. and his colleagues have also used in situ methods based on changes in fluorescence properties taking place during cocrystallisation to monitor directly events within the grinding jar [29].

References

1. James SL, Adams CJ, Bolm C, Braga D, Collier P, Friščić T, Grepioni F, Harris KDM, Hyett G, Jones W, Krebs A, Mack J, Maini L, Orpen AG, Parkin IP, Shearouse WC, Steed JW, Waddell DC (2012) Mechanochemistry: opportunities for new and cleaner synthesis. Chem Soc Rev 41:413–447
2. Cave GWV, Raston CL, Scott JL (2001) Recent advances in solventless organic reactions: towards benign synthesis with remarkable versatility. Chem Commun:2159–2169
3. Tullberg E, Peters D, Frejd T (2004) The heck reaction under ball-milling conditions. J Organomet Chem 689:3778–3781
4. Trask AV, Shan N, Motherwell WDS, Jones W, Feng S, Tan RBH, Carpenter KJ (2005) Selective polymorph transformation via solvent-drop grinding. Chem Commun:880–882
5. Boldyrev VV (2006) Mechanochemistry and mechanical activation of solids. Russ Chem Rev 75:177–189
6. Baláž P (2008) Mechanochemistry in nanoscience and minerals engineering. Springer-Verlag, Berlin
7. Takacs L (2013) The historical development of mechanochemistry. Chem Soc Rev 42:7649–7659
8. Trask AV, Jones W (2005) Crystal engineering of organic cocrystals by the solid-state grinding approach. In: Topics in current chemistry (Organic solid state reactions), vol 254. pp 41–70
9. Daurio D, Medina C, Saw R, Nagapudi K, Alvarez-Nunez F (2011) Application of twin screw extrusion in the manufacture of cocrystals, part I: four case studies. Pharmaceutics 3:582–600
10. Halasz I, Puškarić A, Kimber SAJ, Beldon PJ, Belenguer AM, Adams F, Honkimäki V, Dinnebier RE, Patel B, Jones W, Štrukil V, Friščić T (2013) Real-time in situ powder X-ray diffraction monitoring of mechanochemical synthesis of pharmaceutical cocrystals. Angew Chem Int Ed 52:11538–11541
11. Eddleston MD, Bithell EG, Jones W (2010) Transmission electron microscopy of pharmaceutical materials. J Pharm Sci 99:4072–4083
12. Descamps M, Willart JF, Dudognon E, Caron V (2007) Transformation of pharmaceutical compounds upon milling and comilling: the role of T_g. J Pharm Sci 96:1398–1407
13. Cassidy AM, Gardner CE, Auffret T, Aldous B, Jones W (2012) Decoupling the effects of surface chemistry and humidity on solid-state hydrolysis of aspirin in the presence of dicalcium phosphate dihydrate. J Pharm Sci 101(4):1496–1507
14. Toda F, Tanaka K, Sekikawa A (1987) Host-guest complex formation by a solid-solid reaction. J Chem Soc Chem Commun:279–280.
15. Etter MC, Reutzel SM, Choo CG (1993) Self-organization of adenine and thymine in the solid state. J Am Chem Soc 115:4411–4412
16. Hollingsworth MD, Santarsiero BD, Oumar-Mahamat H, Nichols CJ (1991) New series of 1:1 layered complexes of α,ω-dinitriles and urea. Chem Mater 3:23–25
17. Pedireddi VR, Jones W, Chorlton AP, Docherty R (1996) Creation of crystalline supramolecular arrays: a comparison of co-crystal formation from solution and by solid-state grinding. Chem Commun 987-988:18
18. Kuroda R, Imai Y, Tajima N (2002) Generation of a co-crystal phase with novel coloristic properties via solid state grinding procedures. Chem Commun:2848–2849
19. Caira MR, Nassimbeni LR, Wildervanck AF (1995) Selective formation of hydrogen bonded cocrystals between a sulfonamide and aromatic carboxylic acids in the solid state. J Chem Soc Perkin Trans 2:2213–2216
20. Almarsson O, Zaworotko MJ (2004) Crystal engineering of the composition of pharmaceutical phases. Do pharmaceutical co-crystals represent a new path to improved medicines?, Chem Commun:1889–1896
21. Trask AV (2007) An overview of pharmaceutical cocrystals as intellectual property. Mol Pharm 4:301–309

22. Trask AV, van de Streek J, Motherwell WDS, Jones W (2005) Achieving polymorphic and stoichiometric diversity in Cocrystal formation: importance of solid-state grinding, powder X-ray structure determination, and seeding. Cryst Growth Des 5:2233–2241
23. Fucke K, Myz SA, Shakhtshneider TP, Boldyreva EV, Griesser UJ (2012) How good are the crystallisation methods for co-crystals? A comparative study of piroxicam. New J Chem 36:1969–1977
24. Chiarella RA, Davey RJ, Peterson ML (2007) Making co-crystals - the utility of ternary phase diagrams. Cryst Growth Des 7:1223–1226
25. Childs SL, Rodríguez-Hornedo N, Reddy LS, Jayasankar A, Maheshwari C, McCausland L, Shipplett R, Stahly BC (2008) Screening strategies based on solubility and solution composition generate pharmaceutically acceptable cocrystals of carbamazepine. Cryst Eng Comm 10:856–864
26. Shan N, Toda F, Jones W (2002) Mechanochemistry and co-crystal formation: effect of solvent on reaction kinetics. Chem Commun:2372–2373
27. Nguyen KL, Friščić T, Day GM, Gladden LF, Jones W (2007) Terahertz time-domain spectroscopy and the quantitative monitoring of mechanochemical cocrystal formation. Nat Mater 6:206–209
28. Bučar D-K, Day GM, Halasz I, Zhang GGZ, Sander JRG, Reid DG, MacGillivray L, Duer MJ, Jones W (2013) The curious case of (caffeine)·(benzoic acid): how heteronuclear seeding allowed the formation of an elusive cocrystal. Chem Sci 4:4417–4425
29. Frenette M, Cosa G, Friščić T (2013) Characterisation of organic solid forms and real-time in situ monitoring of their transformations using solid-state fluorescence. Cryst Eng Comm 15:5100–5106

Chapter 22
Innovative Spontaneous Chiral Resolution Phenomenon: Preferential Enrichment

Rui Tamura

Abstract The mechanism and requirements of 'preferential enrichment (PE)' are presented. This is a unique spontaneous chiral resolution phenomenon induced by the solvent-assisted solid-to-solid transformation of a metastable polymorphic form into another more stable one during crystallization from the highly supersaturated solution. This is applied to certain kinds of racemic mixed crystals and common racemic compound crystals as well.

Keywords Preferential enrichment • Regular chiral fluctuation or symmetry-breaking • Chiral resolution • Chemical complexity phenomenon

22.1 Introduction

Polymorphism is one of the subjects of great concern in the pharmaceutical and fine-chemicals industries, because polymorphs exhibit different physical and chemical properties [1, 2]. Particularly, the study on structure-activity relationships is of special importance in the pharmaceutical industry; an unwanted polymorph can have different bioavailability in the target organism, making the drug ineffective or rising its potency to a dangerous limit [2, 3]. Usually polymorphic transformation of a metastable polymorphic form into another more stable one, which very often occurs during crystallization from solvents or in the solid state [4–8], has generally been considered as an intractable phenomenon [1, 2]. To avoid such an undesired transformation, a lot of strategies such as the use of seed crystals and additives, and the choice of solvents and interfaces have been devised for controlling crystallization of polymorphs [9, 10]. In this account, I describe an example of advantageous use of polymorphic transformation, i.e. an unusual spontaneous chiral resolution

The contents of this chapter is organized mainly on the basis of the review articles published in Refs. [11–14].

R. Tamura (✉)
Graduate School of Human and Environmental Studies, Kyoto University, Sakyo-ku, Kyoto 606-8501, Japan
e-mail: tamura.rui.8c@kyoto-u.ac.jp

K.J. Roberts et al. (eds.), *Engineering Crystallography: From Molecule to Crystal to Functional Form*, NATO Science for Peace and Security Series A: Chemistry and Biology, DOI 10.1007/978-94-024-1117-1_22

phenomenon that is initiated by a solvent-assisted solid-to-solid polymorphic transition during crystallization from the supersaturated solution of a certain kind of racemic crystals [11–14]. Development of economically and environmentally acceptable chiral resolution methods has long been the subject of considerable interest in connection with the rationalization of the origin of biomolecular homochirality on Earth [15–21] as well as the industrial and pharmaceutical needs for chiral organic substances [22–26]. Among them, the methods for chiral resolution of racemates by crystallization are classified into two categories [27]; one is a straightforward method to separate enantiomers in the absence of an external chiral element and the other is an indirect method using an external chiral element, such as a diastereomeric salt formation followed by fractional crystallization [23, 28–30] or a diastereoselective host-guest inclusion complexation (Fig. 22.1) [31]. As the former example, since Pasteur's first manual sorting of mirror-image D and L crystals (known as 'racemic conglomerates') of a tartrate salt [32], a lot of modified methods have been devised up to now for the resolution [preferential crystallization (PC)] of racemic conglomerates composed of nonracemizable enantiomers (Figs. 22.2a and 22.3a) [32–35] and for the deracemization of ordinary or epitaxial racemic conglomerates of racemizable molecules [36–44]. However, racemates existing as racemic conglomerates occupy only less than 10% of the characterized crystalline racemates, and more than 90% of them are supposed to be racemic crystals [45] which are further classified into (a) a 'racemic compound' consisting of a regular packing of pairs of *R* and *S* enantiomers in the crystal (Fig. 22.3b) or (b) a 'racemic mixed crystal' (in other words, a pseudoracemate or a solid solution) composed of a random to highly ordered arrangement of equal amounts of two enantiomers in the defined positions (Fig. 22.3c) [11–14, 27]. A notable difference between a racemic compound and a racemic mixed crystal is that racemic crystals of 0% *ee* are exclusively formed from a racemic solution in the former type while nonracemic crystals are easily formed in the latter type. It had been believed for over a century that there was no way for resolution of these racemic crystals by simple crystallization in the absence of an external chiral element.

In 1993 we discovered an unprecedented spontaneous chiral resolution phenomenon that can be applied to certain kinds of racemic mixed crystals (Fig. 22.2b) [46]. In 1998, this new phenomenon was referred to as 'preferential enrichment (PE)' [47], which is completely opposed to PC of racemic conglomerates; in PE it is in the mother liquor that substantial enantiomeric enrichment occurs by recrystallization, and at the same time slight enrichment of the opposite enantiomer always takes place in the deposited crystals (Fig. 22.2b). In 2002, we proposed the mechanism of PE in terms of a unique polymorphic transition during crystallization [48, 49]. Namely, PE was found to be initiated by the solvent-assisted solid-to-solid transformation of a metastable polymorphic form into a thermodynamically more stable one during crystallization from the supersaturated solution of a certain kind of racemic mixed crystals composed of two enantiomers without the aid of any external chiral element [11–14, 48, 49]. Most notably, this polymorphic transition process was followed by selective redissolution of the excess one enantiomer from the incomplete hydrogen bonding sites inside the transformed disordered crystal

Enantiomeric Resolution by Crystallization

(1) Without an external chiral element:

- **a. Preferential Crystallization (1866~)---Conglomerates**
- **b. Deracemization (1980~) ---Conglomerates**
  - **Attrition-enhanced Ostwald ripening (2008~)**
- **c. Preferential Enrichment (1996~)---Racemic mixed crystal**

(2) With an external chiral element:

- **a. Diastereomer salt formation (1853~)---Eutectic mixture**
  - **Dutch resolution (1998~)**
  - **Dielectrically controlled resolution (DCR) (2003~)**
- **b. Diastereoselective inclusion complexation (1980~)**

Fig. 22.1 Classification of enantiomeric resolution methods by crystallization

lattice into solution until the deposited crystals were slightly enriched with the opposite enantiomer ($<10\%$ *ee*), with full reproducibility (Fig. 22.4) [11–14, 48, 49]. Thus, the interplay of (i) such a solvent-assisted solid-to-solid polymorphic transition, (ii) the subsequent selective redissolution process of the excess one enantiomer from the deposited crystals, and (iii) the deposition of mixed crystals, which can retain the resulting fairly-random alignment of two enantiomers inside the crystal lattice to memorize the regular chiral fluctuation or symmetry-breaking, turned out to be responsible for this unique chiral resolution phenomenon (Fig. 22.4) [11–14]. Recently, based on the mechanism of PE with respect to the first-generation of chiral organic compounds **1**, we have successfully expanded the scope of PE to common racemic compounds such as α-amino acids, the cocrystals of α-amino acids with achiral dicarboxylic acids, and chiral drugs [50–53]. These results strongly support the propriety of our proposed mechanism of PE [11–14, 48, 49]. At the same time, we have recognized that PE is ascribed to a ‘complexity’ phenomenon; (a) molecular aggregation in solution, (b) formation of metastable crystalline phases, (c) polymorphic transition, and (d) substantial enrichment of one enantiomer in PE correspond to (a′) dissipative self-organization, (b′) emergence of new order and chaos, (c′) phase transition between chaos and/or order, and (d′) increasing returns in complexity, respectively [54–56].

Here I summarize the features, mechanism, requirements and nonlinear thermodynamics with respect to PE, and briefly introduce the recent successful application of PE to racemic compound crystals.

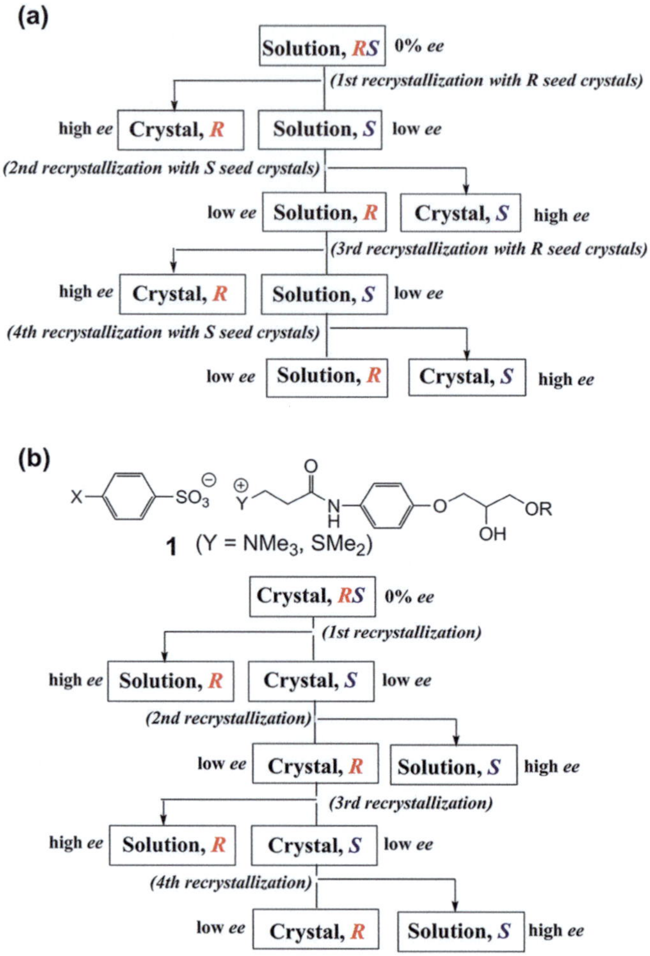

Fig. 22.2 Principles of (**a**) preferential crystallization (PC) of racemic conglomerates and (**b**) preferential enrichment (PE) of a racemic mixed crystal in the case of substantial enrichment of *R* enantiomer in solution after first recrystallization of the first-generation of chiral organic compounds 1

22.2 Mechanism and Requirements

The mechanism of PE and the requirements for its occurrence, which we unveiled with respect to the first-generation of chiral organic compounds **1**, are as follows (Fig. 22.4) [11–14, 48, 49]:

1. *Sufficient Solubility Difference*: The solubility of the enantiomerically enriched sample should be much higher than that of the corresponding racemic one; this implies the high possibility of preferential formation of homochiral

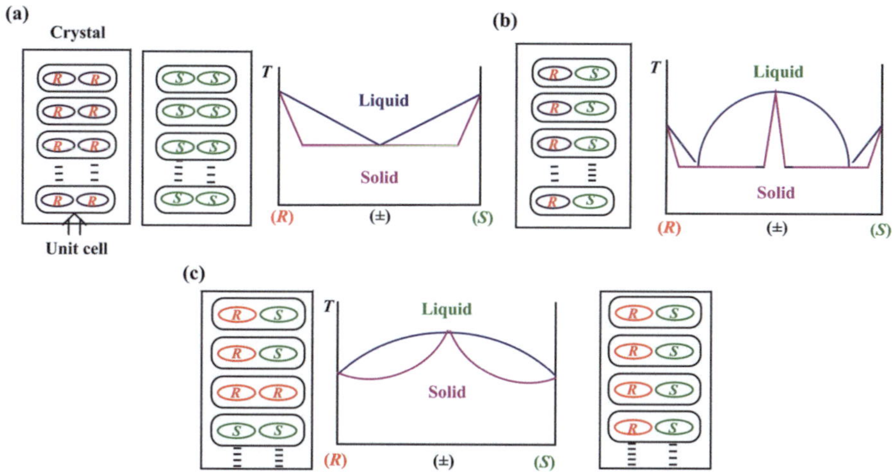

Fig. 22.3 General packing modes and binary melting-point phase diagrams of two enantiomers in (**a**) racemic conglomerates, (**b**) a racemic compound, and (**c**) a fairly-ordered or highly-ordered racemic mixed crystal (racemic compound type). *Lower pink* and *upper blue lines* represent the temperatures of the beginning and the end of fusion, respectively

one-dimensional (1D) *R* and *S* chains even in a racemic solution (Fig. 22.4a). Such a solubility difference substantially affects the *ee* value reached in the mother liquor at the end of crystallization; the larger the difference is, the higher the reached *ee* value in the mother liquor is. Furthermore, the combined use of supersolubility (a solubility obtained by dissolving the sample in a solvent on heating followed by being cooled) measurements under various conditions and number-averaged molecular weight measurement by vapor pressure osmometry, together with the molecular dynamics simulation of the oligomer models, supported a homochiral 1D assembly in preference to a heterochiral one in the racemic solutions of **1** showing PE [48].

2. *Unique Crystal Structure & Polymorphic Transition*: A solid-to-solid polymorphic transition of the initially-formed metastable mixed crystals composed of the random alignment of the homochiral 1D R and S chains into the stable mixed crystals mainly comprising heterochiral RS cyclic dimer chains should occur during crystallization (Fig. 22.4b, c). This polymorphic transition can be monitored by *in situ* ATR-IR spectroscopy; if there is a distinct difference in the vibrations of hydrogen bond forming groups between the highly supersaturated solution just before crystallization and the deposited crystals at the end of crystallization, polymorphic transition is most likely to occur [48]. The occurrence of polymorphic transition during crystallization was further supported by two different types of unique melting point diagrams obtained experimentally for two typical first-generation of chiral organic compounds (**1a** and **1b**) showing PE. In the case of **1a**, two polymorphic curves were found to intersect at two points around 35% *ee* (Fig. 22.5a) [48]. In the case of **1b**, there are two

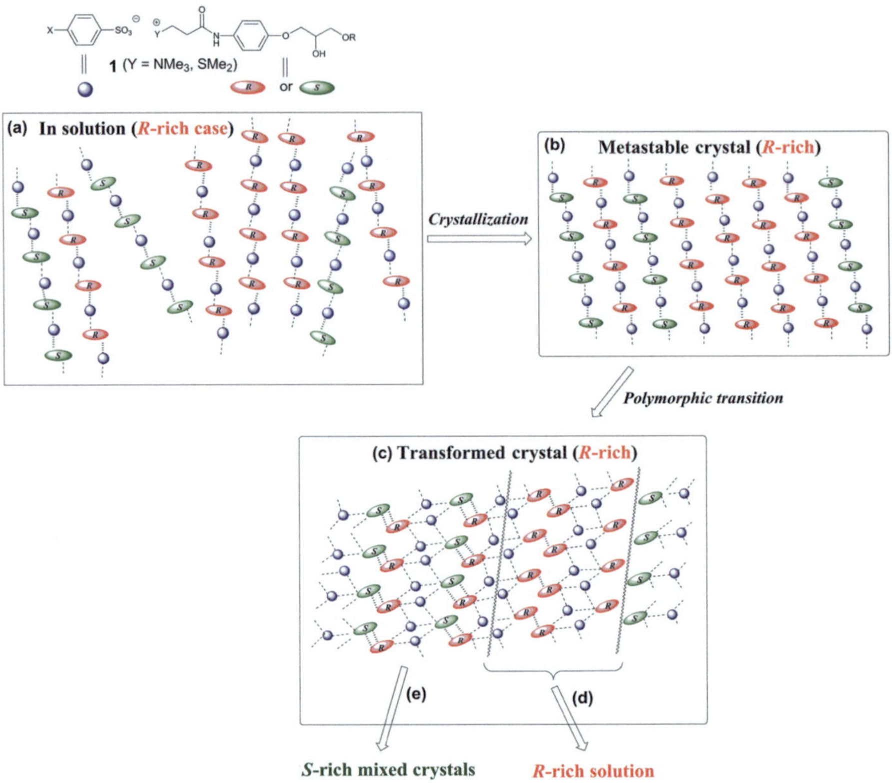

Fig. 22.4 Mechanism of PE in the case of crystallization from the supersaturated solution of slightly *R*-rich compounds **1**. Homochiral 1D molecular association (**a**) in solution and (**b**) in the metastable crystal, (**c**) polymorphic transition, followed by (**d**) selective redissolution of the excess *R* enantiomer into solution, and (**e**) deposition of nonracemic mixed crystals slightly enriched with the opposite *S* enantiomer. (Ref. [14] – Reproduced by permission of The Royal Society of Chemistry)

polymorphic convex curves which do not intersect but are located closely to each other (Fig. 22.5b) [57]. From these two diagrams, the following three conclusions were obtained; (i) their crystalline nature falls into a racemic mixed crystal, (ii) there must be two polymorphic phases in both cases, and (iii) the free energy difference between the two polymorphs is small enough to allow a polymorphic transition to proceed at a moderate rate during crystallization.

3. *Selective Redissolution of Excess One Enantiomer*: At the end of the above polymorphic transition, the transformed disordered crystals become fragile due to the incomplete hydrogen bond formation which results from the irregular alignment of *R* and *S* chains in the metastable crystal lattice. Consequently, selective redissolution of the excess one enantiomer occurs from the fragile transformed crystals into solution until the deposited crystals were slightly enriched with the opposite enantiomer (Fig. 22.4d).

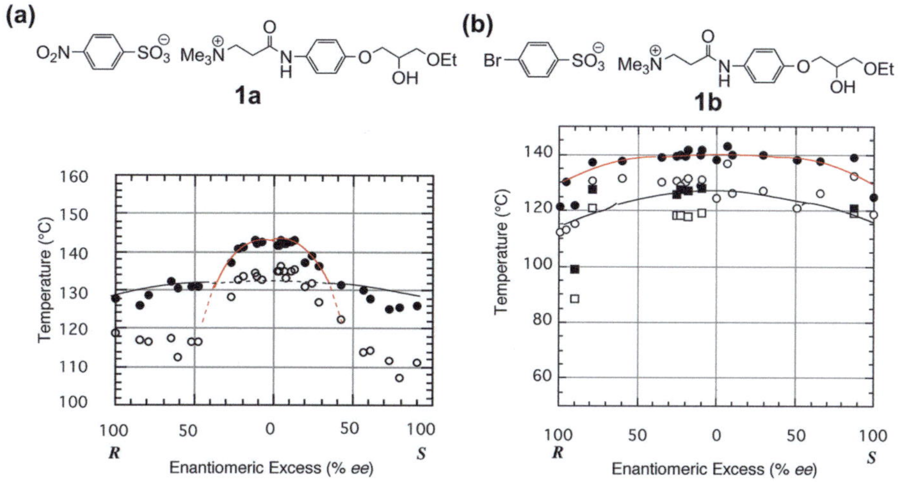

Fig. 22.5 Two types of melting point diagrams of (**a**) **1a** and (**b**) **1b** showing PE. *White* and *black circles* as well as *white and black squares* represent the temperatures of the beginning and the end of fusion, respectively. *Upper red curves* correspond to the stable polymorph, while *lower black curves* belong to the metastable one

4. *Deposition of Nonracemic Mixed Crystals*: The resulting deposited nonracemic mixed crystals should not undergo the subsequent solvent-mediated polymorphic transition into racemic crystals of 0% ee, memorizing the regular chiral fluctuation or symmetry-breaking (Fig. 22.4e).

Accordingly, the technical and phenomenal features of PE are summarized as follows [11–14, 48].

1. Usual standard recrystallization conditions with neither vigorous stirring nor abrasive grinding can be applied to the PE experiment, except that approximately 4–25-fold supersaturated solutions are employed because the supersolubility of the racemates showing PE is considerably higher than that of the solubility at 25 °C. The attainment of such a high supersolubility is closely associated with the preferential formation of homochiral 1D *R* and *S* chains even in the racemic solution. At the supersaturated concentrations lower than four-fold, PE does not efficiently occur.
2. Racemic or nonracemic samples of less than 10% *ee* are more suitable for the PE experiment than those of higher *ee* values to achieve a very efficient resolution, resulting in the deposition of nonracemic crystals slightly enriched (around 5% *ee*) with the opposite enantiomer.
3. When the original supersaturated solution used for crystallization is strictly racemic (0.0% *ee*), the probability for either the *R* or the *S* enantiomer to be enriched in the mother liquor after crystallization is 0.5. This is because initial capricious formation of the first formed nonracemic metastable crystal nucleus should determine which enantiomer is enriched in the mother liquor thereafter.

4. Only racemic or nonracemic samples have only to be crystalline to implement the PE experiment efficiently. It does not matter whether the enantiomerically enriched samples with high *ee* values exist as solids or oils, in sharp contrast to PC of racemic conglomerates.
5. Seed crystals are not necessary at all.

22.3 Application to Common Racemic Compound Crystals

We speculated that a racemic compound with a high eutectic *ee* value might behave like a racemic mixed crystal under nonequilibrium crystallization conditions (Fig. 22.6). Furthermore, it was anticipated that if the four requirements (sufficient solubility difference, unique crystal structure and polymorphic transition, partial crystal dissolution, deposition of nonracemic mixed crystals) for the occurrence of PE were satisfied, PE might occur for a racemic compound with a high eutectic *ee* value under nonequilibrium crystallization conditions. Such a situation was indeed found to be the case. Thus, PE turned out to be applicable to racemic compound crystals by carefully choosing kinetic crystallization conditions so as to induce a polymorphic transition.

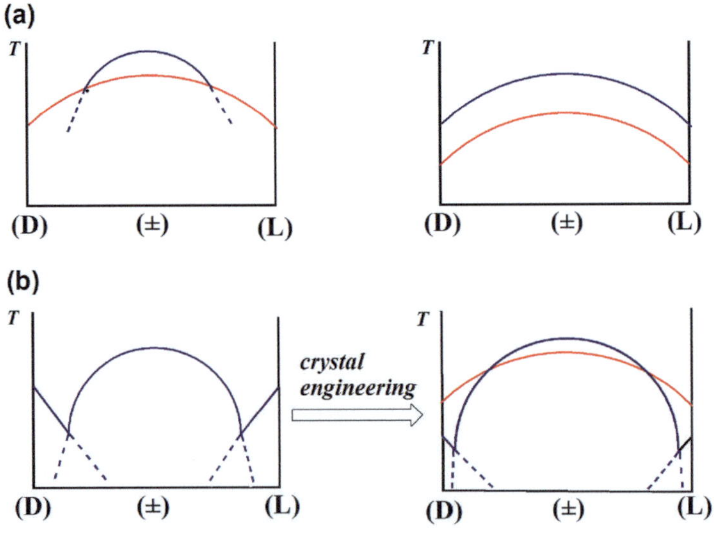

Fig. 22.6 Melting point diagrams with a polymorphism essential to induce PE [50]. (**a**) Two typical diagrams of a racemic mixed crystal capable of showing PE with respect to compounds **1** [11–14]. (**b**) A concept of transforming of a common racemic compound into a racemic mixed crystal (or another racemic compound with a high eutectic *ee* value) with an appropriate polymorphism by crystal engineering. For clarity, only the temperatures of end of fusion are shown. *Lower red* and *upper blue curves* correspond to the metastable and stable polymorphs, respectively, near the racemic composition [14] (Reproduced by permission of The Royal Society of Chemistry)

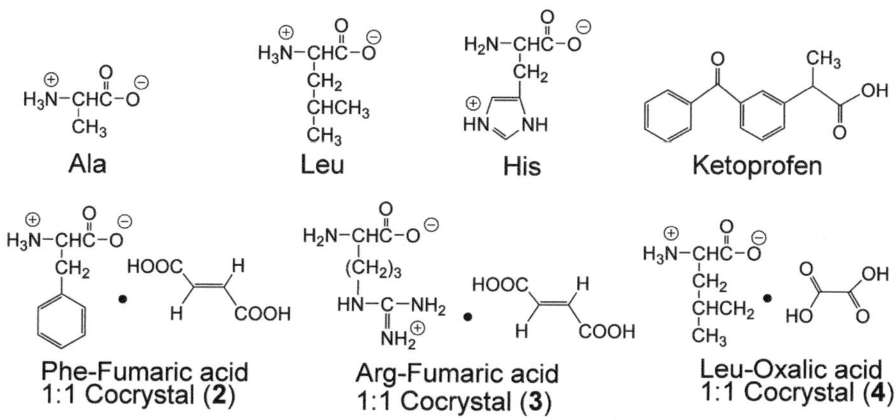

Fig. 22.7 Racemic single-component crystals and cocrystals showing PE

Since a two-component cocrystal has a high potential of showing a polymorphism and inducing polymorphic transition during crystallization [58], we carried out an extensive search from the Cambridge Structural Database (CSD) with respect to the cocrystal of a racemic α-amino acid and an achiral dicarboxylic acid as the model compound, which has the desired crystal structure for the occurrence of PE. Consequently, we found that the 1:1 cocrystal (**2**) of DL-Phe with fumaric acid (Fig. 22.7), which has a eutectic *ee* value of as high as 99% in water [59], satisfied all of the above requirements and indeed showed an efficient PE phenomenon [51], whereas free DL-Phe failed to show a polymorphic transition and thereby PE. Similarly, the 1:1 cocrystals (**3** and **4**) of DL-Arg and DL-Leu with fumaric acid and oxalic acid, respectively (Fig. 22.7), exhibited excellent and good PE phenomena [52, 53]. In addition, the free racemic compound ketoprofen (eutectic point: 82% *ee*) which satisfied the above four requirements showed a good PE phenomenon without cocrystallization [60].

22.4 Conclusions

The mechanism of PE can be explained in terms of a symmetry-breaking complexity phenomenon including two major processes; (1) the polymorphic transition of an incipient metastable crystalline form into the more stable one and (2) the redissolution of the excess enantiomer from the transformed disordered crystals into solution, resulting in a considerable enrichment of the excess one enantiomer in the mother liquor and a slight enrichment of the opposite enantiomer in the deposited crystals. Therefore, PE is not only an excellent chiral separation phenomenon but also a good research model for a complexity phenomenon. Furthermore, PE turned out to be applicable to the common racemic compound crystals, if the five requirements (high eutectic *ee* value, sufficient solubility difference, unique crystal structure & polymorphic transition, partial crystal disintegration, deposition of nonracemic mixed crystals) were satisfied. For example, the racemic compound crystals of free DL-Phe or DL-Arg, which did not show PE, could be persuaded to exhibit PE by co-crystallization with fumaric acid; the crystal engineering principles greatly contribute to produce the requisite molecular arrangement which can induce an appropriate polymorphic transition during crystallization. Therefore, it is of great advantage to use a two-component cocrystal to induce PE, because a solid-to-solid polymorphic transition involving a molecular rearrangement or a rearrangement of hydrogen bonds in the crystal lattice is essential. We believe that PE should become a potent chiral resolution method for other racemic compound crystals such as chiral drugs in concert with the further development of kinetic crystallization technologies.

References

1. Bernstein J (2002) Polymorphism in molecular crystals. Oxford University Press, Oxford
2. Brittain HG (ed) (1999) Polymorphism in pharmaceutical solids. Marcel Dekker, New York
3. Lu J, Rohani S (2009) Polymorphism and crystallization of active pharmaceutical ingredients (APIs). Curr Med Chem 16:884
4. Croker D, Hodnett BK (2010) Mechanistic features of polymorphic transformation: the role of surfaces. Cryst Growth Des 10:2806
5. Mangin D, Puel F, Veesler S (2009) Polymorphism in processes of crystallization in solution: a Practical review. Org Process Res Dev 13:1241
6. Wishkerman S, Bernstein J (2008) Polymorphism and structural mechanism of the phase transformation of Phenyl Carbamate (PC). Chem Eur J 14:197
7. Beckham GT, Peters B, Starbuck C, Variankaval N, Trout BL (2007) Surface-mediated nucleation in the solid-state polymorph transformation of Terephthalic Acid. J Am Chem Soc 129:4714
8. Herbstein FH (2006) On the mechanism of some first-order enantiotropic solid-state phase transitions: from Simon through Ubbelohde to Mnyukh. Acta Cryst B62:341
9. Weisbuch I, Leiserowitz L, Lahav M (1995) In: Mersmann A (ed) Crystallization technology handbook. Marcel Dekker, New York, p 401
10. Kitamura M (2009) Strategy for control of crystallization of polymorphs. CrystEngComm 11:949
11. Tamura R, Ushio T (2004) In: Toda F (ed) Enantiomer separation: fundamentals and practical methods. Kluwer, Dordrecht, p 135
12. Tamura R, Takahashi H, Fujimoto D, Ushio T (2007) Mechanism and scope of preferential enrichment, a symmetry-breaking enantiomeric resolution Phenomnon. Top Curr Chem 269:53
13. Tamura R, Iwama S, Takahashi H (2010) Chiral symmetry breaking phenomenon caused by a phase transition. Symmetry 2:112
14. Tamura R, Iwama S, Gonnade RG (2011) Control of polymorphic transition inducing preferential enrichment. CrystEngComm 13:5269
15. Blackmond DG, Klussmann M (2007) Spoilt for choice: assessing phase behavior models for the evolution of homochirality. Chem Commun 43:3990
16. Blackmond DG (2007) Chiral amnesia as a driving force for solid-phase homochirality. Chem Eur J 13:3290
17. Cintas P (2002) Chirality of living systems: a helping hand from crystals and oligopeptides. Angew Chem Int Ed 41:1139

18. Quack M (2002) How important is parity violation for molecular and biomolecular chirality ? Angew Chem Int Ed 41:4618
19. Avalos M, Babiano R, Cintas P, Jimenez JL, Palacios JC (2000) Chiral autocatalysis: where stereochemistry meets the origin of life. Chem Commun 36:887
20. Buschmann H, Thede R, Heller D (2000) New developments in the origins of the homochirality of biologically relevant molecules. Angew Chem Int Ed 39:4033
21. Podlech J (1999) New insight into the source of biomolecular homochirality: an extraterrestrial origin for molecules of life? Angew Chem Int Ed 38:477
22. Gubitz G, Schmid MG (eds) (2004) Chiral separations: methods and protocols. Humana Press, Totowa
23. Kozma D (ed) (2001) CRC handbook of optical resolutions via diastereomeric salt formation. CRC Press, Boca Raton
24. Subramanian G (ed) (2001) Chiral separation techniques. Wiley-VCH, Weinheim
25. Ager DJ (1999) Handbook of chiral chemicals. Marcel Dekker, New York
26. Collins AN, Sheldrake GN, Crosby J (1997) Chirality in industry II. Wiley, New York
27. Jacques J, Collet A, Wilen SH (1994) Enantiomers, racemates and resolutions. Krieger, Malabar
28. Kinbara K, Saigo K (2003) Chiral discrimination during crystallization. Top Stereochem 23:207
29. Sakai K, Sakurai R, Nohira H (2007) New resolution technologies controlled by chiral discrimination mechanism. Top Curr Chem 269:199
30. Kellogg RM, Kaptein B, Vries TR (2007) Dutch resolution of racemates and the roles of solid solution formation and nucleation inhibition. Top Curr Chem 269:159
31. Toda F (2004) In: Toda F (ed) Enantiomer separation: fundamentals and practical methods. Kluwer, Dordrecht, p 1
32. Pasteur L (1848) Recherches sur les Relations qui Peuvent Exister Entre la Forme Crystalline, la Composition Chimique et le Sens de la Polarisation Rotatoire. Ann Chim Phys 24:442
33. Collet A (1996) In: Reinhoudt DN (ed) Optical resolution, in comprehensive supramolecular chemistry, vol 10. Pergamon, Oxford, p 113
34. Kinbara K, Hashimoto Y, Sukegawa M, Nohira H, Saigo K (1996) Crystal structures of the salts of chiral primary amines with achiral carboxylic acids: recognition of the commonly-occurring Supramolecular assemblies of hydrogen-bond networks and their role in the formation of conglomerates. J Am Chem Soc 118:3441
35. Levilain G, Coquerel G (2010) Pitfalls and rewards of preferential crystallization. CrystEngComm 12:1983
36. Yoshioka R (2007) Racemization, optical resolution and crystallization-induced asymmetric transformation of amino acids and pharmaceutical intermediates. Top Curr Chem 269:83
37. Noorduin WL, Izumi T, Millemaggi A, Leeman M, Meekes H, van Enckevort WJP, Kellogg RM, Kaptein B, Vlieg E, Blackmond DG (2008) Emergence of a single solid chiral state from a nearly racemic amino acid derivative. J Am Chem Soc 130:1158
38. Noorduin WL, Meekes H, van Enckevort WJP, Millemaggi A, Leeman M, Kaptein B, Kellogg RM, Vlieg E (2008) Complete deracemization by attrition-enhanced Ostwald ripening elucidated. Angew Chem Int Ed 47:6445
39. Kaptein B, Noorduin WL, Meekes H, van Enckevort WJP, Kellogg RM, Vlieg E (2008) Attrition-enhanced deracemization of an amino acid derivative that forms an epitaxial racemic conglomerate. Angew Chem Int Ed 47:7226
40. Noorduin WL, Bode AA, van der Meijden M, Meekes H, van Etteger AF, van Enckevort WJP, Christianen PCM, Kaptein B, Kellogg RM (2009) Complete chiral symmetry breaking of an amino acid derivative directed by circular polarized light. Nat Chem 1:729
41. Noorduin WL, Meeks H, van Enckevort WJP, Kaptein B, RKellogg RM, Vlieg E (2010) Enantioselective symmetry breaking directed by the order of process steps. Angew Chem Int Ed 49:2539

42. Green BS, Knossow M (1981) Lamellar twinning explains the nearly racemic composition of chiral, single crystals of hexahelicene. Science 214:795
43. Berfeld M, Zbaida D, Leiserowitz L, Lahav M (1999) Tailor-made polymers for the removal of lamellar twinning: resolution of α-amino acids by entrainment. Adv Mater 11:328
44. Zbaida D, Lahav M, Drauz K, Knaup G, Kottenhahn M (2000) A cyclic continuous process for converting conglomerates into optically pure enantiomers by crystallization and dissolution with the assistance of tailor-made polymers. Tetrahedron 56:6645
45. Eliel E, Wilen SH, Mander LN (1994) Stereochemistry of organic compounds. Wiley, New York, p 297
46. Ushio T, Tamura R, Takahashi H, Yamamoto K (1996) Unusual Enantiomeric resolution phenomenon observed upon recrystallization of a racemic compound. Angew Chem Int Ed 35:2372
47. Tamura R, Takahashi H, Hirotsu K, Nakajima Y, Ushio T, Toda F (1998) Unusual disordered crystal structure of a racemate exhibiting a novel enantiomeric resolution: preferential enrichment. Angew Chem Int Ed 37:2876
48. Tamura R, Fujimoto D, Lepp Z, Misaki K, Miura H, Takahashi H, Ushio T, Nakai T, Hirotsu K (2002) Mechanism of preferential enrichment, an unusual enantiomeric resolution phenomenon caused by polymorphic transition during crystallization of mixed crystals composed of two enantiomers. J Am Chem Soc 124:13139
49. Fujimoto D, Tamura R, Lepp Z, Takahashi H, Ushio T (2003) Mechanism of a new type of solvent-assisted solid-to-solid polymorphic transition causing preferential enrichment: prominent influence of $C(sp^2)H$---O interaction on the control of a crystal structure. Cryst Growth Des 3:973
50. Iwama S, Horiguchi M, Sato H, Uchida Y, Takahashi H, Tsue H, Tamura R (2010) Observation of the preferential enrichment phenomenon for essential α-amino acids with a racemic crystal structure. Cryst Growth Des 10:2668
51. Gonnade RG, Iwama S, Mori Y, Takahashi H, Tsue H, Tamura R (2011) Observation of efficient preferential enrichment phenomenon for a cocrystal of (DL)-phenylalanine and fumaric acid under nonequilibrium crystallization conditions. Cryst Growth Des 11:607
52. Iwama S, Kuyama K, Mori Y, Manoj K, Gonnade RG, Suzuki K, Hughes CE, Williams PA, Harris KDM, Veesler S, Takahashi H, Tsue H, Tamura R (2014) Highly efficient chiral resolution of DL-arginine by cocrystal formation followed by recrystallization under preferential enrichment conditions. Chem Eur J 20:10343
53. Manoj K, Takahashi H, Morita Y, Gonnade RG, Iwama S, Tsue H, Tamura R (2015) Preferential enrichment of DL-leucine using cocrystal formation with oxalic acid under nonequilibrium crystallization conditions. Chirality 27:405
54. Mainzer K (2005) Symmetry and complexity: the sprit and beauty of nonlinear science. World Scientific, Singapore
55. Kauffman S (2000) Investigations. Oxford University Press, Oxford
56. Waldrop MM (1992) Complexity. Simon & Schuster, New York
57. Takahashi H, Tamura R, Fujimoto D, Lepp Z, Kobayashi K, Ushio T (2002) Preferential enrichment: full crystallographic analysis of the unusual phenomenon in the mixed crystals' version. Chirality 14:541
58. Porter WW III, Elie SC, Matzger AJ (2008) Polymorphism in carbamazepine cocrystals. Cryst Growth Des 8:14
59. Klussmann M, Izumi T, White AJP, Armstrong A, Blackmond DG (2007) Emergence of solution-phase homochirality via crystal engineering of amino acids. J Am Chem Soc 129:7657
60. Gonnade RG, Iwama S, Sugiwake R, Manoj K, Takahashi H, Tsue H, Tamura R (2012) Occurrence of spontaneous resolution of Ketoprofen with a racemic crystal structure by simple crystallization under nonequilibrium preferential enrichment conditions. Chem Commun 48:2791

Part III
Function

Chapter 23
Pharmaceutical Solid-State Characterisation Techniques

Thomas D. Turner, Peter J. Halfpenny, and Kevin J. Roberts

Abstract The solid state of a pharmaceutical material impacts on every aspect of its formulation; from solubility and thermodynamic stability to tabletability and flowability. Due to this fundamental connection characterising the solid state is key to providing the information necessary to reduce possible future manufacturing or formulation issues, which critically cuts drug product development costs and time. In this chapter a summary of the importance of the solid state and solid form screening in the pharmaceutical industry is presented. This is followed by an introduction to some of the solid state characterisation techniques routinely utilised in the pharmaceutical industry together with examples of the information provided by each.

Keywords Analytical techniques • DSC • TGA • XRD • DVS • Infrared spectroscopy • Raman spectroscopy • Particles size and shape characterisation • Solid form selection • Solid state characterisation • Quality by design

23.1 Introduction

Characterisation of the solid state is of particular importance to the pharmaceutical industry due to the rigorous selection and testing of not only the active pharmaceutical ingredient (API), but also the final drug product. The stability and processing characteristics of the formulation can impact tremendously on a number of key aspects for industry, most significantly: development time, cost and regulatory procedure.

Due to this, the pharmaceutical industry has moved towards a quality by design approach, whereby monitoring, measurement and characterisation of key mechanical and physical properties of the product throughout its processing, allows continuous optimisation of the manufacturing process. This covers a range of parameters from fundamental characteristics of the API such as polymorphic

---

T.D. Turner (✉) • P.J. Halfpenny • K.J. Roberts
School of Chemical and Process Engineering, University of Leeds, Leeds LS2 9JT, UK
e-mail: T.D.Turner@leeds.ac.uk; p.j.halfpenny@btinternet.com; k.j.roberts@leeds.ac.uk

K.J. Roberts et al. (eds.), *Engineering Crystallography: From Molecule to Crystal to Functional Form*, NATO Science for Peace and Security Series A: Chemistry and Biology, DOI 10.1007/978-94-024-1117-1_23

form stability to prevent unwanted phase transformations, to inter-relationships between critical quality attributes such as the particle size and their impact on solubility and flowability of the formulation.

This need to characterise a number of critical attributes of the drug product and API is becoming increasingly difficult for the pharmaceutical scientist. This is due to the development of progressively more complex API molecules which include salts, hydrates, solvates and co-crystals. This molecular complexity leads to increased conformational flexibility and as such challenging solid forms and particle morphologies. This increases the importance of accurate and reliable methodologies and techniques for characterising the solid state, and moreover the ability to take the results and inter-relate them between different processes to predict the impact one will have on the other.

In this chapter the impact of the solid-state properties on the characteristics of pharmaceutical materials is addressed together with a review of some of the fundamental tools the pharmaceutical scientist can deploy in the characterisation of the solid state.

23.2 Quality by Design

Quality in the pharmaceutical industry can be defined by a product which is fit for use; this is generally true when it meets key quality criteria including potency, purity, identity, bioavailability and packaging. The safety of a drug is also of upmost importance and so the material must perform as described on the labelling and also demonstrate the same clinical performance as determined in clinical testing [1]. The quality of clinical performance for a given pharmaceutical product can be summarised as:

> '*The delivery of efficacy and safety as described in the label, derived from the clinical trials*' [2, 3]

The pharmaceutical industry has until very recently operated by a quality by inspection methodology, whereby products are tested against a set of pre-defined standards after processing. This approach results in measurements which have no direct impact on the development and manufacturing processes; if batches do not meet the defined standards the batch is discarded. More recently the industry has moved towards a quality by design approach (QBD) whereby a more direct monitoring, measurement and characterisation of the quality attributes which define the product, takes place throughout the key processes of dev-elopment and manufacture to ensure improvement and optimisation of these processes.

23.3 Solid Form Screening

QBD has led to improvements in solid-form screening of the API during development to enhance the understanding of the fundamental properties of the crystalline form in an attempt to reduce processing or performance issues in later stages of testing. The solid-form screening process, then, should address the above quality criteria, safety, efficacy, processability and performance. The safety of the drug is reliant on the bioavailability and hence its concentration in the bloodstream. This concentration must always fall within the therapeutic range and below the toxicity limit. The safety and efficacy of drug products are dependent on a number of factors including:

- API content uniformity; where the correct mass of the API is required in each tablet/product formulation.
- Drug release upon administration; where tablet disintegration and particle size of the API/excipient play a role in bioavailability/dissolution profiles.
- Stability; whereby polymorphic phase transformations or loss of water from a hydrated form could lead to crystalline forms with very different API properties.

The selection of the solid form is of fundamental importance during the drug development process and once the criteria above are met, this selection is dominated by solubility and dissolution, which can both be modified by salt or co-crystal formulation. These will include a number of polymorphic forms of the compound together with pseudomorphs such as hydrates, solvates and co-crystals [4]. Polymorphs exist due to the fine balance of intermolecular interactions such as formation of specific H-bonding or the ability to form other non-covalent interactions such as π-π stacking of aromatic ring systems. A balance of these interactions together with conformational optimisation of the drug molecule can lead to the appearance other polymorphs [5–7] or pseudomorphs [8]. This is a particular frustration within the pharmaceutical industry as variation of the solid-state structure to other poly or pseudomorphs can result in undesirable solid state properties [9]. The famous example of Ritonavir is a good example where conformational polymorphism existed between forms I and II, which significantly changed the solubility and hence bioavailability of the drug [10]. According to Ostwald's rule of stages, a metastable phase will be first to crystallise followed by the most stable form [11]. The metastable form can be a solvate or a polymorph and its stability varies widely with some metastable phases being stable for seconds and some for months. The appearance of different polymorphs or metastable phases is largely due to the small energy differences between the molecular conformations and intermolecular packing of an API molecule, hence accessibility of one phase with respect to another is often easily achieved through only small changes to crystallisation conditions [12, 13]. Hence a thorough understanding of the intermolecular interactions in the solid state and also the conformational energies and solvent effects in solution are required in order to control the polymorphism of a given system.

Industrial screening of possible API candidates aims to reduce some of the risk involved with the problems associated with stability and polymorphism. This screening process attempts to improve the understanding of possible phase transformations and to identify all possible polymorphs of a solute system [14]. Additionally, the screening process will also aim to recognise not only polymorphic conversions but also to assess any possible hydrate or solvate formation but also to overcome any of the undesirable properties of the API associated with their appearance. These could include a poor solubility/dissolution profile, poor chemical/environmental stability and poor mechanical properties such as powder compaction and flow.

Table 23.1 Multi-tier evaluation of API candidates [8]

	Properties	Techniques
Tier 1	Crystallinity	Optical microscopy
	Crystal form	Powder X-ray diffraction (XRD)
Tier 2	Thermal properties and thermal behaviour (phase transitions, decomposition, de-solvation)	Differential scanning calorimetry (DSC) thermogravimetric analysis (TGA)
	Hygroscopicity, hydrate / solvate formation	Dynamic vapour sorption (DVS)
Tier 3	Polymorph and hydrate screening	Powder XRD, Raman microscopy, DSC
	Aqueous and pH solubility	Powder XRD of solubility residues, HPLC
	Chemical stability testing (e.g. hydrolysis, oxidation, photolysis)	
	Accelerated physical stability testing	Powder XRD, DSC
Tier 4	Humidity/temperature induced changes in crystal form	Powder XRD, DSC, DVS
	Influence of processing conditions(e.g. milling, micronisation, compaction) on solid form	
	Compatibility with excipients	HPLC

Initially, in the industrial development process, only small amounts of the target API will be available and this limits the capability to screen an API. The progression from an initial target molecule to a desired form, be it a salt or co-crystal, is a long process and to limit experimental time screening all possible forms of an API, a multi-tiered system is used, highlighted in Table 23.1. In this, initial samples, crystallised from a small selection of solvents, will be analysed for polymorphic form, crystallinity and habit usually using Powder XRD and optical microscopy; this will provide initial information as to the crystallographic characteristics of the material. Following acceptable results from Tier 1 analysis, further batches of the API will be tested for thermal behaviour using DSC to characterise any undesirable phase transitions which may occur in the solid state; this can also provide useful information regarding the possible future appearance of any meta-stable phases during stability testing. Establishment of any hydrate and/or solvate structures would also be characterised at this stage using DVS and the thermal de-solvation of any such species can be followed using DSC/TGA.

Once candidate API's have reached Tiers 3 and 4, larger scale batch manufacture is required for a more rigorous analysis approach. This will include full polymorph and solvate/hydrate screening, together with solubility testing for each form. This would be coupled with longer term stability testing to analyse the effect of environmental conditions such as temperature, pressure and relative humidity on a sample's tendency to phase change or chemically degrade. Finally the last candidates for the API will be tested for excipient compatibility and their processing behaviour to understand the effect of mechanical stress on the physicochemical properties of the crystalline material [15].

23.4 Thermal and Gravimetric Techniques

23.4.1 Differential Scanning Calorimetry (DSC)

DSC measures the difference between the heat flow to a sample and that to an inert reference. However, heat flow cannot easily be measured directly and hence differences in temperatures of the sample and reference are measured. Due to thermal events such as phase changes, chemical reactions etc. [16–18] the temperature of the sample does not follow that of the inert reference as shown in Fig. 23.1. The difference between the sample and reference temperatures can be converted into heat flow by calibration, using data from metals of known melting points and enthalpies of fusion, in this case sapphire is used as the reference.

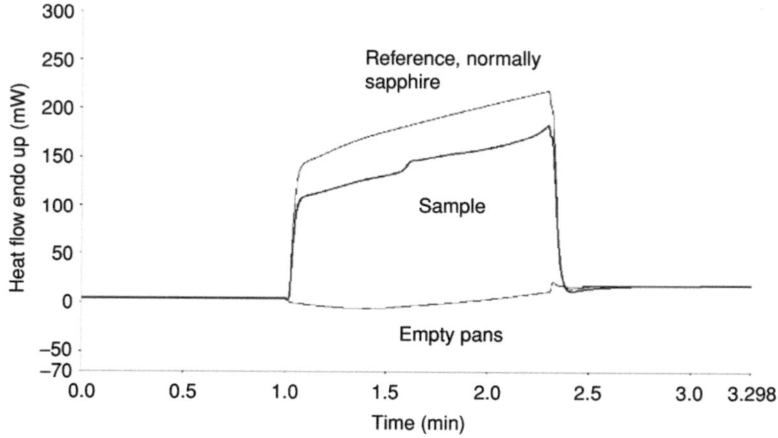

Fig. 23.1 Reference temperature heat flow, empty pan heat flow and the measured sample temperature heat flow [19]

For any thermal event, DSC yields both the temperature at which the event occurs and the corresponding enthalpy change. Some of the more common aspects of materials properties and behaviour on which DSC provides information include:

- Melting
- Crystallisation
- Solid-solid phase transformations
- Chemical reactions
- Measurement of heat capacity (Cp)
- Changes in Cp e.g. accompanying glass transition

Figure 23.2 shows typical heat flux DSC data for a crystalline solid. The integral of heat flow with respect to time, i.e. area under the peak, yields the energy (enthalpy) of a thermal change. An initial small endotherm indicates the glass transition event from the amorphous state, this is followed by a large exotherm indicating crystallisation of the material from the melt. Melting of the sample is revealed by a sharp endotherm at ~180 °C [20].

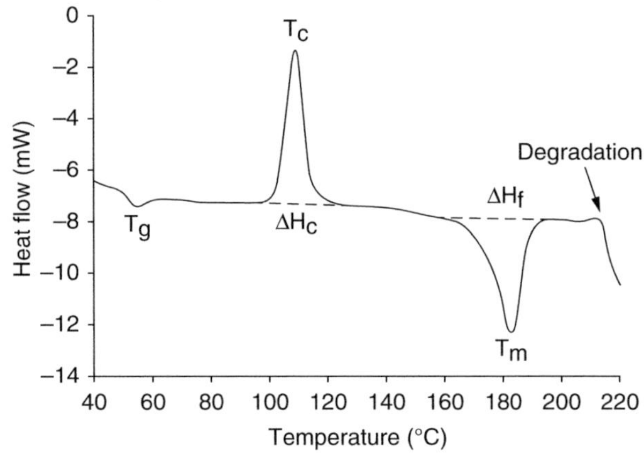

Fig. 23.2 Typical heat flux DSC data for crystallisation of an amorphous material followed by melting of the crystalline solid [20]

The application of DSC to the characterisation of pharmaceutical materials and their processing are numerous. These include measurement of the melting point of crystalline solids, determination of the phase diagram for polymorphic materials, polymorph/solvate characterisation, excipient compatibility and thermal/oxidation stability of API's [21–29]. One of the most important applications to the pharmaceutical industry is in the characterisation of polymorphs and pseudomorphs, and their stability and interconversion. Different crystal forms typically exhibit marked differences in chemical and mechanical properties and so solid form identification in batches is a key step in the development of an API. Figure 23.3 shows DSC data for a drug molecule which exists as a monohydrate and as an anhydrous form.

Figure 23.3a highlights an overlay of the typical DSC curves for the monohydrate and anhydrous forms of the drug compound. Upon heating the

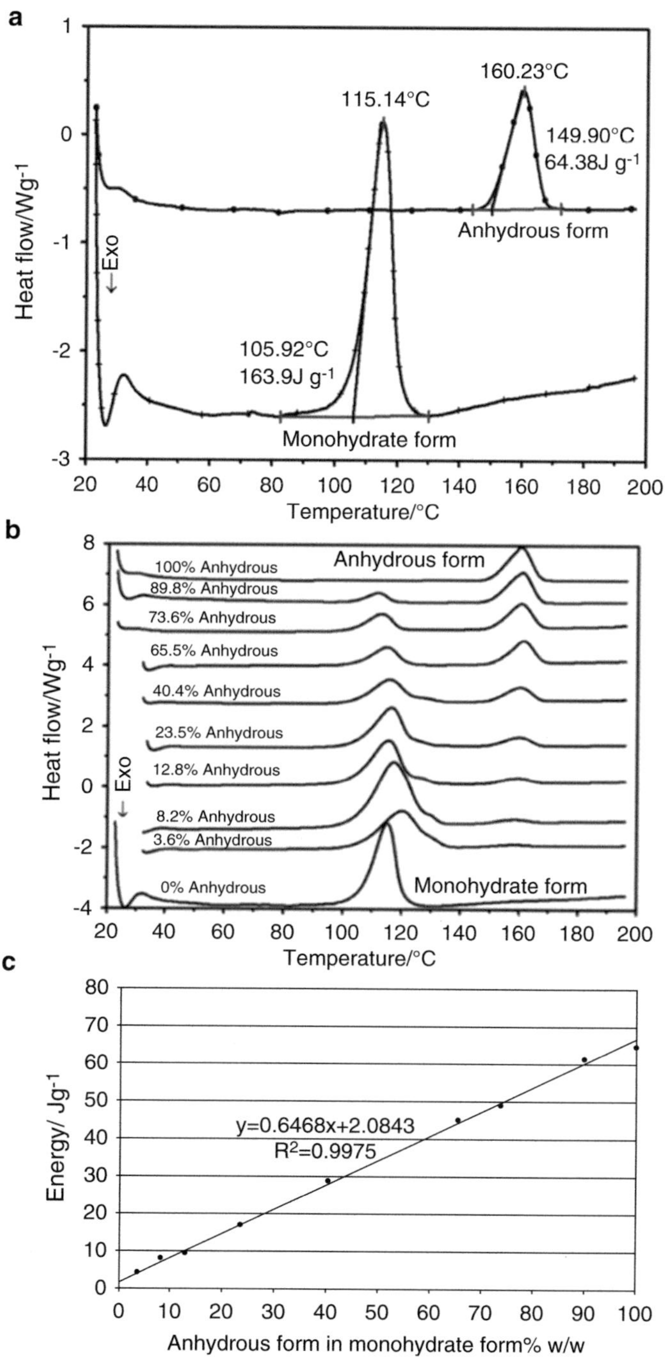

Fig. 23.3 (**a**) DSC data for conversion of a monohydrated form of a drug compound to its anhydrous form, (**b**) DSC data for mixtures of the monohydrate and anhydrous forms and (**c**) calibration curve for quantitative determination of the monohydrate and anhydrous form [30].

monohydrate melting onset is seen at 105 °C with the complete melt endotherm occurring at 115 °C, the area under the endotherm is related to the enthalpy of fusion for the monohydrate. In comparison the anhydrous form is seen to display a melting endotherm at 160 °C. During the development of this drug substance it was noted that the monohydrate had the more desirable chemical and solid-state performance properties. Further development of the monohydrate led to the requirement of a reproducible quantification technique for measuring solid concentrations of the respective forms within test samples. Various solid mixtures of the crystal forms (0–100 w/w% anhydrous form) were prepared and subsequent DSC traces were recorded; these are highlighted in Fig. 23.3b. Due to sufficient separation of the melting endotherms in the DSC traces, a quantitative model relating the measured heat of fusion energy (Jg^{-1}) of the anhydrous form to its mass concentration was constructed. There proved to be a linear correlation between the w/w% concentration of the anhydrous form and its measured fusion energy (Fig. 23.3c), this allowed concentration determination down to the detection limit of the DSC instrument which was ~4 w/w% [30]. This is a common application of DSC methods in the quantification of one phase within another. Thermal methods, in this way, have been extensively applied during the development of inhalation products.

23.4.2 Thermogravimetric Analysis (TGA)

TGA involves the measurement of changes in sample mass as a function of temperature and/or time. The sample atmosphere can be controlled during measurements, specifically pressure, humidity, inert atmosphere or oxidative atmosphere, etc. Thus a wide range of processes and interactions can be investigated, ranging from adsorption to loss of solvent or thermal decomposition [31]. Furthermore, volatile materials evolved during measurements can be quantified and characterised by ancillary techniques such as mass spectrometry (TG-MS) or FTIR spectroscopy (TG-FTIR). The information obtainable from TGA is, again, extensive. A few examples are:

- Volatile materials content
- Drying behaviour
- Thermal stability
- Decomposition behaviour

A TGA plot of copper sulphate pentahydrate is shown in Fig. 23.4, illustrating many of the common features of TGA data. The first weight loss step corresponds to loss of volatiles such as unbound or weakly bound solvent, in this case this is removal of two H_2O molecules. The second elimination step corresponds to more strongly bound solvent; this also corresponds to elimination of 2 H_2O molecules. The third step at 200 °C indicates elimination of the final H_2O molecule. At higher temperatures, thermal decomposition or other chemical reactions with the atmosphere may occur producing a substantial weight loss, this occurs at ~650 °C in Fig. 23.4 leaving a final residue which remains effectively constant in weight with temperature at ~900 °C.

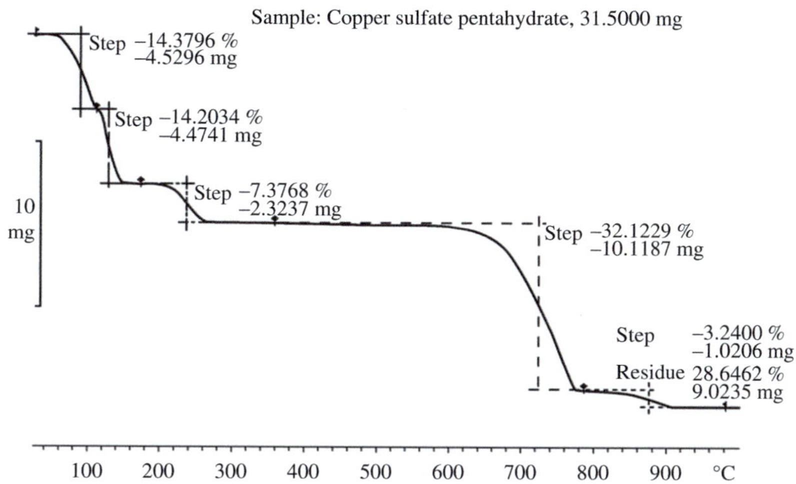

Fig. 23.4 TGA trace for copper sulphate pentahydrate; indicating the important molecular elimination steps [19]

TGA is frequently used in the pharmaceutical industry for the investigation of drying behaviour, solvate or hydrate transformations, composition analysis and thermal stability/decomposition studies [32]. Many crystalline materials crystallise with solvent molecules (frequently water) incorporated into the crystal structures. Such solvates (hydrates in the case of water) exhibit substantial differences in properties from the anhydrous compound. Furthermore, they can be substantially influenced by thermal processing. TGA enables characterisation of the stability, inter-conversion and de-solvation of such compounds, allowing quantification of the solvate/hydrate stoichiometry.

Fig. 23.5 TGA data for the three structurally related drug molecules HPCHC, TCHC and SLCHC [33]

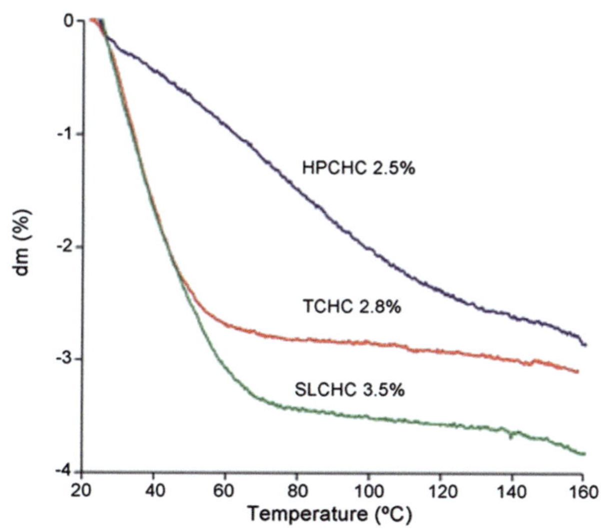

Figure 23.5 compares TGA data for three structurally related anaesthetic drug molecules hydroxyprocaine hydrochloride (HPCHC), tetracaine hydrochloride (TCHC) and hydroxytetracaine hydrochloride (SLCHC), which are known to be hydrated structures, however, their stoichiometry was unknown. The TGA mass change indicated on the plots shows a mass loss between 2.5 and 3.5%, this relates to a molar equivalence of 0.4 – 0.6 mol of water confirming all three structures to be hemihydrates. Further to this the TGA data highlighted the relative thermal stability of the three hydrates where decomposition of HPCHC and SLCHC started around 160 °C whereas TCHC showed a higher thermal stability whereby decomposition did not begin until ~220 °C [33].

23.4.3 Dynamic Vapour Sorption (DVS)

Interaction of pharmaceutical solids with moisture influences a wide range of material properties, as well as their processability and stability. During processing adsorption (surface) and absorption (bulk) of moisture can lead to changes in flow characteristics, structural changes and formation of hydrates. During storage, interaction with water vapour affects chemical stability (via hydrolysis) and particle properties (via caking). Because of the similarities between TGA and DVS, a number of the applications of the two techniques are similar. However, the key difference between the techniques is the programmable humidity which can be achieved with DVS. Some representative applications are listed below:

- Adsorption isotherms, e.g. Fig. 23.6
- Characterisation of amorphous content
- Characterisation of hydrate formation
- Characterisation of glass transition and crystallisation behaviour

Quantifying the extent to which a solid is amorphous or crystalline is of considerable importance in controlling product properties, particularly in the pharmaceutical industry. This stems from the marked differences between amorphous and crystalline solids. Amorphous solids are disordered structures whereas crystalline solids are three-dimensionally ordered arrays of atoms. The properties of amorphous solids differ substantially from the crystalline forms of the same compound, e.g. the solubility and hence bioavailability of the amorphous form is higher than the corresponding crystalline solid [35]. Also, amorphous solids typically convert readily to their crystalline forms. Amorphous material may arise from, e.g. very rapid crystallisation/precipitation or milling with extensive mechanical deformation of the solid [36, 37]. Given the differences in properties and the

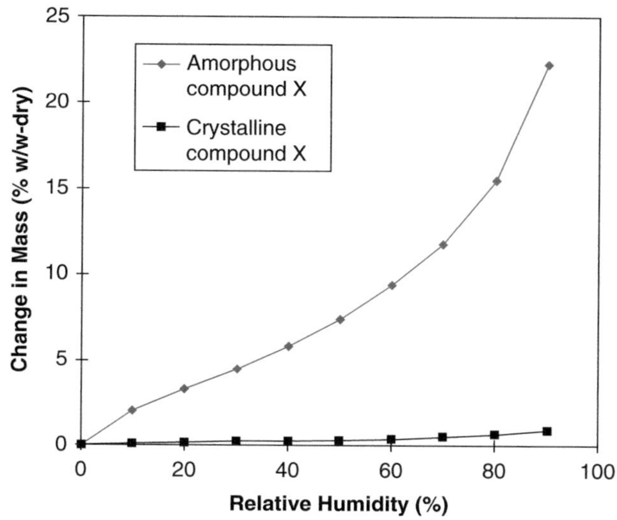

Fig. 23.6 Typical sorption profile for an amorphous compound in comparison to a crystalline material [34]

potential for conversion to crystalline forms, it is important to know the proportion of amorphous material present.

Figure 23.7 shows the change in mass profiles for a sample of a pharmaceutical compound with a 4.8% amorphous content. The sample is subjected to a sequence of high and low humidity pulses. After exposure to high humidity, subsequent changes in mass with humidity are lower. This is as a consequence of an amorphous to crystalline transition in the solid at high humidity. The changes allow calculation of the amorphous content of the sample [34].

A large proportion of pharmaceutical solid materials exist as hydrates and understanding the phase behaviour of these materials when under differing humidity conditions is important for the pharmaceutical industry. Many solids undergo phase transitions due to absorption of water into the bulk structure or adsorption onto the surface providing opportunity for glass transitions or recrystallization to a more stable form of the drug. These can include stable crystalline polymorphs or hydrates with varying stoichiometric proportions of water depending on the humidity and temperature. These changes can result in profound variations in the solid-state and chemical-state properties of the material such as compressibility and solubility [39].

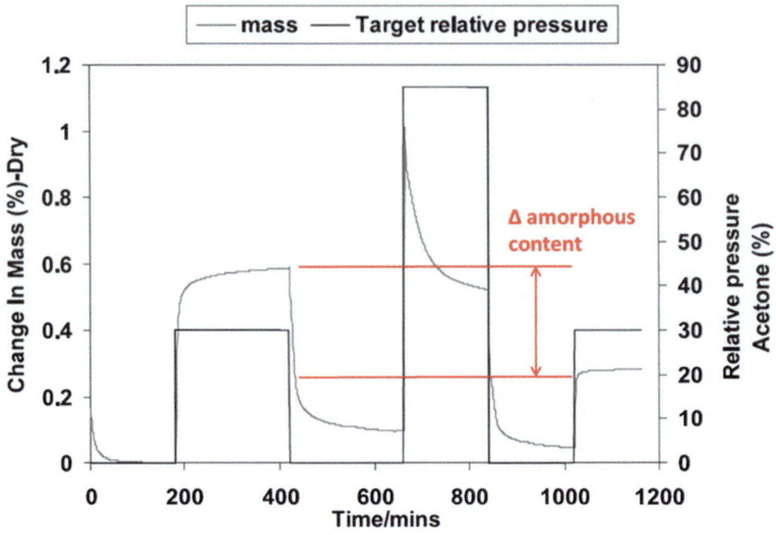

Fig. 23.7 Determination of amorphous content using DVS [34]

Figure 23.8 highlights the sorption and desorption profile of the di-hydrate drug compound Naloxone HCl. The isotherms highlight the change of mass of the sample relative to its dry mass as a function of relative humidity. The isotherms in Fig. 23.8 show a large difference between the sorption and desorption of water from the sample; this is very indicative of hydrate formation as the uptake of water generally occurs at higher relative humidity compared to the loss of water from the structure. The uptake isotherm in Fig. 23.8 allows calculation of the stoichiometry of the hydrate species. The equilibrium mass levels out at a 9.6% uptake of the samples dry mass. This result combined with the molecular weight of the drug material and water, allows calculation of the molar stoichiometry of 1.9, confirming that the sample is indeed in a di-hydrated state [38].

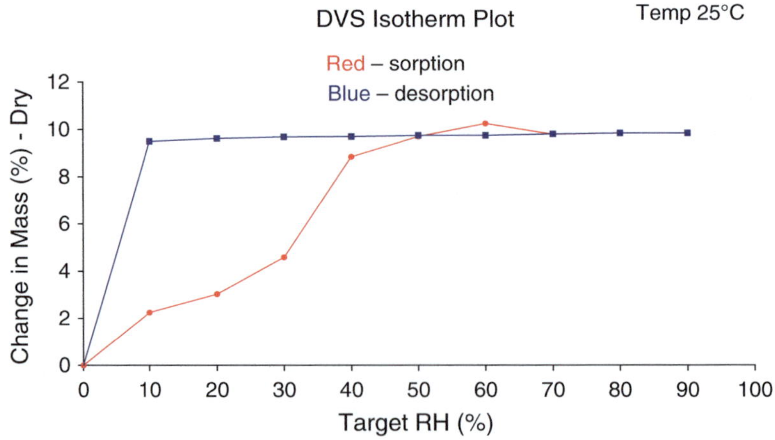

Fig. 23.8 Water sorption and desorption isotherms of Naloxone HCl [38]

23.5 Vibrational Spectroscopy

23.5.1 Infrared Spectroscopy

The two main applications of mid-infrared spectroscopy are (i) identification of functional groups and (ii) structure identification by fingerprinting. The frequency at which a bond absorbs depends upon the strength of the bond and the masses of the bonded atoms. The bonds in a given functional group should, therefore, absorb at characteristic frequencies allowing their identification. This is indeed true, though several factors complicate the situation. Firstly, the absorption frequency of a bond is affected by the environment of the bond, including factors such as conjugation within the molecule. The second complication arises from the overlapping of absorption bands associated with different functional groups. As a consequence, absorption by a given functional group may occur over a range of frequencies depending upon the nature of the molecule in which the group is located. From examination of the infrared spectra of a wide range of compounds, these absorption ranges have been identified and can be summarised in correlation charts, such as that shown in Fig. 23.9.

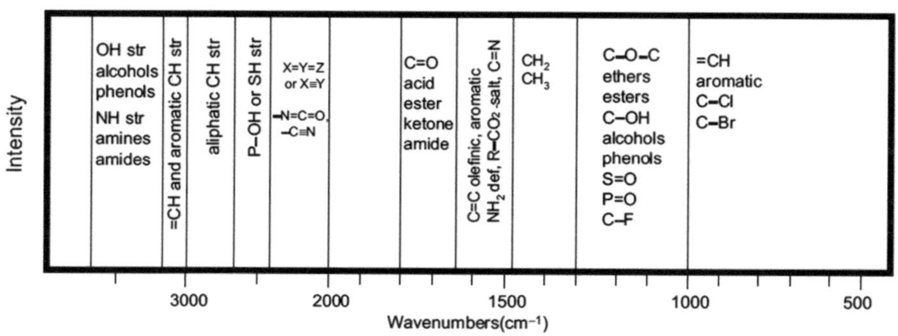

Fig. 23.9 Infrared spectroscopy correlation chart [40]

Functional groups which absorb at a number of characteristic frequencies are easier to identify from infrared spectra than those which have only one. For example: amides exhibit C = O stretching, N-H stretching, N-H bending, etc. this allowing easy identification. Slightly more difficult are esters with C = O stretching and C-O stretching absorption bands. While ketones can be more difficult to identify unambiguously on the basis of C = O stretching alone. In many cases, identification of absorption bands may be difficult due to overlapping or multiple possible assignments. Consequently, infrared spectroscopy is seldom employed as the sole structural tool. The following example illustrates the application of infrared spectroscopy to functional group identification.

The solid state spectra of alpha para-aminobenzoic acid (PABA) are shown in Fig. 23.10. The crystal structure of alpha PABA is characterised by the centrosymmetric H-bonding COOH dimer as well as N-H---C = O H bonding interactions.

The N-H region (3000 – 3500 cm^{-1}) of the spectrum is well characterised in the solid state, with the two peaks at 3459.13 and 3360.37 cm^{-1} corresponding to the symmetric and anti-symmetric N-H stretch of the non H-bonded amino group. The third peak at 3228.51 cm^{-1} is the H-bonded N-H stretch; this interaction is labelled as bond 2 inset in Fig. 23.10. The O-H stretching region is usually visible in the region of 2000 – 3500 cm^{-1}, however, in the case of alpha PABA this region is distorted with no characteristic peaks visible. This is usual for a carboxylic acid molecule such as PABA [42].

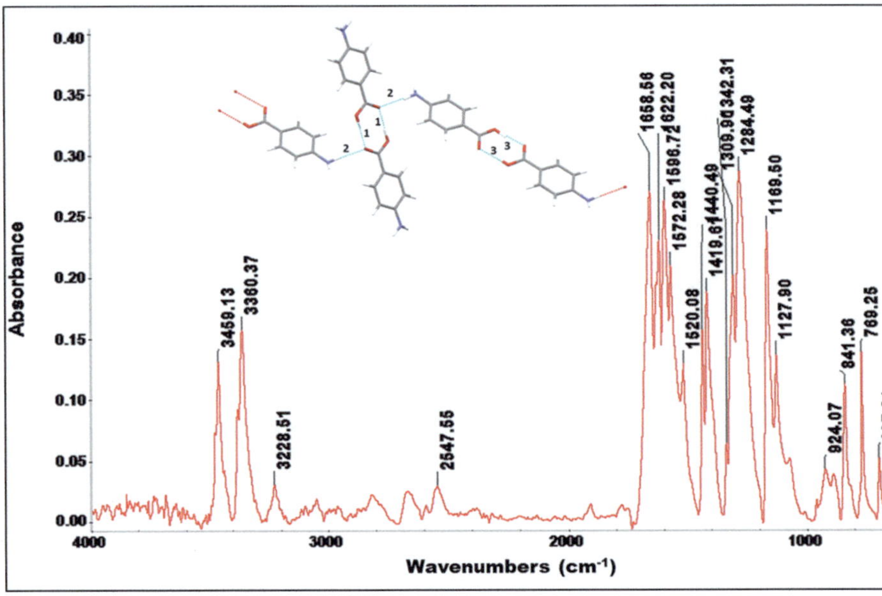

Fig. 23.10 Solid state infrared spectrum of alpha para-aminobenzoic acid, inset the bonding environment within the unit cell of the alpha form highlighting the important H-bonding interactions [41]

The region centred around 1700 – 1500 cm^{-1} highlights the existence of a carbonyl group, the band at 1658.56 cm^{-1} and the shoulder at ~1632 cm^{-1} indicate the symmetrical and anti-symmetrical vibrations of the carboxylic acid dimer, bond 1, in Fig. 23.10. The band at 1622.20 cm^{-1} appears due to the carbonyl group participating in the N-H---C = O H bonding interaction, highlighted as bond 2 inset in Fig. 23.10. The peak at 1596.72 cm^{-1}, indicates the presence of the aromatic ring in PABA.

The region which follows with the two sets of peaks 1440.49, 1419.61 cm^{-1} and 1309.9 and 1284.49 cm^{-1} are common bands indicative of either C-O in-plane stretching or O-H deformation vibrations, both of which are present in PABA [42]. The broad doublet centred around 924.07 cm^{-1} is due to the O-H wag; this region is usually used to indicate the presence of carboxylic acid dimers, with a broader band characteristic of an acid dimer. The position is also important and has been extensively studied with the usual position in the region of 960 – 875 cm^{-1}, dependant on the intermolecular interactions in the crystal structure [43–47].

The fingerprint region of the infrared spectrum ranges approximately from 900 to 1400 cm^{-1}. This region contains a large number of unassigned vibrations which result in a complex pattern of absorption peaks. As a consequence, even very similar molecular structures yield different infrared spectra. Polymorph identification involves examination of, in particular, the fingerprint region of spectra and comparison with reference spectra recorded under the same conditions. Despite the fact that polymorphs have the same molecular structure, substantial differences in infrared spectra arise because of the differing environments and intermolecular forces in the different polymorphic structures.

The antihypertension drug Nifedipine exists as three polymorphic forms α, β and γ, and one glassy amorphous phase. This degree of polymorphism is produced by the degree of conformational rotations within the large flexible Nifedipine molecule. The structural characteristics of these crystalline and glassy phases of Nifedipine were characterised by IR spectroscopy by Chan et al. [48]. It was found that specific conformational rotations of the Nifedipine molecule resulted in the various crystalline forms of the drug, whereas the amorphous glassy phase contained both stable and metastable conformations of Nifedipine. The IR spectra highlighted in Fig. 23.11 indicate the various fingerprint peaks of the crystalline and glassy forms of the material where the wavenumber of the specific vibrational mode is changed to italics. The specific region of interest in resolving the various forms of Nifedipine by their IR spectra is the carbonyl stretching region; 1200 cm^{-1} – 1800 cm^{-1}. This example of polymorphism in the case of Nifedipine highlights the significance of IR spectroscopy in fingerprint identification of the various forms of a pharmaceutical material simply from deviations in the vibrational spectra of solid forms.

23.5.2 *Raman Spectroscopy*

As Raman and mid-IR techniques are complimentary, Raman spectroscopy is often used for fingerprint identification of different forms of a pharmaceutical material such as polymorphs, hydrates or co-crystals [49–51]. Raman spectroscopy is effective as a fingerprinting tool both for closely related compounds and for polymorphs (in a similar manner to mid-infrared spectroscopy fingerprinting). It is also suitable for analysis of complex samples such as drugs within their formulations, e.g. by subtraction of the formulation matrix background. Raman has also been used for *in-situ* characterisation of drug forms during processing and stability testing and also for quantitative studies of polymorph transformations [49]. The main advantage Raman has over infrared techniques is that aqueous sample environments can be analysed due to the weak intensity bands observed in the Raman spectrum of water. This provides a particularly useful tool for qualitative and quantitative analysis of drug formulations, which are often studied in aqueous solvation environments.

The example in Fig. 23.12 highlights the Raman spectra for forms A and D of the drug molecule entacapone in comparison to the surface enhanced Raman spectra of entacapone crystals grown at a gold templating surface to direct the nucleation of

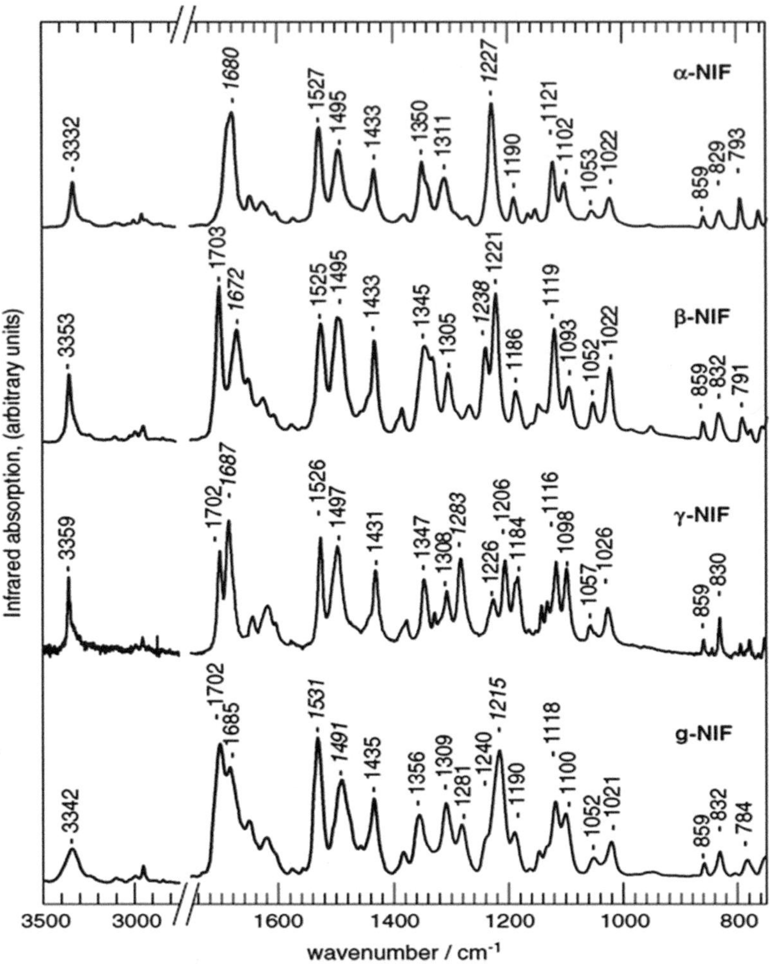

Fig. 23.11 FT-IR spectra of Nifedipine Forms α, β, γ and the glassy form [48]

stable Form A [52]. The Raman probe allows polymorph identification by specific fingerprinting of the spectral region of 1000 – 1500 cm^{-1} which indicates clear differences between the two polymorphic forms. The surface enhanced Raman spectroscopy (SERS) at a gold templating surface indicates the presence of a self-assembled layer of entacapone after adsorption. The SERS indicates that Form A adsorbs onto the gold templating surface by identification of the C = C and C = O vibrational fingerprint peaks at 1500 – 1600 cm^{-1}.

Raman spectroscopy is routinely used in the pharmaceutical industry not only in polymorph identification but also for the evaluation of drug products. Figure 23.13 highlights recorded Raman spectra for a developmental drug product with varying concentration in comparison to a placebo material for that specific formulation. The spectra indicate increasing peak intensity as a function of API concentration relative to the placebo. Using this relationship a calibration curve can be

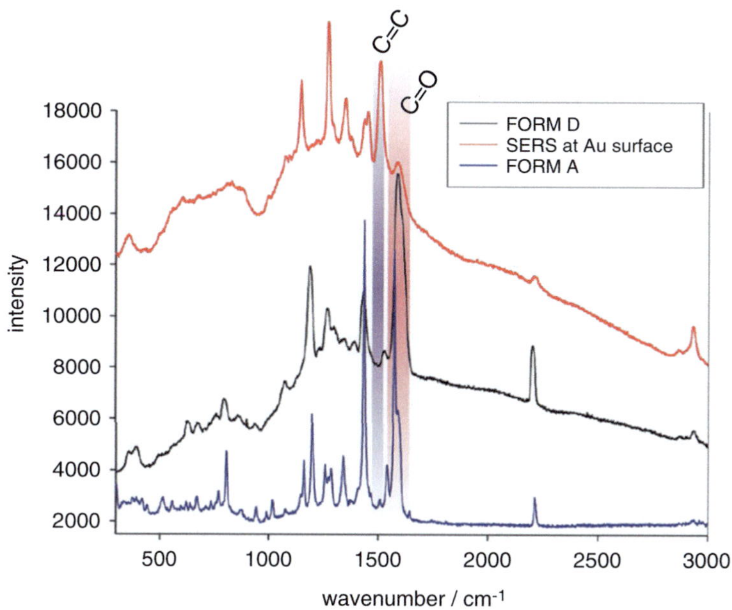

Fig. 23.12 Raman spectra of entacapone form A (*blue*) and form D (*black*) compared with entacapone crystals grown at a gold templating surface [52]

constructed relating fitted peak area to API concentration. This is particularly useful during product development testing and allows a fast, easy and non-destructive analysis of drug products whilst still in packaging [52].

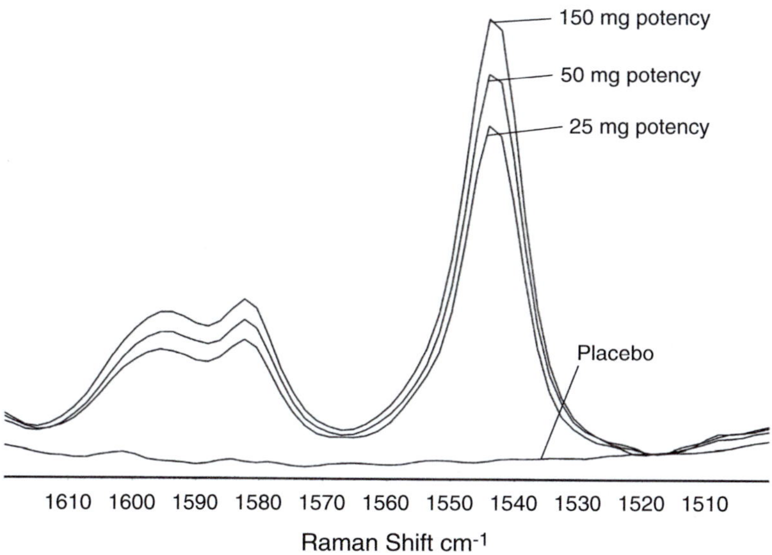

Fig. 23.13 Raman spectra of a developmental compound at 25 mg, 50 mg and 150 mg concentration compared to a placebo formulation [53]

23.6 Powder X-ray Diffraction

A typical diffraction pattern contains many peaks, each due to diffraction from a given set of crystal planes. Bragg's law [54], Eq. 23.1 states that a beam of X-rays of wavelength λ is incident to the crystal planes at an angle of θ, the Bragg angle, measured between the incident beam and the crystal planes. When diffracted, by the crystal planes, the beam will scatter in all directions but only a portion of the scattered beam will be in phase and able to reinforce each other to be a detectable diffracted beam.

$$n\lambda = 2dsin\theta \tag{23.1}$$

n here is the order of diffraction and if the values of λ and d are fixed then multiple values of θ can be obtained at which diffraction may occur, corresponding to $n = 1$, 2, 3 [55]. The number of peaks resulting from the diffraction of X-rays depends upon the crystal structure and the wavelength employed. Structural information is contained in both the positions (angles) and intensities of the peaks.

An array of X-ray diffraction techniques exist for characterisation of a wide range of structural and chemical properties of crystalline solids, including:

- Crystal structure determination
- Compound identification
- Determination of crystallographic orientation
- Particle size analysis
- Crystal strain measurement
- Degree of crystallinity/crystal quality
- Defect density and characterisation

In addition to crystal structure, a number of sample and instrumental factors influence the characteristics of diffraction peaks [56]. Some of these can yield valuable information on material properties. Among the most important effects are:

Preferred Orientation
The orientation of crystallites in a powder is rarely totally random. The shape of the individual crystals is a key factor in determining how particles pack and whether their orientation is random. Plate-like or tabular crystals tend to pack with their largest faces parallel. Needle-like crystals frequently pack with the needle axes aligned. Only in the case of equi-dimensioned crystallites is random orientation approached. The effect of preferred orientation is to change the intensities of diffraction peaks relative to that expected for a completely random sample. This complicates the analysis of diffraction patterns and their use in fingerprinting.

Particle Size
As the particle size decreases, diffraction peak width increases. For a thin crystal, the extent of destructive interference near the Bragg angle is less than for a thick crystal. Hence diffraction occurs over a wider range of angles either side of the

exact Bragg angle and the peak width increases. This effect is only significant for particle sizes around 0.1 mm and below. Peak broadening can also mask features in diffraction patterns and hinder interpretation. Analysis of the peak broadening, however, can enable determination of particle size. The key problem is separation of the size related broadening from the 'normal' peak width. This can be achieved by including with the sample during analysis, an internal standard with a particle size greater than 0.1 mm.

Strain and Loss of Crystallinity

The characteristics of a diffraction peak can be altered by crystal strain. The effect of uniform strain within the crystals is to shift the peak position since the inter-planar spacing is altered by the strain. In the case of non-uniform strain, a range of inter-planar spacing's are present. Diffraction can therefore occur over a wider range of angles resulting in peak broadening.

Figure 23.14 presents powder diffraction patterns of β succinic acid of a plate like morphology and also of the same sample after ball milling of the powder [57]. Ball milling of the sample results in peak broadening; particularly of the diffraction peak at ~26 2theta, this is evidence of a reduction of particle crystallinity due to the mechanical stress of milling. Since both crystal size and non-uniform strain influence the diffraction peak width, they can be difficult to distinguish. Their influences on peak shape, however, are different. Fourier analysis of peak profiles can separate and quantify the size and strain effects.

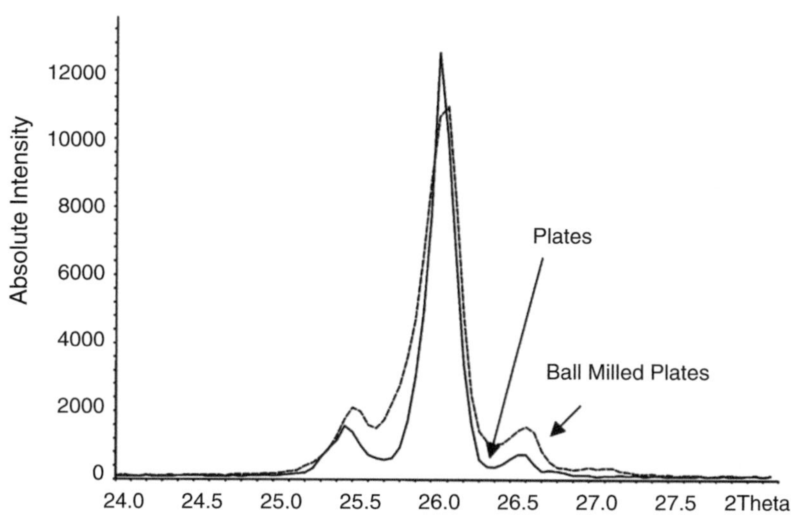

Fig. 23.14 Powder X-ray diffraction patterns of two samples of β succinic acid; plate like crystals and ball milled powder of the same sample indicating a loss of crystallinity upon milling [57]

Since polymorphs, by definition, have different crystal structures, their diffraction patterns are different. Through a combination of differing unit cell size/shape and different atomic positions, substantial differences in patterns are typically observed. Figure 23.15 shows the powder diffraction patterns for two polymorphic forms of paracetamol. Thus powder diffraction patterns can be used to identify polymorphs by fingerprinting, determination of characteristic peak positions and comparison with reference patterns for each polymorph. In addition, the peak intensities are related to the amount of diffracting material. Under favourable circumstances, therefore, powder diffraction patterns can be used to determine the relative proportions of different forms in mixtures of polymorphs.

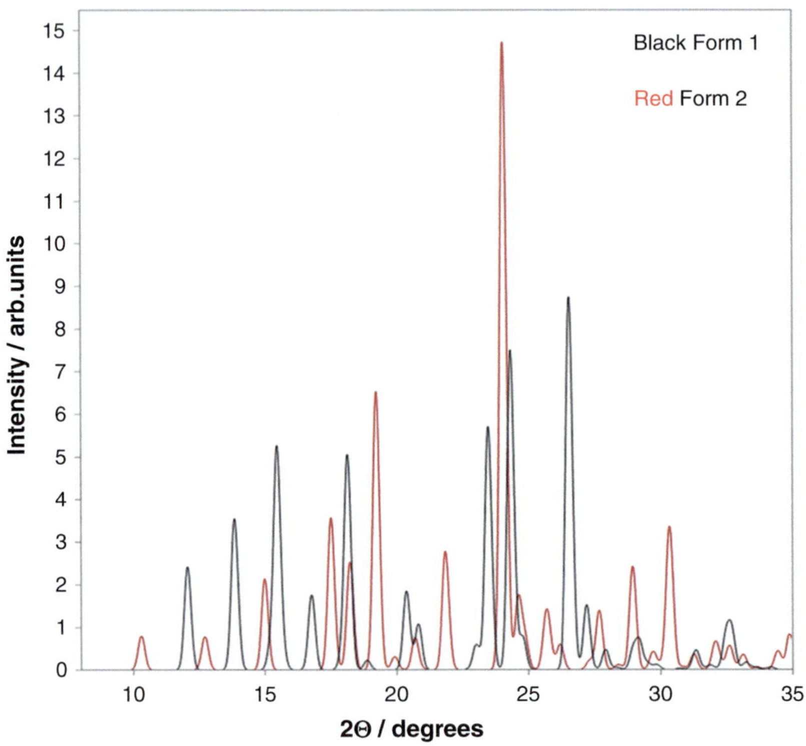

Fig. 23.15 Simulated X-ray diffraction patterns from the crystal structures for the two polymorphs of paracetamol; form 1 and form 2 [58, 59]

23.7 Particle Sizing

Particle size influences a wide range of physical and chemical properties. Among the most important to pharmaceutical material behaviour are solubility, dissolution rate and bioavailability. In addition to affecting fundamental material properties such as these, many aspects of the formulation process are strongly affected by the particle size of active components and excipients. e.g. flow characteristics, packing behaviour, the tendency of solids to cake during storage and segregation behaviour of powder mixtures.

The commonly used techniques for particle size measurements include:

- Sieving
- Microscopy (with or without image analysis) – optical microscope or electron microscope
- Electric stream sensing zone method (Coulter counter)
- Laser diffraction
- Sedimentation – by gravity or centrifugal
- Ultrasonic attenuation

23.8 Laser Diffraction

Small particles scatter light at a small angle to the direction of the incident beam. The intensity patterns depend upon the diameter of the particles involved. In a powder sample, the resulting diffraction pattern is a combination of the scattering from different sized particles. There are two main approaches to extracting size information from diffraction patterns: Fraunhofer diffraction theory and Mie scattering theory. The type of analysis used depends upon the size of the particles in the powder.

Dry powder inhalers deliver drug particulates to the respiratory tract for the treatment of a number of ailments, these particles are required to have a particular size distribution for effective deposition in the lungs [60, 61]. Crystalline drug material often has a tendency to agglomerate due to surface chemistry or crystallite morphology, this is often overcome by formulation with excipient particles which act as carriers for the API [62]. As such it is important that during the delivery of the drug, successful de-agglomeration of the powder to its individual constituent particles occurs in order to prevent poor aerosolisation of the product and hence reduce patient uptake of the API through lower bioavailability [63].

Laser diffraction is a tool which is routinely used in the pharmaceutical industry to analyse particle size distributions (PSD) of dry powders to measure the extent to which a formulation agglomerates during manufacture. In addition to this, recent pressure titration experiments have been used in combination with laser diffraction measurements to assess the strength of agglomerates and its effect on de-agglomeration by measuring the PSD of a powder under varying shear stress from a static powder bed [64, 65]. This has been shown to give a measure of dispersibility for a powder sample by relating measured PSDs and the dispersion pressure exerted upon the sample [66, 67].

Recently, Jafarri et al. have utilised this methodology of dry dispersion laser diffraction to study the degree of agglomeration of seven inhalation drug/excipient powders and ultimately determine their cohesitivity [68]. A range of dispersion pressures were applied to the powders (beclometasone dipropionate, budesonide, fluticasone propionate, lactohale 300, salbutamol base, salmeterol xinafoate and tofimilastand) and the PSD was recorded during the de-agglomeration process. The degree of de-agglomeration was assessed by calculating a de-agglomeration parameter DA, which is the ratio between the primary particle size of the powders and the measured particle size following pressure shear. The study found that the de-agglomeration of all the powders increased with increasing primary pressure, however, the degree of de-agglomeration differed between powders. Fig. 23.16 shows the de-agglomeration profiles of all the powders studied; this indicates the critical primary pressure required to produce the primary particle size (i.e. DA = 1) was much lower in salbutamol base than it was in salmeterol xinafoate. This indicates that the inter-particle cohesitivity is much higher in salmeterol xinafoate than other samples. Thus the use of dry dispersion laser diffraction provides a direct measure of the particle cohesion and de-agglomeration behaviour of bulk pharmaceutical powders and highlights the practicality of such measurements for the development of solid state pharmaceutical materials.

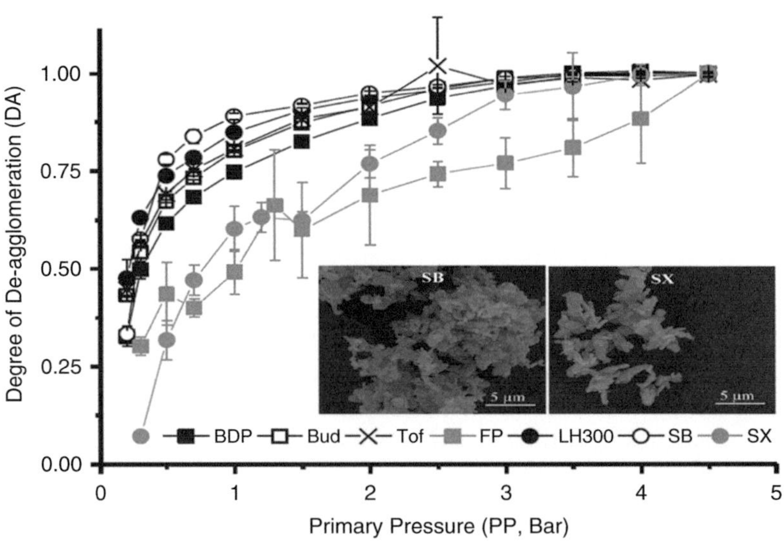

Fig. 23.16 The de-agglomeration profiles of beclometasone dipropionate (BDP), budesonide (Bud), fluticasone propionate (FP), lactohale 300 (LH300), salbutamol base (SB), salmeterol xinafoate (SX) and tofimilast (Tof), with SEM images of the SB and SX samples inset [68]

23.9 Image Analysis

Image analysis offers a number of additional capabilities compared to diffraction techniques.

- It can readily measure particle shape.
- The greater resolution yields number-based distributions and volume-based data.
- Images of individual particles can be observed and recorded

The advantages of image-based characterisation are well illustrated by Fig. 23.17. The four shapes shown could appear identical when considering volume equivalent diameter or circle equivalent diameter, however from a particle processing perspective they have very important differences in their external morphology.

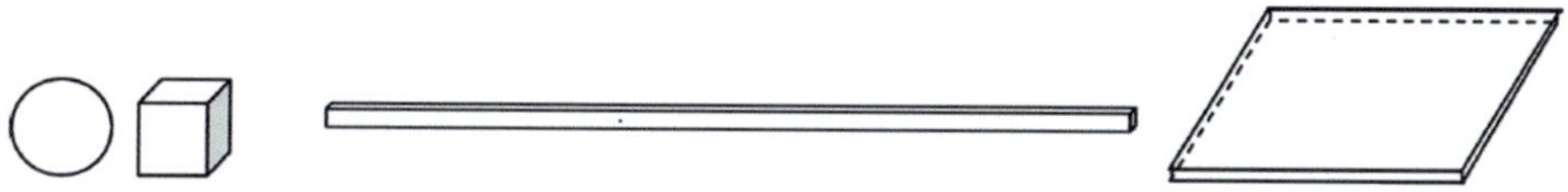

Fig. 23.17 Theoretical particulate shapes having the same volume equivalent diameter but very different external morphology [69].

Figure 23.18. shows a three-dimensional particle illuminated and magnified; the image is captured in two dimensions using a digital CCD. This two dimensional image is converted to a circle with equivalent area of the original image. (Alternative equivalent diameters may be used). The diameter of the circle is reported as a circle equivalent (CE) diameter. Statistics can be generated on the sample as a whole providing mean, standard deviation, median, percentiles, etc. of the CE diameter. The steps are summarised in Fig. 23.18.

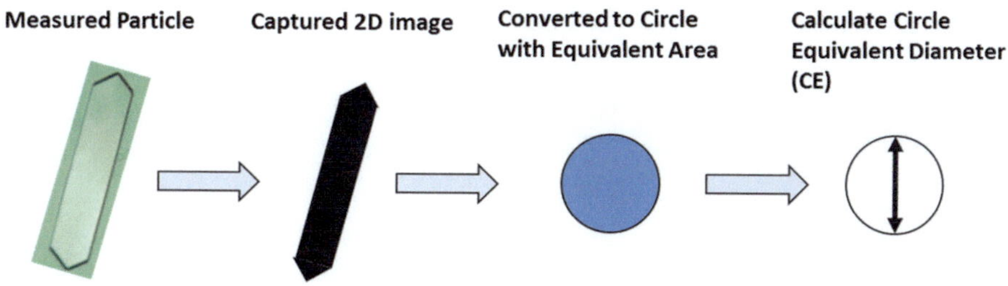

Fig. 23.18 Stages in particle image analysis (measured particle image from Nguyen et al. 2014 [69])

Image analysis is routinely used for characterisation of particle morphology, however, with modern image analysis software combined with powerful microscopy tools, the particle size distributions and, more specifically, morphology differences in a sample can be analysed. This is particularly important when designing a crystallisation process for an API as the crystalline morphology and size distribution impacts upon dissolution and mechanical processing properties such as tabletability.

As an example Fig. 23.19 presents the optical image analysis for samples of para aminobenzoic acid, a small organic material, crystallising from three solvents: ethanol, acetonitrile and water. The micrographs (top left, top right, bottom left) indicate that the crystals grown from ethanol and acetonitrile seem to be the largest and contain crystallographically well defined facets, whereas those grown from water are much smaller with crystallographically poorly defined facets. Optical image analysis using a Malvern Morphologi G3 [70] (bottom right) reveals the length distributions of the needle-like crystals which agree well with the observations of the micrograph analysis where the crystallite length distributions in ethanol and acetonitrile are much larger than the distributions in water [71].

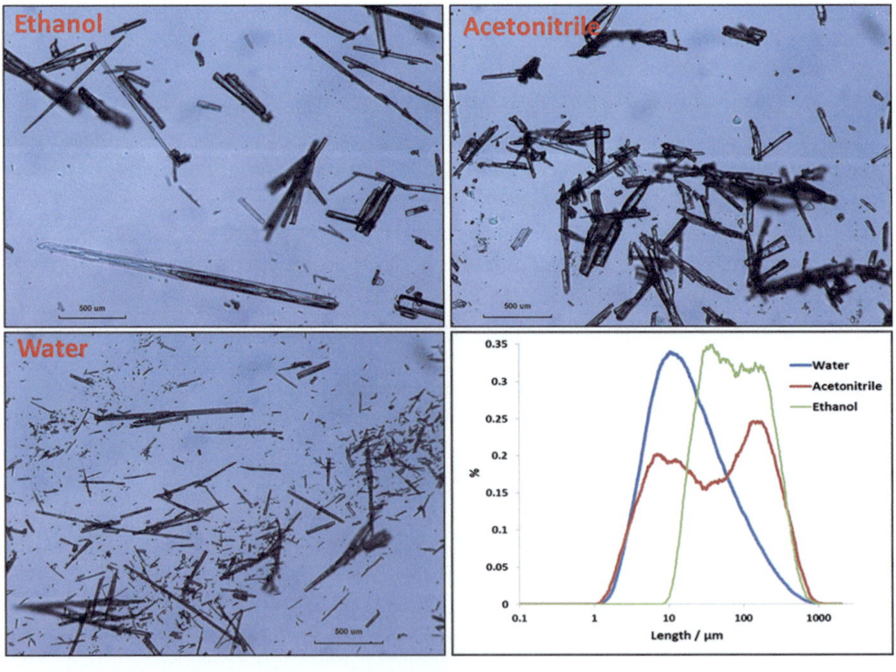

Fig. 23.19 Micrographs of crystals recovered from poly-thermal cooling experiments from ethanol, acetonitrile and water solutions together with length distributions of recovered crystallites from cooling crystallisations at supersaturations =1.2 using optical image analysis [71]

References

1. Fed. Reg. (2005) 70, 134
2. Massa T (2003) Summary of FDA/PQRI Workshop. at the ACPS Manufacturing Subcommittee 17 Sept
3. Woodcock J (2004) The concept of pharmaceutical quality. Am Pharm Rev 7(6):10–15
4. Byrn SR, Pfeiffer RR, Stowell JG (1999) Solid state chemistry of drugs, 2nd edn. Press, SSCI
5. Docherty R, Jones W (1997) Organic Molecular Solids: Properties and Applications. CRC Press: Boca Raton ch 3
6. Etter MC (1990) Encoding and decoding hydrogen-bond patterns of organic compounds. Acc Chem Res 23(4):120–126
7. Desiraju GR (1997) The crystal as a Supramolecular entity. Wiley, Chichester
8. Roberts KJ, Docherty R, Taylor S (2011) Pharmaceutical Process Development: Current Chemical and Engineering Challenges. Royal Society of Chemistry ch. 13
9. Haleblian J, McCrone WC (1969) Pharmaceutical applications of polymorphism. J Pharm Sci 58:911–929
10. Bauer J, Spanton S, Henry R, Quick J, Dziki W, Porter W, Morris J (2001) Ritonavir: an extraordinary example of conformational polymorphism. Pharm Res 18(6):859–866
11. Ostwald W (1897) Studien über die Bildung und Umwandlung fester Körper. Z Phys Chem 22:289–330
12. Ostwald W (1899) Grundriss der allgemeinen Chemie. W. Engelmann, Leipzig
13. Ostwald W (1900) Uber die vermeintliche Isomerie des roten und gelben Quecksilberoxyds und die Oberflachenspannung fester Korper. Z Phys Chem 34:495–503
14. Berge SM, Bighley LD, Monkhouse DC (1977) Pharmaceutical salts. J Pharm Sci 66:1–19
15. Brittain HG (1995) Physical characterization of pharmaceutical solids. Dekker, New York
16. Le Chatelier H (1887) C R Acad Sci Paris, 104:1443–1446
17. Le Chatelier H (1887) De l'action de la chaleur sur les argiles. Bull Soc Fr Minéral Cristallogr 10:204–211
18. Roberts-Austen WC (1899) Report 5. Proc Inst Mech Eng:35–102
19. Gabbott P (2008) Principles and applications of thermal analysis. Wiley, Oxford
20. Clas SD, Dalton CR, Hancock BC (1999) Differential scanning calorimetry: applications in drug development. PSTT 2(8):311–320
21. Giron D (1995) Thermal analysis and calorimetric methods in the characterization of polymorphs and solvates. Thermochim Acta 248:1–59
22. Giron D, Goldbronn C (1995) Place of DSC purity analysis in pharmaceutical development. J Therm Anal 44:217–251
23. Van Rompay J (1986) Purity determination and evaluation of new drug substances. J Pharm Biomed Anal 4:725–732
24. Mura P, Manderioli A, Bramanti G, Furlanetto S, Pinzauti S (1995) Utilization of differential scanning calorimetry as a screening technique to determine the compatibility of ketoprofen with excipients. Int J Pharm 119:71–79
25. Durig T, Fassihi AR (1993) Identification of stabilization effects of excipient-drug interactions in solid dosage form design. Int J Pharm 97:161–170
26. Mura P, Faucci MT, Manderioli A, Furlanetto S, Pinzauti S (1998) Thermal analysis as a screening technique in preformulation studies of picotamide solid dosage forms. Drug Dev Ind Pharm 24:747–756
27. Harris R, Yeung R, Lamont RB, Lancaster RW, Lynn SM, Staniforth SE (1997) Polymorphism in a novel anti-viral agent: lamivudine. J Chem Soc Perkin Trans 2:2653–2654
28. Burger A, Henck J, Hetz S, Rollinger J, Weissnicht A, Stottner H (2000) Energy/temperature diagram and compression behaviour of the polymorphs of D-mannitol. J Pharm Sci 4:457–468
29. Griesser U, Burger A, Mereiter K (1997) The polymorphic drug substances of the European pharmacopoeia. Part 9. Physicochemical properties and crystal structure of acetazolamide crystal forms. J Pharm Sci 86(3):352–358

30. Vitez IM (2004) Utilization of DSC for pharmaceutical crystal form quantitation. J Therm Anal Calorim 78(1):33–45
31. Galwey A, Craig D (2007) Thermogravimetric analysis: basic principles. In: Craig D, Reading M (eds) Thermal analysis of pharmaceuticals. CRC Press, Boca Raton/London/New York
32. Han J, Suryanarayanan R (1999) A method for the rapid evaluation of the physical stability of pharmaceutical hydrates. Thermochim Acta 329:163–170
33. Schmidt AC, Schwarz I (2006) Int J Pharm 320:4–13
34. Mackin L (2002) Int J Pharm 231:227–236
35. Chiou WL, Kyle J (1979) Differential thermal, solubility and aging studies on various sources of digoxin and digitoxin powder: biopharmaceutical implications. J Pharm Sci 68 (10):1224–1229
36. Ward GH, Schultz RK (1995) Process induced crystallinity changes in albuterol sulfate and its effect on powder physical stability. Pharm Res 12(5):773–779
37. Saleki-Gerhardt A, Ahlneck C, Zografi G (1994) Assessment of disorder in crystalline solids. Int J Pharm 101:237–247
38. Azo Materials, Dynamic vapour sorption for determination of hydrate formation and loss. http://www.azom.com/article.aspx?ArticleID=5182. Accessed May 2016
39. Khankari RK, Grant DJW (1995) Pharmaceutical hydrates. Thermochim Acta 248:61–79
40. Larkin P (2011) Infrared and Raman spectroscopy; principles and spectral interpretation. 1st Ed Elsevier
41. Turner TD (2015) PhD thesis, Molecular self-assembly, nucleation kinetics and cluster formation associated with solution crystallisation. School of chemical and process engineering University of Leeds
42. Bellamy L (1980) The infrared spectra of complex molecules. 2:3rd Ed Springer
43. Davies MM, Sutherland GBBM (1938) The Infra-red absorption of carboxylic acids in solution I. Qualitative features. The Journal of Chemical Physics 6:755–766
44. Hadzi D, Sheppard N (1953) The Infra-red absorption bands associated with the COOH and COOD groups in Dimeric carboxylic acids. I. The region from 1500 to 500 cm^{-1}. Proc R Soc London. A Math Phys Sci 216:247–266
45. Flett MSG (1962) Intensities of some group characteristic Infra-red bands. Spectrochim Acta 18:1537–1556
46. Sinclair RG, McKay AFk Jones RN (1952) The infrared absorption spectra of saturated fatty acids and esters. J Am Chem Soc 74:2570–2575
47. RJK D, Dent GK, RKK M, Parveen S (2006) Concerning the relationship between structural and growth synthons in crystal nucleation: solution and crystal chemistry of carboxylic acids as revealed through IR spectroscopy. Cryst Growth Des 6:1788–1796
48. Chan KLA et al (2004) Polymorphism and devitrification of nifedipine under controlled humidity: a combined FT-Raman, IR and Raman microscopic investigation J. Raman Spectrosc 35(5):353–359
49. Tudor AM, Melia CD, Binns JS, Hendra PJ, Church S, Davies MC (1990) The application of Fourier-transform Raman spectroscopy to the analysis of pharmaceuticals and biomaterials. J Pharm Biomed Anal 8(8-12):717–720
50. Hendra PJ (1993) Industrial value of Fourier transform Raman spectroscopy. Vib Spectrosc 5 (1):25–32
51. Compton DA, Compton SV (1991) Examination of packaged consumer goods by using FT-Raman spectrometry. Appl Spectrosc 45(10):1587–1589
52. Kwokal A, Nguyen TTH, Roberts KJ (2009) Surface adsorbed templates for directing the crystal growth of Entacapone as monitored using process analytical techniques. Cryst Growth Des 9(10):4324–4334
53. Findlay WP, Bugay DE (1998) Utilization of Fourier transform-Raman spectroscopy for the study of pharmaceutical crystal forms. J Pharm Biomed Anal 16(6):921–930
54. Bragg WH, Bragg WL (1913) The reflection of X-rays by crystals. Proc R Soc Lond A 88 (605):428–438

55. Cullity BD, Stock SR (2001) Elements of X-ray diffraction, 3rd edn. Prentice-Hall, International
56. Klug HP, Alexander LE (1974) X-ray diffraction procedure for polycrystalline and amorphous materials, 2nd edn. Wiley, New York
57. Chikhaliaa V, Forbes RT, Storey RA, Ticehurst M (2006) The effect of crystal morphology and mill type on milling induced crystal disorder Eur. J Pharm Sci 27(1):19–26
58. Haisa M, Kashino S, Kawai R, Maeda H (1976) The monoclinic form of p-Hydroxyacetanilide. Acta Cryst B32:1283–1285
59. Singh TP, Bhat TN, Vijayan M (1973) Crystallization and crystal data of acetoaminophen and metamizol. Curr Sci 42:384
60. Usmani OS, Biddiscombe MF, Barnes PJ (2005) Regional lung deposition and bronchodilator response as a function of beta(2)-agonist particle size. Am J Respir Crit Care Med 172 (12):1497–1504
61. Shekunov BY, Feeley JC, Chow AH, Tong HHY, York P (2003) Aerosolisation behaviour of micronised and supercritically-processed powders. J Aerosol Sci 34:553–568
62. Telko MJ, Hickey AJ (2005) Dry powder inhaler formulation. Respir Care 50(9):1209–1227
63. Begat P, Morton DAV, Staniforth JN, Price R (2004) The cohesive-adhesive balances in dry powder inhaler formulations. II: influence on fine particle delivery characteristics. Pharm Res 21(10):1826–1833
64. Adi H, Larson I, Chiou H, Young P, Traini D, Stewart P (2006) Agglomerate strength and dispersion of salmeterol xinafoate from powder mixtures for inhalation. Pharm Res 23 (11):2556–2565
65. Ghoroi C, Han X, To D, Jallo L, Gurumurthy L, Davé RN (2012) Dispersion of fine and ultrafine powders through surface modification and rapid expansion. Chem Eng Sci 85:11–24
66. Chiou H, Chan HK, Heng D, Prud'homme RK, Raper JA (2008) A novel production method for inhalable cyclosporine a powders by confined liquid impinging jet precipitation. J Aerosol Sci 39:500–509
67. Kaye RS, Purewal TS, Alpar HO (2009) Simultaneously manufactured nanoin- micro (SIMANIM) particles for dry-powder modified-release delivery of antibodies. J Pharm Sci 98:4055–4068
68. Jaffari S et al (2013) Rapid characterisation of the inherent dispersibility of respirable powders using dry dispersion laser diffraction. Int J Pharm 447:124–131
69. Nguyen TTH, Hammond RB, Roberts KJ, Marziano I, Nichols G (2014) Precision measurement of the growth rate and mechanism of ibuprofen {001} and {011} as a function of crystallization environment. Cryst Eng Comm 16:4568–4586
70. Malvern Instruments; details http://www.malvern.com/en/products/product-range/morphologi-range/morphologi-g3/
71. Turner TD, Corzo DMC, Toroz D, Curtis A, Dos Santos MM, Hammond RB, Lai X, Roberts KJ (2015) The influence of solution environment on the nucleation kinetics and crystallisability of para-aminobenzoic acid. Phys Chem Chem Phys 18:27507–27520

Chapter 24
Techniques for Crystal Optical Characterisation: Chiroptical Spectroscopy

Reiko Kuroda

Abstract Solid state can provide unique and novel chemistry which is different from traditional solution chemistry. However, very few solid-state chirality measurements have been reported to date, as chiroptical spectra are inevitably accompanied by artifact signals originated from macroscopic anisotropies of a sample which are unique to the solid state. We have developed chiroptical spectrophotometers which overcome these problems. In this article, principle and instrumentation of Universal Chiroptical Spectrophotometers (UCS-1, UCS-2 and UCS-3) are described. These instruments measure artifact signals as well, which are then removed to obtain true chirality information. UCS-2 and -3 can measure both diffuse reflectance and transmittance CD (circular dichroism) as they have a horizontal sample stage, ideal for soft materials such as gels or powder crystallines. Some examples of the application are presented. The principle of multichannel (MC) CD which is based on an entirely new idea to detect directly true CD, free from artefact signals, is also briefly described.

Keywords Chirality • Solid state • Circular Dichroism Spectroscopy • Defuse reflectance CD • Circularly Polarized Luminescence • Instrumentation • β-amyloid

24.1 Introduction

Solid-state chemistry is now flourishing as one of the frontier areas, because it does not require organic solvents and hence is environmentally friendly. Further, molecular interactions are much stronger in the solid state where molecules/ions are densely packed as compared with solution where molecules/ions are surrounded by solvent molecules. Thus, unique chemistry can be developed in the solid state. For example, crystal-crystal interactions such as co-grinding of multiple kinds of crystals, or crystal-vapour interactions often produce compounds/crystals which are

R. Kuroda (✉)
Research Institute for Science and Technology, Tokyo University of Science, 2641 Yamazaki, Noda-shi, Chiba 278-8510, Japan
e-mail: rkuroda@rs.tus.ac.jp

K.J. Roberts et al. (eds.), *Engineering Crystallography: From Molecule to Crystal to Functional Form*, NATO Science for Peace and Security Series A: Chemistry and Biology, DOI 10.1007/978-94-024-1117-1_24

different from those obtained from solution [1–9]. Photoreactions in the solid state sometime achieve what solution chemistry cannot afford in terms of stereoselectivity and reaction yield [10–12]. It is particularly interesting in the field of chiral chemistry. Aggregation of proteins/peptides is relevant to some neurodegenerative diseases, and valuable information on the aggregation process in solution and during the solution to solid-state transition can be investigated using chiroptical spectroscopy [13–15].

Thus, characterization of solid state is important, and solid-state spectroscopy provides indispensable information on the structures, supramolecular properties formed in solution-solid transition as well as dynamics of the structural changes, which are not obtainable from the conventional solution spectroscopy. However, very few solid-state chirality measurements have been reported to date [16–23]. This is because chiroptical spectra are necessarily accompanied by artifact signals which originate from the interaction between the macroscopic anisotropies of a sample such as linear birefringence (LB) and linear dichroism (LD) which are unique to the solid state, and the non-ideal characteristics of polarization–modulation instruments [24–26]. Thus, they cannot be in general measured on commercially available circular dichroism (CD) and circular birefringence (CB = optical rotatory dispersion (ORD)) spectrophotometers [24, 25, 27].

24.2 CD Measurement Along the Crystal Unique Optic Axis

The measurements of a crystal chirality along its unique optic axis are free from anisotropic effects described above [16, 17, 21, 22, 25, 26]. Taking the advantage of the fact, Mason [21], Judkins [23] and Kuroda [16, 17] independently studied chirality of transition metal complexes with D_3 symmetry which often crystallizes in the trigonal space groups with their molecular C_3 axis parallel to the crystal optic axis. Kuroda has developed techniques of CD measurement in a KBr matrix so that it can be used for quantitative analyses [16, 17, 28]. By combining single crystal CD and microcrystalline CD in a KBr matrix, Kuroda assigned the two components of the first *d-d* transition band of Co(III) complexes and evaluated the rotatory strengths. These cannot be achieved with solution spectra, as the two transition components are opposite in sign and overlap substantially resulting in severe cancellation of the CD peaks [16, 17, 25].

24.3 Biaxial Crystals

Uniaxial crystals where optic and crystal axes are common are exceptions. In the case of biaxial crystals in which most compounds are crystallized, there is no simple relationship between the two axes, and thus it is almost impossible to find optic axis for the CD measurement. Thus, to obtain true CD and CB spectra of solid samples, we must either take away the parasitic artifact signals from the observed spectra, or devise a special technique to detect artifact-free signals. We have worked on both approaches and made novel spectrophotometers, UCS-1, -2 and -3 (UCS = Universal Chiroptical Spectrophotometer) for the first method, and MC-CD (MC = Multi Channel) for the second approach. In this article, I shall describe the theoretical background and instrumentation of UCSs and their applications to a variety of samples in detail, and only briefly for the MC-CD spectrophotometer and CPL (circularly polarized luminescence) measurement.

24.4 Theoretical Background in General Cases

For designing spectrophotometers and analyzing observed signals, we employed the Stokes-Mueller matrix method [29]. The signals observed of a sample having CD, CB, LD and LB can be expressed as:

$$
\begin{aligned}
&\mathrm{Signal}_{50\mathrm{kHz}} = \\
&\quad G_1\left(P_x{}^2 + P_y{}^2\left[CD + 1/2\left(LD'LB - LB'LD\right) + \left(LD'\sin 2\theta - LD\cos 2\theta\right)\sin\alpha\right]\right. \\
&\quad + G_1\left(P_x{}^2 - P_y{}^2\right)\sin 2a\left\{LB'\sin 2\theta - LB\cos 2\theta\right. \\
&\quad + \left[\text{-}CB + 1/2\left(LD^2 + LB^2 - LD'^2 - LB'^2\right)\sin 4\theta\right. \\
&\quad \left.\left. + \left(LD'LD + LB'LB\right)\cos 4\theta\right]\sin\alpha\right\} + G_1\left(P_x{}^2 - P_y{}^2\right)\cos 2a\left\{LB'\cos 2\theta\right. \\
&\quad - LB\sin 2\theta - \left[1 + 1/4\left(LD^2 - LB'^2 - LD'^2 + LB^2\right)\cos 4\theta\right. \\
&\quad \left.\left. + 1/2\left(LD'LD + LB'LB\right)\sin 4\theta\right]\sin\alpha\right\}
\end{aligned}
\tag{24.1}
$$

Here, G_1 is the apparatus constant related to the sensitivity of the spectrometer, LD' and LB' are 45-degree linear dichroism and 45-degree linear birefringence, respectively. $P_x{}^2$ and $P_y{}^2$ are the transmittance of the photomultiplier along the x and y directions and "*a*" is the azimuth angle of its optical axis with respect to the x axis. θ is the rotation angle of the sample, and α is the residual static birefringence of the photoelastic modulator (PEM). Terms multiplied by $\sin\alpha$ are negligibly small as a PEM having a small residual static birefringence ($\alpha = 0.2$) was selected in our CD spectrophotometer. We can also neglect the contribution of the term containing cos2a, as we set the photomultiplier (PM)'s azimuth angle so as to make cos2a ≈ 0 in the baseline calibration. Thus, 50 kHz signal is written as:

$$\begin{aligned}\mathrm{Signal}_{50\mathrm{kHz}} &= \mathrm{G}_1\left(\mathrm{P_x}^2 + \mathrm{P_y}^2\right)\left[\mathrm{CD} + 1/2\left(\mathrm{LD}'\mathrm{LB} - \mathrm{LB}'\mathrm{LD}\right)\right] \\ &\quad + \mathrm{G}_1\left(\mathrm{P_x}^2 - \mathrm{P_y}^2\right)\sin 2a\left(\mathrm{LB}'\sin 2\theta - \mathrm{LB}\cos 2\theta\right)\end{aligned} \tag{24.2}$$

Similarly, the 100 kHz component of the photocurrent detected by the lock-in amplifier is given as:

$$\begin{aligned}\mathrm{Signal}_{100\mathrm{kHz}} &= \mathrm{G}_2\left(\mathrm{P_x}^2 + \mathrm{P_y}^2\right)\left(\mathrm{LD}'\sin 2\theta - \mathrm{LD}\cos 2\theta\right) \\ &\quad + \mathrm{G}_2\left(\mathrm{P_x}^2 - \mathrm{P_y}^2\right)\sin 2a\left\{-\mathrm{CB} + 1/4\left(\mathrm{LD}^2 + \mathrm{LB}^2 - \mathrm{LD}'^2 - \mathrm{LB}'^2\right)\sin 4\theta \right. \\ &\quad \left. + \left(\mathrm{LD}'\mathrm{LD} + \mathrm{LB}'\mathrm{LB}\right)\cos 4\theta\right\}\end{aligned} \tag{24.3}$$

Here G_2 is the apparatus constant related to the sensitivity of the spectrometer at 100 kHz.

When an analyzer with 45° optical axis is inserted into optical path, the light intensity detected by photomultiplier can be expressed as:

$$\mathrm{Signal}_{50\mathrm{kHz}} = \mathrm{G}_3\left\{\mathrm{CD} + 1/2\left(\mathrm{LD}'\mathrm{LB} - \mathrm{LDLB}'\right) - \mathrm{LB}\cos 2\theta + \mathrm{LB}'\sin 2\theta\right\} \tag{24.4}$$

and when LD is much smaller than LB as is often the case.

$$\mathrm{Signal}_{100\mathrm{kHz}} = \mathrm{G}_4\left\{\mathrm{CB} - 1/2\left(\mathrm{LB}^2 + \mathrm{LB}'^2\right)\sin\left(4\theta + \gamma\right)\right\} \tag{24.5}$$

where $\gamma = \tan^{-1}[\mathrm{LBLB}'/1/2(\mathrm{LB}^2 + \mathrm{LB}'^2)]$.

If there is no macroscopic anisotropy, Eqs. (24.2) and (24.5) give CD and CB, respectively, and this is the case for solution. In other cases, polarization phenomena, LB, LD, CB and CD, are intermingled with each other in several detecting modes. Based on these analyses, chiroptical spectrophotometers UCS-1, 2 and 3 have been developed.

24.5 Development of New Instruments

24.5.1 Universal Chiroptical Spectrophotometer UCS-1

Figure 24.1 shows the block diagram of UCS-1 [30] together with the axis orientations of the optical and electric components. Briefly, it has two lock-in amplifiers (50 and 100 kHz), and an analyzer can be inserted and taken out of the light path,

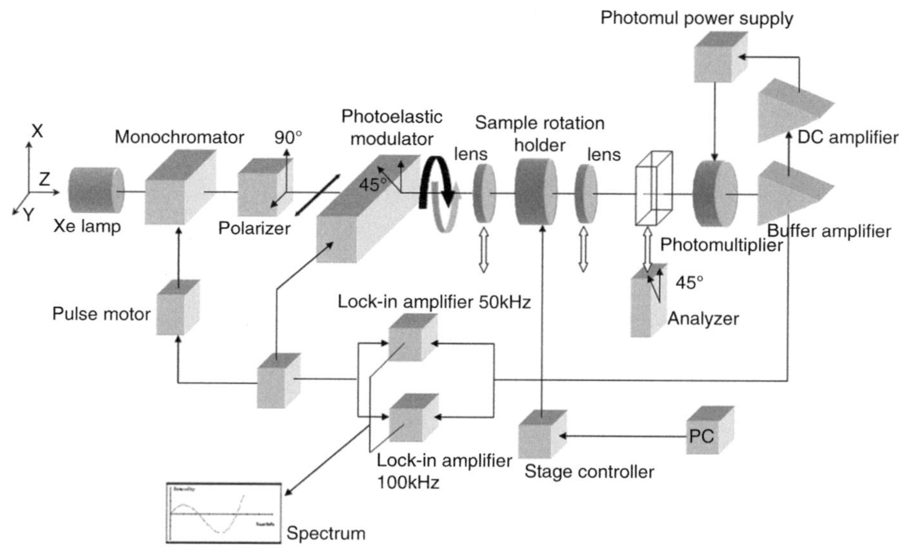

Fig. 24.1 Block diagram of UCS-1 (The figure is adapted from Ref. [30])

and spectra can be recorded while rotating a sample around the axis perpendicular to the optical axis under computer control. We have devised a special sample holder to measure 'front' and 'back' spectra to cancel the artefact signals (see below).

A set of procedures for the measurement of true CD is as follows.

1. With an analyzer, LB measurement is carried out by rotating the sample 360° in the (X-Y) plane at the wavelength of an absorption maximum. Then, the LB spectrum is obtained by the wavelength scanning at the LB_{max} position.
2. Without an analyzer, LD is measured by rotating the sample at the absorption maximum. Then, the LD spectrum is obtained by the wavelength scanning at the LD_{max} position.
3. Similarly, CD measurement is carried out by rotating the sample at the absorption maximum. If the CD value changes on rotating the sample 360° around the z axis, it is clear that the macroscopic anisotropies contribute to the CD spectrum. From the measurements, we can tell the extent of contribution of LD and LB signals to the observed CD signals. If the two 50 kHz spectra obtained with and without the analyzer are different, the 50 kHz signal detected without the analyzer contains an artefact CD signal and that with the analyzer is an LB signal.
4. From the data obtained in step (1), we can locate the LB_{max} and LB_{min} positions. The sample is rotated 45° from the LB_{max} position, wherein the LB value becomes 0 and the LB' value maximum. The wavelength scan is then carried out without an analyzer. From Eq. (24.2), the apparent CD signal of the face side is given as:

$$[\text{appCD}]_{\text{face}} = G_1\left\{(P_x{}^2+P_y{}^2)\left(\text{CD} - 1/2\text{LB}'\text{LD}\right) + (P_x{}^2-P_y{}^2)\sin 2a\left(\text{LB}'\right)\right\} \tag{24.6}$$

5. The sample is then rotated by 180° about the Y axis and the wavelength scan is carried out. This corresponds to the back side measurement. By this rotation, the CD and LD do not change their signs, but LB' becomes –LB'. Hence, the apparent CD signal of the back side becomes:

$$[\text{appCD}]_{\text{back}} = G_1\left\{(P_x{}^2+P_y{}^2)\left(\text{CD} - 1/2\text{LB}'\text{LD}\right) + (P_x{}^2-P_y{}^2)\sin 2a\left(\text{LB}'\right)\right\} \tag{24.7}$$

The average of Eqs. (24.6) and (24.7) gives true CD as in Eq. (24.8)

$$G_1\left(P_x{}^2+P_y{}^2\right)\text{CD} \tag{24.8}$$

The method is best applied when the LB value is not too large, i.e., less than ~30°.

24.5.2 Dual-Purpose Transmittance CD and Diffuse Reflectance CD Spectrophotometers (UCS-2 and -3)

UCS-2 [31] and UCS-3 [32] are designed as dual-purpose spectrophotometers to measure DRCD as well as transmittance CD to achieve high-quality *in-situ* chirality measurements with high sensitivity over a wide wavelength range. Chiroptical spectra of microcrystallines are usually measured either the KBr disk or the nujol-mull method. However, the methods sometime suffer from reactions with the matrix material [33] or dissolution of samples in nujol. We have noticed the crystallines become amorphous or change to other crystal forms by grinding [34]. Thus, it is ideal to measure CD spectra of microcrystallines *in situ*. For this purpose, diffuse reflectance (DR) spectroscopy is the most suited. The DRCD spectrophotometer was first developed by Biscarini, Kuroda et al. in 2002 [35] however, due to the arrangement of the optical trains in the instrument and the low grade of the optical elements used, the CD measurement was limited. UCS-2/3 we have recently developed a unique design distinct from that of commercially available CD instruments. A block diagram of UCS-2/3 is shown in Fig. 24.2. This has the following characteristics:

- they can measure both DRCD and transmittance CD, and for this purpose they house two photomultipliers,

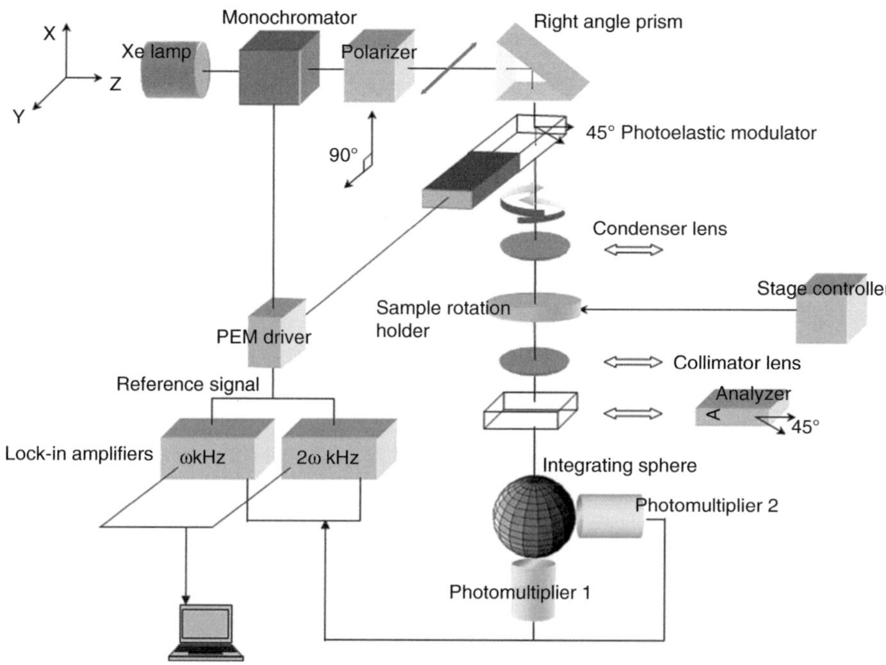

Fig. 24.2 Block diagram of dual-purpose UCS-2/3 (The figure is adapted from Ref. [31])

- they are equipped with two lock-in amplifiers (50 and 100 kHz) and an analyzer so that they are capable of measuring all polarization phenomena, i.e., LB, LD, CB and CD, simultaneously in the transmittance mode, as is UCS-1,
- they are equipped with an integrating sphere for the measurement of DRCD,
- *In situ* chirality measurement without any pretreatment can be carried out as a right angle prism is installed in the instrument, which makes it possible to set a sample on a horizontal plane.

This is ideal to carry out not only *in situ* measurements of solid samples including loose powders without any pretreatment, but also time-dependent measurements of liquid, mesophase and fluid-type condensed phases. We have improved the DRCD spectral range from 250–800 nm in UCS-2 to 190–800 nm with higher efficiency in UCS-3 [32].

24.5.3 Multichannel (MC) CD: A Novel Method for Direct True CD Measurement

So far we have succeeded in obtaining true CD and CB spectra of solid samples, by designing/making a series of UCSs and devising a set of measurement/analytical

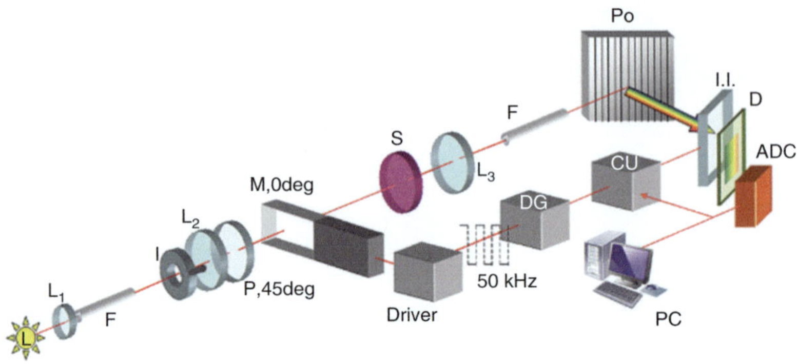

Fig. 24.3 Block diagram of MC-CD spectrometer (The figure is adapted from Refs. [38, 36b])

procedures. With UCSs, many interesting results have been obtained including structures of organic, inorganic and bio samples in the condensed phase and the dynamics of relatively slow structural change of peptides from solution to solid phase transition. UCSs adopt the method of taking away the parasitic artifacts signals from the observed spectra, and hence cumbersome procedures are necessary.

An alternative method adopts entirely novel concept/technique for measuring CD spectra, i.e. to detect only the artifact-free signals. We have developed a new MC-CD spectrophotometer [36–38]. This was designed so that only left- and right-handed circularly polarized components modulated at 50 kHz can be accurately triggered out of the emerging light from the PEM, thus the spectra will not be affected by the macroscopic anisotropies which are intrinsic to solid samples. It also employs non-monochromated white light and measures spectra like a snap shot with a multichannel detector to provide simultaneous detection of whole wavelength ranges. Figure 24.3 shows the block diagram [38]. As wavelength scan is unnecessary, high-speed data acquisition is possible; ca. 50 times faster than conventional spectrophotometers. Thus, this is ideal for studying dynamics of moderately fast structural changes of chiral compounds. Accuracy and S/N ratio of UCSs are expected to surpass those of MC-CD spectrophotometer, and both techniques are complementary to each other.

24.6 Examples of Solid-State CD Spectra

24.6.1 True CD Spectra of Achiral Films [30, 39] and Crystals [40–42]

PVA (Polyvinyl alcohol) film dyed with Congo red is achiral but it shows strong CD spectra due to artifact resulting from the strong LB. True CD spectra obtained with

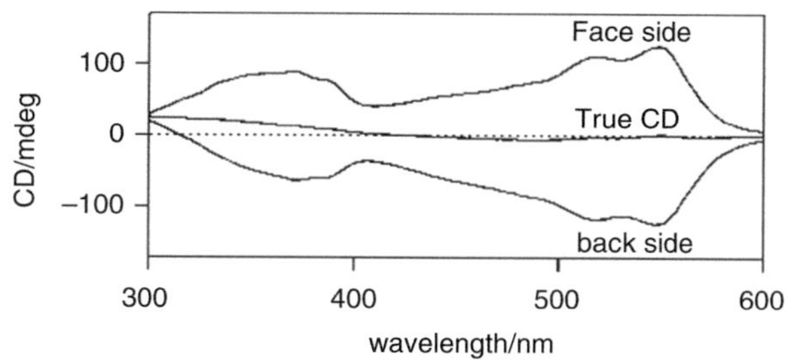

Fig. 24.4 CD spectra of a stretched PVA film dyed with Congo *Red*. True CD spectrum is obtained by averaging the face and back sides apparent CD spectra. (The figure is adapted from Ref. [30])

UCS-1 by the method described above showed almost zero signal, proving that our instrumentation and analyses are correct (Fig. 24.4).

24.7 Cast Film CD of Proteins/Peptides

CD spectroscopy of proteins in dry thin films provides useful information on their conformation in an aggregate or in the condensed phase, which is particularly relevant to some neurodegenerative disorders such as Alzheimer and prion diseases [43, 44]. Based on the different CD spectra of BSA (Bovine Serum Albumin) in solution and in dry films, it was claimed that these proteins undergo structural transformation from native structures in solution to β-sheet predominant structures in the solid state [43]. However, our measurements on UCS-1 have clearly shown that the structural change does not occur in the process of film formation, and the spectral difference is simply due to the artifacts arising from the substantial macroscopic anisotropies [45].

In the case of β-amyloid (1–40) and (1–42) peptides, we could show a clear structure transition from α-helix dominant structure in solution to β-sheet dominant one in cast films [15] (Fig. 24.5). Using UCS-3, we could show that transition from α-helix to β-sheet conformation occurs rather fast [46]

24.8 DRCD Measurement of 1:1 BQ – PYR Complex

The 1:1 complex of BQ (benzoquinone) and PYR (pyrene) exhibits optical activity only in the crystalline state due to the chiral supramolecular arrangement of the non-chiral components [47]. It crystallizes either in the space group of $P4_1$ or $P4_3$. Strong absorption of the crystals ($\varepsilon = 10^4 \sim 10^5$) in the UV-Vis wavelength range and sublimation of BQ hampered the CD measurements by ordinary methods. DRCD measurement is most suitable for this type of samples [47]. A single crystal was co-ground with KBr (separately pre-ground) to microcrystalline powders (grain size: 20–53 μm, dilution: 10–20 wt%) in order to decrease the specular reflection [48–51] and the parasitic signals originating from the intrinsic

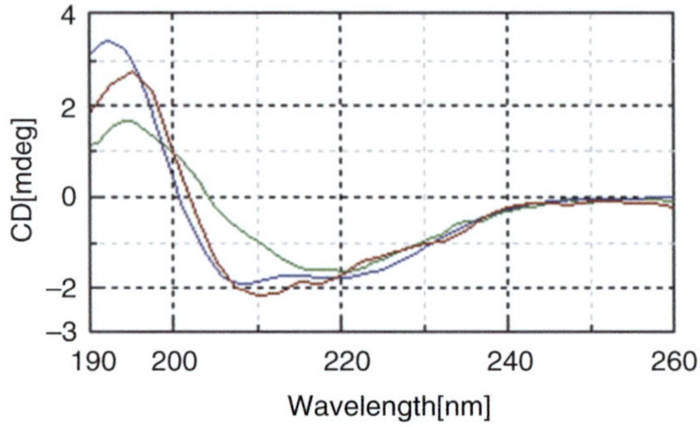

Fig. 24.5 True CD spectra of β-amyloid (1–40) in aqueous solution (*blue*) and in a cast film (*red*: immediately after casting, and *green*: after a short time) showing the secondary structure change

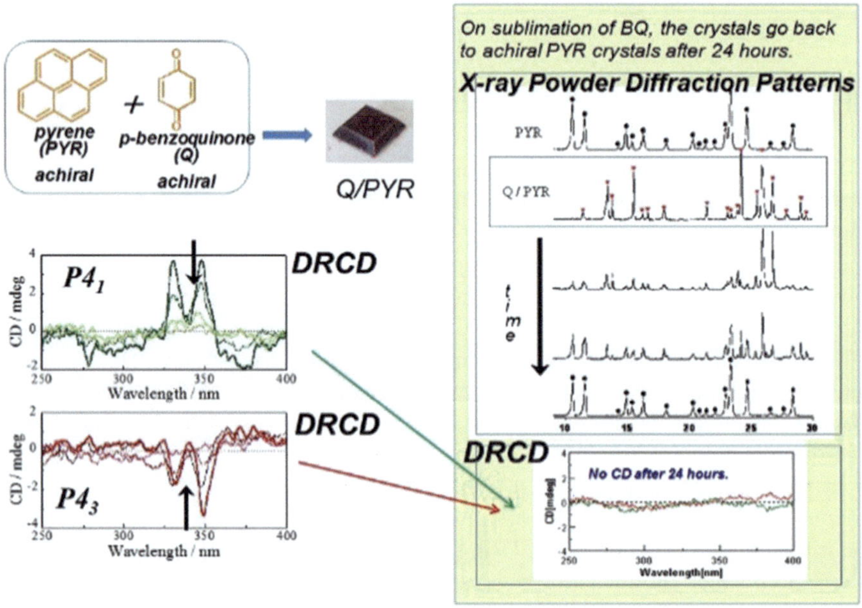

Fig. 24.6 DRCD spectra of enantiomeric supramolecular crystals and the time changing pattern together with that of PXRD (The figures are adapted from Ref. [47])

macroscopic anisotropies (LB and LD) as well as to reduce the large absorption coefficient [47]. Figure 24.6 shows DRCD spectra measured on UCS-3 for both of the enantiomeric crystals, which are almost mirror images of each other. We observed time-dependent change of the DRCD spectra over 24 h on exposing the sample to air. The DRCD spectra decreased the absolute intensities, and after 24 h all the CD peaks disappeared completely. The absorption spectra after 24 h exposure coincided with that of PYR crystal. PXRD also showed that the 1:1 complex crystals reverted to pyrene crystals in the solid state, as a result of sublimation of BQ.

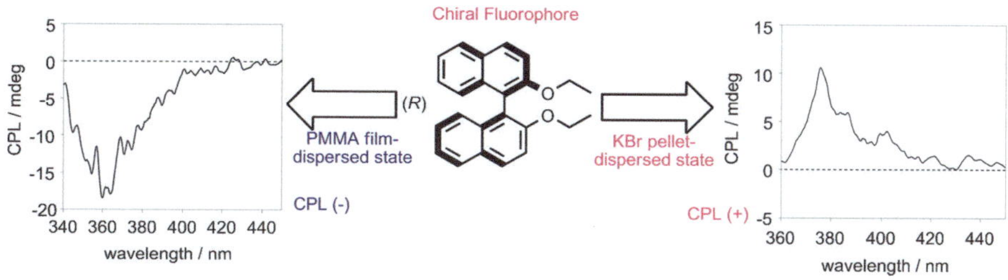

Fig. 24.7 The sign of CPL of (R)-2,2′-diethoxy-1,1′-binaphthyl is opposite depending on the environment (The figure is adapted from Ref. [53])

24.9 CPL (Circularly Polarized Luminescence)

We have studied the crystals of supramolecular organic fluorophores which are composed of chiral 1-phenylethylamine as the chiral component molecule and 4-(2-arylethybyl)-benzoic acid as the π-conjugated fluorescent component molecule by solid state CD and CPL. We have found that the sign of solid-state CD and CPL can be controlled by subtle change in the achiral acid structure due to differences in the crystal packing of these two structures. For example, by changing 4-methylbenzen moiety to 4-fluorobenzen, the sign of the solid-state CD and CPL spectra were reversed, although the amine of the same chirality was used [52]. Further and remarkably, we have shown that the solid-state CPL of axially chiral binaphthyl organic fluorophores exhibit opposite sign depending on the environment; a poly(methyl methacrylate) (PMMA) film-dispersed state and a KBr pellet (Fig. 24.7) [53].

References

1. Braga D, Giaffreda SL, Grepioni F, Pettersen A, Maini PL, Curzi M, Polito M (2006) Mechanochemical preparation of molecular and supramolecular organometallic materials and coordination networks. Dalton Trans:1249–1263 and references in the review articles
2. Friscic T, Trask AV, Jones W, Motherwell WDS (2006) Screening for inclusion compounds and systematic construction of three-component solids by liquid-assisted grinding. Angew Chem Int Ed 45:7546–7550
3. Toda F, Tanaka K, Miyamoto H, Koshima H, Miyahata I, Hirotsu K (1997) Formation of racemic compound crystals by mixing of two enantiomeric crystals in the solid state. Liquid transport of molecules from crystal to crystal. J Chem Soc Perkin Trans 9:1877–1885
4. (a) Pedireddi VR, Jones W, Chorlton AP, Docherty R (1996) From solution and by solid state grinding. Chem Commun:987–988; (b) Trask AV, Motherwell WDS, Jones W (2004) Solvent-drop grinding: green polymorph control of cocrystallisation. Chem Commun 7:890–891
5. Nichols PJ, Raston CL, Steed JW (2001) Engineering of porous pi-stacked solids using mechanochemistry. Chem Commun 1062–1063
6. (a) Braga D, Maini L, Polito M, Mirolo L, Grepioni F (2002) Mechanochemical assembly of hydrogen bonded organic-organometallic solid compounds. Chem Commun 24:2960–2961;

(b) Braga D (2003) Crystal engineering, Where from? Where to? Chem Commun:2751–2754; (c) Braga D, Maini L, Polito M, Mirolo L, Grepioni F (2003) Assembly of hybrid organic–organometallic materials through mechanochemical acid–base reactions. Chem Eur J 9:4362–4370; (d) Braga D, Maini L, de Sanctis G, Rubini K, Grepioni F, Chierotti MR, Gobetto R (2003) Mechanochemical preparation of hydrogen-bonded adducts between the Diamine 1,4-Diazabicyclo[2.2.2]octane and Dicarboxylic acids of variable chain length: an X-ray diffraction and solid-state NMR study. Chem Eur J 9:5538–5548; (e) Braga D, Giaffreda SL, Grepioni F, Polito M (2004) Mechanochemical and solution preparation of the coordination polymers $Ag[N(CH_2CH_2)_3N]_2[CH_3COO].5H_2O$ and $Zn[N(CH_2CH_2)_3N]Cl_2$. CrystEngComm 6:458–462; (f) Braga D, Curzi M, Grepioni F, Polito M (2005) Mechanochemical and solution reactions between $AgCH_3COO$ and $[H_2NC_6H_{10}NH_2]$ yield three isomers of the coordination network $\{Ag[H_2NC_6H_{10}NH_2]+\}$. Chem Commun:2915–2917
7. Chadwick K, Davey R, Cross W (2007) How does grinding produce co-crystals? Insights from the case of benzophenone and diphenylamine. Cryst Eng Com 9:732–734
8. (a) Kuroda R, Imai Y, Tajima N (2002) Generation of a co-crystal phase with novel coloristic properties via solid state grinding procedures. Chem Commun 23:2848–2849; (b) Kuroda R, Imai Y, Sato T (2001) Chirality recognition in solvent-free solid-state crystallization: chiral adduct formation by Bis-β-naphthol derivatives and Benzoquinone crystals. Chirality 13:588–592; (c) Imai Y, Tajima N, Sato T, Kuroda R (2002) Molecular recognition in solid-state crystallization: colored chiral adduct formations of 1,1′-Bi-2-naphthol derivatives and Benzoquinone with a third component. Chirality 14:604–609; (d) Kuroda R, Higashiguchi K, Hasebe S, Imai Y (2004) Crystal to crystal transformation in the solid state. Cryst Eng Comm 6:463–468; (e) Imai Y, Tajima N, Sato T, Kuroda R (2006) Visualization of molecular Recognition: a novel system based on charge-transfer complexes composed of 1,1′-Bi-2-naphthol derivatives and *p*-Benzoquinone. Organic Lett 8:2941–2944; (f) Cheung EY, Kitchin SJ, Harris KDM, Imai Y, Tajima N, Kuroda R (2003) Direct structure determination of a multicomponent molecular crystal prepared by a solid state grinding procedure. J Am Chem Soc 125:14658–14659; (g) Kuroda R, Sato T, Imai Y (2008) Varied charge-transfer complex crystals formed between diols and benzoquinone in the solid and solution states. Cryst Eng Comm 10:1881–1890
9. Nakamura A, Sato T, Kuroda R (2004) Formation of racemic crystals of transition metal complexes by grinding 1:1 mixtures of enantiomeric crystals. Chem Commun 24:2858–2859
10. (a) MacGillivray LR, Reid JL, Ripmeester JA (2000) Supramolecular control of reactivity in the solid state using linear molecular templates. J Am Chem Soc 122:7817–7818; (b) Gao X, Caronna T, Friscic T, MacGillivray LR (2004) Supramolecular construction of molecular ladders in the solid state. Angew Chem Int Ed 43:232–236
11. Tanaka K, Toda F (2000) Solvent-free organic synthesis. Chem Rev 100:1025–1074
12. Sekiya R, Kuroda R (2011) Controlling stereoselectivity of solid-state photoreactions by co-crystal formation. Chem Commun 47:10097–10099
13. Harada T, Kuroda R (2002) Circular Dichroism measurement of a protein in dried thin films. Chem Lett 31:326–327
14. Babenko V, Harada T, Yagi H, Goto Y, Kuroda R, Dzwolak W (2011) Chiral superstructures of insulin amyloid fibrils. Chirality 23:638–646
15. Harada T, Kuroda R (2011) CD measurements of β-amyloid (1-40) and (1-42) in the condensed phase. Biopolymers 95:127–134
16. Kuroda R (1975) Stereochemistry of transition metal complexes. Ph D thesis, University of Tokyo
17. Kuroda R, Saito Y (1976) Solid-state circular dichroism spectra of Tris(diamine)cobalt(III) complexes: decomposition into E and A_2 components. Bull Chem Soc Japan 49:433–434
18. Livolant F, Mickols W, Maestre MF (1988) Differential polarization microscopy (cd and linear dichroism) of polytene chromosomes and nucleoli from the dipteran sarcophaga footpad. Biopolymers 27:1761–1769

19. Safar J, Roller PP, Ruben GC, Gajdusek DC, Gibbs CJ Jr (1993) Secondary structure of proteins associated in thin films. Biopolymers 33:1461–1476
20. Lang J, Liu M (1999) Layer-by-layer assembly of DNA films and their interactions with dyes. J Phys Chem B 103:11393–11397
21. McCaffery AJ, Mason SF (1963) The oxalate and tartrate complexes of chromium(III). Trans Faraday Soc 59:1–11
22. Norden B, Grenthe I (1972) Circular Dichroism of dihedral rare earth carboxylates Chirally stabilized in single crystal. Acta Chem Scand 26:407–409
23. Judkins RR, Royer DJ (1974) Optical rotatory strength of tris-bidentate cobalt(III) complexes. Inorg Chem 13:945–950
24. Shindo Y (1995) Application of polarized modulation technique in polymer science. Opt Eng 34:3369–3384
25. Kuroda R (2000) Solid-state CD: application to inorganic and organic chemistry. in Circular dichroism: principles and applications 159–184, Second Edition, Ed. Berova N, Nakanishi K, Woody RW, Wiley: New York
26. Kuroda R (2004) Circular Dichroism in the solid state. In: Inoue Y, Ramamurthy V (eds) Chiral photochemistry. Marcel Dekker Inc, New York, pp 385–414
27. Shindo Y, Nishio M, Maeda S (1990) Comments on differential polarization microscopy (CD and linear dichroism). Biopolymers 30:405–413
28. Kuroda R, Honma T (2000) CD spectra of solid-state samples. Chirality 12:269–277
29. (a) Schellman JA (1988) Polarization modulation spectroscopy. In: Samori B, Thulstrup EW (eds) Polarized spectroscopy of ordered systems. Kluwer Academic, Dordrecht, pp 231–274; (b) Jensen HP, Schellman JA, Troxell T (1978) Modulation techniques in polarization spectroscopy. Appl Spectrosc 32:192–200; (c) Schellman J, Jensen HP (1987) Optical spectroscopy of oriented molecules. Chem Rev 87:1359–1399; (d) Shindo Y, Nakagawa M, Ohmi Y (1985) On the problems of CD spectropolarimeters.II: artifacts in CD spectrometers. Appl Spectosc 39:860–868; (e) Shindo Y, Ohmi Y (1985) New polarization-modulation spectrometer for ... Instrument design, analysis, and evaluation. Rev Sci Instrum 56:2237–2242
30. (a) Kuroda R, Harada T, Shindo Y (2001) A solid-state dedicated circular dichroism spectrophotometer: development and application. Rev Sci Instrum 72:3802–3810; (b) Japanese Patent No. 3,942,800 28 April 2000 and Japanese Patent No. 4,010,760 13 October 2000
31. Harada T, Hayakawa H, Kuroda R (2008) Vertical-type chiroptical spectrophotometer (I): instrumentation and application to diffuse reflectance circular dichroism measurement. Rev Sci Instrum 79:373103
32. Harada T, Miyoshi Y, Kuroda R (2009) High performance diffuse reflectance circular dichroism spectrophotometer. Rev Sci Instrum 80:046101
33. Braga D, Maini L, Polito M, Grepioni F (2002) Unexpected solid–solid reaction upon preparation of KBr pellets and its exploitation in supramolecular cation complexation. Chem Commun 20:2302–2303
34. Kuroda R (2014) Formation and interconversion of crystals in the solid state, ISCD-26, Prague
35. Bilotti I, Biscarini P, Castiglioni E, Ferranti F, Kuroda R (2002) Reflectance circular Dichroism of solid state chiral coordination compounds. Chirality 14:750–756
36. (a) Kuroda R (2010) Chirality recognition, generation, enhancement and measurement in the solid state. ISCD-22, Sapporo; (b) Harada T, Takahashi H, Kuroda R, to be submitted
37. Kuroda R, Harada T (2012) Solid-state Chiroptical spectroscopy. Principles and applications. In: Berova N, Polavarapu P, Nakanishi K, Woody RW (eds) Comprehensive chiroptical spectroscopy. Wiley, Hoboken, pp 91–114
38. Japanese patent, No. 2008–98471 and WO 2009-JP56950; US Patent No.12/935,853
39. Shindo Y, Kani K, Horinaka J, Kuroda R, Harada T (2001) The application of polarization modulation method to investigate the optical homogeneity of polymer films. J Plast Film Sheeting 17:164–183
40. Harada T, Shindo Y, Kuroda R (2002) Crystal chirality of the non-chiral inorganic salt, α-Ni $(H_2O)_6 \cdot SO_4$. Chem Phys Lett 360:217–222

41. Harada T, Sato T, Kuroda R (2005) Intrinsic birefringence of a chiral sodium chlorate crystal: is cubic crystal truly optically neutral? Chem Phys Lett 413:445–449
42. Harada T, Sato T, Kuroda R (2008) Inversion of the sign of the solid-state circular dichroism at low temperature. Chem Phys Lett 456:268–271
43. Koo EH, Lansbury PT, Kelly JW (1999) Amyloid diseases: abnormal protein aggregation in neurodegeneration. Proc Natl Acad Sci U S A 96:9989–9990
44. Tran PB, Miller RJ (1999) Aggregates in neurodegenerative disease: crowds and power? Trends Neurosci 22:194–197
45. Hu HY, Li Q, Cheng HC, Du HN (2001) β-sheet structure formation of proteins in solid state as revealed by circular dichroism spectroscopy. Biopolymers 62:15–21
46. Kuroda R, Harada T, to be submitted
47. Asano N, Harada T, Sato T, Tajima N, Kuroda R (2009) Supramolecular chirality measured by diffuse reflectance circular dichroism spectroscopy. Chem Commun 8:899–901
48. Kortüm G, Braun W, Herzog G (1963) Principles and techniques of diffuse-reflectance spectroscopy. Angew Chem Int Ed 2:333–341
49. Childers JW, Röhl R, Palmer RA (1986) Direct comparison of the capabilities of photoacoustic and diffuse reflectance spectroscopies in the ultraviolet, visible and near-infrared regions. Anal Chem 58:2629–2636
50. Castiglioni E, Albertini P (2000) An integrating sphere to measure CD from difficult samples. Chirality 12:291–294
51. Lagorio MG (2004) Why do marbles become paler on grinding? Reflectance spectroscopy, color, and particle size. J Chem Educ 81:1607–1611
52. Nishiguchi N, Kinuta T, Nakano Y, Harada T, Tajima N, Sato T, Fujiki M, Kuroda R, Matsubara Y, Imai Y (2011) Control of solid-state chiral optical properties of a supramolecular organic Fluorophore containing of 4-(2-arylethynyl)-benzoic acid. Chem Asian J 6:1092–1098
53. Kimoto T, Amako T, Tajima N, Kuroda R, Fujiki M, Imai Y (2013) Control of solid-state circularly polarized luminescence of binaphthyl organic fluorophores through environmental changes. Asian J Org Chem 2:404–410

Chapter 25
Unique Ferromagnetic Properties Observed in All-Organic Radical Liquid Crystals

Rui Tamura

Abstract We discovered that all-organic rod-like liquid crystalline (LC) compounds (1) with a stable nitroxide radical unit in the central core position exhibit unique intermolecular ferromagnetic interactions, which was referred to as 'magneto-LC effects', induced by low magnetic fields in the various LC phases. The origin of magneto-LC effects was interpreted in terms of the generation of a sort of spin glass-like inhomogeneous ferromagnetic interactions (the average spin-spin exchange interaction constant $\bar{J} > 0$). By measuring the electric field dependence of EPR spectra of the ferroelectric LC phase which can simultaneously show 'positive magneto-LC effect', two magnetic bistable states, anisotropy in spin-spin dipole interactions, and 'magneto-electric effect' were observed for the first time in a surface-stabilized liquid crystal cell.

Keywords Magneto-LC effects • Organic radical liquid crystals • Ferromagnetic liquid crystals • Ferroelectric liquid crystals • Magneto-electric effect

25.1 Introduction

Liquid crystals (LCs), which are defined as a thermal mesophase between crystalline and isotropic phases and can also be regarded as high temperature polymorphs of crystals, are unique soft materials that combine fluidity and anisotropy (Fig. 25.1). From another viewpoint, liquid crystalline (LC) phases are considered to be a sort of 'complexity' system consisting of non-equilibrium dynamic states due to the molecular motion and the coherent collective properties of molecules in the LC state. Accordingly, they are so sensitive to external stimuli, such as heat, light, temperature, pressure, electric or magnetic field, and added chiral dopants, that LC superstructure can be easily altered [1–4].

R. Tamura (✉)
Graduate School of Human and Environmental Studies, Kyoto University, Sakyo-ku, Kyoto 606-8501, Japan
e-mail: tamura.rui.8c@kyoto-u.ac.jp

K.J. Roberts et al. (eds.), *Engineering Crystallography: From Molecule to Crystal to Functional Form*, NATO Science for Peace and Security Series A: Chemistry and Biology, DOI 10.1007/978-94-024-1117-1_25

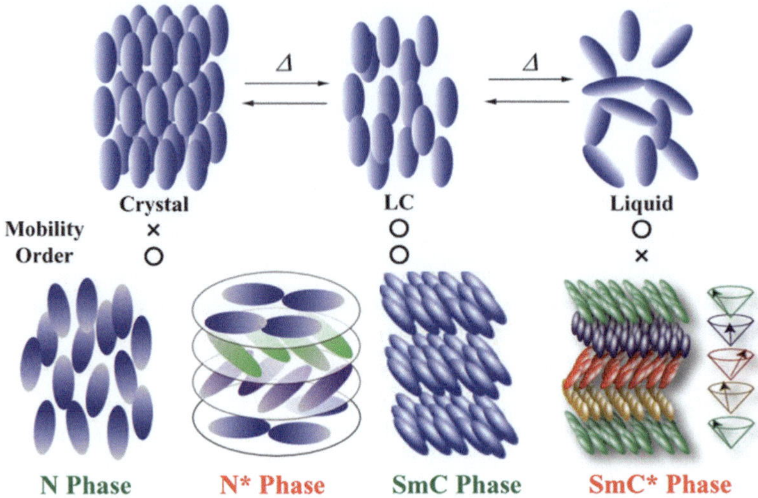

Fig. 25.1 Classification of chiral and achiral rod-like liquid crystalline phases

Magnetic LC materials have attracted great interest as soft materials to enhance the effect of magnetic fields on the electric and optical properties of LCs [5–9]. For example, they were anticipated to exhibit unique magnetic interactions and thereby unconventional magneto-electric [10–13] or magneto-optical [14–16] properties in the LC state. However, there had been no prominent study on this interesting topic, because the majority of magnetic LCs were highly viscous transition or lanthanide metal-containing metallomesogens. These were not always appropriate for investigating the swift molecular motion and reorientation in the LC phases in low applied magnetic fields [5, 6, 8, 9]. Therefore, we have investigated if metal-free magnetic LC materials with low viscosity can afford the following two possibilities; (i) formation of magnetic domains (or order) in applied magnetic fields and (ii) occurrence of coupling between magnetic dipole and electric dipole moments (magneto-electric effect) in the LC phases [17–23]. As a consequence, we could discover that all-organic rod-like LC compounds (**1**) with a stable nitroxide radical unit in the central core portion (Fig. 25.2) [24, 25] exhibit unique intermolecular ferromagnetic interactions induced by low magnetic fields in the various LC phases [26, 27]. This observation was interpreted in terms of the generation of a sort of spin glass (SG)-like inhomogeneous ferromagnetic interactions (the average spin-spin exchange interaction constant $\bar{J} > 0$) and proved to have nothing to do with the molecular reorientation effects arising from simple molecular magnetic anisotropy ($\Delta\chi$) [27]. This unique magnetic phenomenon was referred to as 'positive magneto-LC effect' ($\bar{J} > 0$) [28].

Of various all-organic magnetic LC compounds synthesized, (2*S*,5*S*)-**1** (n > 10) exhibited a chiral smectic C (SmC*) phase and showed both excellent ferroelectricity in a surface stabilized liquid crystal cell and explicit ferromagnetic interactions (positive magneto-LC effect) in the bulk SmC* phase [25, 27, 30]. Therefore,

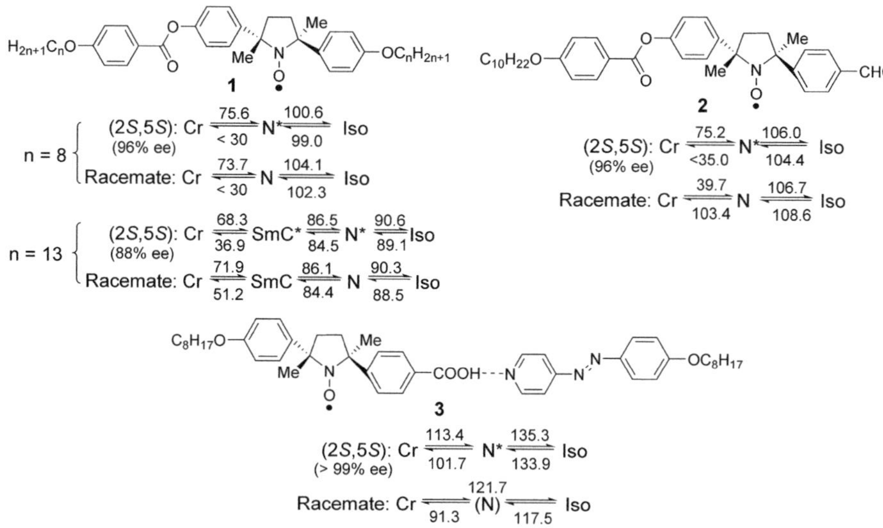

Fig. 25.2 Second-generation of rod-like all-organic radical LC compounds

it was expected that the unique magneto-electric coupling, which was observed only for inorganic multiferroic materials possessing both ferroelectricity and magnetic order (ferromagnetism or antiferromagnetism) at cryogenic temperatures [10–13], might occur in the ferroelectric LC (FLC) phase of (2*S*,5*S*)-**1** showing the positive magneto-LC effect at high temperatures. In this lecture, I introduce some of our works on the molecular design of (2*S*,5*S*)-**1** [24, 25] and their ferroelectricity [25, 29–31], magneto-LC effects [26–28, 33, 34] and magneto-electric effect [34–36].

25.2 Molecular Design and Synthesis

Amongst stable organic free radicals such as nitroxides, verdazyls, thioaminyls, a certain hydrazyl, phenoxyls, and carbon-centered radicals, nitroxide radicals show outstanding thermodynamic stability ascribed to the delocalization of the unpaired electron over the N—O bond and thereby no dimerization [37]. However, only a few all-organic nitroxide radical LC compounds were prepared before 2004, because the geometry and bulkiness of the radical-stabilizing substituents were believed to be detrimental to the stability of LCs which needs molecular linearity and planarity. Although several first-generation of achiral rod-like all-organic LCs with a stable cyclic nitroxide radical unit (DOXYL or TEMPO group) as the spin source were prepared [38–43], their molecular structures were limited to those containing a nitroxyl group (N–O·) in the peripheral position, away from the rigid central core (Fig. 25.3), and thereby allowed the free rotation of the nitroxyl moiety

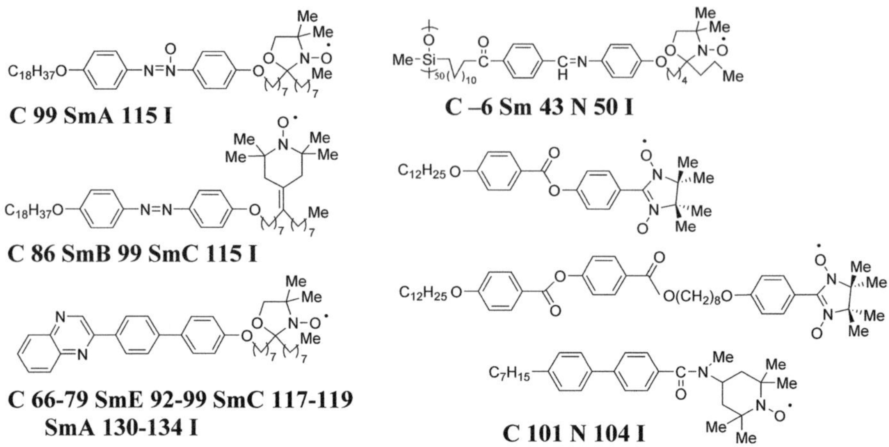

Fig. 25.3 First-generation of rod-like all-organic radical LC compounds

inside the molecule, resulting in a considerable decrease in both paramagnetic anisotropy ($\Delta\chi_{para}$) and dielectric anisotropy ($\Delta\varepsilon$) of the whole molecule.

With this situation in mind, we designed and synthesized the second-generation of chiral rod-like all-organic radical LC compounds **1** as the prototype which could satisfy the following three mandatory requirements (Fig. 25.2):

1. Spin source; a nitroxyl group with a large electric dipole moment (ca. 3 Debye) and known principal *g*-values (g_{xx}, g_{yy}, g_{zz}) should be used as the spin source, because i) the dipole moment is large enough for the source of spontaneous polarization (*P*s) and ii) the principal *g*-values are useful to determine the direction of molecular alignment in the LC phase by EPR spectroscopy.
2. High thermal stability; a 2,2,5,5-tetraalkyl-substituted pyrrolidine-1-oxy (PROXYL) radical structure, which is stable enough for repeated heating and cooling cycles below 150 °C in the air, is the most suitable.
3. Molecular structure; i) to avoid the free rotation of the nitroxyl group inside the molecule so as to make use of the $\Delta\chi_{para}$ and $\Delta\varepsilon$, a geometrically fixed chiral PROXYL radical unit should be incorporated into the rigid core of LC molecules and ii) to obtain a slightly zigzag molecular structure and a negative $\Delta\varepsilon$ advantageous for the appearance of an SmC* phase, a *trans*-2,5-dimethyl-2,5-diphenyl PROXYL radical skeleton in which the electric dipole moment orients to the molecular short axis is the best choice.

We could synthesize both optically active (2*S*,5*S*)-**1** and racemic **1** [(2*S*,5*S*):(2*R*:5*R*) = 1:1] which showed chiral and achiral nematic (N* and N) phases and/or chiral and achiral smectic C (SmC* and SmC) phases, respectively.

25.3 Ferroelectricity [25, 29–31]

Although stable nitroxide radical structures have been used as the spin sources or the redox species to develop metal-free solid-state magnetic materials and spintronic devices, or polymer battery devices, respectively, the large electric dipole moment (ca. 3 Debye) of a nitroxyl group has never been utilized in these nitroxide radical-based materials. Therefore, we have prepared organic LC compounds which can benefit from the unique magnetic and electric properties intrinsic to the nitroxide radical structure.

It is well known that an SmC* phase undergoes deformation of the helical superstructure in a thin sandwich cell to show ferroelectric switching by application of electric fields (Fig. 25.4). In fact, the optically active (2*S*,5*S*)-**1** (n = 11–15) which showed an SmC* phase indeed exhibited ferroelectricity in a planar anchoring thin sandwich cell (4 μm thickness) (Table 25.1). Among them, (2*S*,5*S*)-**1** (n = 13) showed the best ferroelectric properties with a *P*s value of 24 nC cm^{-2}, an optical response time ($\tau_{10–90}$) of 213 μs, and an ideal layer tilt angle (θ)) of 29°. Furthermore, second-harmonic generation (SHG) was clearly observed by Kogo and Takezoe *et al.* under a phase-matching condition in the SmC* phase of (2*S*,5*S*)-**1** (n = 13) loaded into an LC cell (20 μm thickness), validating the existence of ferroelectricity [44].

25.4 Magneto-LC Effects [26–28, 32, 33]

The possibility of a ferromagnetic rod-like LC material had been considered unrealistic due to the inaccessibility of long-range spin-spin interactions between rotating molecules in the LC state [7]. Interestingly, however, a nonlinear relationship between the applied magnetic field (H) and the observed magnetization (M) in chiral and achiral LC phases of **1** (Fig. 25.5) implies the generation of unusual magnetic interactions in the LC phases under weak magnetic fields. Such a nonlinearity was not observed in the crystalline phases of the same compounds which showed a usual linear relationship indicating a paramagnetic behavior and no contamination of magnetic impurities in the sample. The Arrott-Belov-Kouvel plots (M^2 vs H/M) [45, 46] using the same data strongly suggested the generation of a sort of SG-like inhomogeneous magnetic interactions (the average spin-spin exchange interaction constant $\bar{J} > 0$) induced by low magnetic fields in various LC phases of **1** (Fig. 25.6). In fact, these radical LC droplets floating on water were attracted by a permanent magnet and moved quickly and freely on water under the influence of this magnet (Fig. 25.7), whereas the crystallized particles of the same compounds did not respond to the same magnet.

To directly prove that the magnetic behavior in the LC phase is different from that in the crystalline phase, the temperature dependence of molar magnetic susceptibility (χ_M) was measured for (2*S*,5*S*)-**1** and racemic **1** (n = 8 and 13 each) at a

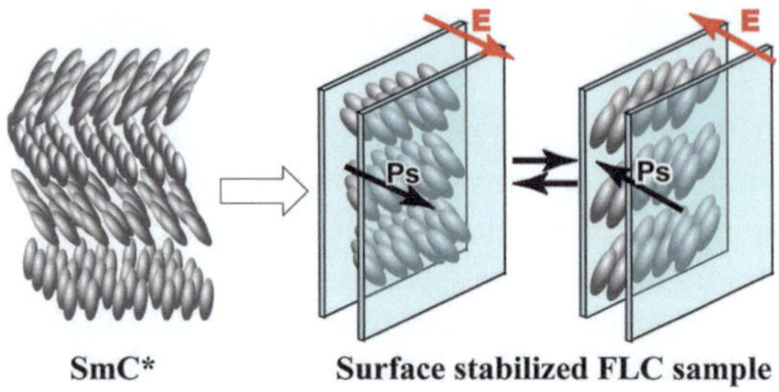

Fig. 25.4. Ferroelectric switching in a thin sandwich cell

Table 25.1 Ferroelectric properties of (*S*,*S*)-**1** (n = 11–15)

(*S*,*S*)-**1**	Spontaneous polarization *P*s (−10°)	Response time $\tau_{10\text{–}90}$ (−10°)	Viscosity η (−10°)	Tilt angle θ (−10°)
n	[nC cm−2]	[μs]	[m Pa s]	[deg]
11	15 (74 °C)	850	182.1	41
12	18 (64 °C)	334	85.9	37
13	24 (74 °C)	213	73.0	29
14	16 (79 °C)	159	36.3	36
15	17 (79 °C)	146	35.5	34

magnetic field of 0.05 T. The χ_M –*T* plots obeyed the Curie–Weiss law in the crystalline phase, while a considerable χ_M increase at the Cr-to-LC phase transition was observed for all LC phases examined (Fig. 25.8). This unique magnetic phenomenon was referred to as 'positive magneto-LC effect'.

We reported that EPR spectroscopy is a much better means than SQUID magnetization measurement to analyze the temperature dependence of the χ_{para} for organic nitroxide radical LC materials at high temperatures because of the following three reasons [27]: (1) Treatment of the χ_{dia} term is unnecessary. (2) The experimental error is very small even at such high temperatures. (3) The analysis of microscopic magnetic interactions such as spin–spin dipole and exchange interactions is also feasible. For example, on the basis of the EPR spectra for (2*S*,5*S*)-**1** (n = 13) measured between 25 and 115 °C at a magnetic field of 0.33 T by using a quartz tube, the temperature dependence of relative paramagnetic susceptibility (χ_{rel}), which is defined as:

$$\chi_{rel} = \chi_{para}/\chi_0 \tag{25.1}$$

where χ_0 is the standard paramagnetic susceptibility at 30 °C in the heating run, as shown in Fig. 25.9a. Again a considerable χ_{rel} increase at the Cr-to-SmC* phase transition was observed, indicating the generation of positive magneto-LC effect. Furthermore, to gain an insight into the origin of the positive magneto-LC effect operating in the SmC* phase, the temperature dependence of line-width (ΔH_{pp}) was compared with that of χ_{rel} (Fig. 25.9). As a consequence, the χ_{rel} increase was accompanied by the large ΔH_{pp} increase at the crystal-to-SmC* phase transition in

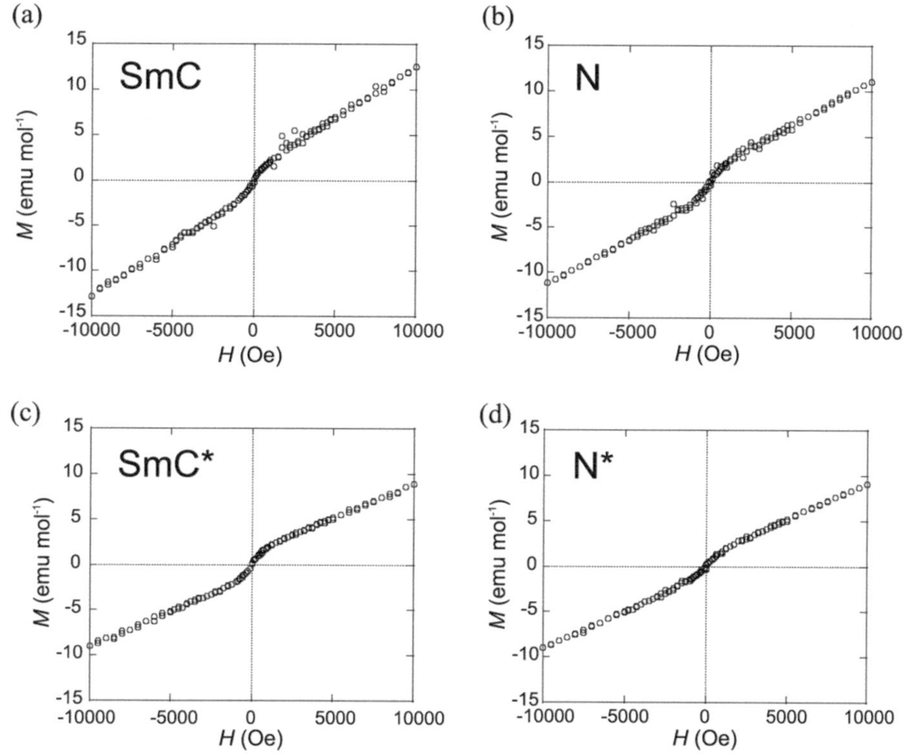

Fig. 25.5 Magnetic field (H) dependence of molar magnetization (M) at 77 °C for **(a)** the SmC phase of (±)-**1** (n = 13), **(b)** the N phase of (±)-**1** (n = 8), **(c)** the SmC* phase of (2*S*,5*S*)-**1** (n = 13), and **(d)** the N* phase of (2*S*,5*S*)-**1** (n = 8). (Reprinted with permission from Ref [27]: Copyright 2015 American Chemical Society)

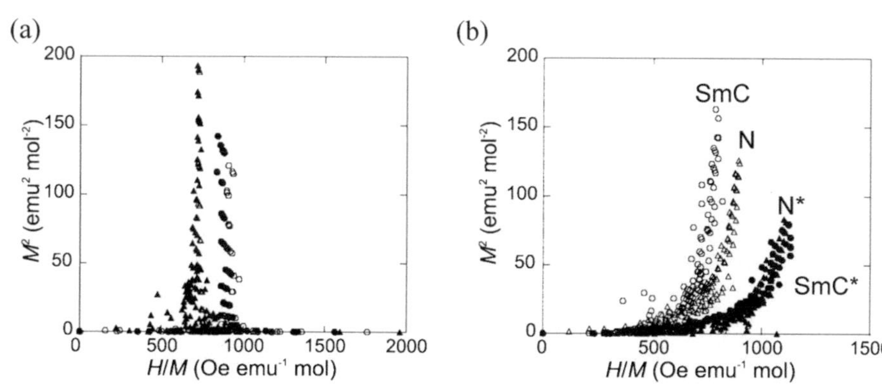

Fig. 25.6 The Arrott-Belov-Kouvel plots (M^2 vs H/M) 77 °C for (a) the crystalline phases and (b) the LC phases of **1** (n = 8, 13). (Reprinted with permission from Ref [27]: Copyright 2015 American Chemical Society)

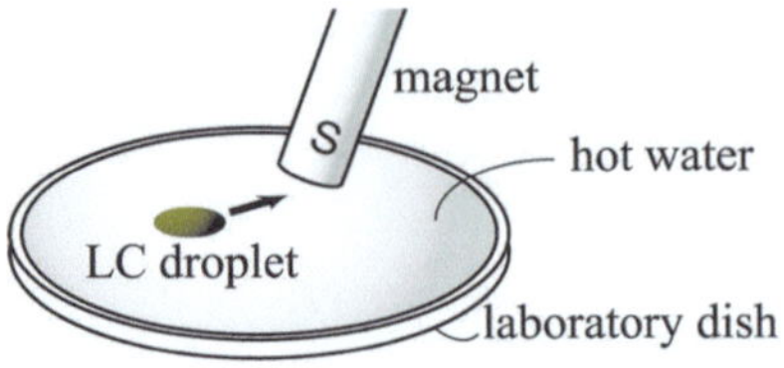

Fig. 25.7 Schematic representation of the experimental set-up for observing the attraction by a permanent magnet (<0.5 T) of a magnetic LC droplet on water in a shallow laboratory dish. (Reprinted with permission from Ref [27]: Copyright 2015 American Chemical Society)

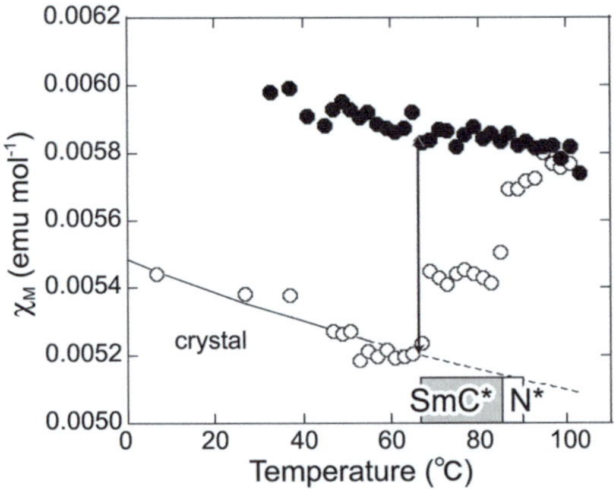

Fig. 25.8 Temperature dependence of molar magnetic susceptibility (χ_M) of (2*S*,5*S*)-**1** (n = 13, 88% *ee*) at a field of 0.05 T in the temperature ranges of 0 to +105 °C. *Open and filled circles* represent the first heating and cooling runs, respectively. The solid lines show the Curie-Weiss fitting curves, while the dotted lines represent the extrapolated curves. An arrow indicates the difference in the χ_M between the crystalline and LC states

the heating run. Definitely, strong spin-spin dipole interactions as well as exchange interactions turned out to be responsible for the positive magneto-LC effect. Fig. 25.10 shows the relative stability in spin-spin dipole interactions. There are two types; one is a linear type and the other is a side-by-side type. Since the ferromagnetic head-to-tail dipole interaction is energetically more stable than the antiferromagnetic side-by-side dipole interaction and other two interactions, the origin of positive magneto-LC effects can be interpreted in terms of the resulting ferromagnetic spin-spin dipole interactions induced by magnetic fields in the anisotropic LC superstructure.

To examine the relationship between the signs of $\Delta\varepsilon$ and magneto-LC effects, (2*S*,5*S*)-**2** and racemic **2** with a terminal formyl group which have a positive $\Delta\varepsilon$ were synthesized (Fig. 25.1) [28]. Under weak magnetic fields, positive magneto-LC effect ($\bar{J} > 0$) operated in the chiral N* phase of (2*S*,5*S*)-**2**, whereas negative magneto-LC effect ($\bar{J} < 0$) was observed in the achiral N phase. The origin of such negative magneto-LC effect operating in the N phase of racemic **2** was interpreted

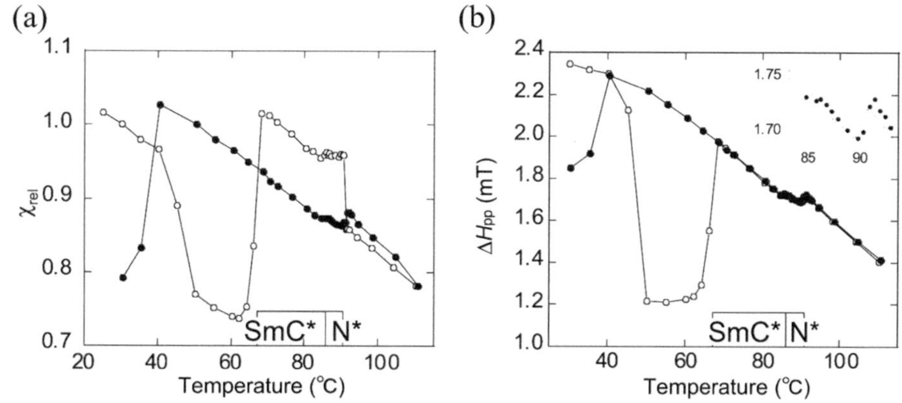

Fig. 25.9 Temperature dependence of (**a**) relative paramagnetic susceptibility (χ_{rel}) and (**b**) ΔH_{pp} for (2*S*,5*S*)-**1** (n = 13, 88% *ee*) at a magnetic field of 0.33 T. *Open and filled circles* represent the heating and cooling runs, respectively. The LC temperatures shown in a *box* refer to the heating process. The *inset* in panel **b** indicates the magnification in the temperature range of 83–93 °C. (Reprinted with permission from Ref [27]: Copyright 2015 American Chemical Society)

Fig. 25.10 The origin of strong positive magneto-LC effect. (**a**) Relative stability of four possible spin-spin dipole interactions between two spins. (**b**) Spin-spin interaction between spins localized in radical moieties in the SmC or SmC* phase of **1** (n = 13)

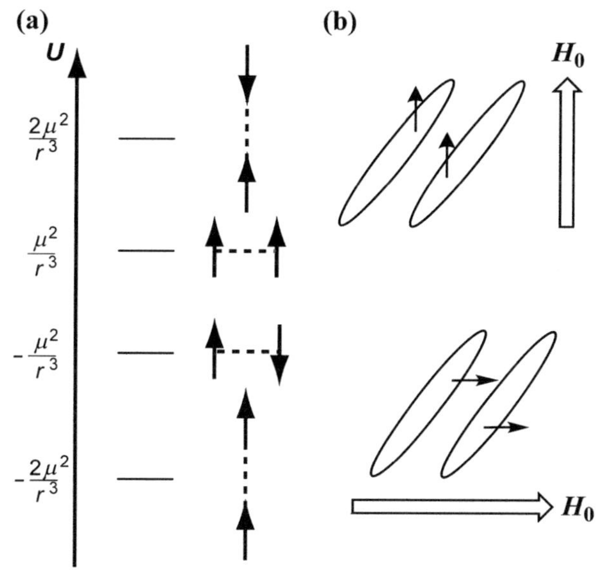

in terms of the occurrence of antiferromagnetic interactions due to the local SOMO-SOMO overlapping and thereby the side-by-side spin-spin dipole interaction, which is associated with the strong racemic (2*R*,5*R*)/(2*S*,5*S*) dimer formation to cancel the electric dipole moments, while the energetically favored ferromagnetic head-to-tail spin-spin dipole interactions should operate in the N* phase because the same enantiomers cannot form a strong dimer (Fig. 25.11). In contrast, only positive magneto-LC effect was observed for **1** with a negative $\Delta\varepsilon$, irrespective of the

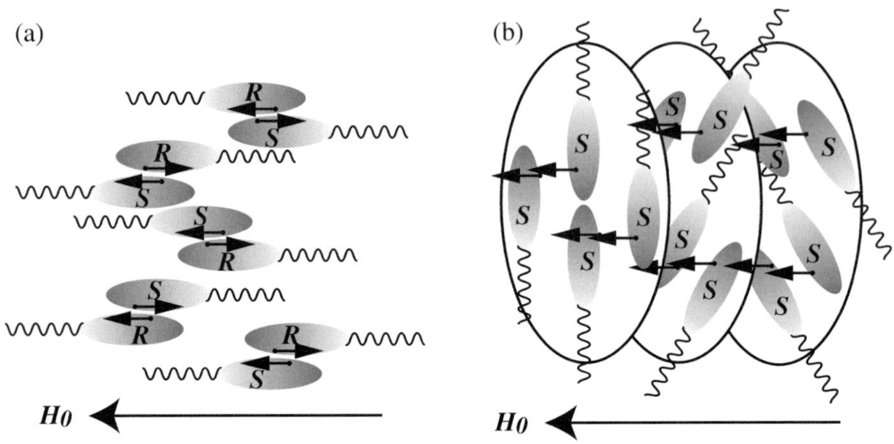

Fig. 25.11 Schematic illustration of spin-spin dipole interactions in LC phases. **(a)** The N phase of (±)-**2** and **(b)** the N* phase of (2*S*,5*S*)-**2** (96% *ee*). (Ref. [28] – Reproduced by permission of The Royal Society of Chemistry)

chirality of LC phases [27]. This result can be explained in terms of the occurrence of electric dipolar repulsion between the *R* and *S* enantiomers, which results in ferromagnetic head-to-tail spin-spin dipole interactions (Fig. 25.12).

25.5 Magneto-Electric Effect [34–36]

To clarify the relationship between the ferroelectric properties and positive magneto-LC effect, i.e., the existence of magneto-electric effect, we measured the electric field dependence of EPR spectra of (2*S*,5*S*)-**1** (n = 13) confined in a surface-stabilized liquid crystal cell and thereby evaluated the EPR parameters such as *g*-value, χ_{rel}, and the ΔH_{pp}.

The sample of (2*S*,5*S*)-**1** (n = 13) of 65% *ee* was introduced by capillary action into the lower tip of a handmade 4 μm-thick sandwich cell (50 mm x 5 mm) in which the inner surface of two glass substrates with indium tin oxide (ITO) electrodes were coated with polyimide polymer and then only one inner surface in the cell was rubbed with a velvet roller (Fig. 25.13). Then the cell tip was inserted into the EPR cavity.

Next, the existence of ferroelectric bistable state was verified by evaluating the electric field dependence of *g*-value under the conditions in which the magnetic field (0.33 T) was applied perpendicular to the electric field and parallel to the rubbing direction (Fig. 25.14). The experimental *g*-values (g_{exp}) exhibited a hysteresis loop between +25 V and −25 V. Since parallel and perpendicular *g*-values ($g_{//}$ and $g_{\perp}$) of molecule **1** (n = 13) were previously determined to be 2.0054 and 2.0068, respectively [47], the g_{exp} (2.0057) at −25 V reflects a large contribution of

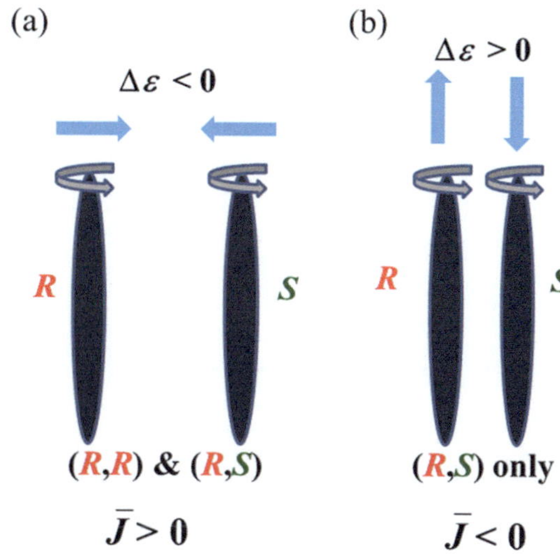

Fig. 25.12 Relationship between the signs of magneto-LC effects ($\bar{J}$) and dielectric anisotropy ($\Delta\varepsilon$). **(a)** Positive magneto-LC effect ($\bar{J} > 0$: ferromagnetic interactions) was exclusively observed in the case of $\Delta\varepsilon < 0$, irrespective of the chirality of LC phases. **(b)** Negative magneto-LC effect ($\bar{J} < 0$: antiferromagnetic interactions) was observed in the case of racemic samples with $\Delta\varepsilon > 0$

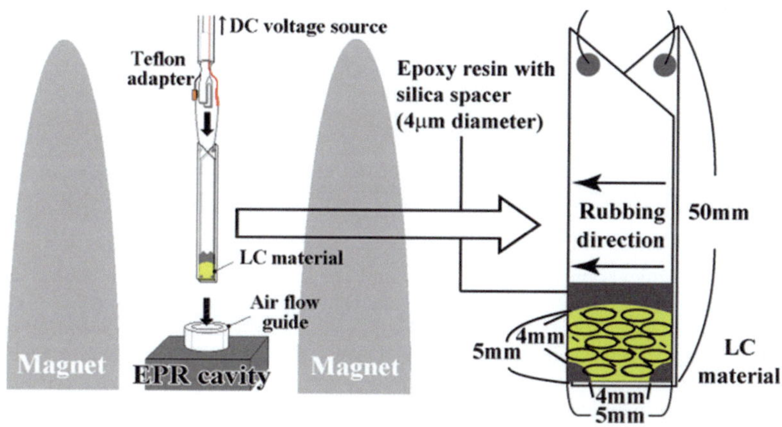

Fig. 25.13 (**a**) Experimental setup to monitor the variable-temperature or electric field-dependent EPR spectra of (2*S*,5*S*)-**1** (n = 13) confined in a long 4 μm thick sandwich cell. (Ref. [34] – Reproduced by permission of The Royal Society of Chemistry)

$g_{//}$, suggesting that molecular long axis of (2*S*,5*S*)-**1** (n = 13) aligns almost parallel to the magnetic field (Fig. 25.14a). On the other hand, the g_{exp} (2.0063) at +25 V indicates that the molecular long axis of (2*S*,5*S*)-**1** (n = 13) is tilted by 58° (2θ) from the direction of the magnetic field. Thus, (2*S*,5*S*)-**1** (n = 13) has proved to take a surface stabilized ferroelectric bistable state between +25 V and −25 V.

The electric field dependence of $\chi_{rel,E}$, which is defined as:

$$\chi_{rel,E} = \chi_{para}/\chi_1 \tag{25.2}$$

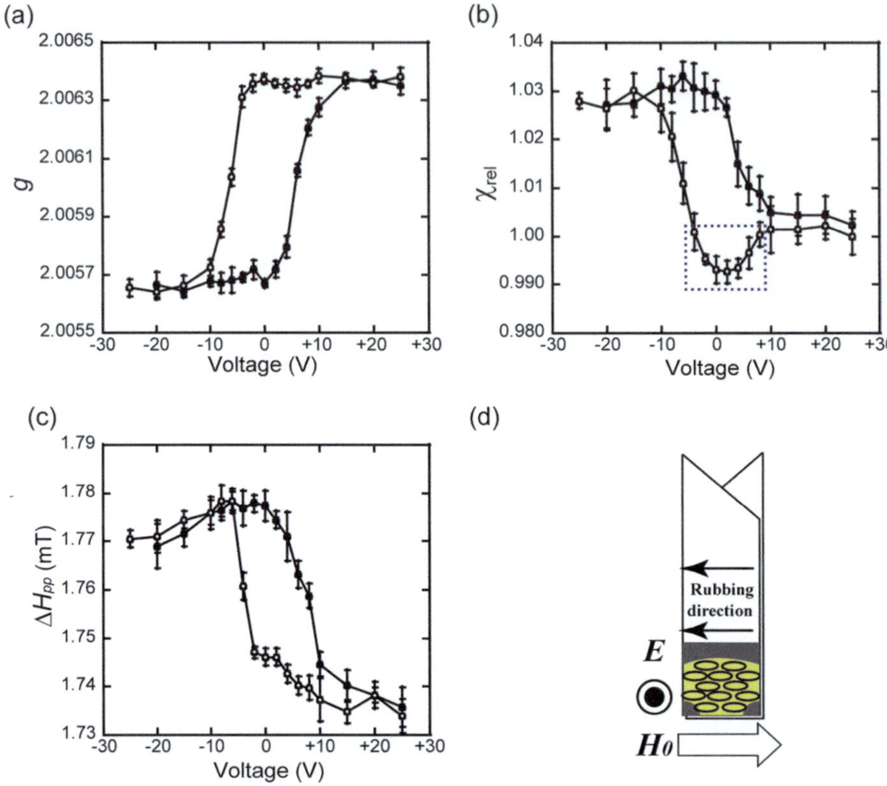

Fig. 25.14 Electric field dependence of **(a)** *g*-value, **(b)** $\chi_{\mathrm{rel,E}}$, and **(c)** ΔH_{pp} for the ferroelectric phase of (2*S*,5*S*)-**1** (n = 13, 65% ee) confined in a thin rubbed sandwich cell at 75 °C by EPR spectroscopy at a magnetic field of 0.33 T. **(d)** The magnetic field was applied perpendicular to the electric field and parallel to the rubbing direction. Open and filled circles represent the application of electric fields from +25 V to −25 V and from −25 V to +25 V, respectively. Magneto-electric effect was observed in the range from +10 V to −8 V surrounded by a dotted square in the panel **(b)**. (Ref. [34] – Reproduced by permission of The Royal Society of Chemistry)

where χ_1 is the standard value at the initial potential of +25 V and at 75 °C, was plotted for (2*S*,5*S*)-**1** (n = 13), showing a hysteresis loop between +25 V and −25 V (Fig. 25.14b). Since the molecular long axis is parallel to the magnetic field at −25 V, the spin easy axis in the ferroelectric LC phase of (2*S*,5*S*)-**1** (n = 13) seemed to exist along the molecular long axis. Then, to gain an insight into the origin of the anisotropic positive magneto-LC effect operating in the ferroelectric LC phase, the electric field dependence of ΔH_{pp} was compared with that of $\chi_{\mathrm{rel,E}}$ for (2*S*,5*S*)-**1** (n = 13) (Fig. 25.14c). The ΔH_{pp} exhibited a hysteresis similar to that of $\chi_{\mathrm{rel,E}}$. The ΔH_{pp} is known to reflect the following two competing factors: (1) the spin-spin exchange interaction and (2) the spin-spin dipole interaction. If the $\chi_{\mathrm{rel,E}}$ change results from the spin-spin exchange interaction, the experimental ΔH_{pp} would decrease with increasing $\chi_{\mathrm{rel,E}}$. However, the observed ΔH_{pp} increased and decreased with increasing and decreasing $\chi_{\mathrm{rel,E}}$, respectively, suggesting that the electric field dependence of $\chi_{\mathrm{rel,E}}$ should primarily arise from the change in the spin-spin dipole interaction. Thus, as the spin-spin dipole interactions contribute to the formation of the spin easy axis, these results proved the existence of a spin easy axis along the molecular long axis in the ferroelectric LC phase of (2*S*,5*S*)-**1** (n = 13).

Interestingly, however, the $\chi_{\mathrm{rel,E}}$ decreased from +10 V to 0 V and increased from 0 V to −8 V in Fig. 25.14b. This anomalous behavior cannot be explained simply by either the molecular reorientation or the spin-spin dipole interaction, because during the same electric field the *g*-value was almost constant and the ΔH_{pp} slightly increased (Fig. 25.14c). Accordingly this ΔH_{pp} increase from +10 V to 0 V should be ascribed to the decrease in the spin-spin exchange interaction and that form 0 V to −8 V to the increase in both the spin-spin exchange and dipole interactions; we are most likely to observe 'magneto-electric effect' in the range from +10 V to −8 V in the ferromagnetic and ferroelectric liquid crystal at as high as 75 °C.

25.6 Conclusions

We observed the generation of spin glass-like inhomogeneous ferromagnetic interactions ($\bar{J} > 0$) induced by applied magnetic fields at high temperatures in the various LC phases of the second-generation of chiral rod-like all-organic radical LC compounds such as **1** and referred to this phenomenon as 'positive magneto-LC effect'. The origin of this phenomenon was interpreted in terms of inhomogeneous intermolecular contacts to give ferromagnetic spin-spin dipole and exchange interactions. Then we evaluated the influence of electric fields on the positive magneto-LC effect in the FLC phase of (2*S*,5*S*)-**1** (n = 13) confined in a surface-stabilized liquid crystal cell by measuring the electric field dependence of the EPR spectra. By application of electric fields between +25 V and −25 V under the conditions where the magnetic field (0.33 T) was applied perpendicular to the electric field and parallel to the cell surface and rubbing direction, for the first time we observed (1) two magnetic bistable states, (2) anisotropy in spin-spin dipole interactions, and (3) 'magneto-electric effect' in the range of +10 V to −8 V at as low as 75 °C in the ferromagnetic and ferroelectric liquid crystal. The electric field dependence of ΔH_{pp} indicated that the anisotropy in spin-spin dipole interactions was responsible for the magnetic bistable states; strong spin-spin dipole interactions were observed when the magnetic field was applied parallel to the molecular long axis. Such unique magnetic properties may be applicable to the development of magnetic data storage materials operable at room temperature.

References

1. Bilinov LM (1983) Electro-optical and magneto-optical properties of liquid crystals. John Wiley & Sons, New York
2. Chamdrasekhar S (ed) (1992) Liquid crystal, second edn. Cambridge University Press, Cambridge
3. Demus D, Goodby J, Gray GW, Spiess HW, Vill V (eds) (1999) Physical properties of liquid crystals. Wiley-VCH, Weinheim
4. Dierking I (2003) Textures of liquid crystals. Wiley-VCH, Weinheim
5. Serrano JL (1996) Metallomesogens: synthesis, properties, and applications. Wiley-VCH, Weinheim
6. Dunmur D, Toriyama K (1999) In: Demus D, Gooby J, Gray GW, Spies HW, Vill V (eds) Physical properties of liquid crystals. Wiley-VCH, Weinheim, p 102
7. Kaszynski P (1999) In: Lahti PM (ed) Magnetic properties of organic materials. Marcel Dekker, New York, p 305
8. Griesar K, Haase W (1999) In: Lahti PM (ed) Magnetic properties of organic materials. Marcel Dekker, New York, p 325
9. Binnemans K, Gröller-Walrand C (2002) Lanthanide-containing liquid crystals and surfactants. Chem Rev 102:2303
10. Eerenstein W, Mathur ND, Scott JF (2006) Multiferroic and magnetoelectric materials. Nature 442:759
11. Rao CNR, Serrao CR (2007) New routes to multiferroics. J Mater Chem 17:4931
12. Felser C, Fecher GH, Balke B (2007) Spintronics: a challenge for materials science and solid-state chemistry. Angew Chem Int Ed 46:668
13. Seki S (2012) Magnetoelectric response in low-dimensional frustrated spin systems. Springer, Tokyo
14. Rikken GLJA, Raupach E (1997) Observation of magneto-chiral dichroism. Nature 390:493
15. Rikken GLJA, Raupach E (2000) Enantioselective magnetochiral photochemistry. Nature 405:932
16. Train C, Gheorghe R, Krstic V, Chamoreau LM, Ovanesyan NS, Rikken GLJA, Gruselle M, Verdaguer M (2008) Strong magneto-chiral dichroism in enantiopure chiral ferromagnets. Nat Mater 7:729
17. Likhtenshtein GI, Yamauchi J, Nakatsuji S, Smirnov AI, Tamura R (2008) Ntroxides: application in chemistry, biomedicine, and materials science. Wiley-VCH, Weinheim, p 303
18. Tamura R, Uchida Y, Ikuma N (2008) Paramagnetic all-organic chiral liquid crystals. J Mater Chem 18:2872
19. Tamura R, Ikuma N, Shimono S (2009) In: Nalwa HS (ed) Soft nanomaterials, vol 1. American Scientific Publishers, Stevenson Ranch, p 257
20. Tamura R, Uchida Y, Suzuki K (2012) In: Li Q (ed) Liquid crystals beyond display. Wiley, Hoboken, p 83
21. Tamura R, Uchida Y, Suzuki K (2012) In: Kokorin AI (ed) Nitroxides: theory experiment and application. INTECH, Rijeka, p 191
22. Tamura R, Suzuki K, Uchida Y, Noda Y (2013) EPR characterization of diamagnetic and magnetic organic soft materials using nitroxide spin probe techniques. In: Electron Paramag Reson, vol 23, p 1
23. Tamura R, Uchida Y, Suzuki K (2014) In: Gooby JW, Collings PJ, Kato T, Tschierske C, Gleeson HF, Raynes R (eds) Handbook of liquid crystals, vol 8, 2nd edn. Wiley-VCH, Weinheim, p 837
24. Ikuma N, Tamura R, Shimono S, Kawame N, Tamada O, Sakai N, Yamauchi J, Yamamoto Y (2004) Magnetic properties of all-organic liquid crystals containing a chiral five-membered cyclic nitroxide unit within the rigid core. Angew Chem Int Ed 43:3677

25. Ikuma N, Tamura R, Shimono S, Uchida Y, Masaki K, Yamauchi J, Aoki Y, Nohira H (2006) Ferroelectric properties of paramagnetic, all-organic, chiral nitroxyl radical liquid crystals. Adv Mater 8:477
26. Uchida Y, Ikuma N, Tamura R, Shimono S, Noda Y, Yamauchi J, Aoki Y, Nohira H (2008) Unusual intermolecular magnetic interaction observed in all-organic radical liquid crystal. J Mater Chem 18:2950
27. Uchida Y, Suzuki K, Tamura R, Ikuma N, Shimono S, Noda Y, Yamauchi J (2010) Anisotropic and inhomogeneous magnetic interactions observed in all-organic nitroxide radical liquid crystals. J Am Chem Soc 132:9746
28. Suzuki K, Uchida Y, Tamura R, Shimono S, Yamauchi J (2012) Observation of positive and negative magneto-LC effects in all-organic nitroxide radical liquid crystals by EPR spectroscopy. J Mater Chem 22:6799
29. Ikuma N, Tamura R, Masaki K, Uchida Y, Shimono S, Yamauchi J, Aoki Y, Nohira H (2006) Paramagnetic FLCs containing an organic radical component. Ferroelectrics 343:119
30. Ikuma N, Uchida Y, Tamura R, Suzuki K, Yamauchi J, Aoki Y, Nohira H (2009) Preparation and properties of C_2-symmetric organic radical compounds showing ferroelectric liquid crystal properties. Mol Cryst Liq Cryst 509:108
31. Uchida Y, Tamura R, Ikuma N, Yamauchi J, Aoki Y, Nohira H (2008) Synthesis and characterization of novel radical liquid crystals showing ferroelectricity. Ferroelectrics 365:158
32. Vorobiev AK, Chumakova NA, Pompgailo DA, Uchida Y, Suzuki Y, Noda Y (2014) Determination of structural characterization of all-organic radical liquid crystals based on analysis of the dipole-dipole broadened EPR spectra. J Phys Chem B 118:1932
33. Uchida Y, Suzuki K, Tamura R (2012) Magneto-LC effects in hydrogen-bonded all-organic radical liquid crystal. J Phys Chem B 116:9791
34. Suzuki K, Uchida Y, Tamura R, Noda Y, Ikuma N, Shimono S, Yamauchi J (2013) Influence of applied electric fields on the positive magneto-LC effects observed in the ferroelectric liquid crystalline phase of a chiral nitroxide radical compound. Soft Matter 9:4687
35. Suzuki K, Uchida Y, Tamura R, Noda Y, Ikuma N, Shimono S, Yamauchi J (2013) Electric field dependence of molecular orientation and anisotropic magnetic interactions in the ferroelectric liquid crystalline phase of an organic radical compound by EPR spectroscopy. Adv Sci Tech 82:50
36. Tamura R, Uchida Y, Suzuki K (2015) In: Tamura R, Miyata M (eds) Advances in organic crystal chemistry: comprehensive reviews. Springer, Tokyo, p 689
37. Aurich HG (1989) In: Breuer E, Aurich HG, Nielsen A (eds) Nitrones nitronates and nitroxides. Wiley, Chichester, p 313
38. Dvolaitzky M, Billard J, Polydy F (1974) Note des membres et correspondants et notes presentees ou transmises par leurs soins. C R Acad Sci Paris Ser C 279:533
39. Dvolaitzky M, Taupin C, Polydy F (1976) Nitroxides piperidineques: synthesis de nouvelles sondes paramagnetiques. Tetrahedron Lett 18:1469
40. Dvolaitzky M, Billard J, Polydy F (1976) Smectic E, C and A free radicals. Tetrahedron 32:1835
41. Allgaier J, Finkelmann H (1994) Synthesis and magnetic properties of mesogenic side-chain polymers containing stable radicals. Macromol Chem Phys 195:1017
42. Greve S, Vill V, Friedrichsen W (2002) Novel Nitronyl Nitroxides: Synthesis and Properties. Z Naturforsch 57b:677
43. Nakatsuji S, Mizumoto M, Ikemoto H, Akutsu H, Yamada J (2002) Preparation and properties of organic radical compounds with mesogen cores. Eur J Org Chem 2002:1912
44. Kogo R, Araoka F, Uchida Y, Tamura R, Ishikawa K, Takezoe H (2010) Second harmonic generation in a paramagnetic all-organic chiral smectic liquid crystal. Appl Phys Express 3:041701
45. Mandel SK, Nath TK, Das A, Kremer RK (2006) Magnetic glassy phase in Zn0.85Fe0.15O diluted magnetic semiconducting nanoparticles. Appl Phys Lett 89:162502

46. Schneider J, Handstein A, Zaveta K (1984) Temperature and field dependence of the magnetization in crystalline and amorphous metallic alloys. J Magn Magn Mater 42:73
47. Uchida Y, Tamura R, Ikuma N, Shimono S, Yamauchi J, Shimbo Y, Takezoe H, Aoki Y, Nohira H (2009) Magnetic-field-induced molecular alignment in an achiral liquid crystal spin-labeled by a nitroxyl group in the mesogen core. J Mater Chem 19:415

Chapter 26
Mechanical Deformation Chemistry of Crystals: Designing Mechanical Performance

C. Malla Reddy

Abstract Crystal engineering approaches can be useful to understand, predict and design mechanical properties of the active pharmaceutical ingredients (APIs) for their improved performance in various stages of production. For example, the understanding of correlation among structure, mechanical property and powder compaction would allow prediction and design of powder tabletability of APIs. A design approach to achieve mechanically flexible plastic and elastic molecular crystals has recently been proposed. This involves the introduction of active slip planes (with minimal ruggedness) into the crystal structure by making different non-interfering weak interactions such as van der Waals (vdW), π-stacking and hydrogen bonding. By analyzing the reported crystal structures of plastically flexible crystals it can be hypothesized that the spherical hydrophobic groups will assemble via shape complementarity (shape synthons) to reliably form low energy slip planes. As these groups do not interfere with the π-stacking or hydrogen bonding groups, they can pack in a predictable manner and thus form slip planes to facilitate mechanical flexibility, as successfully demonstrated in a series of naphthalene diimide derivatives. Such studies can allow the preparation of exotic plastic crystals by design and through this demonstrate the potential for using soft interactions for tuning mechanical behaviour of ordered molecular materials. A comment is made on the prospects and ramifications of this emerging field, in the context of pharmaceutical solids.

Keywords Active pharmaceutical ingredients • Crystal engineering • Mechanical properties • Tabletability • Shape synthons

C.M. Reddy (✉)
Department of Chemical Sciences, Indian Institute of Science Education and Research (IISER) Kolkata, Mohanpur Campus, Mohanpur 741246, India
e-mail: cmreddy@iiserkol.ac.in

K.J. Roberts et al. (eds.), *Engineering Crystallography: From Molecule to Crystal to Functional Form*, NATO Science for Peace and Security Series A: Chemistry and Biology, DOI 10.1007/978-94-024-1117-1_26

26.1 Introduction

The crystal engineering approach has been used very effectively in recent times to advance the understanding of structure-mechanical property correlation in molecular crystals which is relevant to the pharmaceutical technology aspects such as manufacturability, handling of solids at various production processes, etc. Recently there have been growing reports pertinent to the mechanical properties of molecular crystals studied by application of external stress using a range of techniques. This includes both qualitative (shearing, bending and brittle deformation) and quantitative (nanoindentation, powder compaction and high-pressure) studies. This has allowed the establishment of correlation of anisotropic mechanical behaviour with underlying crystal structure of the materials. Hence, structure-mechanical property correlation studies, which reveal the spatial arrangement of intermolecular interactions and overall packing of molecules that influence the crystal deformation, can significantly help in designing various pharmaceutical processes to handle the solids efficiently.

The study of mechanical properties of organic materials which has increasingly gained the attention of researchers across the disciplines owing to their potential applications in various fields [1–4]. For instance, the consideration of mechanical properties and controlling dislocations is important in organic semiconductor devices [5–9], formulation of APIs in pharmaceuticals [10, 11], mechanical actuators, etc. [12–14]. In the pharmaceutical industry, bulk properties of a drug like hardness, strain rate sensitivity, granularity and flowability depend on its mechanical properties that are vital for easy tabletting and formulation [15]. Mechanical deformation behavior of single crystals may also influence the particle size distribution, for instance in industrial scale crystallization, filter drying by mechanical agitator, etc. as the breakage of particles can easily occur due to the mechanical agitations involved in such processes. The forces that bind atoms into molecules and/or molecules into crystalline solids, to a great extent, determine the physical and chemical properties of a solid material at a given temperature. It is important to relate the mechanical properties of molecular crystals to their underlying crystal packing. Here the techniques like nanoindentation, powder compaction and high-pressure studies are very helpful for gaining knowledge for the design and synthesis of new materials using crystal engineering approaches [4]. In this context, it is important to understand the precise nature of intermolecular interactions and their role in directing the three dimensional (3D) structures. As the present understanding of structural basis for the mechanical deformation of molecular crystals is not adequate, it would be useful to draw parallels with well studied classes of materials like inorganic crystals [16], with anticipation that such knowledge will guide us in exploring the phenomena systematically in molecular crystals.

Currently, empirical methods are used in pharmaceutical processing to address issues related to powder compaction and caking behavior of solid powders during industrial crystallization, filtration, storage and mechanical milling, etc. The behavior of powder particles is typically analyzed and predicted by considering morphology of crystalline particles and almost no crystal structure information is considered for addressing issues arising from the mechanical aspects in various processes. Recent studies suggest that it is important to establish a good correlation among crystal structure, mechanical behavior of single particles (crystals) and bulk properties, such as compaction behavior for addressing the issues by scientific approaches rather than empiricism. Keeping this in mind, a brief summary of the results on structure-mechanical behavior of single crystals is outlined here.

26.2 Classification of Molecular Crystals Based on Mechanical Response

Molecular crystals are generally considered to be brittle, but some recent studies have shown that certain molecular crystals can be extremely plastic or elastic [3]. Based on the mechanical behaviour, molecular crystals can be broadly classified into (1) brittle, (2) shearing, (3) plastic bending and (4) elastic bending types [3, 13]. This classification is based on the physical morphological changes in crystals upon application of external mechanical stress.

1. *Brittle type crystals*: Majority of organic crystals fall into this category. These crystals are typically very fragile and break easily upon application of a mechanical stress. Crystal packing in these materials is generally isotropic with similar type of interactions in all the dimensions. For example interlocked packing with the hydrogen bonding in all three dimensions (3D network of H-bonding) in crystals makes them brittle. Weak interactions with similar strength in all three dimensions also may lead to brittle fracture (Fig. 26.1c, f [3].
2. *Shearing type crystals*: Crystals with flat two-dimensional layers (2D) with relatively strong *intra*layer and weak *inter*layer interactions show shearing behaviour upon application of a mechanical stress. When stress is applied parallel to the layers, portions of these crystals slide similar to a pack of cards, leading to the permanent deformation in crystals. This is because the relatively strong *intra*layer interactions hold the molecules within a layer tightly, while the *inter*layer interactions mediate smooth sliding of molecular layers over each other (Fig. 26.1a, d [3, 17]. Here the low ruggedness of sheets and weak interplanar interactions are critical for such plastic shearing (irreversible) under an external mechanical stress.
3. *Plastic bending type crystals*: Crystals with active slip planes between thick molecular sheets undergo plastic bending (Fig. 26.1b, e). The slip planes in these crystals are typically constituted by weak van der Waals interactions from close packing of pseudo-spherical functional groups, such as alky, −Cl, −OMe, etc. When such weak interactions are formed to create a smooth surface (with low ruggedness) along a crystallographic plane that is nearly parallel to one of the observed crystal faces, crystals bend on that face, e.g. like a bundle of printing papers [18].

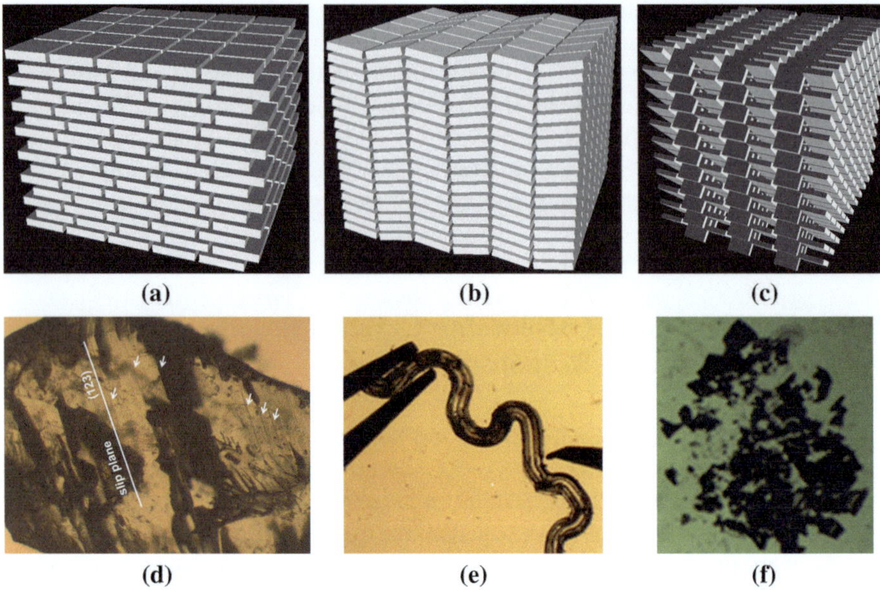

Fig. 26.1 Schematic representation of structure based models for mechanical shearing (**a**), bending (**b**) and brittle (**c**) type molecular crystals, and images of crystals from the three classes, (**d**), (**e**) and (**f**), respectively. The arrangement of the molecules and type of intermolecular interactions in the models are distinct and lead to different types of deformation mechanisms

4. *Elastic bending type crystals*: Some crystals of brittle type, having a combination of isotropic intermolecular interactions and mechanical interlocking via weak van der Waals interactions, show high elastic flexibility (essentially falling into a sub class of brittle crystals, albeit with high elastic limits (see Figs. 26.2 and 26.3) [18]. In these crystals, upon bending, the outer and inner surfaces of the crystal experience large tensile and compressive strains, respectively. This is in contrary to the situation in plastic crystals where the tensile and compressive forces are negligible, but slippage of molecules via weak interaction planes with low ruggedness leads to the changes in interfacial angles (Fig. 26.2). The elastic type crystals may show very poor tabletability properties (yet to be explored by systematic studies). However, crystals with such high elastic properties can be relevant to various other practical applications (e.g. flexible electronics, wave guides, etc.).

The elastic type crystals, not only recover the original shape upon release of the external mechanical stress, but also show no sign of fatigue in their crystal structure. From the X-ray diffraction data, obtained from a straight single crystal that recovered after subjecting to several mechanical bending cycles, it was possible to reproduce the full structure solution (with identical unit cell parameters) with equally good refinement. Such high elastic recovery in pharmaceutical crystalline particles may pose extreme problems in powder compaction.

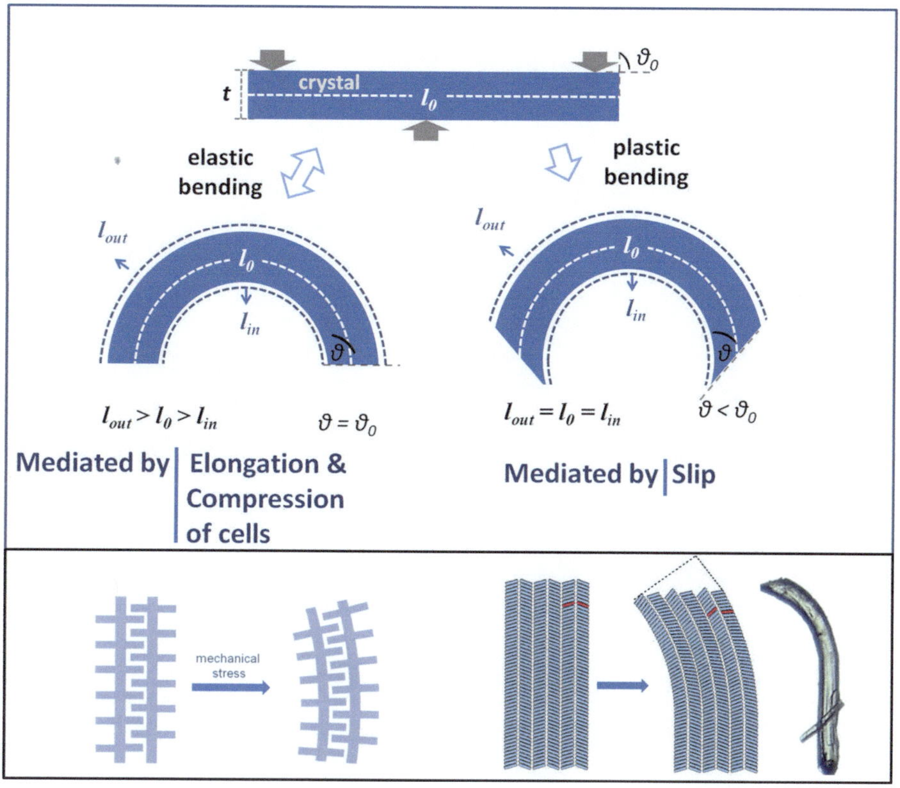

Fig. 26.2 Schematic models showing the comparison of elastic (*top-left*) and plastic (*top-right*) bending in crystals. Notice the absence of change in interfacial angles (*edges*) in elastic crystal, unlike in the plastic bending type crystal. Schematic depiction of possible mechanisms of elastic (*bottom-left*) and plastic (*bottom-right*) bending in crystals (The *top-panel* in this figure is adapted from Ref. [18] and the panel in *bottom-right* is adapted from Ref. [12])

The variety of crystals discussed here demonstrates the possibility of diverse and extreme mechanical behaviors in crystalline organic particles. This arises from the possibility of diverse structural features from the underlying crystal packing in molecular crystals. A careful analysis of crystal structure of a product at an early stage of development may prove to be valuable for predicting the mechanical behavior of pharmaceutical powders beforehand and to design a robust process based on scientific inputs rather than empiricism.

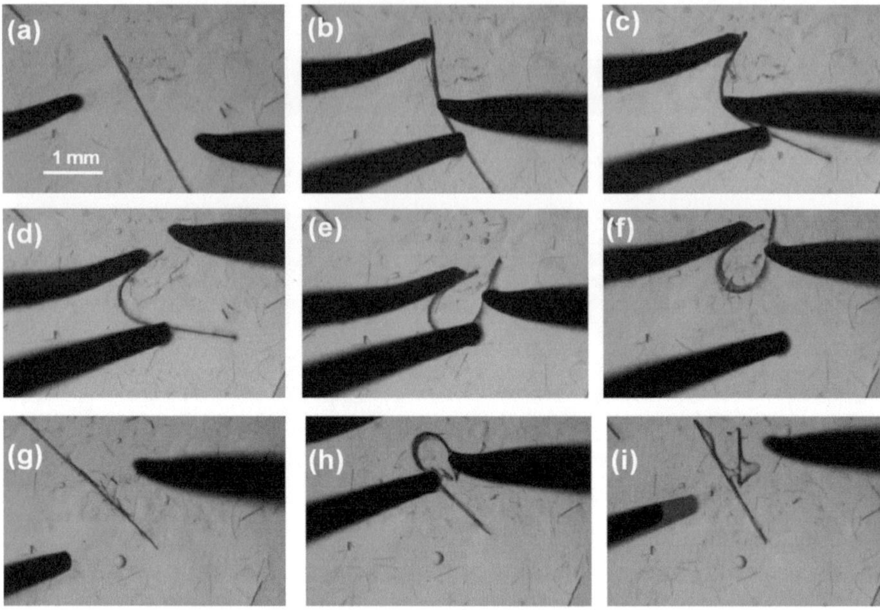

Fig. 26.3 Elastic bending in the co-crystal solvate, caffeine: 4-chloro-3-nitrobenzoic acid:methanol (1:1:1). (**a–f**) First bending cycle by applying mechanical stress, using a pair of forceps and a metal pin and (**g**) the subsequent recovery upon withdrawal of the force. (**h**) Second bending cycle and (**i**) the crystal breakage when the elastic limit is exceeded. Notice the remarkable shape recovery in the two broken halves in (**i**) (This figure is adapted from Ref. [18])

26.3 Impact of Crystalline Particles' Mechanical Behaviour on Powder Compaction

Typically, the percentage of an active substance in a tablet is very low compared to the matrix hence the intrinsic bulk properties of the drug, such as powder compaction, may not have any significant influence on the formulation. However, in the case of formulated dosage forms containing a high drug loading (>750 mg), the size of the tablet for the same ratio of drug/excipient would become unsuitable for oral dosage (too large to swallow) and also may cause issues with packaging and storage. In such cases, the intrinsic adverse mechanical behavior of the drug may pose serious problems in tabletting [15]. For instance, among the polymorphs of paracetamol (PCA), Form II is more suitable for tableting than Form I but the latter is preferred due to its superior stability. So challenges remain to enhance the tabletability of Form I and this necessitates extensive efforts. We recently demonstrated the potential of plastically bending type crystalline forms to achieve better tablets compared to shearing and brittle type forms [15, 17]. Our initial studies strongly support that a detailed knowledge of correlation among structure-mechanical property-powder compaction can potentially allow devising straightforward strategies for powder compaction by simply using the crystal structure knowledge. Such a method will be very beneficial considering the time and costs involved in the trial-and-error methods, especially in the developmental stages. Some of our recent studies on correlation among crystal structure-mechanical behaviour and powder compaction are detailed below. A design strategy for achieving flexible crystals is also presented in the later part.

Our recent experiments on a trimorphic 6-chloro-2,4-dinitroaniline (**cda**) suggest that the plastically bendable molecular crystals are far better suited for pharmaceutical tabletting compared to the brittle or shearing type crystals [15, 17] (Figs. 26.4 and 26.5). The brittle type crystals, due to poor plastic deformation and high elastic recovery, do not enhance the bonding area between particulates in powder compaction and hence do not produce strong tablets. In shearing type crystals, easy shearing of crystalline particles, however, quickly increases the bonding area and reduces the pore volume upon compaction, but the soft nature of the interactions results in poor interparticulate binding energy, hence leads to poor tabletability. For these reasons, shearing type crystals typically also lead to the problems such as caking of particles, poor flowability and poor tabletability. On the other hand, plastic deformation in bending type crystals reduces the pore volume and increases bonding area of powders. In bending type materials, the bonding strength is also reasonably good compared to the shearing type crystals. Hence they show far better powder compaction properties than that of brittle and shearing type crystals.

The blue highlight in Fig. 26.4 is to indicate the thickness of layers (formed via stronger interactions) and interlayer spacing. Notice the comb like interlocking and some C–H···O interactions connecting adjacent combs in Form III packing. The images of the tablets in (d) (f) and (h) made at the compaction pressure of 10 MPa and (e), (g) and (i) at 200 MPa, of Forms I, II and III, respectively. The images correspond to the tablets broken at their respective maximum tensile strengths. Notice the poor shape of the Form III tablet in (h) compared to those in (d) Form I and (f) Form II. Figure 26.5 shows the quantitative assessment of the mechanical performance of the three forms, brittle (black line), shearing (blue line) and plastically bending (red) type forms. This clearly indicates that the plastically bending type forms are more favourable for powder tabletting compared to the other two types, at least in this case. A similar trend was also observed in another study conducted by us recently on the co-crystals of **cda** [17].

26.4 Crystal Engineering: Designing Mechanical Performance via Shape Synthons

By carefully examining various bending type crystals [3, 12–15, 17–19], we proposed a model for achieving flexible plastic crystals using certain crystal engineering principles. According to the model [13], the crystals with weak interactions or active slip planes, formed from van der Waals groups, promote the

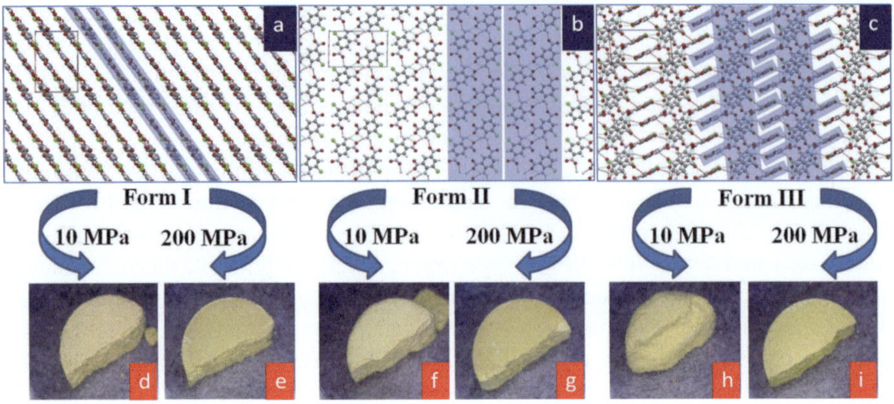

Fig. 26.4 Crystal packing and tablet images of the three forms of 6-chloro-2,4-dinitroaniline (**cda**). (**a**) Thin and flat 2D layer packing in Form I. (**b**) Thick 2D layer packing in Form II. (**c**) Comb-like interlocked 3D structure in Form III (This figure is adapted from Ref. [15])

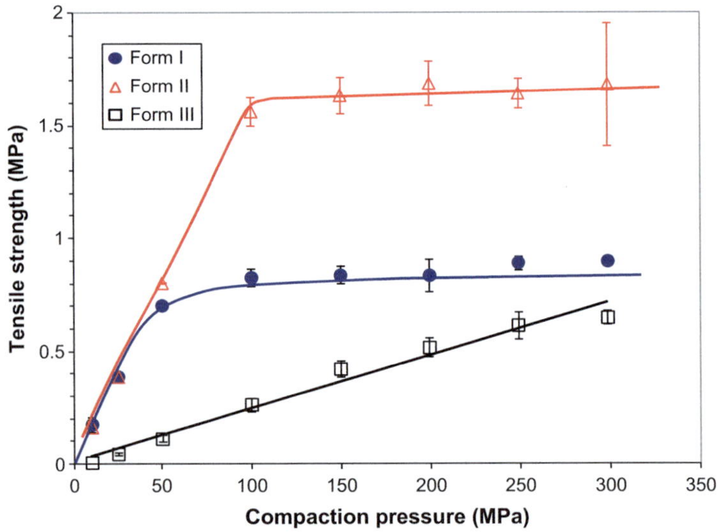

Fig. 26.5 Tabletability curves of **cda** Forms I (o), II (Δ) and III (□) (This figure is adapted from Ref. [15])

plasticity in them. To demonstrate the design strategy we recently exploited some naphthalene diimide (NDIs) derivatives [20]. Although these examples are not pharmaceutically relevant, the self-assemble principles involved in their packing are common to all the molecular materials, including APIs. The study showed that proper utilization of the weak van der Waals interactions can allow us to design softer pharmaceutical crystals [20]. A set of non-interfering groups, spherical vdW groups and aromatic groups with or without hydrogen bonding functionalities

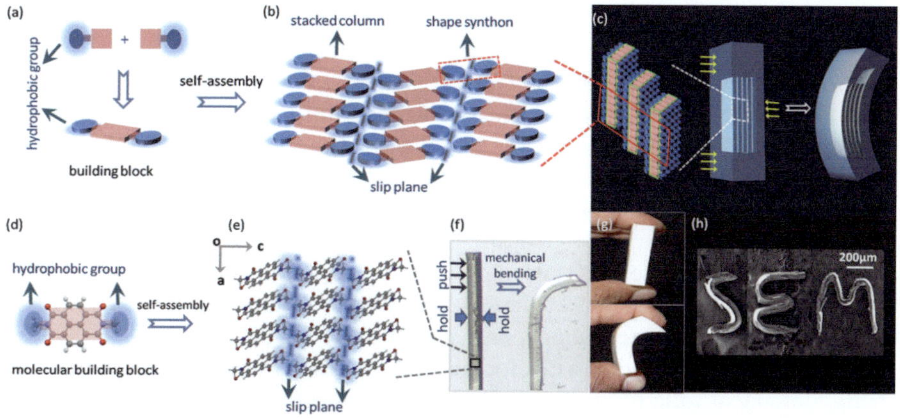

Fig. 26.6 Schematic representation of the supramolecular shape synthon-based model for achieving plastic flexibility in molecular single crystals. (**a**) A building block having a central core (in *red*) and peripheral vdW (in *blue*) groups. (**b**) Packing of building blocks to form slip planes (*broken blue lines* via shape synthons) by vdW groups. (**c**) Flexible molecular sheets which may bend along slip planes upon mechanical deformation. (**d**) Prototype example of N-methyl naphthalene diimide (**N-MNDI**) derivative, and (**e**) its experimentally determined crystal packing, which is consistent with the slip plane model in (b). (**f**) Micrographs of a flexible single crystal of **N-MNDI**, before (*left*) and after (*right*) mechanical bending. (**g**) Bending of sticky *yellow pad* (via slippage of individual sheets) is compared to bending of a plastic crystal (*via* slippage of molecular layers). Notice the similarity of the top ends of the bent crystal (image on *right*, in **f**) and bent sticky note pad in (**g**). (**h**) SEM images of the mechanically bent crystals (This figure is reproduced from Ref. [20])

(Fig. 26.6) were strategically considered. The spherical vdW groups self-sort to close pack with themselves and generally do not interfere with the aromatic or hydrogen bonding groups. Similarly, the aromatic groups of NDI derivatives close pack via π-stacking interactions. This typically leads to the formation of slip planes by the vdW groups while the aromatic groups pack in perpendicular direction. Such an anisotropic packing, if achieved in crystal structures, may facilitate plastic flexibility, as depicted in Fig. 26.6.

With this in mind, a series of naphthalene diimides (NDIs), with alkyl vdW substituents such as methyl (**N-MNDI**), *n*-propyl (**N-PNDI**), *p*-tolyl (**N-PTNDI**) and *tert*-butyl (**N-TBNDI**) on either ends of the central NDI core, were synthesized [20]. All the compounds in this series typically formed long needle type crystals, with lengths of up to 1 cm. In the crystal structures of **N-MNDI** ($P2_1$/c), **N-PNDI** (*Pbca*), **N-PTNDI** ($P2_1$/c) and **N-TBNDI** (*P*-1), as anticipated, the vdW groups close pack with themselves to form the shape synthons (isotropic interactions) while the aromatic groups form π-stacked columns in the perpendicular direction, showing an excellent agreement with the proposed model. For instance, in **M-NDI** the spherical methyl groups self-sort to form slip-planes parallel to (0 0 2) via shape synthons (vdW interactions) while the central NDI aromatic core group forms π-stacked columns (Fig. 26.6e). Similar packing is observed in the other compounds, except in **N-PNDI**, where the elongated *n*-propyl groups from the adjacent stacks facing in the opposite directions.

Mechanical behavior of all the crystals was examined by a simple qualitative method, i.e. applying mechanical stress using a pair of forceps and a metal needle while viewing them under a stereomicroscope [3]. The single crystals of **N-MNDI**, **N-PTNDI** and **N-TBNDI** showed excellent plastic flexibility and could be bent to 360°, while notably **N-PNDI** underwent brittle fracture (due to interlocking of *n*-propyl groups). This impressive plastic flexibility in the former examples is demonstrated by molding their crystals into a variety of shapes, such as letters "S E M". Hence, these examples make obvious the possibility of utilizing the crystal engineering principles for designing plastically flexible single crystals. The similar strategy can be applied to APIs for generating new pharmaceutically relevant multi-component solid forms, or predict the mechanical behavior of single component forms for devising suitable formulation strategies (without resorting entirely to empirical methods).

26.5 Future Directions

Apart from demonstrating the importance of mechanical properties of single crystals for various applications, the above results highlight the importance of establishing a strong structure-mechanical property correlation in molecular materials. The proposed model, which is based on weak interactions, has allowed us to design plastically flexible crystals of diverse materials [3, 14, 20]. The same strategy can be applied to various other molecular materials, including the pharmaceutical solids, for tuning their mechanical performance. The structure-property correlation developed in these crystals may also help to predict the mechanical behavior of single crystals in various solid-state processes involved in the early product development. For instance, the crystal size distribution in bulk crystallization processes, powder compaction and mechanical behaviour (disintegration during mechanical agitation) of a crystal form can be studied and predicted by a careful examination of crystal packing.

References

1. Desiraju GR, Vittal JJ, Ramanan A (2011) Crystal engineering: a text book. World Scientific, Singapore
2. Desiraju GR (2013) Crystal engineering: from molecule to crystal. J Am Chem Soc 135:9952–9967
3. Reddy CM, Krishna GR, Ghosh S (2010) Mechanical properties of molecular crystals—applications to crystal engineering. CrystEngComm 12:2296–2314

4. Varughese S, Kiran MSRN, Ramamurty U, Desiraju GR (2013) Nanoindentation in crystal engineering: quantifying mechanical properties of molecular crystals. Angew Chem Int Ed 52:2701–2712
5. Fratzl P, Barth FG (2009) Biomaterial systems for mechanosensing and actuation. Nature 462:442–448
6. Lv S, Dudek DM, Cao Y, Balamurali MM, Gosline J, Li H (2010) Designed biomaterials to mimic the mechanical properties of muscles. Nature 465:69–73
7. Koshima H, Ojima N, Uchimoto H (2009) Mechanical motion of azobenzene crystals upon photoirradiation. J Am Chem Soc 131:6890–6891
8. Koshima H, Takechi K, Uchimoto H, Shiro M, Hashizumec D (2011) Photomechanical bending of salicylideneaniline crystals. Chem Commun 47:11423–11425
9. Morimoto M, Irie M (2010) A diarylethene cocrystal that converts light into mechanical work. J Am Chem Soc 132:14172–14178
10. Kiran MSRN, Varughese S, Reddy CM, Ramamurty U, Desiraju GR (2010) Mechanical anisotropy in crystalline saccharin: nanoindentation studies. Cryst Growth Des 10:4650–4655
11. Karki S, Friščić T, Fábián L, Laity PR, Day GM, Jones W (2009) Improving mechanical properties of crystalline solids by cocrystal formation: new compressible forms of paracetamol. Adv Mater 21:3905–3909
12. Reddy CM, Gundakaram RC, Basavoju S, Kirchner MT, Padmanabhan KA, Desiraju GR (2005) Structural basis for bending of organic crystals. Chem Commun:3945–3947
13. Reddy CM, Padmanabhan KA, Desiraju GR (2006) Structure-property correlations in bending and brittle organic crystals. Cryst Growth Des 6:2720–2731
14. Krishna GR, Kiran MSRN, Fraser CL, Ramamurty U, Reddy CM (2013) The relationship of solid-state plasticity to mechanochromic luminescence in difluoroboron avobenzone polymorphs. Adv Funct Mater 23:1422–1430
15. Bag PP, Chen M, Sun CC, Reddy CM (2012) Direct correlation among crystal structure, mechanical behaviour and tabletability in a trimorphic molecular compound. CrystEngComm 14:3865–3867
16. Dieter GE (1988) (Adapted by D. Bacon), Mechanical metallurgy, SI Metric Edition. McGraw-Hill, Singapore
17. Krishna GR, Shi L, Bag PP, Sun CC, Reddy CM (2015) Correlation among crystal structure, mechanical behavior, and tabletability in the co-crystals of vanillin isomers. Cryst Growth Des 15:1827–1832
18. Ghosh S, Reddy CM (2012) Elastic and bendable caffeine cocrystals: implications for the design of flexible organic materials. Angew Chem Int Ed Eng 51:10319–10323
19. Panda MK, Ghosh S, Yasuda N, Moriwaki T, Mukherjee GD, Reddy CM, Naumov P (2014) Spatially resolved analysis of short-range structure perturbations in a plastically bent molecular crystal. Nat Chem 7:65–72
20. Krishna GR, Devarapalli R, Lal G, Reddy CM (2016) Mechanically flexible organic crystals achieved by introducing weak interactions in structure: supramolecular shape synthons. J Am Chem Soc 138:13561–13567

Chapter 27
DEM Analysis of the Effects of Die Shape and Orientation on Die Filling Processes

Chunlei Pei and Chuan-Yu Wu

Abstract Powder compaction is widely used to manufacture particulate products (e.g. tablets and pellets), in which die filling is a critical stage governing the performance of the powder compaction process and the quality of finished products. In this paper, how the shape and orientation of the die affect powder flow behaviour during die filling is analysed using the discrete element method (DEM), in which powders are treated as a collection of individual particles. Die filling with dies of various shapes and orientation is considered, and die filling performance is evaluated in terms of the mass flow rate and the critical fill speed. It is found that the die filling efficiency during parallel filling into a die of large aspect ratio is generally low. This study also demonstrates that DEM is a useful tool for modelling die filling processes.

Keywords Discrete element method • Die filling • Critical filling speed • Powder compaction • Fill ratio

27.1 Introduction

Die filling is a common and critical stage of the manufacturing process of particulate products in many industries, such as fine chemicals, powder metallurgy and pharmaceuticals. During the die filling process, the powder is usually fed in a shoe and then the shoe moves over a die and delivers the powder into the die. The performance and efficiency of the die filling process significantly influence the subsequent processes, such as compaction and tableting, and the quality of final products. For instance, an incomplete fill or a non-uniform packing density distribution in the die can lead to breakage and failure in the downstream compaction process, which produces low quality final products. The performance of the die

C. Pei • C.-Y. Wu (✉)
Department of Chemical and Process Engineering, University of Surrey, Guildford, Surrey GU2 7XH, UK
e-mail: C.Y.Wu@surrey.ac.uk

K.J. Roberts et al. (eds.), *Engineering Crystallography: From Molecule to Crystal to Functional Form*, NATO Science for Peace and Security Series A: Chemistry and Biology, DOI 10.1007/978-94-024-1117-1_27

filling depends on many factors, such as the properties of particles and the geometry of the dies [1, 2].

Dies of various shapes are used in practice to produce products of designed geometries. It is anticipated that the die shape and filling system set-up (e.g. orientation) can also affect the powder flow behavior. Bocchini [3] investigated the influence of the width and orientation of die openings on the filling density in rectangular dies. It was found that the filling density decreased when the width of the die decreased. In addition, the filling density at the region close to the die walls is almost zero, indicating that there was a boundary layer near the die walls. This boundary layer depended on the particle properties and the geometry of the die. Rice and Tengzelius [4] suggested that the particles were restricted to rearrange and expend in a relatively smaller die. Therefore, lower density was observed in the narrow dies. In the experiment of Bocchini [3], two different orientations of dies were considered: parallel filling, in which the longest axis of the die aligns with the motion of the shoe, and orthogonal filling, in which the longest axis is perpendicular to the motion of the shoe. It was observed that the orthogonal filling gave a higher filling density than the parallel filling. To characterize the flow behaviour of particles in the die filing system, Wu and Cocks [2] suggested that a critical filling speed can be used to classify the die filling process, below which the die can be completely filled. If the filling speed is higher than the critical filling speed, then the die cannot be completely filled. The fill ratio of the die at these higher filling speeds can be defined as the filled mass of particles in the die divided by the fully filled mass. Wu and Cocks [2] found that, when the filling speed is higher than the critical filling speed, the fill ratio can be treated as a function of the filling speed, which can be written as:

$$\delta = \left(\frac{v_c}{v_s}\right)^m \tag{27.1}$$

where v_c is the critical filling speed; v_s is the filling speed; m is a power index in the range 1.0–1.6. Clearly, the critical filling speed and the power index m vary with the geometry of the die.

Die filling was also investigated using the Discrete element method (DEM). Wu and Cocks [5] investigated the flow of particles into a confined die using DEM and analyzed the flow pattern of 2-D polygonal particles filling into a stepped die with two (narrow and wide) sections divided by a central block. It was observed that a smoother flow of particles was observed in the wide section than in the narrow one. Moreover, arching and bridging were observed in the narrow section, implying that relatively lower packing density can occur in the narrow section. Wu and Guo [6] modelled the suction filling and the gravity filling using DEM-CFD, a coupled DEM with computational fluid dynamics. In the suction filling, the bottom of the die can move downwards and 'suck' particles into the die due to the pressure difference. The suction process can significantly increase the mass flow rate especially for air-sensitive particles [1]. It has been demonstrated that DEM can be a useful tool to explore the die filling process, especially when the detailed dynamic information, e.g. the velocity and position of particles, are required. In this study, the influences of the shape and orientation of the die on the die filling process are investigated using DEM, in terms of the mass flow rate and the critical fill speed.

27.2 Model Set-Up

In the discrete element method used in this study, the contact interactions between objects are determined by classical contact mechanics: Hertz theory for the normal contact and the theory of Mindlin and Deresiewicz for the tangential interaction. The motion of the particle can be computed using the Newton's law of motion.

The die filling system consists of a moving shoe and a die as shown in Fig. 27.1. Initially, the particles are deposited in the shoe. The shoe then moves over the die at a specified speed and deposits the particles into the die. After the die filling, the amount of particles deposited into the die is then determined.

In the current study, different shapes and orientations of the die are chosen as shown in Fig. 27.2. Various dies considered include circle (C0), square (R0), ellipse (C1 & C2), and rectangle (R1–R4). The orientation with respect to the shoe moving direction includes parallel die filling (C1, R1 & R3), in which the shoe moves along the long axis of the die, and orthogonal die filling (C2, R2 & R4), in which the shoe moves perpendicular to the long axis of the die. The geometries of the dies are listed in Table 27.1. For all cases considered, the cross-sectional area and the height of all the dies are identical at 28.27 mm^2 and 12 mm, respectively. Therefore, the volume of all dies is 339.24 mm^3.

In order to represent the difference of die shapes for the same orientation and its influence on the die filling process, the aspect ratio is introduced and defined as the ratio between the long axis and the short axis:

$$\gamma = \frac{d_l}{d_s} \tag{27.2}$$

where d_l is the long axis of the die opening; d_s is the short axis of the die opening. The aspect ratios of all dies are also listed in Table 27.1. For instance, C1 and R3 are of the same aspect ratio and orientation but different shapes. R1 and R3 are rectangular and parallel to the shoe moving direction but of different aspect ratios.

To model the contact interactions between the curved surface (C0–C2) and the particles, the die is divided into triangular meshes to represent the complex shape as shown in Fig. 27.3. The contact detection and interaction between particles and the dies are then implemented between the particles and the triangular meshes [7]. The dies of C0, C1 and C2 are discretised into 442, 792 and 734 triangular meshes, respectively. Monosized elastic spherical particles with the diameter of 0.8 mm are used. The material properties of the particles and the die systems are presented in

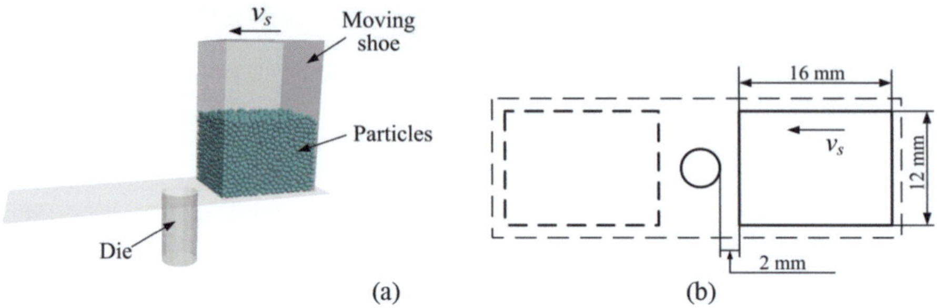

Fig. 27.1 The model setup

Fig. 27.2 The geometry of the dies

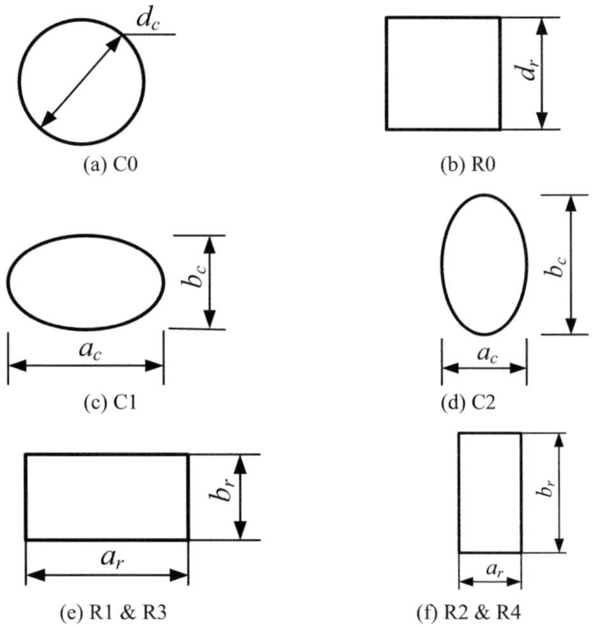

Table 27.1 Details of various dies considered

ID	Dimension (mm)	Aspect ratio	v_c (*mm/s*)	*m*
C0	d_c: 6.0	1.0	116.99	1.023
R0	d_r: 5.3	1.0	115.24	1.043
C1	$a_c \times b_c$: 9.0 × 4.0	2.25	105.57	0.95
C2	$a_c \times b_c$: 4.0 × 9.0	2.25	111.62	1.076
R1	$a_r \times b_r$: 7.070 × 4.07	1.74	107.4	0.96
R2	$a_r \times b_r$: 4.07 × 7.070	1.74	112.17	1.05
R3	$a_r \times b_r$: 7.98 × 3.54	2.25	99.9	0.95
R4	$a_r \times b_r$: 3.54 × 7.98	2.25	105.8	1.04

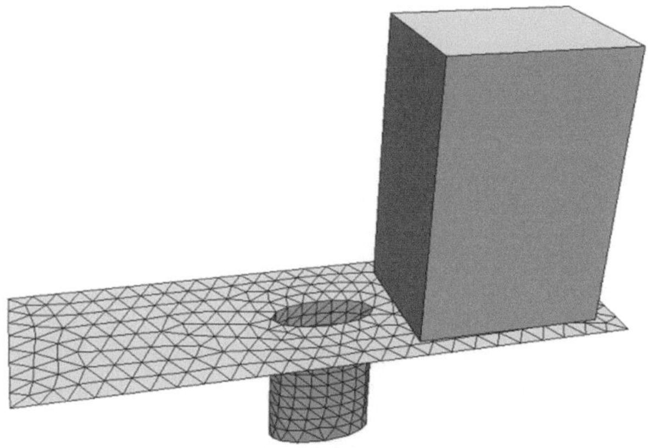

Fig. 27.3 The meshed die system (C1)

Table 27.2 Material parameters of the particle and the die system

	Particle	Die
Elastic modulus, Y (Pa)	8.7×10^9	2.1×10^{11}
Poisson's ratio, ν	0.3	0.3
Density, ρ (kg·m^{-3})	1.5×10^3	7.9×10^3

Table 27.2. In the current study, the shoe velocity is varied in the range from 75 to 300 mm s^{-1}. The mass flow rate and the critical fill speed are then analyzed.

27.3 Results

27.3.1 The Particle Profiles

Figure 27.4 shows the particle profiles of C0 during die filling at a filling speed of 85 mm s^{-1}. Initially, all particles are deposited into the shoe as shown in Fig. 27.4a. Then the shoe moves towards the left side. When the leading (left) side of the shoe starts to pass the die opening, the particles begin to fall into the die under gravity (Fig. 27.4b). This process continues (Fig. 27.4c) until the tailing (right) side of the shoe passes the die opening (Fig. 27.5). It is found that the die is completely filled at the filling speed of 85 mm/s. In addition, an inclined surface of the granular bed in the shoe is observed due to the acceleration provided by the moving shoe. Figure 27.5 presents particle profiles of various die orientations and shapes after die filling at a filling speed of 85 mm s^{-1}. It can be seen that all dies with various shapes and orientations can be completely filled at a filling speed of 85 mm s^{-1}.

Figure 27.6 shows the particle profiles during die filling into a cylindrical die at a filling speed of 200 mm s^{-1}. The particles are moved and delivered from the shoe to the die, which shows similar features to that at 85 mm s^{-1}. However, at a filling speed of 200 mm s^{-1}, the die cannot be completely filled as shown in Fig. 27.6d.

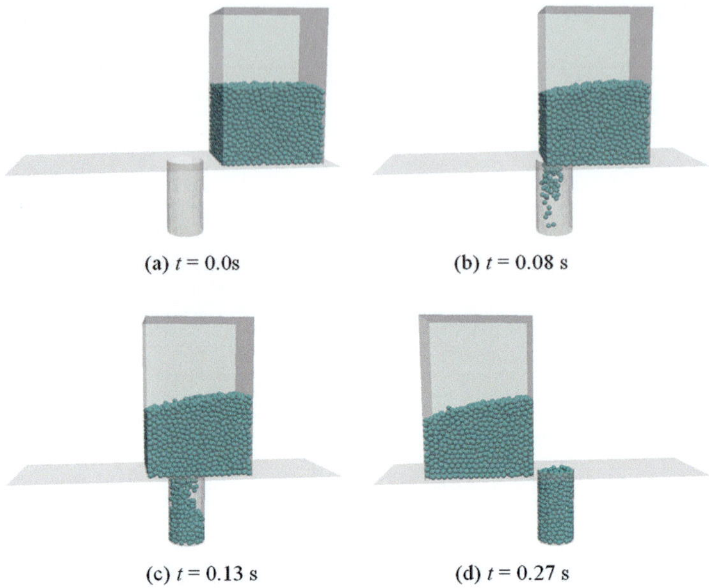

Fig. 27.4 Particle profiles of C0 during die filling at a filling speed of 85 mm s^{-1}

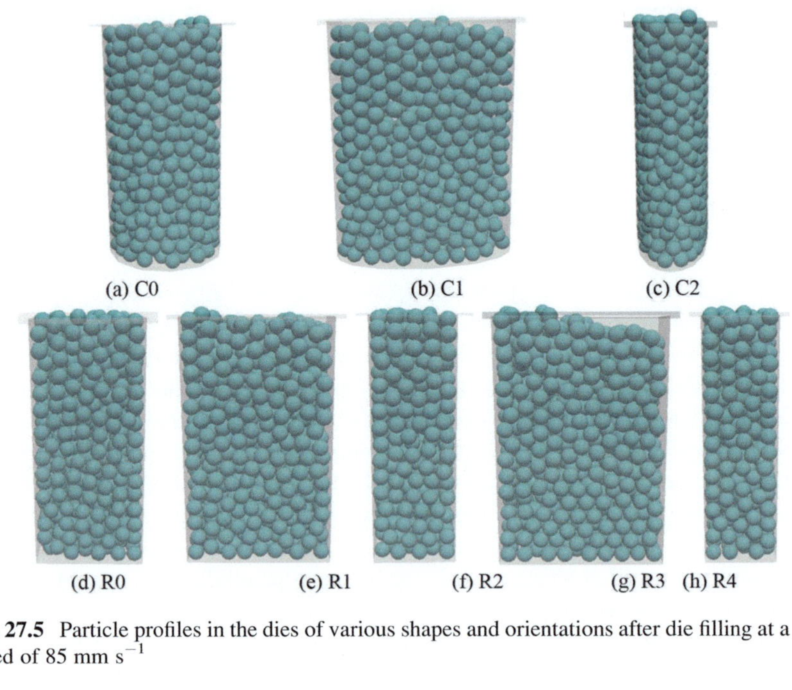

Fig. 27.5 Particle profiles in the dies of various shapes and orientations after die filling at a filling speed of 85 mm s^{-1}

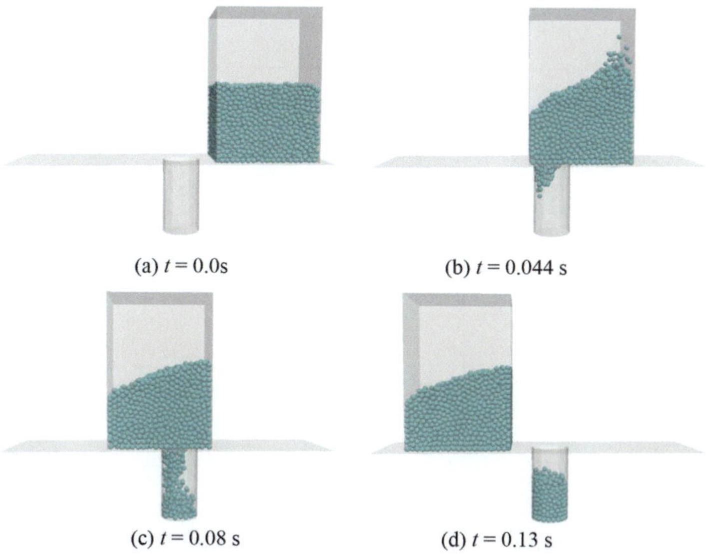

Fig. 27.6 Particle profiles during die filling with a cylindrical die (C0, see Fig. 27.2) at a filling speed of 200 mm s^{-1}

In other words, the filling speed of 200 mm s^{-1} is too high to completely fill the die. Moreover, it is observed that the inclined surface (Fig. 27.6b, d) has a larger inclination angle than that at a filling speed of 85 mm s^{-1} (Fig. 27.3b, d), which is caused by a high inertia. Figure 27.7 shows particle profiles of various die orientations and shapes after die filling at the filling speed of 200 mm s^{-1}. It can be seen that all dies considered cannot be completely filled at a filling speed of 200 mm/s.

27.3.2 The Mass Flow Rate

To determine the mass flow of the die filling process, the mass of particles filled into the die during the process is recorded. Figure 27.8 presents the time evolution of mass of particles deposited into various shaped dies and orientations during the die filling process at a filling speed of 85 mm s^{-1}. A similar pattern is observed for all cases considered here. Initially, the mass of particles deposited in the die keeps zero due to the initial distance (2 mm) between the shoe and the die. Then the mass of the particles in the die starts to increase when the shoe starts to move over the die as shown in Fig. 27.3. Eventually, the mass of the particles in the die achieves a maximum value once the shoe has passed the die. The final (maximum) masses of particles in various dies give nearly equal values (Fig. 27.8a, b), which is the mass of particles in the completely filled die. This corresponds well with the phenomenon observed in Figs. 27.4 and 27.5.

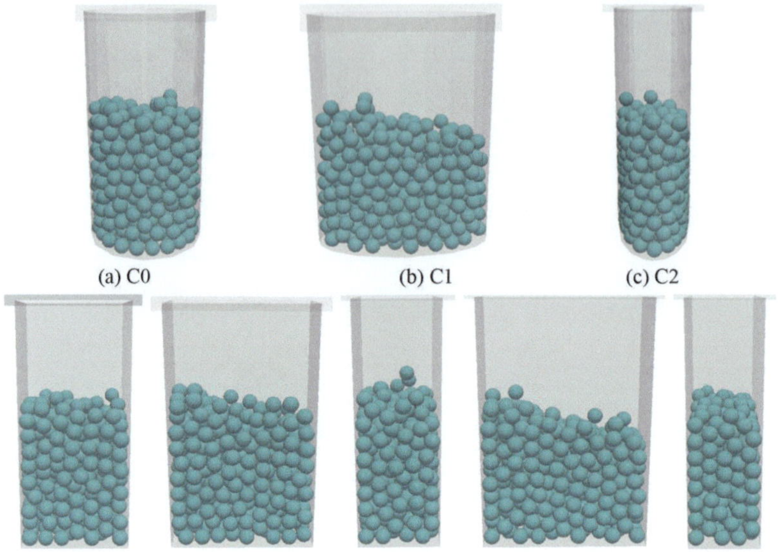

Fig. 27.7 Particle profiles of various die orientations and shapes after die filling at the filling speed of 200 mm s^{-1}

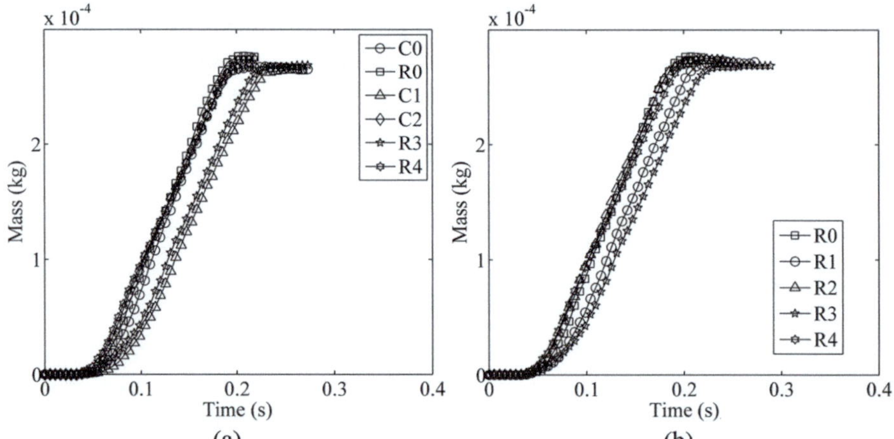

Fig. 27.8 Time evolutions of mass of particles deposited into the die at a filling speed of 85 mm s^{-1}

However, it is observed that in Fig. 27.8a, for the cases of C1 and R3, it takes a longer time to achieve the maximum mass than in other cases. Similarly, in Fig. 27.8b, for R1 and R3, it takes a longer time as well. In addition, R3 needs a longer time than R1. These phenomena indicate that the orientation and the shape of

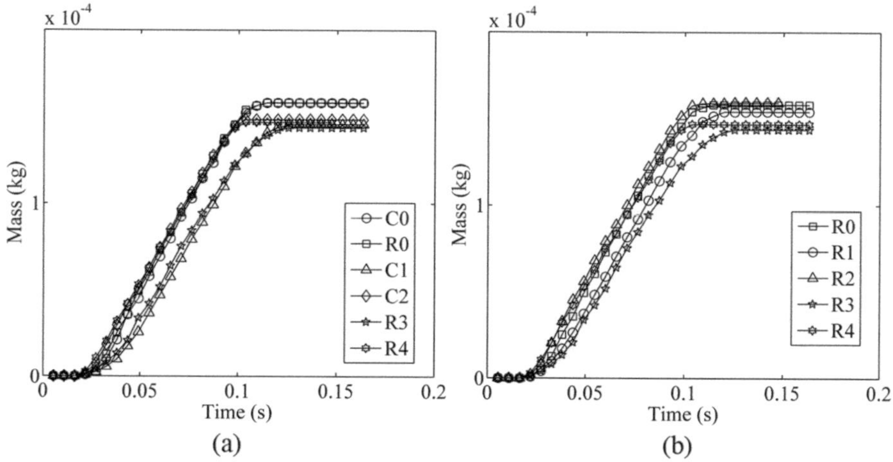

Fig. 27.9 Time evolutions of mass of particles deposited into the die at a filling speed of 200 mm s^{-1}

the die influence the efficiency of the die filling process. Specifically, the parallel filling of cases C1, R1 and R3 induces a lower efficiency of die filling than the orthogonal filling. Meanwhile, a larger aspect ratio and a smaller thickness (i.e. short axis), such as C1 and R3, also lead to a lower efficiency.

Figure 27.9 presents the time evolution of mass deposited into the die at a filling speed of 200 mm s^{-1}. It shows similar patterns as observed at 85 mm s^{-1}. However, it can be seen that the final mass deposited into the die is smaller than the mass in the fully filled die at 85 mm s^{-1} (Fig. 27.8). This indicates that the die is incompletely filled at a filling speed of 200 mm s^{-1}, which corresponds well with Figs. 27.6 and 27.7. The orientation and the shape of the die also have similar effects to that described in Fig. 27.8.

To represent the efficiency (i.e. the flow rate) of the die filling process for various dies, the average mass flow rate is defined as:

$$\tau = \frac{m}{t_c} \tag{27.3}$$

where m is the final (maximum) mass of particles filled into the die; t_c is the time from the time at which the particles start to flow into the die to the time at which the final (maximum) mass is just reached.

Figure 27.10 shows the mass flow rates at various filling speeds. It is clear that, for all filling speeds considered, C1, R1 and R3 have smaller mass flow rates than the other cases. Meanwhile, R1 has a relatively larger mass flow rate than C1 and R3. In other words, the parallel filling and a larger aspect ratio can induce a relatively smaller mass flow rate and hence a lower efficiency of die filling. However, it is noticeable that for the orthogonal filling, the aspect ratio does not have as significant influence as for the parallel filling, especially for the lower filling

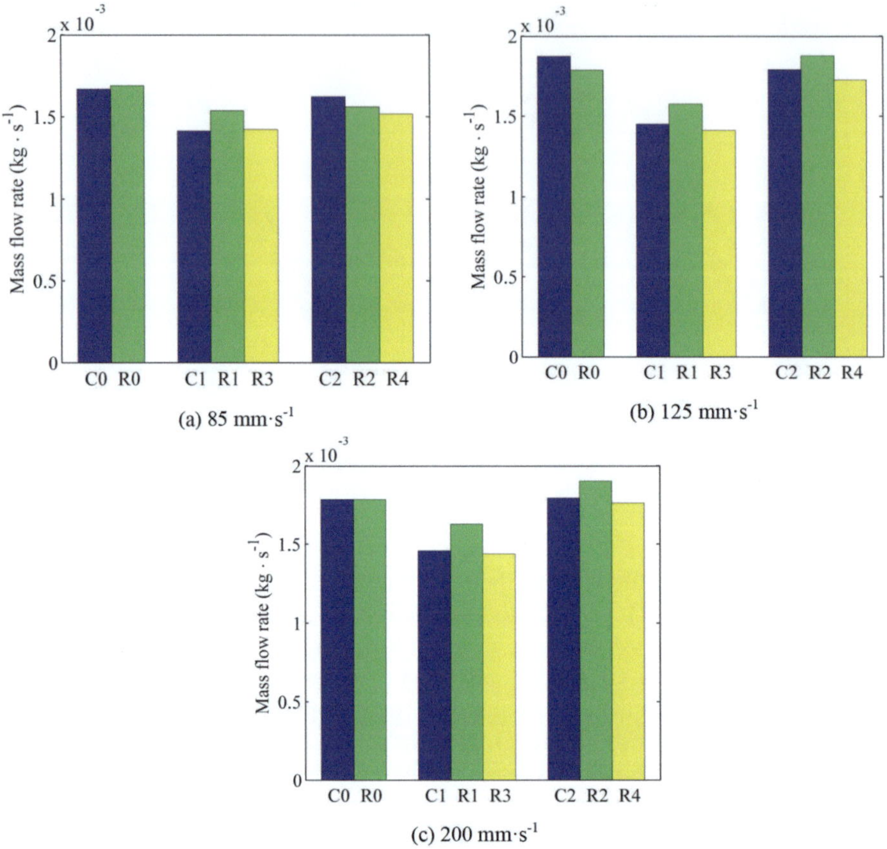

Fig. 27.10 Mass flow rates during die filling at various filling speeds

speed (Fig. 27.9a). In addition, for the same aspect ratio, a higher mass flow rate is induced for the elliptic die opening (C2) than that for the rectangular shape (R4).

27.3.3 The Critical Filling Speed

To characterize the flow of particles during die filling, Eq. (27.1) is used to fit the simulation data to obtain the critical fill speeds and the power index m for various cases considered. Figure 27.11 shows the relationship between the fill ratio and the filling speed for various dies. The solid lines represent the best fit to the simulation data (from 125 to 300 mm s^{-1}) with Eq. (27.1). For all cases, when the filling speed is relatively smaller (75 and 85 mm s^{-1}), the fill ratios for all cases are generally equal to 1, indicating that the die is completely filled. When the filling speed is

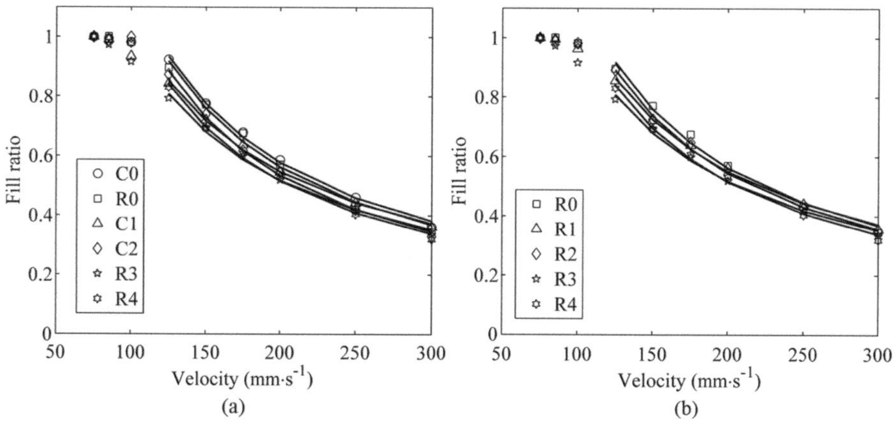

Fig. 27.11 The fill ratio as a function of the filling speed for various cases considered

higher (100 mm s^{-1}), the fill ratios decrease as the shoe velocity increases. This indicates that the die is incompletely filled and the mass filled into the die decreases as the filling speed (>100 mm s^{-1}) increases. Therefore, the data used to fit is in the range from 125 to 300 mm s^{-1}. It can be seen that Eq. (27.1) gives a good fit with the simulation data.

The obtained critical filling speeds for various cases considered are shown in Fig. 27.12 and Table 27.1. It is clear that the critical filling speed varies with the shape and the orientation of the die. C0 and R0 result in the highest critical filling speeds, while C1, R1 and R3 lowest critical filling speeds. Meanwhile, R1 leads to a relatively higher critical filling speed than C1 and R3. This implies that the parallel filling and a larger aspect ratio can result in a lower critical filling speed. It is also noticeable that, for both parallel and orthogonal filling, C1 and C2 result in larger critical filling speeds than R3 and R4, even when their aspect ratios are equal. This phenomenon indicates that the geometrical shape can also influence the die filling process.

27.4 Discussion

The fill ratio decreases as the filling speed increases, especially when the filling speed is larger than the critical filling speed. When the shoe moves faster, fewer particles will flow into the die. According to Wu and Cocks [2], the critical filling speed can be used to characterize powder flow during die filling. The critical filling speed depends on the properties of the particles and the design of the die filling system, such as the shape and the orientation of the die. A higher critical filling speed means that the die can be completely filled at a higher filling speed, indicating that the powder has a better 'flowability' or a more

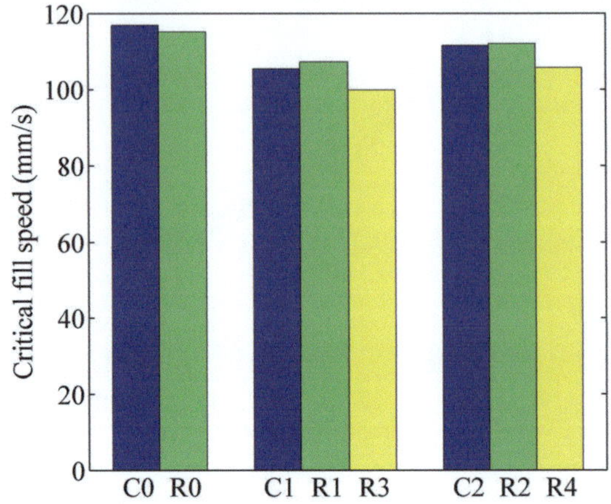

Fig. 27.12 Critical filling speeds for various cases considered

efficient die filling can be performed. Form Fig. 27.11, it can be seen that the relationship between the fill ratio and the filling speed can be well described using Eq. (27.1).

The parallel filling results in a low efficiency of the die filling process. From Figs. 27.10 and 27.12, it is found that the parallel filling induces smaller mass flow rates and critical filling speeds. During die filling, the effective powder discharge can only be achieved when the die is completely covered by the shoe. However, for the parallel filling, the time duration for effective powder discharge is much less than that for the orthogonal filling.

Powder flow behavior during die filling also depends on the aspect ratio. A large aspect ratio means that the die opening is relatively narrow. A narrow die opening can restrict the particles from flowing into the die especially for relatively large particles as used in this study. On the other hand, a smooth transition at the corner of the die as C1 and C2 can improve the particle flow and lead to a higher mass flow rate and a larger critical filling speed than R3 and R4. Therefore, due to the parallel filling, larger aspect ratio and the rectangular shape, R3 results in the lowest mass flow rate and smallest critical filling speed.

In the current study, the particle size (0.8 mm) is relatively large compared to the dimension of the die (3–10 mm). Thus, the shape and orientation of the die may have great influence on the die filling process due to the small size ratio of the particles to the die. In practice, the particle size varies with the industrial processes. For instance, for pharmaceutical processes, the particle size in the range of 100 microns is quite common. For further research, detailed studies should be considered for the influence of the die shape and orientation on the die filling process with large size ratios of the particles to the die.

27.5 Conclusions

In this study, the die filling processes with various die shapes and orientations are investigated using the discrete element method. Various dies of different shapes and orientations are modelled, including circle, square, ellipse and rectangle. The orthogonal and parallel filling are also considered. It is found that the dies with larger aspect ratio and parallel filling can lead to smaller mass flow rates and lower critical filling speeds. In addition, the dies with elliptic openings generally result in higher critical filling speeds than those with rectangular shapes. These results indicate that it is necessary to consider the influence of the geometry of the die on die filling during the product design process.

Acknowledgements This work was sponsored by the International Fine Particle Research Institute (IFPRI).

References

1. Guo Y, Kafui KD, Wu CY, Thornton C, Seville JPK (2009) A coupled DEM/CFD analysis of the effect of air on powder flow during die filling. AICHE J 55:49–62
2. Wu CY, Cocks ACF (2004) Flow behaviours of powders during die filling. Powder Metall 47:127–136
3. Bocchini GF (1987) Influence of small die width on filling and compacting densities. Powder Metall 30:261–266
4. Rice ER, Tengzelius J (1986) Die filling characteristics of metal powders. Powder Metall 29:183–194
5. Wu CY, Cocks ACF (2006) Numerical and experimental investigations of the flow of powder into a confined space. Mech Mater 38:304–324
6. Wu CY, Guo Y (2012) Numerical modelling of suction filling using DEM/CFD. Chem Eng Sci 73:231–238
7. Kremmer M, Favier JF (2001) A method for representing boundaries in discrete element modelling—part II: kinematics. Int J Numer Methods Eng 51:1423–1436

Chapter 28
Finite Element Modeling of Powder Compaction

Alexander Krok and Chuan-Yu Wu

Abstract Powder compaction is widely used to manufacture a wide range of particulate products. Understanding powder compaction behaviour is of practical importance to improve the efficiency of product development and the manufacturing performance. This paper briefly introduces the finite element method (FEM) that has been extensively used in modelling powder compaction, for which the DPC model is introduced. Typical finite element analysis results are presented to illustrate the capability of FEM in modelling powder deformation. It shows that non-uniform density and temperature distributions are generally obtained during powder compaction. In addition, the correlations between the fracture patterns observed experimentally with the stress distribution obtained using FEM are examined. It is shown that the stress distribution could provide useful information on the potential fracture patterns during powder compaction.

Keywords Powder compaction • Tabletting • Dracker Prager Cap model • Thermomechanical modelling • Stress distribution • Capping • Air entrapment • Fracture • X-ray computed tomography

28.1 Introduction

Powder compaction is a very common manufacturing process in the pharmaceutical, food, fine chemicals, powder metallurgy and ceramics industries. It is a net shape technique that is widely used to manufacture products of complex geometries, ranging from simple oblong tablets to complex multi-level ceramic parts. In powder compaction, powders are deposited into a die and compressed under high pressures so that coherent compacts can be formed. There are three distinct process stages during powder compaction: die filling, compaction and ejection. Powder behaviour at each stage plays an important role in controlling the performance of

A. Krok • C.-Y. Wu (✉)
Department of Chemical and Process Engineering, University of Surrey,
Guildford, Surrey GU2 7XH, UK
e-mail: C.Y.Wu@surrey.ac.uk

K.J. Roberts et al. (eds.), *Engineering Crystallography: From Molecule to Crystal to Functional Form*, NATO Science for Peace and Security Series A: Chemistry and Biology, DOI 10.1007/978-94-024-1117-1_28

downstream processes and the quality of final products. Moreover, any defects induced at those process stages will have a detrimental impact on the quality of the products.

Powder compaction is also an energy conversion process: the total work imparted by the mechanical compression (via the compression force) is converted into (1) the energy required to deform the powder material; (2) heat that is induced by plastic deformation and friction (particle-particle and particle-wall). A portion of the generated heat during powder compaction may be dissipated into the tooling through heat conduction, while the remaining heat is absorbed by the compressed powder, leading to the rise of temperature in the powder. The heat generation can cause significant temperature rise, which could be very detrimental to heat sensitive materials, as it can cause material degradation and change in mechanical properties.

A better understanding of powder compaction processes is hence critical in product design and process control. Thus, powder compaction has attracted an increasing interest in the scientific community over the last few decades. Nevertheless, modelling of powder compaction is still very challenging. For example, powder flow behaviour during die filling depends upon geometrical set-up of the tooling [1, 2], whereas the powder is generally in a loose state and its behaviour is determined by the interactions between particles, particle-wall and particle-surrounding media (i.e. air or fluids). This can be well modelled using the discrete element method [2]. In the initial phase of powder compaction, densification takes place but it is dominated by a relative movement and particle re-arrangement. Subsequently, when the relative motions of particles are strongly restricted, elastic and plastic deformations begin to occur. There is a significant reduction of the powder volume, causing a sharp increase in the relative density (or a reduction in the porosity of the material). As the pressure increases further, powder compaction is dominated by plastic deformation for ductile materials and fracture and fragmentation of particles for brittle materials. For this process, the DEM is not applicable, as it cannot model the deformation and fragmentation of each particle effectively. However, the boundary and loading conditions of the compaction process are well defined. Thus, the finite element method has been introduced to model powder compaction processes.

28.2 The Finite Element Method (FEM)

The finite element method (FEM) is proven as a reliable and robust modelling tool for analysing the behaviour of powders during compaction. It provides useful information on the compacted material, such as distributions of density, stress or temperature. A number of studies showed that the behaviour of powders during compaction could be described as an elastic-plastic continuum using phenomenological models, such as the Cam-Clay model or the Dimaggio-Sandler model [3–5]. These models were originally developed for geological materials and soil mechanics, but found to be also suitable for other particulate materials, such as

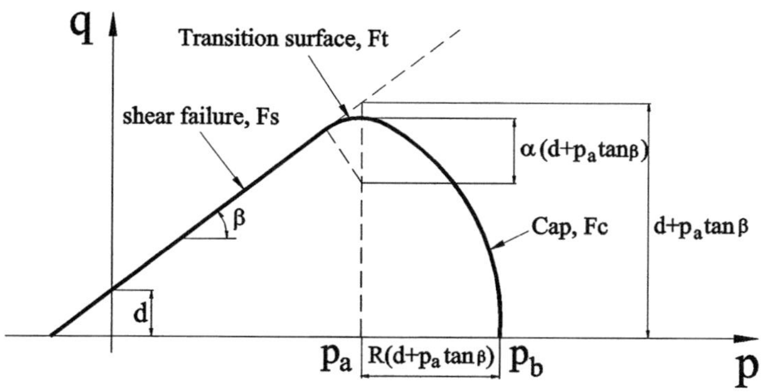

Fig. 28.1 The Drucker-Prager-Cap model in the p–q plane [6]

metallic, ceramic and pharmaceutical powders. Another phenomenological model, the Drucker-Prager-Cap (DPC) model [6–11] was also introduced to model the behaviour of pharmaceutical, metallic and ceramic powders during compaction, The DPC model can also consider the hardening, softening and inter-particle friction.

In the Drucker-Prager-Cap model, the yield surface consists of three segments (see Fig. 28.1): a shear failure segment $F_{s,}$ a cap segment F_c and a transition segment F_t that is introduced primarily to ensure the smooth transition between the first two segments, so that computational stability can be maintained [6]. In the p–q plane, these segments are generally defined using two stress invariants:

$$p = \frac{1}{3}(\sigma_1 + \sigma_2 + \sigma_3) \tag{28.1}$$

$$q = \sqrt{\frac{1}{6}\left[(\sigma_1 - \sigma_2)^2 + (\sigma_2 - \sigma_3)^2 + (\sigma_3 - \sigma_1)^2\right]} \tag{28.2}$$

where p is the equivalent pressure stress, q *is* the Mises equivalent stress, and σ_i $(i = 1, 2, 3)$ are the principal stresses.

The shear failure segment defines a criterion for the occurrence of shear flow that is a function of the cohesion d and the angle of friction β:

$$F_s(p, q) = q - p \tan\beta - d = 0 \tag{28.3}$$

The cap segment is defined as:

$$F_c(p, q) = \sqrt{(p - p_a)^2 + \left(\frac{Rq}{1 + \alpha - \alpha/\cos\beta}\right)^2} - R(d + p_a \tan\beta) = 0 \tag{28.4}$$

It intersects the equivalent pressure stress axis and is an elliptical curve with constant eccentricity. In Eq. (28.4), R and α are parameters describing the shape of the cap segment and the smooth transition surface, p_a is an evolution parameter describing material hardening or softening:

$$p_a = \frac{p_b - Rd}{1 + R\tan\beta} \tag{28.5}$$

where p_b is the hydrostatic yield pressure that defines the position of the cap. p_b is generally a function of the volumetric inelastic strain ε_{vol}^{in}. This represents the hardening or softening of the cap segment: compaction leads to hardening, while dilation develops softening.

The transition segment is always very small and its size is controlled by the parameter α, which typically has a value of 0.01–0.05. The transition segment is mathematically defined as:

$$F_t(p,q) = \sqrt{(p - p_a)^2 + \left[q - \left(1 - \frac{\alpha}{\cos\beta}\right)(d + p_a \tan\beta)\right]^2} - \alpha(d + p_a \tan\beta) = 0 \tag{28.6}$$

Full calibration of the DPC model requires triaxial compression test data. However, the model can also be calibrated using experimental data from uniaxial compression with an instrumented die fitted with radial stress sensors. Detailed discussion on the experimental techniques and the calibration method can be found in Seville and Wu [12].

To model the thermo-mechanical behaviour of powders during compaction, it is essential to determine the amount of energy converted into heat that is responsible for the change in temperature in the material during compaction. In this study, three primary heat generation mechanisms resulting in the change in temperature within the powder are considered: (i) friction between particles; (ii) wall friction along the tooling surfaces; and (iii) elastic/plastic deformation of the particles. The temperature gradient in the powder can be described as:

$$\rho C_p \frac{dT}{dt} = \nabla(k\nabla T) + \dot{q}_{\phi} + \dot{q}_{\varphi} \tag{28.7}$$

where ρ (kg.m^{-3}), C_p (J kg^{-1} K^{-1}), T (K) and k (W m^{-1} K^{-1}) are the density, specific heat, temperature and thermal conductivity, respectively. $\dot{q}_{\varphi}$ is the total amount of heat generated from the inter-particle friction and plastic deformation of the powder. It can be determined from stresses and strains as follows:

$$\dot{q}_{\phi} = \xi \sum_{i,j=1}^{3} \sigma_{ij} d\varepsilon_{ij}^{pl} \tag{28.8}$$

where ξ (−) is the inelastic heat fraction.

$\dot{q}_{\varphi}$ in Eq. (28.7) represents the heat generated from wall friction and can be given as:

$$\dot{q}_{\varphi} = \int_{A} \nu \mu \sigma_{nn} dA \tag{28.9}$$

It is clear that $\dot{q}_{\varphi}$ is dependent on the wall friction coefficient μ (−), the interacting area dA (m^2) between powders and walls, the local normal stress σ_{nn} (Pa) and the relative velocity v (m s^{-1}).

28.3 FE Analysis of Powder Compaction

For modelling powder compaction using FEM, a range of input parameters are needed, which include material properties and process parameters. For a given powder, it is important to choose an appropriate material model that can well represent the deformation behaviour of the powder, but also to perform model calibration so that the input parameters are specific to the powder considered. In addition, process parameters are also needed to specify the geometry and process conditions in the boundary and loading conditions. Using these essential input parameters, FEM modelling can be performed using a FE solver to explore various responses of the systems, which includes the response of the powder during the process, e.g. compaction behaviour of the powder during compaction processes, properties of final products, such as density and temperature distributions, as well as the response of the tooling.

As an example, let us consider compaction of a microcrystalline Cellulose (MCC) Avicel PH 102 powder. The powder was compressed into cylindrical tablets with two different surface curvatures. All the tablets had a diameter of 10 mm. The powder had an initial height of 6 mm and was compressed using the upper and lower punches of different surface curvatures. The die wall and punches were modelled as rigid bodies since their deformation is negligible compared to that of the powder. To make the shallow convex tablets, the radius of the punch surface curvature was 16.5 mm and the punch depth was 0.776 mm. As it is an axisymmetrical problem, the powder was hence modelled as a 2D axisymmetric deformable continuum. The initial FE mesh is shown in Fig. 28.2a for making the shallow convex tablets. The powder was discretised using a non-adaptive mesh composed of elements with displacement – temperature degrees of freedom. The initial temperatures of the powder die and punches were set as 22 °C. All three stages of a typical compaction process, i.e., compression, decompression and ejection, were modelled. The interaction between the powder and the tooling surfaces were modelled as master-slave contacts with finite sliding. The nodes on the symmetrical axis (AB) were restrained to move in horizontal direction and the nodes at boundary BC could only move in the horizontal direction. The upper punch could only move

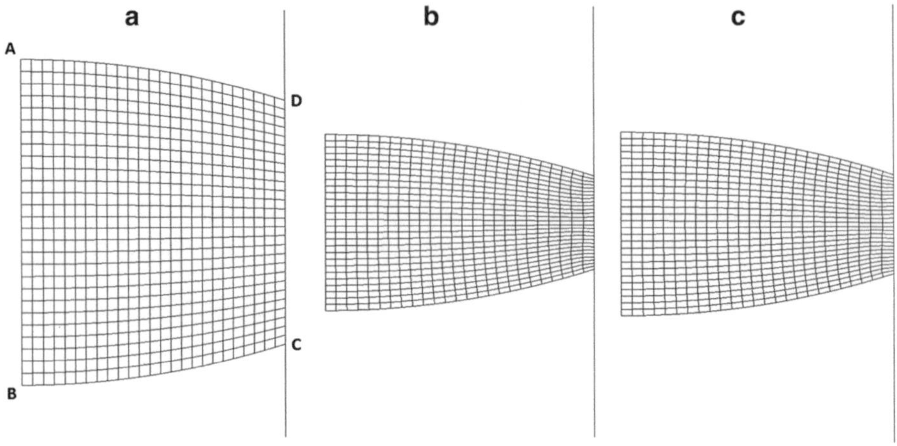

Fig. 28.2 Finite element meshes: (**a**) before compaction; (**b**) at the end of compression and (**c**) at the end of decompression

vertically at a specified compression speed. The compaction speed was set to 100 mm s^{-1}, which was also the speed employed in the physical experiments reported here. It was assumed that the powder had a uniformly distributed initial relative density.

Figure 28.2b shows the FE meshes at maximum compression and Fig. 28.2c shows the meshes at the end of decompression. By examining the FE meshes at different instants, it is clear that, during compression, the volume of the powder bed reduces, which manifests in the deformation or distortion of the finite element meshes. While during decompression, dilation takes place as the volume of the powder bed increases. It is clear that the powder bed was compressed during loading and springs back (i.e. relaxes) during the unloading. Furthermore, the relative density and stress under the applied pressures were also determined, which are shown as contour plots in Figs. 28.3 and 28.4, respectively.

The density distributions were produced by mapping the relative density in each element. It is clear that the density distributions are not uniform. For double ended compression considered here, at the upper and lower edges, a higher density was obtained, which is primarily due to the effect of wall friction, as pointed out by many researchers [6, 8, 13–15]. During compaction, both punches were moved towards each other so that the powder near the upper and lower edges was compressed more (with higher von Mises stresses as shown in Fig. 28.4b) than that in other areas. Consequently, higher relative densities near the upper and lower edges were obtained. Similar patterns were also obtained from X-ray computed tomography [16].

It is also apparent that the stress distribution of the shallow convex tablet at the end of compression (Fig. 28.4b) has a similar pattern to the relative density distribution (Fig. 28.3b). The highest stresses appeared near the upper and lower edge, which was caused due to the combined effects of the shear stress induced by

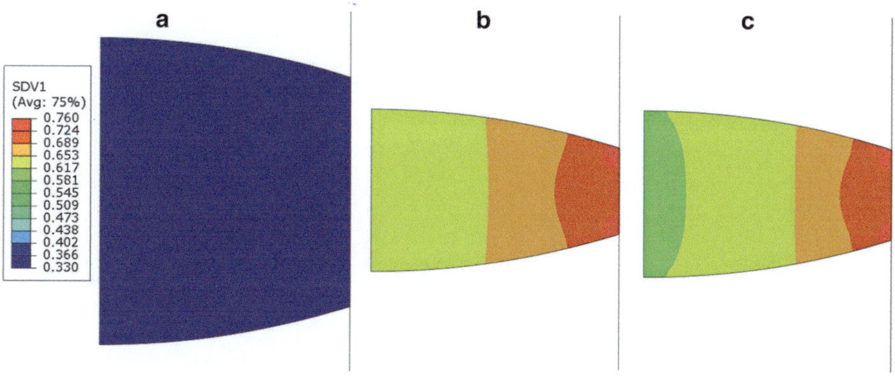

Fig. 28.3 Relative density distributions: (**a**) before compaction; (**b**) at the end of compression and (**c**) at the end of decompression

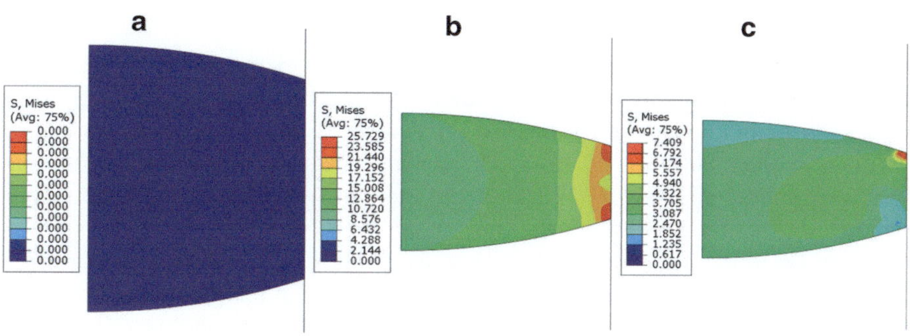

Fig. 28.4 Von Mises Stress distribution: (**a**) before compaction; (**b**) at the end of compression and (**c**) at the end of decompression

wall friction and compressive stresses as the punches move towards each other. From Fig. 28.4c it can be observed, that during unloading, the stress decreased and was redistributed in the tablet. At this stage, only the upper punch withdrew, resulting in a relaxation of material due to the elastic recovery.

The FEM can also be used to predict the temperature distribution (Fig. 28.5) on the surface and inside the tablet during the compression, decompression and ejection. Figure 28.5b shows the temperature distribution of the tablet at the maximum compression. At a high compression speed, part of the generated heat near the die wall is transferred by conduction into the die tools and punches, therefore it can be seen that the highest temperature was induced inside the tablet. After unloading, the temperatures inside the tablet decreased slightly (Fig. 28.5c). While the upper surface has a slightly higher temperature than the lower surface or the wall of the tablet, because some heat in these areas was dissipated by conduction.

Physical experiments were also carried out to measure the temperature distributions using an infrared (IR) camera, from which the temperature distributions on the

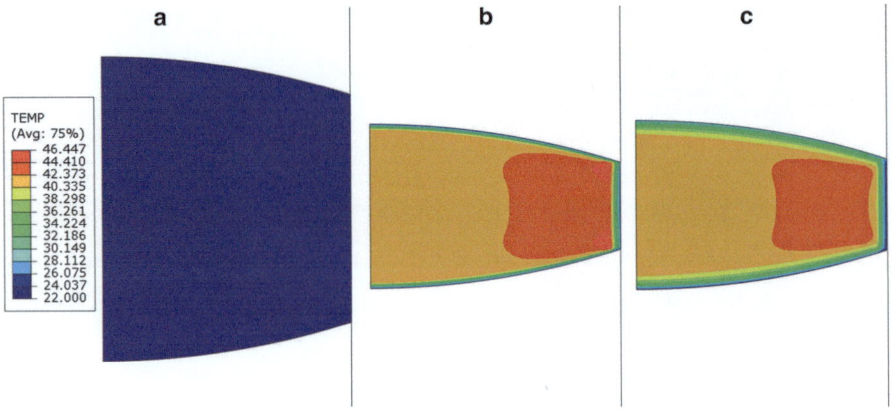

Fig. 28.5 Temperature distributions: (**a**) before compaction; (**b**) at the end of compression and (**c**) at the end of decompression

surface of the flat-face (FF) and the shallow convex (SC) tablets, immediately after ejection, were obtained and presented in Figs. 28.6a, b. It was observed from the experimental measurement that for both cases the highest temperatures were obtained on the centreline of the upper surface. The same compaction processes were also modelled using FEM and the temperature distributions obtained from the FE simulation are shown in Figs 28.6c, d. It was clear that the surface temperature in the top centre was the highest, which is in broad agreement with the experimental observation. It was also found that the temperatures inside the tablets for both cases are much higher than the surface temperatures. The temperature of tablets changed with the tablet shape: as the tablet surface curvature increases, the irreversible amount of work increases, generating a greater quantity of heat in the tablet.

28.4 FE Analysis of Tablet Fracture During Powder Compaction

Wu et al. showed that there was a strong correlation between the stress distributions with the potential fracture patterns during powder compaction. This is illustrated in Fig. 28.7, which shows the internal cracks identified using X-ray computed tomography (left), photographs of broken/capped tablets made of lactose (insert), and the distribution of shear stress at the early stage of the unloading/decompression (right). It is clear that that there is an intensive localised shear band induced, running from the top edge towards the mid-centre. Shear deformation in powders generally leads to powder dilation. Consequently, the strength in this zone is relatively weak and the von Mises stress in this region is low [17]. This is further confirmed by experimental observations of the tablet failure during powder compaction. It was found that cracks were developed from the top edge towards

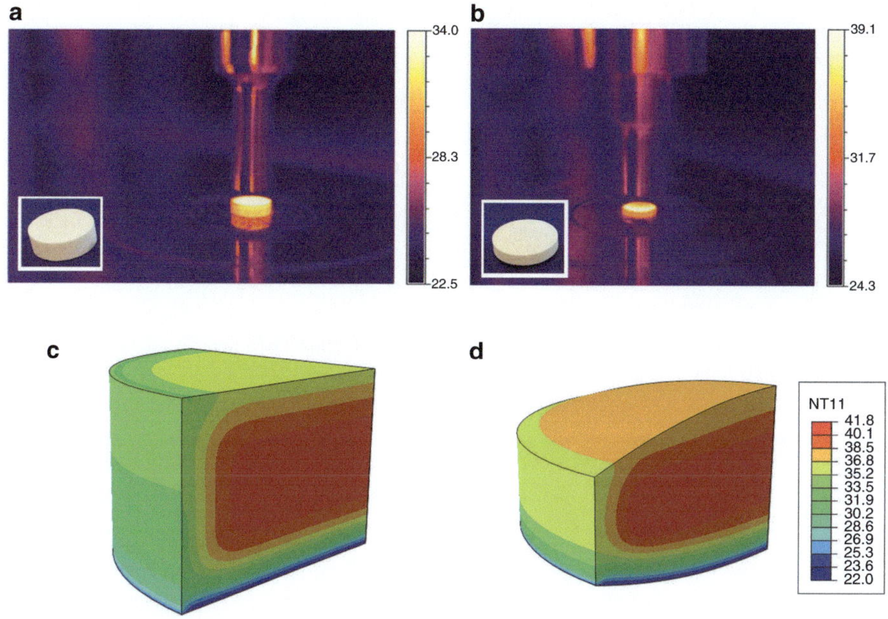

Fig. 28.6 Temperature distributions obtained immediately after ejection from experiments and simulations for MCC Avicel PH 102 tablets with a compression speed of 100 mm/s

the bottom centre, similar to the pattern of shear band obtained from FE analysis. This suggests that the tablet cracks or capping can be attributed to the formation of the shear band or the highly localised stress distributions.

In order to further demonstrate the capability of FE modelling, powder compaction with an entrapped air pocket was also considered. It is assumed that the permeability of the powder is very poor so that it is impossible for the entrapped air to escape during compaction. Both experimental and numerical analyses are performed. In the experimental study, the entrapped air was approximated using a soft elastic ball of different sizes that will deform under high pressure; once the pressure is removed, the compressed ball will expand to restore its original size. This can be used to mimic the expansion of compressed air pocket upon unloading. The initial position of the ball in the powder mass was altered to explore its effect on the powder compaction behaviour. Corresponding numerical analysis using FEM was also performed, in which the powder was modelled using the DPC model while the entrapped air was modelled as a highly elastic material. The results are shown in Fig. 28.8. It can be seen that micro-cracks in the powder compacts were observed, which was induced by the expansion of the simulated air pocket during unloading. FEM modelling results clearly indicate that intense stress localisation takes place. It is interesting to notice that the predicted stress localisation patterns are consistent with the micro-cracks observed experimentally. This again illustrates that the finite element analysis can be performed to predict the potential fracture patterns during powder compaction.

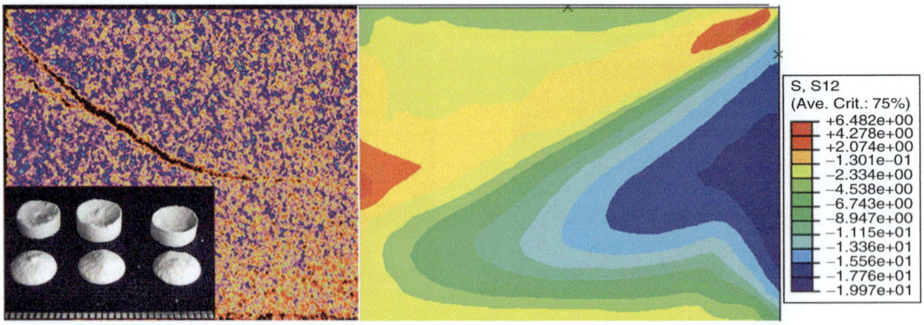

Fig. 28.7 The distribution of shear stress (*right*) obtained using FEM and visual examination of X-ray computed tomographic images of ejected lactose tablets (*left*)

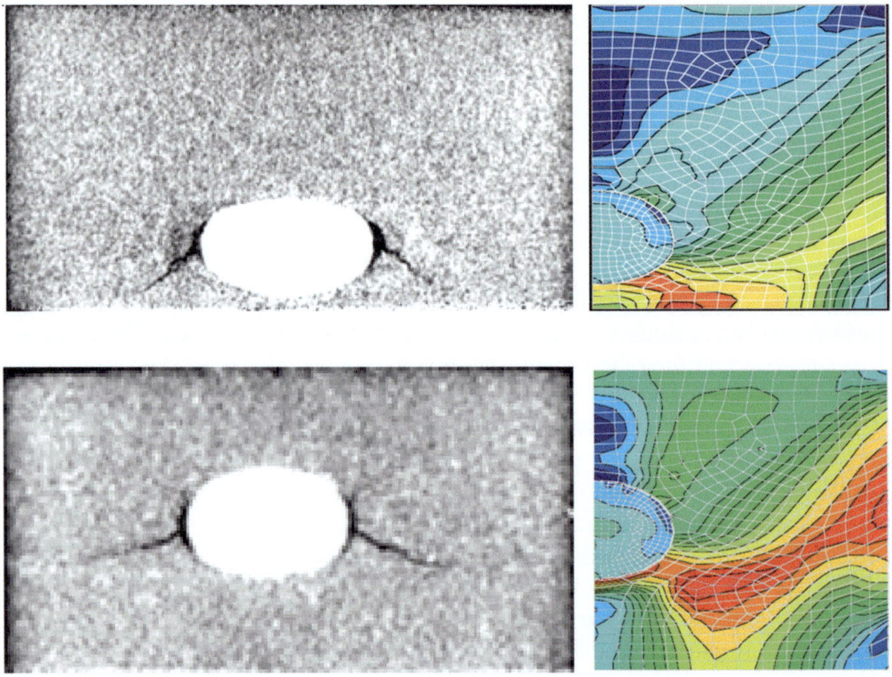

Fig. 28.8 Tablet fracture induced by entrapped air analysed using FEM (*right column*) and X-ray computed tomographic imaging (*right column*): the *top row* shows the results when the 'air pocket' is close to the tablet surface; the *bottom row* shows the results when the 'air pocket' is located in the centre of the tablets

28.5 Summary

During powder compaction, particles experience high compression pressures that can cause large deformation even fragmentation of individual particles. This possesses significant challenges to model this process using DEM or other particle based methods. Consequently, continuum modelling using FEM is a viable alternative. In FEM, powders are treated as elastic-plastic materials and their powder compaction behaviour was obtained by solving the boundary value problem. The widely used DPC model is introduced and some typical results from FEM modelling are discussed. FEA results showed that powder compaction results in a non-uniform temperature distribution in the tablets, whereas the highest temperature was induced inside the tablets. It was also shown that there are good correlations between the fracture patterns and the predicted stress distributions obtained using FEM. This implies that FEA can be performed to indicatively predict possible failure patterns during powder compaction. Nevertheless, it is still a challenging task to model and predict the propagation of the cracks, hence deserving further investigation.

Acknowledgments This work was supported by the Marie Curie Intra-European Fellowships, funded through the People Programme (Marie Curie Actions) of the European Union's Seventh FP7 under REA grant agreement No. 622874 (acronym: ThermoPC).

References

1. Wu CY, Cocks ACF, Gillia OT, Thompson DA (2003) Experimental and numerical investigations of powder transfer. Powder Technol 138:214–226
2. Pei C, Wu CY (2016) DEM analysis of the effects of die shape and orientation on die filling processes. In: Roberts KJ, Tamura R (eds) Molecules to crystals to powders: understanding structure versus function. Springer, Dordrecht. in press
3. Brewin PR, Coube O, Doremus P, Tweed JH (2008) Modelling of powder die compaction. Springer, London
4. Khoei RA (2005) Computational plasticity in powder forming processes. Elsevier, London
5. Chtourou H, Guillot M, Gakwaya A (2002) Modeling of the metal powder compaction process using the cap model. Part I Experimental material characterization and validation. Int J of Solids Struct 39:1059–1075
6. Wu CY, Ruddy O, Bentham AC, Hancock BC, Best SM, Elliott JA (2005) Modelling the mechanical behaviour of pharmaceutical powders during compaction. Powder Technol 152:107–117
7. Cunningham JC, Sinka IC, Zavaliangos A (2004) Analysis of tabler compaction. I. Characterization of mechanical behavior of powder and powder/tooling friction. J Pharm Sci 93(8):2022–2039
8. Sinka IC, Cunningham JC, Zavaliangos A (2003) The effect of wall friction in the compaction of pharmaceutical tablets with curved faces: a validation study of the Drucker-Prager cap model. Powder Technol 133:33–43
9. Michrafy A, Ringenbacher D, Techoreloff P (2002) Modelling the compaction behaviour of powders: application to pharmaceutical powders. Powder Technol 127:257–266
10. Coube O, Riedel H (2000) Numerical simulation of metal powder die compaction with special consideration of cracking. Powder Metall 43:123–131
11. Aydin I, Briscoe BJ, Sanlitürk KY (1996) The internal form of compacted ceramic components: a comparison of a finite element modelling with experiment. Powder Technol 89:239–254
12. Seville JPK, Wu CY (2016) Particle technology and engineering. Elsevier, London
13. Train D (1957) Transmission of forces through a powder mass during the process of pelleting. Trans Inst Chem Eng 35:258–266

14. Kim HS (2003) Densification modelling for nanocrystalline metallic powders. J Mater Process Technol 140:401–406
15. Han LH, Elliott JA, Bentham JC, Bentham AC, Mills A, Amidon BE, Hancock BC (2008) A modified Drucker-Prager Cap model for die compaction simulation of pharmaceutical powders. Int J Solids Struct 45:3088–3016
16. Klinzing GR, Zavaliangos A, Cunningham J, Macaro T, Winstead D (2010) Temperature and density evolution during compaction of a capsule shaped tablet. Comput Chem Eng 34:1082–1091
17. Frenning G (2008) An efficient finite/discrete element procedure for simulating compression of 3D particle assemblies. Comput Methods Appl Mech Eng 197:4266–4272

Chapter 29
From Molecules to Crystals to Functional Form: Science of Scale

Robert Docherty, Garry O'Connor, Radoslav Y. Penchev, Jonathan Pickering, and Vasuki Ramachandran

Abstract An understanding of the materials science of a new active pharmaceutical ingredient (API) is crucial at the interface of the chemical synthesis and drug product development. The selection of the crystallisation process and particle attributes during development is a key milestone in the conversion of a new API into a drug product. The physical and chemical properties of an API can impact product performance and process robustness and are strongly influenced by the solid state structure of the API. Product performance can only be assured when the API is delivered to the patient in a chemically and physically stable solid form. In this chapter we will attempt to integrate progress with cutting edge computational tools in academia to the best current industrial practices.

Keywords Particle attributes • Morphology • Surface chemistry • Cambridge structural database • Crystal engineering • Particle engineering and quality-by-design

29.1 Introduction

Selection of the commercial crystallization process and the associated particle attributes is one of the key steps in the development of any new drug product. Particle attributes connect the final unit operations of the active pharmaceutical

R. Docherty • G. O'Connor • R.Y. Penchev (✉)
Pharmaceutical Sciences, Pfizer Global R&D, Ramsgate Road, Sandwich, Kent CT13 9NJ, UK
e-mail: Robert.Docherty@pfizer.com; Garry.O'Connor@pfizer.com; Radoslav.Y.Penchev@pfizer.com

J. Pickering • V. Ramachandran
School of Chemical and Process Engineering, University of Leeds, Leeds LS2 9JT, UK
e-mail: j.h.pickering@leeds.ac.uk; v.ramachandran@leeds.ac.uk

K.J. Roberts et al. (eds.), *Engineering Crystallography: From Molecule to Crystal to Functional Form*, NATO Science for Peace and Security Series A: Chemistry and Biology, DOI 10.1007/978-94-024-1117-1_29

ingredient (API) manufacturing process with the drug product (tablet, capsule, etc.) manufacturing process and are critical to both API manufacturing and to drug-product functionality attributes; i.e., performance, processing, and stability. The regulatory landscape associated with the solid form and particle attributes of the API and dosage form development has been well documented [1, 2]. The progress of automation and computational technologies, which allow pharmaceutical scientists to search and identify the solid form and particle with optimal properties, has also been reported [3, 4].

More recently the importance of the Materials Science tetrahedron, depicting the relationships between internal structure, particle properties, material processing and performance of a drug product, has been described [5]. The industrial perspective on engineering pharmaceutical materials has also been highlighted [6, 7]. Pharmaceutical Materials Science has emerged as a foundation of Quality by Design (QbD) with solid form, crystallization and particle engineering being core elements linking the drug product functional form to the final steps of the API manufacturing process [8].

Whilst increasing interest in the crystallisation of pharmaceutical materials within academia has resulted in substantial progress over the last decade, the challenge for the pharmaceutical scientist in tackling the crystallisation of highly complex APIs remains a significant one. These complex organic structures are exacting because the crystallisation process is, typically, required to successfully navigate a complex polymorphic landscape. Different crystal faces exhibit different surface chemistry and hence interact differently with solvents, process impurities and excipients. For a given API the combination of different particle sizes, habits and surface chemistry can lead to different chemical and physical stabilities, biopharmaceutical properties and processing behaviour (API and drug product).

This chapter seeks to outline some of the recent progress on the application of emerging computer modelling technologies as foundation elements of the modern design strategy for the development, and manufacture, of advanced functional particulate products, see Fig. 29.1. The chapter highlights the opportunity for a molecule to crystal to functional form philosophy, with the aim of bridging across the chemical, analytical and formulation disciplines. These relationships, combined with institutionalised corporate knowledge of formulation design practices provide a route map to a fully integrated, holistic product design approach consistent with QbD philosophy.

29.2 Particle Morphology and Surface Structure

The crystal chemistry of drug molecules has already been described in Chap. 5 of this publication and a detailed analysis of morphology prediction has also been described in Chap. 9. The external shape of a crystal is referred to as the crystal morphology. Early crystallographers were fascinated by the flat, and symmetry

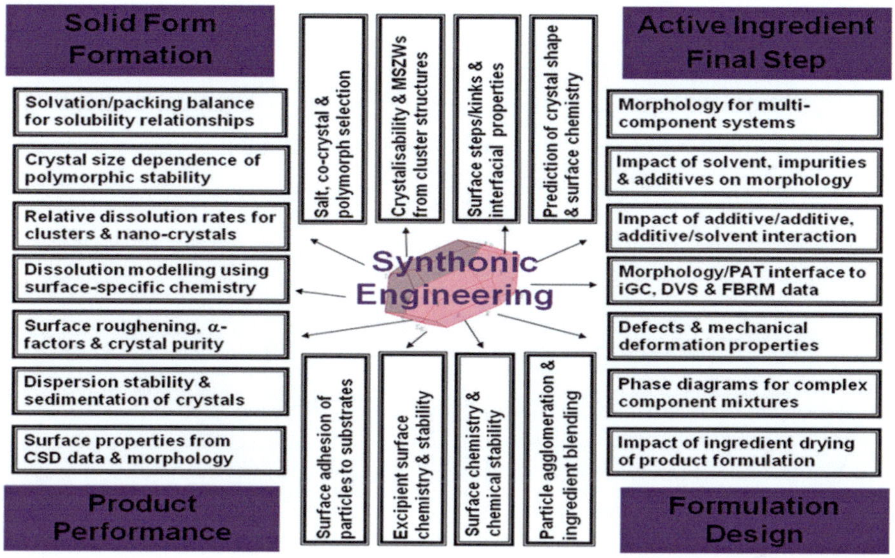

Fig. 29.1 The importance of particle design tools in impacting product safety, efficacy and quality. Distilled from discussions at a cross-industry workshop at Institute of Physics 2nd March 2012 [9]

related external faces in both natural and synthetic crystallised solids. This led to the postulate that the ordered, external arrangement was a result of an ordered internal arrangement.

Crystals are bounded by the slowest growing faces and the crystal habit is determined by the relative growth rates of the various faces. Crystal habit has been traditionally described using a variety of qualitative terms such as plate-like, prismatic and needle-like. In our current vision for this Summer School we are moving to more quantitative descriptions of the shape utilising Miller indices to allow a greater understanding of the different surface chemistries being exposed. The crystallographic planes that define the external growth morphology of the 'as grown' crystal can be described by the Law of Rational Indices as expressed through their Miller indices [10].

Miller indices are important to the crystallisation scientist as they provide a link between the modern structural crystallography of X-ray diffraction and classical morphological crystallography of shape and habit. This allows the process chemist or pharmaceutical scientist to link the internal molecular structure to the chemical functionality of the external surface structure. Figure 29.2 shows the observed morphology for ibuprofen with the Miller indices labelled.

Morphological simulations based on crystal-lattice geometry were initially proposed [10, 11]. Subsequent work focused on quantifying the crystal morphology in terms of the interaction energies between crystallising units [12, 13]. Attachment and slice energies can be calculated directly from the crystal structure by partitioning the lattice energy in certain crystallographic directions. The calculated

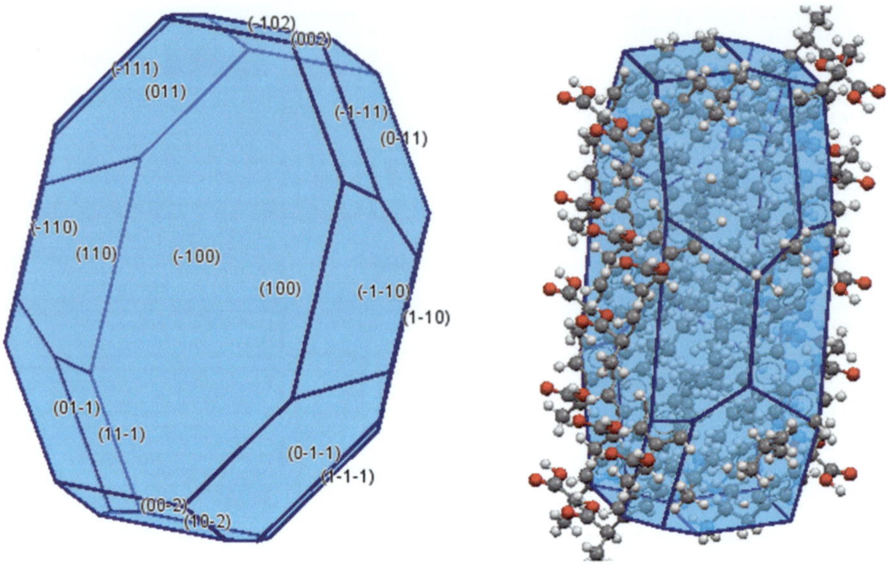

Fig. 29.2 The image shows the crystal habit of ibuprofen as predicted by the Bravais Friedel Donnay-Harker [11] method. The *left hand image* shows the predicted crystal faces, labelled with the corresponding Miller indices. The image on the *right* shows the habit packed with molecules, note the large (100) and (−100) faces are dominated by hydrogen bonding carboxylic groups, while the smaller faces are dominated by aliphatic and aromatic groups

attachment energies for crystal faces can be used as a measure of relative growth rates and so the theoretical morphology may be computed by determining the smallest polyhedron that can be enclosed by these faces and their relative growth rates. This has been exemplified in both academic and industrial case studies [14, 15].

A computer program HABIT [16] has been developed to allow such calculations to be carried out. The calculated attachment energies can be used as a measure of relative growth rates and so the theoretical morphology may be computed by determining the smallest polyhedron that can be enclosed by the specified forms given their relative growth rates. This approach has proved successful for a number of compounds and applied to a number of challenges during the development of the APIs and drug products. These will be highlighted in the subsequent sections. An inherent assumption of the attachment energy model is that the surface is a perfect termination of the bulk and little or no surface relaxation takes place.

Continuing with the assumption of perfect termination, it is possible to explore the interactions of a probe molecule, solvent, impurity or host, with a given surface. This is achieved in the systematic search algorithm by scanning the crystal surface with a probe molecule in a grid of positions and orientations (see Chap. 6) [17].

This approach is demonstrated in Fig. 29.3 (left) on the (100) face of ibuprofen crystal with the white triangles indicating the grid points visited by the probe

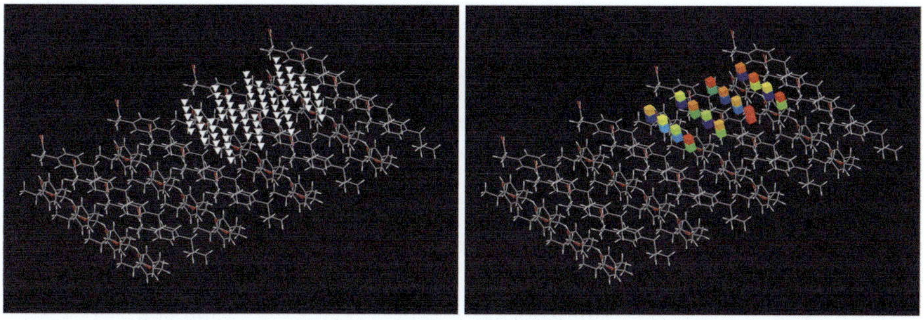

Fig. 29.3 *Left*: Image of a three dimensional grid points defined above a slab of (100) crystal face of ibuprofen as implemented by systematic search algorithm in Visual HABIT. At each position the probe is rotated through a grid of Euler angles. At each hyper-node (position & orientation) the interaction energy with the slab is calculated. *Right*: Image of the potential energy field for 1,2-xylene above (100) face of ibuprofen crystal. Each cube represents a node in a search grid, they are coloured from *red*, the highest energy, through *green* to *blue*, the lowest energy

molecule on a plane parallel to the crystal face under evaluation. The coloured cubes in Fig. 29.3 (right) represent the strength of the binding energies between the probe molecule and the crystal face at each grid point.

29.3 Connecting Form Formation and Formulation: The API-DP Interface

Integrated API and dosage form design to deliver an optimal drug product is at the core of the pharmaceutical scientist mission at the heart of this summer-school. A key part of this work is the definition of API properties for bioavailability, stability and manufacturability of the drug product. An API target attribute profile is designed to help the pharmaceutical scientist deliver this mission. This enables:

The link between API attributes and the drug product profile for clinical performance.
An integrated API & drug product design approach exploring and expanding API attribute acceptable ranges.
The rationale for establishing API target attributes to be captured.
The evolution of the API physico-chemical specification to be archived.
The knowledge transfer of API target attributes into both API and drug product manufacturing.

It has been recognised that the API particle size distribution descriptor, which has an impact on the material behaviour, does not take into account the anisotropic nature of crystalline materials. This has led to a broadening of the description of the important API attributes, from particle size and internal structure, to include particle shape and surface properties. To allow assessment of these properties, a

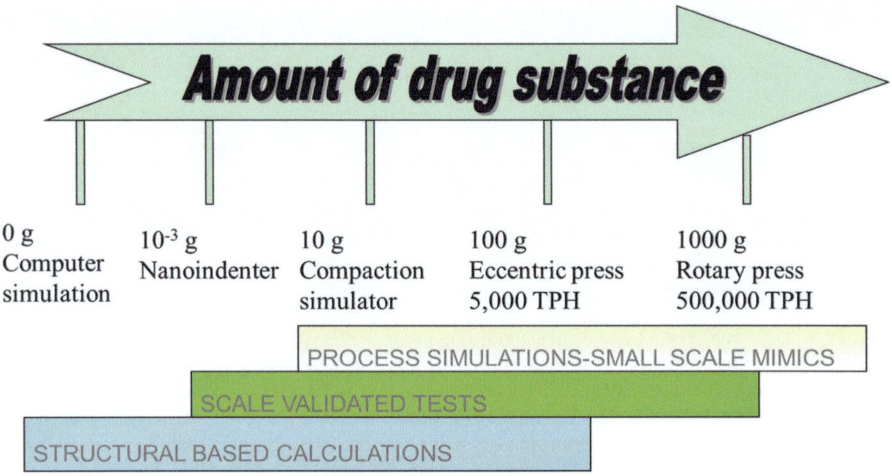

Fig. 29.4 The evolution from process mimics, through science of scale tools to materials and process modelling

range of analytical techniques has been integrated into pharmaceutical material characterisation including a number of high throughput dynamic image analysis techniques to assess shape [18, 19] and surface characterisation techniques including atomic force microscopy [20] and inverse gas chromatography [21]. These analytical techniques have been complemented by advances in approaches to access surface chemistry from the crystal structure of materials [22]. To justify the increasing amount of resource for API characterisation, to include the shape (habit), surfaces and mechanical properties, it is important to be able to link those properties to the critical attributes of the drug product. Therefore, it is interesting to consider the literature in respect to their potential to impact on drug product processing and performance. This evolution is summarised in Fig. 29.4.

Over the last decade, through embracing both academic advances and technology initiatives, significant progress has been made in defining relationships between API properties and drug product attributes, thereby enhancing the formulation design aspects of new products. Examples of established progress include:

API particle size distributions and content uniformity [23].
API particle size distributions and flow [24].
API particle size distributions, surface area and mechanical properties [25].
API particle size and dissolution [26].
Crystal brittleness and milling behaviour [27, 28].

Models have been built that allow the pharmaceutical scientist to explore the impact of particle size variation on dissolution rate and bioavailability. The Biopharmaceutics Classification System (BCS) [29] is used to define classes of compounds based on the solubility and permeability of the compounds. Permeability is a molecular property, but solubility and dissolution rate are related to the

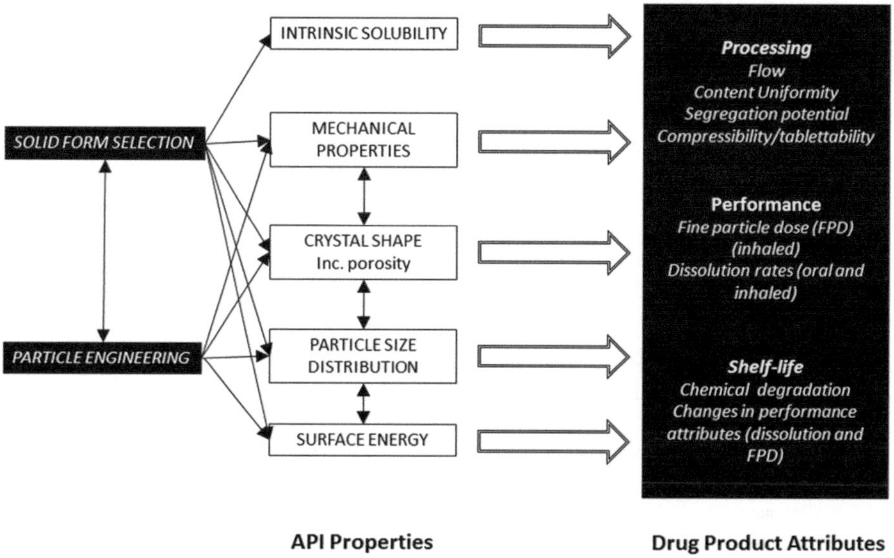

Fig. 29.5 The link between solid form selection (internal structure), particle engineering and key API properties and drug product attributes reproduced from [30]

internal structure (salt and polymorph) and particle size distribution/surface area/ surface chemistry. More recently this concept was built upon where the interplay between permeability and solubility and the impact of using simulated gastric fluids in measuring solubility has brought greater definition to the impact of API particle attributes impacting on drug efficacy [31]. Additionally, a recent cross-industry attempt to develop a Manufacturing Classification System (MCS) highlights the intense interest in this area from the manufacturing sector [32]. The MCS provides a common template for understanding risk, aiding product and process development and facilitating the transition between clinical and commercial manufacturing sites. This is summarised in Fig. 29.5 and Tables 29.1 and 29.2.

29.4 The API Perspective

Characterizing the detailed structural nature of the nano-particles associated with the formation of a distinct polymorph post-nucleation is experimentally challenging due to the small particle sizes involved. Hence, computational techniques are now being used to provide understanding of the inter-relationship between crystal form and surfaces at the nano-scale.

For glutamic acid, the observed crystal structures of the two polymorphic forms were optimized and their respective morphologies simulated using the technologies described earlier [33]. The predicted morphologies of both forms were in excellent

Table 29.1 Towards a particle passport, Part I: Modelling the bulk properties [34]

Description of property (API attributes)	Input	Output	Drug product manufacturability/ performance
Description of main inter-molecular and inter-atomic interactions	Crystal structure, charges	Lattice energy convergence graph Coordination shell (structure and energy)	Links the structure to the stability of the API and the drug product composite
Bulk-probe interactions	Crystal structure, impurity molecular structure	Segregation coefficient local site coordination of impurity Binding energy	Segregation tendency based on particle attributes of API and excipients
Prediction of particle morphology	Crystal structure	Figures of morphology	Input for the particle shape diversity
Evaluate potential particle shape diversity (e.g., solvent effects)	Crystal structure Solvent structure Concentration data	Crystal shapes	Impact on isolation, filtration, flow, size reduction (milling/micronisation)
Standard mechanical properties of API	Crystal structure	Identification of: Slip planes, likely burgers vectors, slip system, elastic tensor & dislocation energy	Links the structure to the milling and compaction behaviour. Plasticity/brittleness assessment

agreement with experimental results and observations from the literature. The meta-stable, α-form, crystal showed a prismatic shape whilst the stable, β-form, crystal was predicted to have an elongated needle-like shape. For the α-form morphology, the dominant face was calculated to be the (002) face followed by the smaller faces (101), (111), and (−111). For the β-form morphology, the dominant face was calculated to be the (020) face followed by the smaller faces (101) and (021).

Molecular clusters of different sizes were created by overlaying the predicted morphological shape and the optimized crystal structure. The energies of the facetted molecular clusters of different sizes were plotted as a function of cluster size for both polymorphic forms of glutamic acid. The results revealed that the meta-stable, α-form, is the more thermodynamically stable form compared to the "stable," β- form, at cluster sizes smaller than 240 molecules.

The influence of the growth solvent on crystal morphology (and habit) has been well documented [35]. It is known that the underlying mechanism involves face-specific, solvent-solute interactions resulting in different degrees of solvent binding which inhibits crystal growth on selective crystal-habit planes. Recently, authors successfully demonstrated the application of a surface-specific, grid-based search method, as outlined schematically in Fig. 29.3, to predict both the solute and solvent

Table 29.2 Towards a particle passport, Part II: Modelling the surface properties/features [34]

Description of property (API attributes)	Input	Output	Drug product manufacturability/ performance
Description of all dominant crystal surfaces and relative face area distribution.	Crystal morphology	Figures of surfaces with detailed description of chemistry, fractional surface area	Useful for assessing API wettability in dissolution or stomach media and particle adhesion/ cohesion
Surface energy prediction	Morphology & attachment energy	Figures of predicted crystal morphology & surface energy	Impact of surface energies on dissolution rates and stability
Probe-surface interactions. The required probes are:	Crystal structure	The statistics (mean, minimum, total and number of binding sites)	Predict probe-surface & surface-surface interactions with common excipients – calculate adhesion/cohesion energy
List of all standard excipients	Solvent/ impurity/ excipient structure	Distribution plot	
List of potential impurities + any known impurities for the synthesis of the specific API			
List of commonly used organic solvents			
Effect of particle shape (plate, needle, prism & matched to experimental data morphology) and size on dissolution.	Crystal structure	Dissolution profile of crystals with different shape	Predict impact of shape on dissolution
	Solubility data		
	Diffusion coefficient		
	Dose and volume of the dissolution medium		

binding to crystal habit surfaces of aspirin associated with crystallisation from aqueous ethanol solution [36].

The results from these simulations were representative of experimentally observed crystal habits for aspirin growth from ethanol and water. A comparison of simulations reveals that the unadjusted attachment energy model provides a good general match to the experimental morphology but, that the crystal habit, as predicted, manifests a much thicker and more tabular shape when compared to the experimentally observed shape of the crystals grown from solution. In contrast the crystal habit prediction, adjusted to allow for the effect of surface wetting and was a much better match to the experimental data. This is consistent with this modified model providing a more appropriate method for predicting the crystal habit of solution-grown crystals.

29.5 The Drug Product Perspective

A key challenge in product design is that during drug formulation, one or more active pharmaceutical ingredients (APIs) are combined with non-active excipients to produce a marketable product. The excipients provide a range of features from bulking to optimising the mechanical and processing properties of the active ingredient. Excipients can also have a major impact on the stability of the product and so appropriate selection is critical to the success of product design.

The optimum excipient selection process requires a thorough understanding of the components, their interactions with the API and the proposed manufacturing process. Computational tools have been utilised to define particle properties in terms of surface chemistry [36], to examine dissolution [37] and mechanical properties [28] and the agglomeration/crystal-crystal interactions [33, 38]. The prediction of adhesive strength at the API/excipient interface using computer simulation of API-excipient interactions has also been carried out [39].

Ketoprofen was chosen as a model API and a range of commonly used excipients were used as the probe systems for the study. The molecular modelling has two parts, prediction of crystal morphology followed by the prediction of binding energy of a probe molecule on the surfaces of the ketoprofen crystal. Three conformers for each of the excipients (probe molecules) were taken into consideration. For those excipients that are polymers, a representative monomer unit was used as a probe molecule.

It was found in this study that the broad range of excipients gave favourable interactions compared to the base line value for the ketoprofen molecule itself. β-D-glucose has a similar binding energy to ketoprofen whilst palmitic acid has a stronger, and similar, binding energy on both (100) and (110) faces. For D-mannitol the binding on the (100) face is stronger than the binding to (110) whilst the opposite trend is true for β-lactose. This output shows how modelling techniques can be used to examine surface/surface intermolecular interactions. This is summarised in Table 29.3.

A challenge in inhalation formulation is that it is comprised of micronised particles of API and excipient(s) which tend to form agglomerates. These agglomerates need to be dispersed effectively upon delivery for effective drug deposition to the lungs. Therefore, in studying the de-agglomeration and aerosolisation behaviour, it is essential to have a complete understanding of the inter-particle interactions between the API and excipients, or in other words, the cohesive and adhesive forces between particles of homogeneous and heterogeneous species respectively. HABIT has been successfully applied to predict the cohesive behaviours of an excipient, α-lactose monohydrate, and three APIs, salbutamol, fluticasone propionate and budesonide [40]. The prediction has been validated through independent experimental de-agglomeration studies using laser diffraction techniques [41].

The interaction energies between a probe molecule and all the surfaces of their respective crystal morphology were calculated using HABIT. The cohesive energies are the average of the minimum interaction energy of the probe molecule on

Table 29.3 Minimum interaction energy (kcal/mol) of some of the excipients on the ketoprofen (100) and (110) surfaces compared with the interaction energies of a ketoprofen molecule with itself [39]

Excipient name	Minimum interaction Energy (kcal/mol)	
	(100)	(110)
Ketoprofen	−6.35	−9.18
β-D-glucose	−5.84	−6.30
β-lactose	−13.74	−19.35
D-mannitol	−11.05	−7.01
Palmitic acid	−10.05	−10.67

every crystal surface of its respective morphology. The cohesive strength is stronger if the interaction energy is more negative. On that basis, it is predicted that fluticasone propionate is the most cohesive followed by salbutamol and budesonide. One of the interesting features of the binding energy range for salbutamol is that the range is very tight, within 1 kcal/mol, across all the different faces. In contrast, the binding energy range span for fluticasone propionate is broader, 3 kcal/mol, and reflects the different range of binding modes accessible on the different crystal faces.

An additional challenge for the pharmaceutical scientist is to deliver consistent dissolution rates of API particles used in formulations. Additionally, as the chemistry route is optimised and the final step isolation is refined, we need to be able to define the potential impact of changes of API size and shape on product efficacy. The models currently used to predict the dissolution kinetics of crystalline particles are based on Noyes-Whitney equation which relates dissolution rate to the surface area of particles [42]. Recent advances have shown the possibility of exploring both the impact of particle shape on surface area and the impact of surface chemistry on the stability of molecules in different environments [43]. The use of molecular dynamics to explore the crystal chemistry of acetaminophen has recently shown the potential of such tools [44].

The vast majority of achievable physical attributes of APIs are set at the point of solid form selection. The molecular and the crystal structure are a canvas, and particle engineering enables the manufacture of particles exhibiting the preferred properties from those allowed by the structural features. Over the last decade, through embracing both academic advances and technology initiatives, significant progress has been made in defining relationships between the API properties of new and existing chemical entities and the formulation design aspects of drug products.

29.6 Particle Passport and the Future of Drug Product Design

In an analogous journey to that of solid form design, structural and computational tools are now being integrated with experimental workflows to help define a 'particle passport' that connects surface chemistry attributes to product performance.

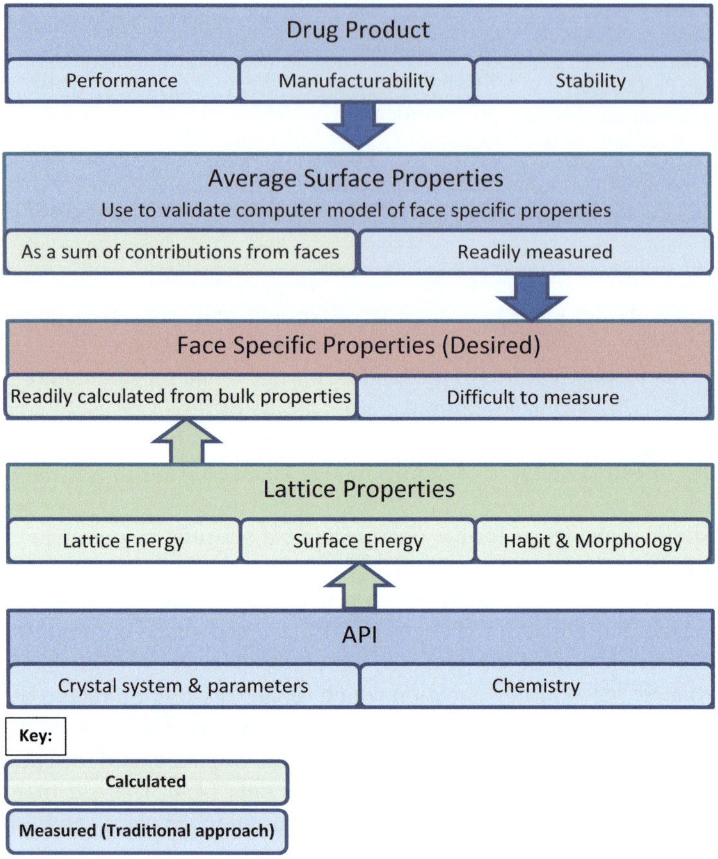

Fig. 29.6 Integration of experimental and computational workflows

To achieve an advanced definition of particle attributes at the API-DP interface the surface properties of crystals are required. The average surface properties of a crystal can be readily measured experimentally but it is difficult to measure the properties of individual faces. Conversely, using computer modelling, the face specific properties can be calculated, but a means is needed of validating the calculation. This can be provided by calculating the average surface properties using the face specific properties weighted by the fractional surface area, as in crystal habit prediction calculations, see Fig. 29.6.

Whilst, traditionally, the solid-form selection process has focused on two main factors, namely achieving an appropriate degree of product stability and bio-availability, increased emphasis is also being focused on selection of solid forms at the pre-formulation stage which have optimal physical properties such as mechanical behaviour, surface properties and particle shape. Given this perspective, this chapter has, briefly, outlined some recent research on the application of

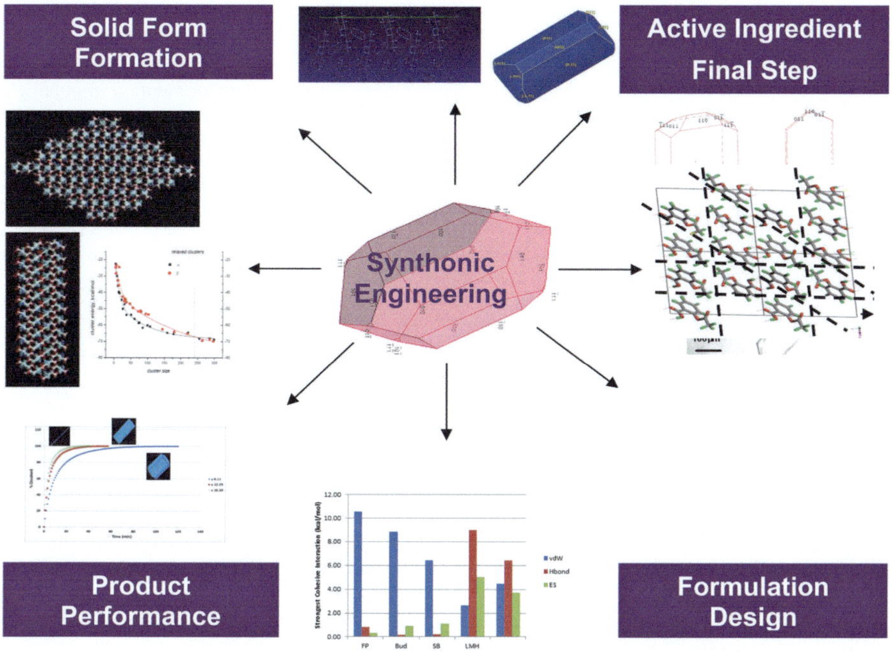

Fig. 29.7 The particle passport vision highlighted in Fig. 29.1 translated onto the current computational models

emerging computational technologies as foundation elements of the modern paradigm (Fig. 29.7).

- Predicting crystal surface-solvent interactions and the solvent-mediation of the crystal habit [36]
- Understanding inter-particle interactions associated with polymorphic transformation [38]
- Estimating the enhancement of solubility as a function of particle size reduction and morphological change [37].
- Visualising the impact of crystal packing motifs on the potential for mechanical deformation processes [28].
- Understanding the cohesive/adhesive balance between API and excipients [39, 40]

These relationships, combined with institutionalised corporate knowledge of formulation design practices plus computational methodologies of the sort highlighted in this chapter open up the potential of a fully integrated holistic product design process consistent with a Quality by Design strategy and aligned with the "form to formation to function" philosophy which was at the heart of the 48th International School on Crystallography upon which this publication is based.

References

1. DeCamp WH (2001) The impact of polymorphism on drug development: a regulator's viewpoint. Am Pharm Rev 4:75–77
2. Byrn SR, Pfeiffer RR, Stowell JG (2002) Polymorphs, regulations, crystallization, and solid-state chemistry. Am Pharm Rev 5:96–99
3. Storey RA, Docherty R, Higginson PD (2003) Integration of high throughput screening methodologies and manual processes for solid form selection. Am Pharm Rev 6:104–105
4. Ticehurst M, Docherty R (2006) From molecules to pharmaceutical products – the drug substance/drug product interface. Am Pharm Rev 9:34–36
5. Sun CC (2009) Materials science tetrahedron – a useful tool for pharmaceutical research and development. J Pharm Sci 98:1671–1687
6. Hancock BH, Elliot J (2006) Pharmaceutical materials sciences. MRS Bull 31:869–900
7. Chow K, Tong HHY, Lum S, Chow AHL (2008) Engineering of pharmaceutical materials: an industrial perspective. J Pharm Sci 97:2855–2877
8. Shekunov BY, Chattopadhyay P, Tong HHY, Chow AHL (2007) Particle size analysis in pharmaceutics: principles, methods and applications. Pharm Res 24:203–227
9. Roberts KJ, Docherty R, Marshall A (2012) Cross-industry synthonic engineering workshop. Institute of Physics, London
10. Friedel MG (1907) Etudes sur la loi de Bravais. Bull Soc Franc Miner 9:326
11. Donnay JDH, Harker D (1937) A new law of crystal morphology extending the law of Bravais. Am Min 22:446–467
12. Hartman P, Perdok WG (1955) The relations between structure and morphology of crystals. II. Acta Crystallogr 8:521–524
13. Berkovitch-Yellin Z (1985) Toward an ab initio derivation of crystal morphology. J Am Chem Soc 107:8239–8253
14. Docherty R, Clydesdale G, Roberts KJ, Bennema P (1991) Application of Bravais-Friedel-Donnay-Harker attachment energy and Ising models to predicting and understanding the morphology of molecular crystals. J Phys D Appl Phys 24:89–99
15. Black SN, Williams LJ, Davey RJ, Moffatt F, McEwan DM, Sadler DE, Docherty R, Williams DJ (1990) Crystal chemistry of 1-(4-chlorophenyl)-4,4-dimethyl-2-(1H-1,2,4-triazol-1-yl) pentan-3-one, a paclobutrazol intermediate. J Phys Chem 94:3223–3226
16. Clydesdale G, Docherty R, Roberts KJ (1991) HABIT – a program for predicting the morphology of molecular crystals. Comput Phys Commun 64:311–328
17. Hammond RB (2015) Modelling route map: from molecule through the solution state to crystals, chapter 6. In: Roberts KJ, Docherty R, Tamura T (eds) Engineering crystallography: from molecule to crystal to functional form. Springer Advanced Study Institute (ASI) series, 2017, in press
18. Yu W, Hancock BC (2008) Evaluation of dynamic image analysis for characterizing pharmaceutical excipient particles. Int J Pharm 361:150–157
19. Gamble JF, Ferreira AP, Tobyn M, DiMemmo L, Martin K, Mathias N, Schild R, Vig B, Baumann JM, Parks S et al (2014) Application of imaging based tools for the characterisation of hollow spray dried amorphous dispersion particles. Int J Pharm 465:210–217
20. Jones MD, Young PM, Traini D, Shur J, Edge S, Price R (2008) The use of atomic force microscopy to study the conditioning of micronised budesonide. Int J Pharm 357:314–317
21. Ho R, Heng JYY (2013) A review of inverse gas chromatography and its development as a tool to characterize anisotropic surface properties of pharmaceutical solids. KONA Powder Part J 30:164–180
22. Modi SR, Dantuluri AKR, Puri V, Pawar YB, Nandekar P, Sangamwar AT, Perumalla SR, Sun CC, Bansal AK (2013) Impact of crystal habit on biopharmaceutical performance of Celecoxib. Cryst Growth Des 13:2824–2832
23. Yalkowsky SH, Bolton S (1990) Particle size and content uniformity. Pharm Res 7:962–966

24. Mullarney MP, Leyva N (2009) Modeling pharmaceutical powder-flow performance using particle-size distribution data. Pharm Technol 33:126–134
25. Narayan P, Hancock BC (2003) The relationship between the particle properties, mechanical behaviour, and surface roughness of some pharmaceutical excipient compacts. Mat Sci Eng A A355:24–36
26. Siepmann J, Siepmann F (2013) Mathematical models of drug dissolution. Int J Pharm 453:12–24
27. Taylor L, Papadopoulos DG, Dunn PJ, Bentham AC, Dawson NJ, Mitchell JC, Snowden MJ (2004) Predictive milling of pharmaceutical materials using nanoindentation of single crystals. Org Process Res Dev 8:674–679
28. Olusanmi D, Roberts KJ, Ghadiri M, Ding Y (2011) The breakage behaviour of Aspirin under quasi-static indentation and single particle impact loading: effect of crystallographic anisotropy. Int J Pharm 411:49–63
29. Amidon GL, Lennernaes H, Shah VP, Crison JR (1995) A theoretical basis for a biopharmaceutic drug classification: the correlation of in vitro drug product dissolution and in vivo bioavailability. Pharm Res 12:413–420
30. Martyn DT, Ivan M (2015) Integration of active pharmaceutical ingredient solid form selection and particle engineering into drug product design. J Pharm Pharmacol 67:782–802
31. Butler JM, Dressman JB (2010) The developability classification system: application of biopharmaceutics concepts to formulation development. J Pharm Sci 99:4940–4954
32. Leane M, Pitt K, Reynolds G (2015) A proposal for a drug product manufacturing classification system (MCS) for oral solid dosage forms. Pharm Dev Technol 20:12–21
33. Hammond RB, Pencheva K, Roberts KJ (2007) Molecular modeling of crystal-crystal interactions between the α- and β- polymorphic forms of L-glutamic acid using grid-based methods. Cryst Growth Des 7:875–884
34. Ramachandran V, Roberts KJ, Docherty R, Synthonic engineering: particle passport and its impact on functional form (unpublished work)
35. Weissbuh I, Leiserowitz L, Lahav M (1995) "Tailor-Made Additives" and impurities. In: Mersmann A (ed) Crystallizaion technology handbook. Marcel Dekker, New York
36. Hammond RB, Pencheva K, Roberts KJ (2006) A structural-kinetic approach to model face-specific solution/crystal surface energy associated with the crystallization of acetyl salicylic acid from supersaturated aqueous/ethanol solution. Cryst Growth Des 6:1324–1334
37. Hammond RB, Pencheva K, Roberts KJ, Auffret T (2007) Quantifying solubility enhancement due to particle size reduction and crystal habit modification: case study of acetyl salicylic acid. J Pharm Sci 96:1967–1973
38. Hammond RB, Jeck S, Ma CY, Pencheva K, Roberts KJ, Auffret T (2009) An examination of binding motifs associated with inter-particle interactions between nano-crystals of acetylsalicylic acid and ascorbic acid through the application of molecular grid-based search methods. J Pharm Sci 98:4589–4602
39. Roberts KJ, Hammond RB, Ramachandran V, Docherty R (2016) Synthonic engineering: from molecular and crystallographic structure to the rational design of pharmaceutical solid dosage forms. In: Abramov YA (ed) Computational approaches in pharmaceutical solid state chemistry. Wiley, Chichester
40. Ramachandran V, Murnane D, Hammond RB, Pickering J, Roberts KJ, Soufian M, Forbes B, Jaffari S, Martin G, Collins E, Pencheva K (2015) Formulation pre-screening of inhalation powders using computational atom-atom systematic search method. Mol Pharm 12:18–33
41. Jaffari S, Forbes B, Collins E, Barlow DJ, Martin GP, Murnane D (2013) Rapid characterisation of the inherent dispersibility of respirable powders using dry dispersion laser diffraction. Int J Pharm 447:124–131
42. Noyes A, Whitney WR (1897) The rate of solution of solid substances in their own solutions. J Am Chem Soc 19:930–934

43. Krzyzaniak JF, Meenan PA, Docherty CL, Pencheva K, Luthra S, Cruz-Cabeza A (2016) Integrating computational materials science tools in form and formulation design. In: Abramov YA (ed) Computational approaches in pharmaceutical solid state chemistry. Wiley, New York
44. Gao Y, Olsen KW (2013) Molecular dynamics of drug crystal dissolution: simulation of acetaminophen form I in water. Mol Pharm 10:905–917

MIX
Papier aus verantwortungsvollen Quellen
Paper from responsible sources
FSC® C105338

If you have any concerns about our products, you can contact us on
ProductSafety@springernature.com

In case Publisher is established outside the EU, the EU authorized representative is:
Springer Nature Customer Service Center GmbH
Europaplatz 3, 69115 Heidelberg, Germany

Printed by Libri Plureos GmbH
in Hamburg, Germany